Geoinformatics for Environmental Management

About the cover page

The cover page illustrates Hyperspectral Image of Fish Lake Valley located in South - Western Nevada to the East of the white mountains that form the CA-NV border and to the West of Silver peak range. The first picture at top of the right hand side shows the output of NOAA satellite with vector overlay. The middle picture shows the digital terrain model where as the bottom picture shows an action plan of Cherial Watershed.

Geoinformatics for Environmental Management

M. ANJI REDDY

Professor of Environmental Science and Technology,
Jawaharlal Nehru Technological University Hyderabad,
Kukatpally, Hyderabad-72 (A.P.) India.
mareddyanjireddy@rediffmail.com

BS Publications
A unit of **BSP Books Pvt., Ltd.**

4-4-309/316, Giriraj Lane, Sultan Bazar,
Hyderabad - 500 095
Phone : 040 - 23445605, 23445688

Published by :

4-4-309/316, Giriraj Lane, Sultan Bazar,
Hyderabad - 500 095
Phone : 040 - 23445605, 23445688
e-mail : info@bspbooks.net

ISBN : 978-93-85433-70-2 (HB)

Dedicated to

---

all those who believe in working together and putting their vision into action.

Jawaharlal Nehru Technological University
Kukatpally

Hyderabad – 500 072, A.P. (India)
Phone: 040-23156109 (O)
23156116 (R)
Fax: 040-23156112
E-mail: jntuadm@hd2vsnl.net.in
www.jntu.ac.in

Dr. Y. Venkatarami Reddy
B.E., M.Sc., (Engg.) Ph.D (I.I.Sc)
Vice-Chancellor

FOREWORD

Environmental awareness is an important subject of today's education. This aspect has been stressed recently by the Supreme Court order that environmental topics must be taught from primary to PhD level. Environment is the source of all resources for all living systems. Advances in technology and zeal for economic progress are responsible for the all round environmental degradation. However, evaluation of the extent of environmental destabilization has been extremely difficult till recently. Fortunately, the subject of GIS is helping in the evaluation of the status of environment using Remote Sensing. While, general information on the environmental topics is known quite widely the subject of Geoinformatics is relatively of nascent origin and is not still comprehended by many people in the environmental management side.

Sensing this gap of knowledge between the GIS and environmental management, a timely, informative and educative as well as a reference book has been prepared by Prof. M. Anji Reddy, Centre for Environment, JNT University. A glance through the book indicates the usefulness of this book not only for students, research scholars but also for other members in the field of Environmental management.

Rightly Dr. Anji Reddy gave information about the basic aspects of remote sensing and GIS including the important topics of Geoinformatics and also explained certain topics like the survey technology, Cartography, Photogrammetry, Geodesy, Global Positioning Systems and Map Projection, almost as introductory topics. He has focused his attention on various specific environmental applications of Geoinformatics in the later chapters. The important cause for Global Warming like the Forest Resource Management, Watershed Management, Water Quality Mapping and Modeling, Solid Waste Management and other alike topics have been specifically dealt with in the later part of the book.

To make it much more impressive and understanding the author has included various figures and color plates making the subject easier for the readers of the book. I am very happy that such an important and timely publication is brought out by one of the faculty of Jawaharlal Nehru Technological University. I wish that all concerned will not only use it as a book of knowledge and information but also for teaching and research purpose.

(Y. Venkatarami Reddy)
Vice- Chancellor

Preface

My students and research scholars are instrumental in bringing out this book : Geoinformatics for Environmental Management. When preparing this book, a major problem was not of finding sufficient material for inclusion, but what could be omitted, if a comprehensive text was to be compiled within reasonable limits. Initially, it was contemplated that more would be written on image analysis and digital photogrammetry. This would, however, have resulted in an imbalance and reduced the contents of other topics. Since in my first book, *Textbook of Remote Sensing and GIS*, I have dealt with all fundamental aspects of Remote Sensing and GIS, therefore, in this book, I emphasized on other important topics of Geoinformatics and its applications to environment.

The main thrust of the book is to provide a comprehensive text in a perspective of Geoinformatics technologies covering Remote Sensing, Geographical Information Systems and Global Positioning Systems and other interrelated subjects like geodesy, map projections along with various environmental applications for its management.

I have been working on a number of projects related to environmental problem and solving them using the technologies of Remote Sensing, GIS and GPS. With 14 years of my experience, it is contemplated to write a book of this kind for the benefit of resource scientists, engineers, decision makers and planners, academic community and my dear students.

The book is divided into two parts : principles of geoinformatics and its applications to environment management. In the first part, topics namely, Surveying Technology, Cartography, Photogrammetry, Remote Sensing, Geodesy, Global Positioning System, Map Projection and GIS are provided along with the introductory chapter.

The principles and the fundamental concepts of these technologies are discussed keeping the application scenario in mind. The second part of this book is mainly focused on the specific environment applications of geoinformatics. The specific application areas that are presented in this book are Forest Resource Management, Watershed Management, Saltwater Intrusion Modelling, Water Quality Mapping and Modelling, Solid Waste Management, Landslides Analysis, Alignment Surveys, Fishing Zone Mapping and Urban Management.

A number of figures and color plates are provided in this book which were developed by my students and research scholars under various projects sponsored by State and Central Government Organizations. This book is very much useful for any application scientist for the execution of their projects. This book can be used as a reference book or textbook.

The author is deeply indebted to Sri CKP Unni, Survey of India (Retd.) for his valuable guidance, suggestions and critical evaluation of chapters of first part of the book. I am grateful to Prof. Y. Anjaneyulu, Director, Institute of Science and Technology, JNTU, for his precious suggestions, technical advice and help throughout the preparation of this book. I want to express my thanks to my students, research scholars and academic assistants at Centre for Environment for their assistance and providing me the relevant data while preparing this book at various stages.

Dr. M. Anji Reddy

(mareddyanjireddy@rediffmail.com)

Professor of Environmental Science and Technology,

JNT University, Hyderabad.

Contents

7 Global Positioning System (GPS)

8 Map Projections

9 Geographical Information Systems

12 Salt Water Intrusion Modelling

13 Water Quality Mapping and Modelling

14 Solid Waste Management

18 Urban Planning and Management

1 Introduction

1.1 Geoinformatics and Geographic Information Science

Several terms such as "geomatics," "geomatic engineering," and "geoinformatics" are now in common use pertaining to activities generally concerned with geographic information. These terms have been adopted primarily to represent the general approach that geographic information is collected, managed, modelled, applied, monitored and mapped. Surveying, Photogrammetry, Remote Sensing, Cartography, Geographical Information Systems (GIS) and Global Positioning Systems (GPS) are important components of Geomatics or Geoinformatics.

When GIS became mature as a branch of information technology, the various disciplines surrounding it gradually converged to form a particular field of scientific study, leading to the proposition of the term "Geographic Information Science" (GIScience). Goodchild (1992) defined geographic information science as the set of basic research issues raised by the handling of geographic information. These issues include, for example, the unique characteristics of geographic data, the distinct nature of research that requires geographic problem solving, the interaction between geographic information research and related academic disciplines, as well as the impacts of using geographic information on society. Geographic information science aims to provide the theoretical and organizational coherence for the scientific study of geographic information. It seeks to redefine geographic concepts and their applications in the context of GIS. Therefore, Geographic Information Science is not intended as a substitute of GIS in terminology and in practice.

The term GIS has now gained universal acceptance which will be difficult to replace it with other alternative names. In this book, we use GIS as the acronym of "geographic information systems" to denote both the approach and the technology pertaining to geographic information. We use the term with the understanding that GIS not only represents the skills and procedures for collecting, managing and using geographic information, but also entails a comprehensive body of scientific knowledge from which these skills and procedures are developed. As we will demonstrate in later chapters, such a liberal view of GIS allows us to provide a balanced treatment of the subject matters between concepts and techniques throughout this book.

1.2 Components of Geoinformatics

Geoinformatics, in alternative words, is defined as the science and technology dealing with the character and structure of spatial information and the infrastructure necessary for the optimal use of spatial information. Broadly, the basic components of Geoinformatics are :

- (i) Conventional Mapping
- (ii) Photogrammetry
- (iii) Digital Photogrammetry
- (iv) Remote Sensing
- (v) Digital Image Processing
- (vi) Cartography and Automated Cartography
- (vii) Global Positioning Systems (GPS)
- (viii) Database Management System
- (ix) CAD/Digitisation
- (x) Geographical Information Systems (GIS)

The major functional elements of Geoinformatics are Capturing, Organization, Classification, Qualification, Analysis, Management, Display and dissemination of spatial information for specific needs of user community.

1.3 Applications of Geoinformatics

Some of the application areas where Geoinformatics can be used with more accurate results are as follows :

- Forestry
- Agriculture
- Geology and Geomorphology
- Terrain Investigation and Evaluation
- Environmental Studies and impact assessment

- Ground Water investigations, targeting, mapping and management
- Surface Water mapping and inventory and drainage system
- Transportation Engineering and planning
- Disaster Management and mitigation
- Land use planning and land suitability
- Soil mapping
- Oceanography and coastal zone management

The specific applications in each of the above disciplines are provided in the following sections.

1.3.1 Geoinformatics in Forestry

- Vegetation type and density mapping.
- Vegetation monitoring and change detection.
- Wildlife census using TIR Remote Sensing.
- Wildlife habitat mapping, corridor analysis.
- Tree height, crown diameter measurements, inventory and stock mapping, regeneration status using Microwave Remote sensing and biomass estimation.
- Nutrient deficiency, water stress and disease detection and assessing plant vigour.
- Functional and Structural analysis of vegetation.
- Fire risk zonation mapping.
- Biodiversity characterization at landscape level.
- Mapping areas suitable for afforestation and reforestation.
- Forest road planning.
- Estimation of lignin and protein using hyper spectral data.

1.3.2 Geoinformatics in Agriculture

- Crop type classification and mapping.
- Crop condition assessment – crop diseases, insect damage, plant stress, integrating with soil conditions to use as a basis for programming microprocessor controlled equipment such as fertilizer spreaders and irrigators.
- Crop yield estimation and modelling.
- Using color infrared aerial photographs to determine plowing planting progress, detecting delayed emergence and low plant density, check on stand growth and development through the growing season, check stand condition and acreage to be harvested and total area harvested etc.

1.3.3 Geoinformatics in Geologic and Soil Mapping

- Identification of landforms, rock types and rock structure.
- Mineral resource exploration – surface and near surface deposits.
- Location of deep potential deposits in combination with geophysical methods.
- Mapping of lineaments, faults and lithological units.
- Geobotanical approach to ascertain the availability of nutrients to vegetation.
- Soil type mapping.
- Land capability classification by integrating with relevant information.

1.3.4 Geoinformatics in Urban Planning

- Identification of landforms, rock types and rock structure.
- Mapping transportation network – roads, railways and signal points.
- Mapping transmission lines – electric and telephone.
- Mapping water pipe and sewage lines.
- Planning the best possible routes for roads, transmission lines, drainage network, railways etc.
- Mapping areas suitable for constructing colonies, industries etc.
- Assessment of environmental impact on major projects.
- Traffic and parking studies.
- Urban change detection and growth assessment.

1.3.5 Geoinformatics in Environment

- Assessment of natural resources.
- Study of volcanic eruptions.
- Estimation of coastal sediments.
- Watershed modelling and management.
- Contouring and Leveling - DTM and DEM etc.
- Forest Resource Management.
- Saltwater Intrusion Modelling.
- Water Quality Mapping and Modelling.
- Solid Waste Management.
- Industrial Siting.
- Natural disaster management.
- Coastal zone management.
- Urban planning and management.

1.4 Approache to the Study of Geoinformatics

People now study geoinformatics for different purposes. Many students study geoinformatics simply as an academic pursuit; others study it to prepare for work as a specialist in a rapidly growing industry. There are also practicing professionals in various fields who study geoinformatics in order to learn a new set of software tools that is increasingly used in their workplaces. The basic approache to the study of this subject and to have a solid understanding of the concepts and techniques, this book serves the purpose.

Before the 1980s, GIS was a set of relatively obscure computer applications that were found only in a small number of government agencies and universities. Although articles pertaining to GIS could be found in a wide variety of learned journals in geography, cartography, computer science, and surveying, opportunities in the academic study of GIS in general were rather limited (Unwin, 1991). Today, it is hard to find a department of geography, geology, civil engineering and environment where no GIS course is offered. GIS is also widely taught in programs, in Earth and environmental sciences, computer science, business administration, natural resources management, urban and regional planning, and civil engineering.

The introduction of GIS into the academic curriculum not only means the addition of a new course of study to the degree programs, it also represents a new way of looking at geographic data. It has injected new concepts and techniques into traditional academic disciplines such as those noted here. In geography, in particular, the advent of GIS has been widely regarded as one of the most significant development since the so called quantitative revolution in the 1960s. Given the growing interest among geographers in geoinformatics today, it would be hard to contest the proposition that GIS will be central to the development of geography as an academic discipline.

Until the late 1990s, GIS courses were taught mostly at the undergraduate level as a part of degree programs in geography and other disciplines. Since then many universities started to offer specialist degree programs in GIS at both the undergraduate and post graduate levels (Waters, 1999). These specialised degree courses generally have a strong academic thrust that makes them distinct from technically oriented GIS programs offered by community colleges and institutes of vocational training. Although the contents of different programs tend to vary from one another, the concepts and techniques of Geoinformatics presented in this book constitute the fundamental component. Supplementary courses incorporated in most programs include traditional spatially oriented fields such as cartography and remote sensing, as well as those fields that GIS relies on in its development, such as computer science, information systems, statistics, mathematics and cognitive science. These programs prepare students for an academic and research career in geographic information science and in different disciplines in science and engineering. They also provide the basic education for students who aspire to become professional practitioners in government, business and industry.

1.5 Geoinformatics Products

The history of the topological data structure provides a typical example of how GIS has become less special. It was originally devised to solve two related problems associated with the digital representation of a certain type of map commonly found in environmental and resource management, and it is not surprising that these turn out to be the most successful areas for the nascent Geomatics industry (Prasad M., 2002).

Economies of scale underlie all of today's information technologies. The internet is successful because its packet technology can be used for any type of information irrespective of its meaning, and a single packet might contain codes for text, numbers, music, or images. Digital technology is successful because its binary alphabet can be adapted, using simple coding systems, to represent virtually anything. As a result the unit costs of digital technology are low, and falling lower.

1.5.1 Geographic Information Systems (GIS)

MapInfo released Version7 with new features of direct reading of maps in ESRI shape format and enhanced object editing and processing tools. Notable amongst them being support for the Geographic Markup Language (GML). ESRI launched ArcView 8.2, with features of two new optional ArcGIS extensions: ArcGIS Publisher and ArcGIS StreetMap. ArcGIS Publisher allows GIS users and data suppliers to easily publish and share electronic maps locally, over networks, or via the Internet. Autodesk Map 6 had key additions of functionalities to annotate objects-display textual values and expressions without the need to run a separate query, created highly precise maps using coordinate geometry (COGO) input, simplified topology capabilities and facilities to create, edit, and manipulate multiloop mapping polygons. The GeoMedia5 from Intergraph had features-improvement on aggregation, functional attributes, dynamic buffer zones, grids, improved text and user-defined line styles. The GeoConcept V5 had strong points like object-oriented environment, better ergonomics and open architecture with an engine common to all the product range. The Idrisi32 R2 from Clark Labs included an HDF reader and import/export routines for ERDAS Imagine '.img' files and ER Mapper '.ers' files. Geomatica Version 8.2 from PCI Geomatics offered web tools, added spatial analysis functions and 3D stereo feature editing / extraction capabilities. Some of the new functionalities added in the ERDAS IMAGINE V8.6 are Spectral Analysis tool for Hyper Spectral image processing and ESRI Geodatabase support. Able Software came with new release for raster to vector conversion R2V version 5.0.3 that includes attributes like creation of polygon features from polylines. Surfer 8 from Golden Software promises better rendering of 3D features and raster overlaying.

Hitachi Software Global Technology launched Any*GIS, an enterprise GIS platform, AltaMap4 from GeoMicro provides a collection of OLE/COM objects for building GIS applications with key features of light weight ActiveX control. Caris launched GEMM 4.0, which includes a geology symbol template for digitizer tablets (Sanjay Kumar, 2002).

The new kid on the block was RaveGeo, a vector database storage format, which provides new display methods for opening large vector data files. It is designed for environments ranging from handheld devices to workstations. GeoExpress from LizardTech, enables one to create, disseminate, and access digital imagery, provides lossless encoding and compression, preserving images with pixel-for-pixel fidelity.

1.5.2 Remote Sensing

Remote Sensing segment was relatively quiet, with companies focusing attention more on the delivery and market consolidation. DigitalGlobe commenced the commercial operation for its Quickbird 61cm imagery; this was suitably matched with restructuring of prices for IKONOS 1m imageries. National Remote Sensing Agency, India, (NRSA) with its PAN 5.8m data of IRS continued its flat approach oblivion to the changing equation, with no new scheme or price restructuring to sustain the customer base.

Spot 5 which was placed successfully in orbit the distribution data will be through DigitalGlobe in the U.S. and ImageONE in Japan. SPOT 5's High Resolution Geometric (HRG) sensors will provide improved 2.5 meter and 5 meter resolution imagery over wide swaths (up to 60 x 120 km). High Resolution Stereo (HRS) instrument can now do large area "strip mapping" with stereo imagery.

ENVISAT launched by the European Space Agency in March 2002, is providing measurements of atmosphere, ocean, land and ice. ENVISAT data supports earth science research. DigitalGlobe M5 System, setup in association with Ball Aerospace and Technologies Corporation, plans to launch Remote Sensing system M5, scheduled to be operational in early 2006. The M5 constellation will consist of four satellites, each of which will collect five-meter resolution multi-spectral data over a 185 kilometer-wide area. The first M5 satellite is scheduled to be operation in the first quarter of 2006, and all four by the third quarter of 2007. M5's off-nadir pointing capability – collecting images from an angle – will give DigitalGlobe, the ability to revisit any point on the Earth's surface multiple times per day, providing a valuable resource of information for time-sensitive applications.

1.5.3 Mobile Mapping

Year 2k2 also saw the increased interest about mobile mapping technology amongst the user segment. Still treading with caution and not any landslide sale of M2 product to vindicate the absolute faith of the community in this technology. IntelliWhere OnDemand by Intergraph, ESRI's ArcPad and MapX Mobile from MapInfo's GIS for hand helds were some of the Mobile Mapping products released.

1.6 Legal Implications

GIS methodology using digitized maps is now changing the way digital spatial data are used, revised and assessed. Maps as tangible property and as tools are capable of good and harm. Maps as property raise issues of intellectual property generally and copyright in particular whereas maps as tools raise liability issues especially when the data in them are erroneous. Both these issues are interrelated as their interpretations lead to a legal conundrum, maps as factual compilations are incapable of copyright protection whereas courts require them as 'facts' for evidence (George Cho, 2002).

In the future it is expected that most litigation in this context will revolve around Geographic Information Systems products and spatial data, its collection, analysis, use and third-party use. More often the GIS professional will be in court either as a witness to the development and application of a spatial database or as an expert witness.

1.6.1 Restricted Maps

The very classification of a map as 'restricted' evokes a fearful response from the custodians of these documents. It was not very long ago that the GIS data user community heard of a Crime Branch of India (CBI) inquiry in context with the digitization of some restricted maps by a private agency for a Governmental project. None of us would like to be scapegoats for some untimely mishap, which may suggest the use of information from 'restricted' maps (Samant. H, 2002).

A recent addition to data acquisition is the advent of GRACE, which facilitates geoid measurements up to centimeter accuracy. This leads one to doubt the sanctity of the restriction of the Everest Ellipsoid parameters by Survey of India (SoI). Students of Earth Sciences perennially face the problem of data restriction. This results in a biased selection of research project work sites. Further many students in our universities spend almost 80% of their research time in just acquiring data.

1.6.2 Burning Issues

Some of the burning issues confronting us as GIS professionals, include (i) protection for data in a global context (ii) the issues of liability in geographic information science and (iii) the protection of personal privacy in a geographic databases. We need to think seriously about these when we next print a map from our Geoinformatics technology.

1.7 Data : The Soul of Research

The integrity of any research endeavor is no better than the integrity of the data, namely, thematic data, topographic data, field data and other collateral data. By and large they can be defined as measurements, observations or any other primary product of an act of research. Such product provides a factual basis for interference, conclusions and publications. On them depends the integrity of research and it's usefulness to the society. However, empirical research need not involve primary data collection. Till, that is the case, and the scale is small, the academicians can appropriately procure it. But, the deterrents occur when many researchers use data collected by others to conduct new analyses-secondary analyses-designed to answer their own research questions. The major advantage of secondary analysis is that it allows a researcher to bypass the time consuming and costly steps of sample selection, data collection and initial data processing. Besides, what is the need too? Moreover, why should the data be duplicated, especially when existing data elsewhere ensures accuracy, like, the remote sensing agencies where :

- Large area coverage facilitates local, regional, national and global surveys.
- Continuous observations are conducted for disaster mapping and assessment.
- Periodic monitoring of environmental resources like land, water, agriculture, etc., is regularly conducted.
- All weather viewing capability is available.
- Procurement of fully computer compatible data, ensuring fast and accurate results is possible along with analog and digital images of the inaccessible regions.

In case there are other agencies in particular research field that have already procured and analyzed data to a certain extent, then no need emerges for its replication. The researcher can thus concentrate on the conceptual and practical problems of data analysis interpretation and reporting. Like, for instance in the case of land use studies, the following agencies can be a few sources of information.

1.7.1 Sources of Information

- Revenue records compiled by the Directorate / Bureau of Economics and Statistics.
- Topographical maps from the Survey of India.
- Land use Atlas from National Thematic Mapping Organization (NATMO).
- Soil Survey Organization (NBSS & LUP AND AIS&LUS) generate soil and land capability maps.
- National Remote Sensing Agency (NRSA), with access to both analog and digital data.

The question that arises is that if an academic researcher is aware of the data access problems then what is the issue? The main issue is data sharing which can only be approached through the bureaucratic maze of procedures enshrined in government rule books. The data generating departments overlook the social commitment of an academician and are unmindful of the serious hurdles that come in his way. What purpose does the data generated by the funds collected by the common man's money serve? Somewhere the procedure of accruing data is a hurdle, whereas in others the lack of finances, despite educational concessions, becomes a cause (Seema M. Parihar, 2002). Academic community can definitely become a successful data provider if individual researchers collaborate in contributing towards common goals. In contrast when an academician works in isolation, the society loses the benefits of improving and thus we remain unheard in policy forums.

As IT awareness increases, GIS will also become more and more mass based. The right to Spatial Information should be recognized as a fundamental part of the right to information. In fact, if the market becomes truly demanding and lively, then it will take care of access to information, whether or not the government has a policy. Market needs to play a key role in influencing policy. People who intent to misuse information get it by whatever means. The sufferers are the genuine users. In fact, we need to look at the issue of stimulating demand rather than getting too obsessed with policy. Information markets are still not fully developed (Amitabha Pande, 2002).

1.8 Geoinformatics and Environmental Modelling

Geoinformatics are not well designed for handling either time or the interactions of continuously changing variables. By contrast, environmental models have been effective for many years at handling time and variable interactions – especially systems-based models-but such models rarely include spatial information either for the purpose of analysis or even for the display of their results. The ongoing challenge has been to integrate these two vastly different approaches that seek to provide an understanding of our environment.

Both GIS and environmental modeling are relatively new areas of research that have made extraordinary progress in the last few decades. Both are likely to make a significant impact on how we can manage our scarce and dwindling resources. Using these tools, and integrating these approaches with the analysis of socioeconomic systems, we should be able to stem the tide of environmental degradation and pollution and protect endangered habitats.

1.8.1 Developments in Environmental Modelling

Four developments were of particular importance. First, there are the conceptual innovations in overlay modelling popularized by Ian McHarg (1969). Clarke (1999) describes the original contributions of Jacqueline Tyrwhitt to this methodology. Steinitz et al. (1976) provide a complete history of the map overlay approach and suggest that the first explicit discussion of the overlay technique despite McHarg's implied statement that his first use of the method had been novel (McHarg, 1996). Tomlinson (1999,) explains that it was his company, Spartan Air Services of Ottawa, that in 1962 first proposed computerizing the overlay methodology.

Second was the development of the idea that modelling as opposed to 'mere description' was an informative approach that would provide useful insights. The third development involved the introduction of analytical and empirical approaches as an aid in understanding geographical and environmental data. The fourth development was the rise of systems thinking. These four areas were developed largely independently of each other and there was, in addition, a significant divide separating research contributions and communications among nations.

1.8.2 The System Approach of Modelling

The second attempt of define the nascent GIScience discipline mentioned by Goodchild et al. (1999) is that of the U.S.-based University Consortium for Geographic Information Science (UCGIS). In 1996 ten topics are defined : 1. spatial data acquisition and integration 2. distributed computing 3. extensions to geographic representation 4. cognition of geographic information 5. interoperability of geographic information 6. scale 7. spatial analysis in a GIS environment 8. the future of the spatial information infrastructure 9. uncertainty in spatial data and GIS-based analysis and 10. GIS and society. In 1998 each of these topics are refined into a series of "white papers," available at http:/ www.ucgis.org/research98.html and now under reconsideration.

Besides spatial dependence, environmental GIS and often present other difficulties. In much environmental modeling a digital elevation model (DEM) is used. Elevation is useful as an independent variable in its own right for such environmental analyses as habitat modeling, but it can also be used to calculate other independent variables such as slope and aspect. Slope, like variables that represent percentages, is a constrained distribution (values must fall within the range 0 to 90 degrees) and aspect is an example of a variable that forms a circular distribution (values go from 0 to 360 and then back to 0 degrees). Few statistics texts address the problem of calculating statistics for such variables, although Burt and Barber (1996) provide one exception while Zar (1999) provides another in a through and comprehensive review of descriptive and inferential statistics and associated significance tests for this type of data.

1.8.3 Need of Environmental Data

Environmental data are needed for one or more of the following reasons (Harmancioglue 1998, Varma, 1996):

(i) To identify the nature, trends and anomalies and characteristics of environmental processes in order to better understand of these processes.

(ii) To assess the effects of natural and human-made factors upon the general trends in environmental processes.

(iii) To evaluate the effectiveness of pollution control measures.

(iv) To assess the appropriateness of environmental quality standards.

(v) To assure compliance with established quality standards and enforcement of quality control measures.

(vi) To conduct environmental impact assessment.

(vii) To monitor and assess the general quality of the environment at regional to global scales.

(viii) To develop, calibrate and validate models of environmental processes.

1.8.4 Environmental Data Collection Efforts

The availability of appropriate and adequate environmental data and the full extraction of information from collected data, are important concerns. In the past, many data collection efforts are local in nature to meet the objectives of environmental impact assessments for projects, monitoring environmental resources for compliance and enforcement purposes, or for modeling problems at the local level. With the evolution and broadening of environmental problems, the nature and type of data requirements have changed and newer data collection efforts have been undertaken. Moreover, environmental data need to be collected in a way that facilitates understanding the interrelationships and interactions among various earth resources during an environmental disaster (natural or human-made). There is a need to integrate information on all resources of the Earth so that we become "information rich" and not just "data rich".

Spatio-temporal data are collected primarily by means of field surveying, instrumentation, photogrammetry and remote sensing. Field data collection using sampling, surveying and/or GPS instruments may be required for gathering geospatial information. Such field work can help fill gaps in existing data, provide ground reference information for use in calibrating remote sensing instruments, for processing and interpreting remote sensing data and in the establishment of ground control (Pickles, 1995). Field surveying techniques involve the use of precise electronic distance measuring equipment and/or GPS features in the landscape. Instrumentation placed in the field to measure environmental phenomena are increasingly self-supporting with their own solar power and ability to automatically transmit data readings via satellite relay to a reception point for processing and archiving. Weather instruments, seismographs and stream gauges are examples of telemetered equipment. Photogrammetry, the process of using remote sensing imagery to create a precise stereo model and compile geometrically accurate map products, has evolved into a digital process that uses soft-copy input. Remote sensing includes airborne and satellite imaging systems, both passive and active. Aerial photography is a passive airborne source of remote sensing data, whereas, light detection and ranging (LIDAR) and interferometric synthetic aperature radar (IFSAR) technologies employ an active signal to obtain information about the environment. Similarly, multispectral scanners on-board satellites are passive sensors, however, radar systems are an active source of data.

Digital data is also created by the process of converting existing maps or graphic documents into an appropriate digital form using digitizing or scanning methods (Clarke, 1997; Hohl, 1998; Jackson & Woodsford, 1991) for details on digitizing and scanning. Data conversion is still an important task in GIS projects and constitutes a major portion of the cost and time of projects. Map conversion projects need to pay attention to scale, level of generalization, datum, coordinate system, projection, data of source and accuracy (Fisher, 1991) for spatial data sources and data problems. (Flowerdew, 1991) for spatial data integration (Hohl, 1998) for discussion of data conversion.

1.9 A Tour of the Contents

No matter what approach a student wishes to follow when studying Geoinformatics, there is a set of core concepts and techniques that he/she must fully understand and master (Fig. 1.1). The primary purpose of this book is to provide students the concepts and techniques fundamental to studying Geoinformatics from different perspectives and its applications to environmental management. The contents of the book have been designed with the needs of beginning students and Geomatics professionals in mind. Accordingly the contents of this book is divided into two major parts. The first part, that is, from chapters 2 through 9 covers the basics of Geoinformatics. This part also includes the introduction (chapter 1) which provides the broad general view of geoinformatics. The second part of this book deals with various applications of geoinformatics for environmental management along with specific case studies, in 10 chapters from chapter 10 to chapter 19.

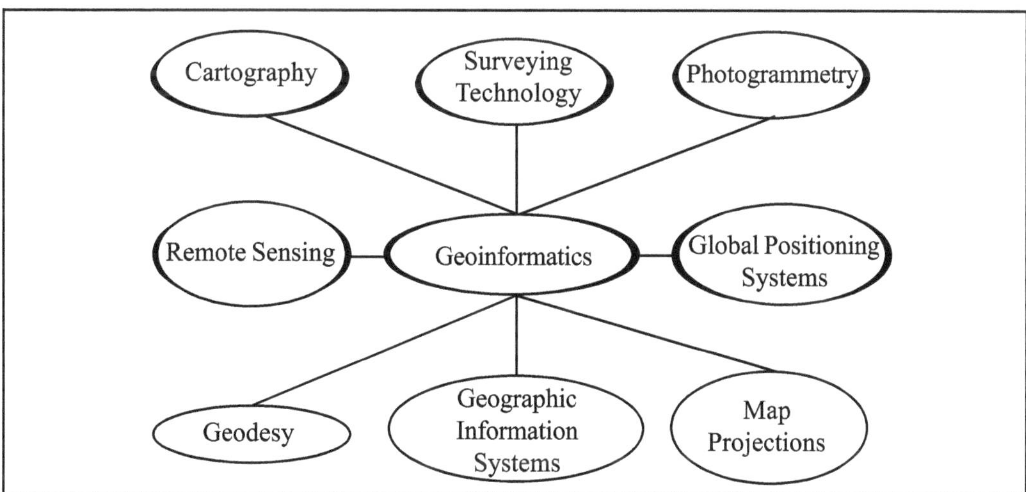

Fig. 1.1 Concepts and Techniques of Geoinformatics

Chapter 2 discusses the concepts of surveying technology including datums and reference systems, classification of surveying methods, stages in surveying, planimetry and height control by traverse method, plain table surveying methods and modern trends in surveying and mapping. Chapter 3 examines the cartographic principles with details of cartographic symbolization and generalization. Cartographic design, thematic and digital cartographic principles are also discussed in this chapter. The history of photogrammetric engineering, classification of aerial photograph, components and types of aerial cameras, photographic scale and geometry along with stereo photogrammetric principles and ground control, aerial triangulation are presented in chapter 4. Chapter 5 highlights remote sensing physics and techniques, remote sensing platforms and sensors. Chapter 6 describes the geodetic surveying principles and mathematical aspects of ellipsoid, expressions for gravity and potential and measurement of gravity on earth are discussed under Geodesy. An emphasis is also given to the conceptual design of satellite geodesy in this chapter.

Chapter 7 describes various components of global positioning system, creation of new data from GPS readings along with the GPS data capture for GIS. Chapter 8 discusses the concepts of map projections including co-ordinate systems, re-projection system, common map projections such as transverse meracator and Universal Transverse Meracator (UTM) along with the related parameters. Chapter 9 focuses fundamentals and roots of GIS, definitions and its terminology, theoretical models of GIS, various data models and GIS operations along with broad view of applications.

The second part of this book starts from 10th chapter and it considers the Forest Resource Management and large variety of data products that use the various methods forest resource surveys for its effective management. Chapter 11 deals with the watershed management and saltwater intrusion modeling is presented in chapter 12. Water quality mapping and modeling which is very new area of research is clearly discussed in chapter 13. Chapter 14 gives an overview on solid waste management and also describes different selection criteria for selecting suitable sites for solid waste dumping using specially designed and developed decision support system. Chapter 15 and 16 stresses both concept and practice of geoinformatics applications to landslides and pipeline alignment surveys respectively. Potential fishing zone mapping and urban planning and management explained and presented in chapters 17 and 18.

This book emphasizes both concept and applications of geoinformatics for environment management. Geoinformatics concept explain the purpose and objectives of operations and the interrelationship among these operations. For most geomatics users GIS is a problem solving tool (Wright etal 1997). To apply the tool correctly and efficiently the GIS user must become proficient in using the tools of geoinformatics.

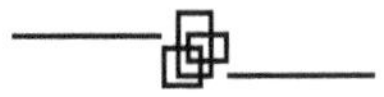

2
Surveying Technology

2.1 Introduction

A topographic map is a graphic representation of selected natural and man made features of a part of the earth's surface on a definite scale and in their correct relative geographic positions and elevations. Symbols, colour differentiations and contours help to show the physical features; mountains, valleys and plains; in their true relationship to land and to man made features; in a way, they are an inventory of the physical features of and on the surface of the earth. They are a living testimony and record of humanity's achievements and a modern blue print for planning the future.

2.2 Men Behind the Making of a Map - "Surveyors"

Surveyors - ever in quest of unknown groping from darkness to light - each step a link in the unraveling of the world around them - alone, often forsaken but always dedicated. For a surveyor exploration of the most beautiful and bountiful land has been always an exciting adventure.

Surveying implies the determination of the relative location of points on, or near the surface of the earth. It is the art of measuring horizontal and vertical distances between objects, determining the direction of lines and in general establishing the geographical position of points by measurement of geometrical quantities like, angles and distances and physical quantities like gravity, atmospheric pressure etc.

In most of the survey operations the ultimate aim is to present the data collected in a form convenient for obtaining the information required without going to the ground. The "form" generally used is the "map" which gives the representation of the terrain on any plane surface to certain scale.

2.3 Datums and Reference Systems

A datum and reference system are any quantity or set of quantities that may serve as a referent of basis for the calculation of other quantities. Generally, datums and reference system can be explained by means geodetic datums. A geodetic datum is a set of constants specifying the coordinate system used for geodetic control i.e., for calculating coordinates of points on the earth. Broadly two types of datums are used in surveying for mapping purposes, namely, horizontal datums and vertical datums. The horizontal coordinates, geodetic latitude and longitude, define a horizontal datum. This is usually accomplished on a country by country bases. A few examples of such datums are the North American datum, European datum, Tokyo datum etc., (Geodetic Glossary, 1986). At least eight quantities are needed to form a complete datum - three to specify the location of the origin of the coordinate system, three to specify the orientation of the coordinate system and two to specify the reference ellipsoid (Moritz, 1980).

Heights of surveyed stations or Bench marks (BMs) which may or may not be part of the horizontal geodetic network are used to adjust and define a vertical datum. For vertical datum, there must be a quantity that serves as a referent, which is the geoid or more practically a surface which approximates the geoid. The geoid is defined as the equipotential surface of the earth's gravity field which best fits in the least squares sense, mean sea level.

Both datums and reference systems are used to determine new coordinates of points needed for boundary determination, navigation, construction, natural resource management and myriad other applications.

2.4 Survey Operations

Survey operations can be divided into two main categories. The first category is the "field work", of taking measurements on the ground. The second category is the "Office Work" of computing, plotting, drawing and any other operations necessary for the purpose of surveying and mapping.

2.4.1 Field Work

For carrying out field work efficiently, the following points are to be borne in mind. A preliminary reconnaissance of the ground to decide the detailed procedure of work. A reconnaissance based on the existing map (if any) covering the area is to be invariably carried out before going to the ground.

Proper administrative arrangements and selection of the right type of instruments consistent with the accuracy is desired. Also clear technical specifications for work is also required.

2.4.2 Office Work

This includes the necessary calculations to transform the field measurements, observations etc., to a form suitable for plotting. The type of map projection required for plotting or drawing have to be decided. Inking and completion of the drawing depends on the need and specifications. Latest drafting methods can be used for better results.

Calculation of quantities shown on the drawings for special purposes such as areas of land etc; for compilation of technical data concerning the map.

If the field work is based on Photogrammetrically compiled drawing maps, the photogrammetric work also forms as a part of office work.

2.5 Classification of Surveys

Broadly speaking surveys may be classified into two categories :

(i) Geodetic Surveys and

(ii) Plane Surveys

2.5.1 Geodetic Surveys

Geodetic Surveying has for its object the precise determination of the positions of some selected points on the earth's surface in a system of control network. The positions are determined both relatively, in terms of lengths and azimuths of the lines joining them and absolutely in terms of coordinates of latitude, longitude and elevation above mean sea level. These geodetic control form the initial frame work for second order control work and topographical control work.

The geodetic surveys can be viewed in terms of geodetic reference systems. Soon after the launch of the first man-made satellite in 1957, geodesists defined geodetic coordinates with respect to a rectangular coordinate system whose origin is located at the centre of mass of the earth and is called as geocentre. Using a mean earth (best fitting) ellipsoid, these coordinates can be converted to geodetic latitude, longitude and height above the ellipsoid using the inverse of equations. The geodetic reference system reflects the evolution of a datum given changes in technology. The introduction of Global Position System (GPS) technologies led the way to a more global, rather than local system.

Geodetic surveys include geodetic astronomy, gravity surveys, magnetic declination surveys, geodetic triangulation and studies connected with the size and shape of the earth.

2.5.2 Plane Surveys and Shape of the Earth

The shape of the earth is that of an oblate spheroid the polar axis being shorter than the equatorial axis. For small areas the curvature of the earth being insignificant, the earth's surface over that area is considered as a plane. Surveys of limited extent, where curvature of earth is neglected are called "Plane Surveys".

It is seen that the Polar axis is shorter than the equatorial axis by about 27 miles. Relative to the diameter of the earth, this is a very small quantity, less than 0.34%. When it is considered that the length of an arc 11 ½ miles long lying along the earth's surface is only 1.5 cm greater than the subtended chord and further that the difference between the sum of the angles in a plane triangle and the sum of those in a spherical triangle is only one second for a triangle at the earth's surface having an area of 75.5 miles, it will be appreciated that the shape of the earth need to be taken into consideration only in surveys of precision covering large areas.

2.5.3 Applications of Plane Surveying

Plane Surveying has a wide scope and public utility and its methods are employed in the vast majority of surveys undertaken for various purpose such as engineering, architectural, legal, commercial, scientific and geographical. The operation of determining elevation is usually considered as a division of plane surveying. Elevations are referred to a spheroidal surface, a tangent at any point in the surface being normal to the plumb line at that point.

2.6 Classification of Plane Surveys

Based on the specifications and applications, plane surveys can be classified into six categories, namely,

- Topographical Surveys
- Photogrammetric Surveys
- Engineering Surveys
- Hydrographic Surveys
- Mine Surveys and
- Cadastral Surveys

2.6.1 Topographical Surveys

On the basis of data obtained in the field, a topographic map can be prepared indicating the relief or elevations and inequalities of natural and cultural details. The field data consists of angular and linear measurements termed as horizontal or plan control and elevation of points by spirit levelling or trigonometrical levelling by what is termed as height control.

2.6.2 Photogrammetric Surveys

Science of measurement by means of photographs, as applied to topographical surveying is called photogrammetric surveying. With specially designed cameras, photographs are taken from aircrafts. The principle of "perspective" is employed in the projection of details in plan and elevation from photographs on to the plotted sheets drawn to scale. In chapter 4, the fundamentals of photogrammetry are presented.

2.6.3 Engineering Surveys

Surveys that are required for location and construction of lines of transportation of communication such as highways, railways, canals, power transmission etc; come under this category. Besides topographical work, such surveys may also include the running of levels and calculation of areas and volumes of earth work.

2.6.4 Hydrographic Surveys

This pertains to surveying bodies of water for navigation, water supply or subaqueous construction. The various operations usually involved are

- Topographic survey of shores and banks.
- Determination of depth of waters and character of the bottom of by angular and linear measurements.
- Plotting of Hydrographic map.
- Observation of fluctuation of Ocean tides.
- Measurement of the discharge of streams.

2.6.5 Mine Surveys

This is a combination of the principles involved in land, topographic and engineering surveys. Both surface and underground surveys are often required to locate on the surface shafts, adits, bore holes, rail, roads, tramways, mills and other details and also delineate fully the mine workings.

2.6.6 Cadastral Surveys

These are the classes of land surveys which are executed for the purpose of systematically recording land rights, producing registers of land-holdings, or an inventory of land areas, land uses and classifications or, of determining tax assessments from the land.

2.7 Principles of Surveying

In all survey work, the aim should be to keep the errors to the minimum and to maintain a uniform standard of accuracy throughout the survey. To achieve this the undermentioned principles are followed.

- Working from whole to part
- Economy of accuracy
- Consistency of accuracy and
- Independent checks

2.7.1 Working From Whole to Part

This is the basic principle that is being followed in all types of survey work. In fairly extensive areas, first a system of control points are to be fixed with a high standard of accuracy. In between these points the work can be carried out by less accurate and less expensive methods. The idea of working in this way is to prevent accumulation of errors. On the other hand if the reverse process is followed and the survey is made to expand outwards, it will generally be found that minor errors become highly magnified and the process becomes uncontrollable at the finish.

2.7.2 Economy of Accuracy

The instruments used and the procedure followed in any survey should be such, as are capable of yielding the accuracy required. It would be a waste of labour and money to adopt elaborate procedure, not required, depending on the purpose for which a survey is meant. That is, there should not be any needless refinement.

2.7.3 Consistency of Accuracy

In extensive surveys, where different portions of survey have to be done by different surveyors or at different times, a uniform standard of accuracy should be maintained. Similar procedure should be followed and instruments used should be similar, in all such cases so that the whole survey can be treated as one and corresponding to the same specifications, throughout.

2.7.4 Independent Checks

An independent check on the work by a surveyor must be always provided to ensure that the work is of required accuracy. For example, in plane-tabling, if the position of an object has been fixed by two ray intersections a third ray from another position gives an independent check. In triangulation, the three angles of a triangle should be observed, though two angles are sufficient for carrying out computations.

2.8 Methods of Surveying

It is always practicable to select two points in the field and to measure the distance between them. These points can be plotted on paper at a distance apart depending upon the selected scale. From these initial points other points can be located by two suitable measurements and can be plotted further on the sheet.

Direct methods of locating the third point C, with respect to two given points A and B are illustrated below, in figure 2.1.

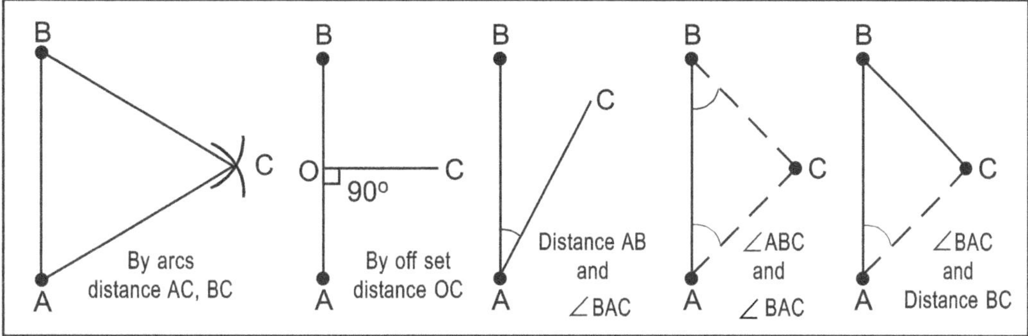

Fig. 2.1. Direct methods of locating point C, with respect to two given points A, B.

A Survey may therefore be executed in several ways by different combinations of instrument and methods. The principal factors to be considered are :

- Purpose of survey
- Nature of the Country
- Degree of accuracy required
- Extent of survey and
- Time available for both field and office work

To select the methods best suited to a particular case demands on the part of a surveyor, a degree of skill which can be acquired only by experience.

2.9 Stages in Surveying

In a normal surveying project, the following stages are involved.

(i) Control Survey

Planimetry and height by the following methods :

- Triangulation (both planimetry and height)
- Traverse (both planimetry and height)
- Height control by spirit levelling.

(ii) Detail Survey by original plane tabling methods

(iii) Contour survey

(iv) Detail Survey by other methods

- Revision survey (updating of existing maps)
- Blue print survey
- Verification survey
- Colour print survey

2.9.1 Planimetry and Height Control by Triangulation

Triangulation is the process of measuring the angles of a chain or net work of triangles formed by stations marked on the surface of the ground. It is an elementary trigonometric proposition that if the angles of a triangle and length of one side are known, the lengths of all the other sides can be computed. Further, if the position of one point and the direction of one side are known with reference to some system of coordinates, it is possible to calculate the coordinate of all the other points.

2.9.2 Data Necessary for Commencing Triangulation

The initial data required for commencing topographical triangulation in an area of which a detail survey is contemplated are :

- base of known length
- an azimuth (direction from true north)
- latitude, longitude and mean sea level height of one station

The above can be made available from the data of previous triangulation or geodetic triangulation series data.

2.9.3 Layout of Triangles in a Series

The arrangement of triangles of a series must be suited to the nature of the country. A series may consist of simple triangles, braced quadrilaterals, centered triangles and polygons. Ideally simple triangles should be equilateral, quadrilateral should be square and other figures should be regular to avoid inaccuracies. Normally the layout chart is planned and drawn on existing maps of the area with the given specifications.

A figure showing the different triangle lay out is given in Fig. 2.2.

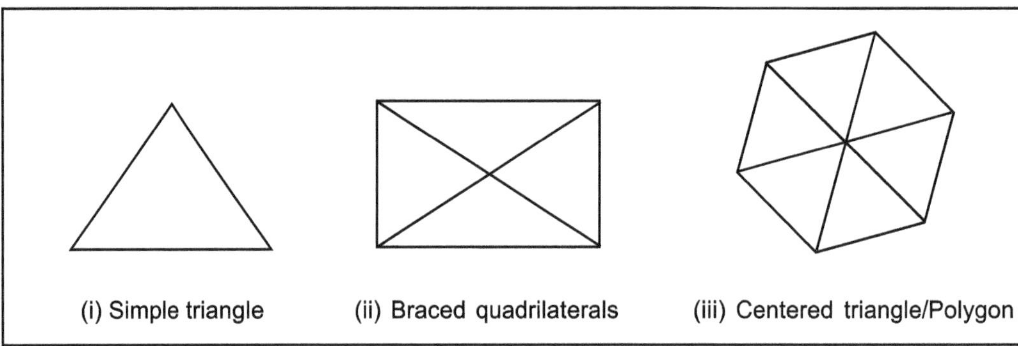

Fig. 2.2 Layout of triangles

2.9.4 Ground Reconnaissance

When a plan for a proposed triangulation has been drawn up it is necessary to fix sites and select stations giving necessary intervisibility. During the reconnaissance, the stations are prepared and also auxiliary control points called as “intersected points” are also selected. During reconnaissance, a reconnaissance chart on a proper scale is prepared and note book is maintained giving necessary particulars of the triangulation stations.

2.9.5 Preparation of Stations

During reconnaissance the control station points are selected and are prepared. Stones on which circle with centre dot are engraved and buried one below the ground and another one exactly above the underneath made in flush with the ground. After preparation of the ground mark, signals are prepared with cross poles and cairns for observation from other stations. While preparing the signals care is taken to keep the signal exactly above the ground mark and also to maintain the verticality.

2.9.6 Observation of Angles at Station

For topographical triangulation optical glass arc theodolites are used for observation of angles (horizontal and vertical). At every station the horizontal and vertical angles of the required and specified stations are observed as per the reconnaissance chart. All the specifications and rules regrading the observations of horizontal and vertical angles are strictly carried out. A programme for visiting and observing all the stations covering the area is prepared and the programme is executed accordingly.

2. 9.7 Computations and Calculation of Coordinates and Heights after Adjustment

After completion of the field observations, the entire figure of triangulation is checked. The coordinates are computed and adjusted. Number of manual and computer based programmes are available for computation of triangulation coordinates, and running down the list of coordinates. A computation volume with list and description of stations is prepared after the completion of the programme. A "history sheet", of the triangulation work is also prepared.

2.10 Planimetry and Height Control by Traverse

2.10.1 Definition of Traverse

A traverse is a course taken when measuring a connected series of straight lines, each joining two points on the ground. The points are called "traverse stations" and the straight line between two traverse stations, "traverse leg". Traverses can be closed or unclosed (open). Traverse starting from a known station and closing at the same station is called closed traverse. Traverse starting from a known station and closing at another known station is called open traverse. The open traverses are preferred over closed traverses because of independent check availability. All traverse coordinates are computed from bearings. A traverse bearing is the angle which the traverse leg makes with the meridian at the origin of survey as measured from north by east.

2.10.2 Traverse Observation

The lengths of the traverse legs are measured by Electronic Distance Measurement (EDM) instruments. Angles are measured by glass arc theodolites. Vertical angles are measured for computation of heights of stations. Number of specifications are given for angular observation for error checks and accuracy improvement. All traverse stations are prepared on ground either permanent, semi permanent before start of observation by digging stones with circle and dot marks or by digging long wooden pegs in the ground with numbering of each station.

2.10.3 Computation of Traverse Lines

Number of manual or computer based programs are available for running down the traverse line angle and bearings and computation of coordinates and heights. Once the field observation books are scrutinised, the traverse lines are "run down" for azimuth bearing checks and computation of coordinates carried out. After completion of the work, a "Traverse Volume" is prepared with all the data and with the list of coordinates and heights. A history sheet is also prepared giving all the details of the traverse work including the accuracy achieved.

2.10.4 Height Control

2.10.4.1 Independent Datums for Planimetric Coordinates and Heights

Datum for planimetric coordinate network can be chosen by the mapping agency or the map designer which depends upon the projection used in the map. But for heights the datum through out the country is "Mean Sea Level" (MSL). Reference made to para (2.3). The heights are measured by the principle of flow of water. Water flow is from a higher elevation to the lower elevation. Hence it is by natural choice, the datum for determination of heights and computation of elevations is the lowest level of the land viz., : the sea level. For computational purposes the datum for the entire country is uniformly determined and is called the "Mean Sea Level". The mean sea level datum is exclusively for measurement and computation of heights/ elevation of control points.

2.10.4.2 Methods of Determination of Elevation of Control Points

The methods are as follows :

- Height determination during Triangulation/Traverse
- Height determination by Spirit Levelling

2.10.4.3 Height Determination During Triangulation / Traverse

During triangulation and traverse, along with horizontal angles, vertical angles of the control points are also observed. The vertical angles along with the distances are used for computation of the height differences between the points. If height of the point of observation is known the heights of all the other observed points can be deduced by observation of the vertical angles and calculation of the distances between the points. Number of specifications are laid down for determination of heights and adjustment of heights in a triangulation/traverse network.

2.10.4.4 Height Control by Spirit Levelling

Spirit levelling or levelling is the term used for determination of heights by an instrument called "level". The levelling instrument is designed to read the exact horizontal direction from a given point of observation. Once the instrument is fixed at a particular point, the observation (horizontal) are taken to levelling staves kept at a known start point (back position) and at another point (normally) forward in the same line. The levelling staves are of 4 meters height and graduated in such a way that horizontal readings can be taken upto three decimal points of a metre by observation through the telescope of the level instrument.

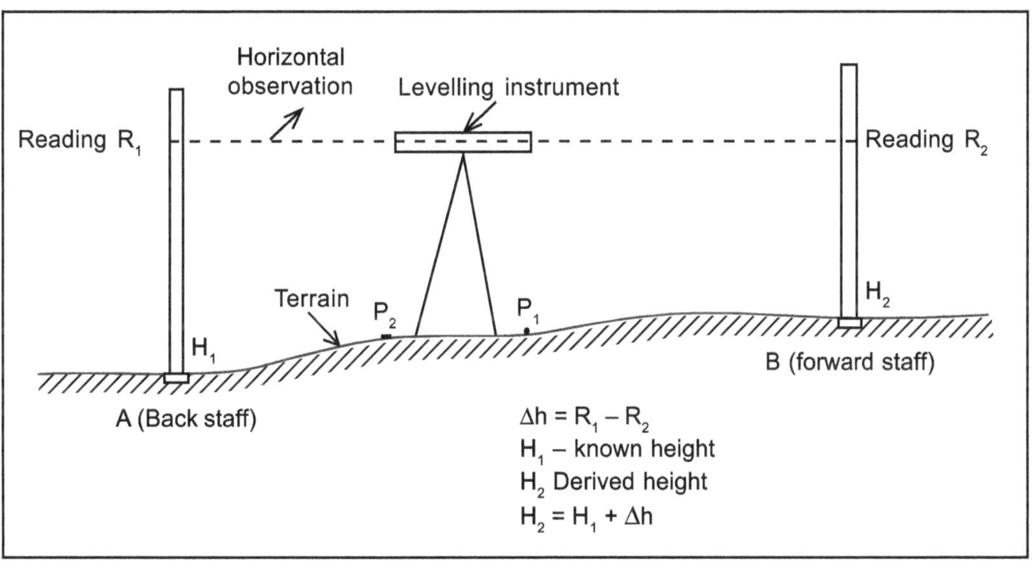

Fig. 2.3 Principle of spirit levelling and determination of heights.

Figure 2.3 shows the basic principle and method for carrying out spirit levelling for height control. 'A', is the point on firm ground or peg or platform where the levelling has to start. It will be known as back station. 'B', is again a firm point on ground where the first level height has to be established. The point B will be known as front station or forward station. The levelling instrument will be placed between the centre of the distance AB to avoid collimation error of instrument. Two levelling staves are placed over points A and B vertically and held by helpers for observation by the level instrument. The observer at position P, will first observe the reading of the level staff positioned vertically at A. Before taking the reading the instrument is levelled exactly horizontal by means of a spirit level bubble fixed on top of the observation telescope of the level instrument. Once the instrument is levelled and pointed to the graduation marks of the levelling staff kept at A the "reading" is recorded from the staff (say R_1). Without disturbing the instrument the telescope is again pointed to the front position staff kept are position 'B'. The instrument is again "levelled" by means of the spirit level bubble and the observation on the front staff is recorded from position P_2 (say R_2). The difference of the reading is taken always from Back-forward i.e., $R_1 - R_2$.

The difference of reading R1 – R2 (say Δh_1) gives the height difference between the points A and B. If back - forward reading is (+) the point B is at a higher altitude and if (–) the point B is at a lower altitude. If the height of point A is known, the height of point B can be calculated by applying Δh (+ or –). For carrying forward the level line, now B becomes the back station and another front point can be selected say 'C', The height difference Δh_2 can be observed with the same procedure of observation carried out earlier, by keeping the level instrument between point's B and C. The same procedure can be carried out for any number of points and Δh established between the back and forward points. The level line can be closed on a known point for check. After closing on a known point are the Δhs are "*run down*" and heights deduced for each and every intermediate points and adjusted at every point.

These are number of already established height points of different accuracy levels through out our country. These established height points are called Bench Marks (B.M.s).

2.11 Detail Survey by Original Plane-Tabling Methods

The plane table measures 750mmx 600mm and length of the tripod are usually 1,200 mm long. The instrument is made entirely of well seasoned wood, except for the metal plate, bolts, nuts and screws which are of brass and shoes to the legs which are of iron. The sight rule is a simple instrument for pointing to the details and giving rays. The instrument consists of a metal bar with straight edges hinged at one ends by "Sight-Vane eye end and at the other end by sight-vane object end. Object is sighted by intersecting it by the sighting thread. For plane table surveys the instruments other than plane table and sight rule are clinometer (for height determination) Spirit level, box compass, umbrella etc.,

Three requirements are met for setting up the plane table at a station.

(i) The table must be levelled (ii) the table must be oriented and (iii) the table be kept vertically over the station (centering).

2.11.1 Methods of Plane Tabling

Methods of surveying with a plane table may be classified under four distinct heads viz. Radiation, Intersection, resection and traversing. The radiation consists in drawing a ray to the point to be surveyed from the plotted P.T. position measuring the distance and plotting it. The method is generally used in large scale surveys in combination with other methods. The scope of the method is widened if distances are obtained tacheometrically. The inter section consists of in locating the position of points by intersection of two rays drawn to them from two different plotted positions on the board. This method is largely used for surveying details. The resection method is used for locating or fixing surveyor's position on board from which the survey of details could be done by radiation or intersection. Resection may be performed from three visible points and their plotted positions. Given three points A, B, C visible from a station P over which the table is set up and a, b, c their plotted positions to determine p the plotted position of P. The following procedure is adopted to fix the position of P on the board.

Orient the board approximately with the help of compass (if available) or by estimating the probable position of the point. Take rays from A, B, C through their respective plotted positions If the board is truly oriented, these three rays will pass through a point which is the required position. If not they will form the sides of a small triangle called, "Triangle of error". The exact position of the point P is determined by applying *Lehmann's rules,* and the board is oriented to the farthest position. The rays Aa, Bb, and Cc will now meet at p. If not the process is repeated till a unique result is obtained.

Lehmann's rules

(i) The point 'p' is so situated that its distances from the three rays are in proportion to the distances of the respective stations.

(ii) Looking at the direction of each station, the point 'p' will lie on the same side of each ray either left or right.

(iii) If P is out side the circumscribing Circle of A, B, C then p lies on that side of the most distant ray as intersection of the other two rays.

The strength of the fix is bad if

(i) P lies close to one of the sides of the triangle ABC

(ii) Both angles subtended at 'p' are small

(iii) P lies near the circumscribing circle of triangle ABC

The strength of the fix is good if

(i) p is well with the triangle ABC

(ii) The middle station is much nearer to P than the other two stations

(iii) One angle is small (tending to zero) when two points are in range and the other is large.

The fixing is indeterminate, if P lies on the circumscribing circle of the triangle ABC. (Danger circle).

Traversing method is used for laying down the survey lines of a closed or unclosed traverse and corresponds to theodolite traversing. The orientation may be done by compass or back ray. The details may be located by offsets. The methods is specially useful for surveying long courses of streams, or in jungle country where visibility is restricted. The closing error in the traverse is adjusted graphically.

2.11.2 Survey and Depiction of Details

By employing the methods of plane tabling described above the details on the ground are surveyed and symbolised on the plane table sections. Symbols should be shown according to the prescribed symbol chart. Separate symbols are available for large scale surveys and small scale surveys. Symbolisation can be classified as (i) Symbols in profile e.g., temples, tree etc. (ii) symbols in plan e.g. huts wells etc. (iii) Symbols depicting particular area e.g. Cultivation limit, tanks etc (iv) Line symbols, e.g. road, footpaths etc. (v) Symbols depciting boundaries e.g. District boundaries, state and international boundaries etc. The symbol chart depicts the details in the following order.

(i) Heights-trigonometrical and plane tabling
(ii) Abbreviations e.g., P. O. Hospital etc
(iii) Water features
(iv) River banks, canals etc
(v) Villages, buildings etc
(vi) Roads and Bridges
(vii) Railways, telegraph lines etc,
(viii) Embankments and cuttings
(ix) Coastal symbols
(x) Boundaries and limits
(xi) Orchards and gardens
(xii) Ornamentation and trees
(xiii) Sand features
(xiv) River features in plains

2.12 Contour Survey and Depiction of Heights

In topographical maps, hill areas are represented by contours. The delineation of the ground by this method, by means of which relief is obtained by relative frequency of the contour lines, gives a faithful representation of its inequalities and slopes. It enhances the value of the map as a guide to engineering and other operations for planning and developmental projects.

2.12.1 Contouring

The survey of contours (lines depicting the same level) are carried out in small scale topographical surveys by providing numerous heights by the help of *clinometer* and in large scale surveys by provision of height mesh by means of levelling instrument. In both cases the contours are inserted/interpolated on the plane table stations which gives a very accurate representation of the relief pattern of the terrain.

2.12.2 Vertical intervals

Normal vertical intervals for 1:25,000, 1:50,000 and 1:250,000 maps are 10,20 and 100 meters respectively. For larger scale of survey (e.g. : engineering, town planning etc) on scales 1:10,000, 1:5000 the vertical interval varies from 0.5 metre, 1 metre and 5 metre depending upon the requirement and specification of the projects.

2.12.3 Depiction of Heights

Numerous heights are provided during contouring. In addition to this, heights will be provided to important details in the map. Such details can be isolated hillock tops, road and river junctions, bridges, distance stones etc. For large scale mapping as per the requirement of the project authorities height mesh (grid pattern) is also provided using levelling instrument, along with contours.

Relative heights are also provided, where more change in heights occur. For example, embankments, cuttings railway, road, bridges etc.

2.13 Completion of Plane Table Section

After carrying out the survey (detail survey and contouring) and inking up in proper colours, the surveyor must complete the responsibility column box; giving starting date, closing date, total area surveyed and other details like inspecting officer's remark and signatures.

2.14 Office Work

Office work consists of two stages. Pre-field office work and post-field office work. The pre-field office work stage consists of preparation of field sections for carrying out field surveys. This includes planning, division of the entire area into field sections, control plotting, checking and mounting on plane table sections. The post-field office work stage starts after completion of the field work. The surveyed sections are received. They are properly checked for any type of errors or omissions, and corrected. Prints of field sections are obtained and mosaiced in the required format. The enhancement of the field drawing is carried out by different sophisticated drafting methods either manually or based on computer based softwares. After this the cartographer takes over. The map and the design and layout of the maps are prepared. After completion of the design which consists of symbology, colour combination, visual balancing and connected specifications, the map is finally printed.

2.15 Detail Survey by Other Methods - Revision Survey

When the existing survey and map have undergone changes due to development and other physical changes, the following methods are adopted for updating the existing survey and maps. This type of survey is called as "Revision Survey". The following methods are adopted for carrying out revision surveys.

2.15.1 Blue-Print Survey

This is carried out on blue prints obtained generally from existing maps and blue prints obtained from photogrammetrically plotted sections.

In the first case blue prints consists of outline (details) and contours both. The details are verified on the ground and the contours are interplated (if required) and then checked on the ground. In the second case the blue prints normally contain only the planimetry (details only). The planimetry is rigorously verified on the ground by plane tabling methods and contouring is carried out as per specifications. On completion of survey the entire surveyed section is inked up in full colours.

2.15.2 Colour Print Survey

This type of survey is carried out when the previous work is accurate and where major portion of the details and hill features has not undergone any changes.

The survey in such cases is carried out on a combined print with outline (details) in dark grey and contours in brown. Corrections and changes in features are effected by erasing the old work and inking up the new surveyed details and contours (if any) in black and brown.

2.15.3 Verification Surveys

This type of surveys is resorted to when a specific item of detail is required to be checked. It is carried out on the same method described in colour print surveys.

If the verification is not heavy, a print of the published map may be used. New details are inked in red and non-existing detail which require deletion are shown in green colour. An insertion and deletion guide is also prepared in red and green.

2.16 Modern Trends in Surveying and Mapping

In the foregoing sections the discussions were mainly on the conventional approach, conventional methodology, instrumentation and specification. The discussions makes it very clear that the conventional methodology of surveying and mapping is based on very sound footing and there is absolutely no scope for error propogation, accuracy degradation or quality deterioration. It has been already discussed that the basic requirements in map making are :

Reference System and datum, Scale, Symbols, Colours, lettering (Text), Instrumentation (both field & office), Base materials, Man power and funds.

The stages in map making comprises of :

- Ground control (X, Y, Z)
- Extension of control
- Photogrammetry
- Remote Sensing data

- Ground Survey/verification
- Cartography
- Final map composition
- Preparation of final output.

In modern map making environment, the basic requirement and the stages are still the same but the approach and methodologies have undergone a total revolutionary transformation.

The factors contributing to the transformation are :

- Sophisticated Instrumentation
- Computer based -

  Mapping Hardware

  Mapping Software.

The modern mapping technology is based on the following major changes in instrumentation and computer aided mapping software and hardware :

- Global Positioning System (GPS)
- "Total Station", based Surveys.
- Digital Photogrammetry
- Remote sensing data availability for mapping (computer Hardware & Software supported)
- GIS Technology

A brief discussion of each will be done in the following sections.

2.16.1 Global Positioning System (GPS)

GPS consists of 24 satellites and their ground stations (Chapter 7). GPS receivers collects signals from these Satellites and calculates the ground positional coordinates within meter accuracy. Advanced methodology and improved receivers can calculate the positional coordinates with centimeter accuracy. (DGPS) GPS and DGPS methodology are now widely used for Alignment Surveys, Large Scale Surveys, Cadastral mapping and Vehicle Tracking.

2.16.2 Total Station System

Very sophisticated Surveying system for on-line survey of ground details. Total station system is called "Electronic Plane Table" as the system can carry out the complete detail survey, just as a conventional plane table.

2.16.3 Digital Photogrammetry

Aerial Photographs and Photogrammetry have come a long way from the conventional mapping to the latest Digital mapping. Starting from Analogue Photogrammetry,

Analytical Photogrammetry, semi analytical/digital Photogrammetry the latest technology which is widely used is Digital Photogrammetry. The basic orientation procedure, data capture and height determination all remain the same. But the entire process is now carried out and executed under digital environment (chapter 4).

The hard copy aerial photographs are replaced by scanned imageries, as input. The following stages are done :

- Creation of 3D stereo imagery from scanned over lapping aerial photographs.
- Computer based interior and exterior orientation procedures
- Data capture by hardware attachment and 3D viewing systems.
- Contouring / DEM generation / orthophoto preparation.

2.16.4 Remote Sensing Technology

This system of mapping is called mapping from space. Remotely sensed geographical data received from space by highly sophisticated sensors fitted to satellites are used as input for mapping (chapter 5). Processing of remote sensing digital data received at earth satellite stations are carried out by modern sophisticated technologies and are used as input for mapping and also for updating of existing maps.

2.16.5 Geographical Information Systems (GIS)

The concept of map as a hard copy for reference by map users have been totally changed by the GIS. The map is now an Information System. The hard copy map having all the details, description and contours have now been transferred to an information system based on layers (chapter 9).

The geographical Information system is a dynamic, on-line Information System of complete map data (in the integrated form or in the layer based form) which is proved to be most convenient for on-line information and monitoring, on-line study and analysis and on-line decision making. Most of the map users now prefers the GIS than the hard copy map.

3
Cartography

3.1 Introduction

The user of Geoinformatics must be aware of the capabilities of the system they are using. What classification methods are provided ? What forms of symbolization are possible ? Can a user customize the map ? What operations for generalisations are available ? For too often, because of poor education in cartography, users default to customised approaches that are inappropriate for their data, and easily mislead the user of the analysis. Since more often than not, the final output of a GIS based analysis is a map, the GIS specialist is well advised to carefully consider the cartographic representation (Dent, B.D. 1996) early in the process. Cartography is witnessing significant changes as modern computer systems allow more sophisticated forms of visualisation, four dimensional cartographies and multimedia representations. Cartography can be defined as the art and Science of map making, chart or map relief models. Cartography comprises of the various elements and techniques that can be used in or can influence the design of maps, how map appears and how its appearance can influence the understanding of the map by the map user. Cartography is concerned with two factors: how features can be represented on maps and how maps can express themselves or communicate to the map user. Cartography is a discipline that is thousands of years old, yet has just in the last several decades seen enormous changes as a result of digital technologies. Comprehensive surveys of the history of cartography are provided Bagrow (1964), and Thrower (1996).

3.2 Task of Cartographer

The art or science of cartography comes into picture at the stage of design and preparation of maps. The task of cartographer in design of maps can be listed as :

- All maps are functional documents.
- Maps have to convey specific information to the map user.
- To ensure that the specific information is conveyed clearly to the map user.
- Cartographic design is very important in the processing and production of a map.

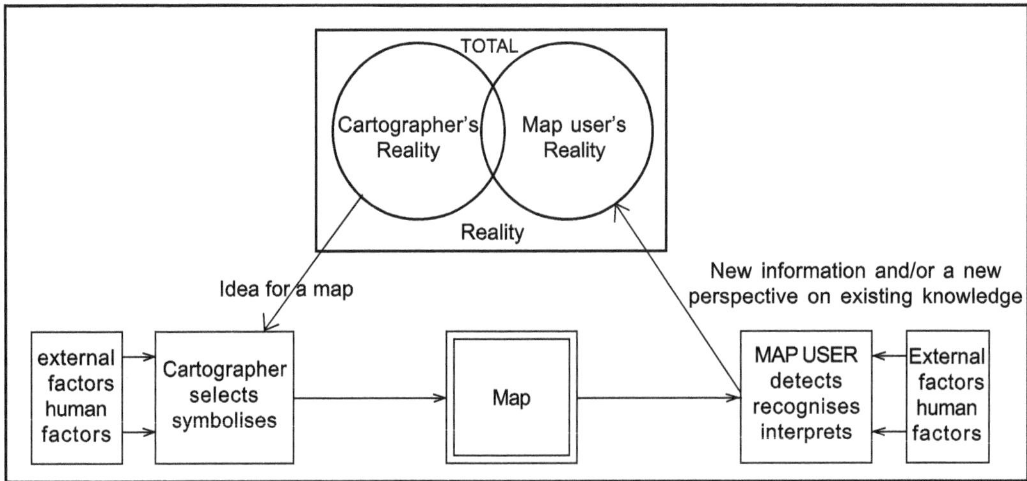

Fig. 3.1 Model of Cartographic communication.

3.3 Model of Cartographic Communication

The following flow chart gives a clear understanding regarding the function of a cartographer in map design for clear communication property by the maps.

3.4 Visual Perception and Communication

Looking at various types of maps, one can safely say all maps have the following in common properties like :

(i) All maps contain information

(ii) this information in the map is presented in a graphical form, and

(iii) a common purpose of all the maps is to give the information represented to the map user. In other words, one can say that all maps contain a certain amount of information which is represented graphically and the representation should be such that the map user can readily understand and appreciate the information.

Communication is the transfer of ideas, knowledge or information from one person to another. The most common method of communication is through the medium of language, which can be spoken or written. Maps are also a form of communication. In other words the map user gets the information from the map through the process of visual perception. The communication process is best understood by comparison of the map with a book. In a book the communication is carried out by language, whereas in a map the communication is effected by symbology. This is depicted in the following flow charts.

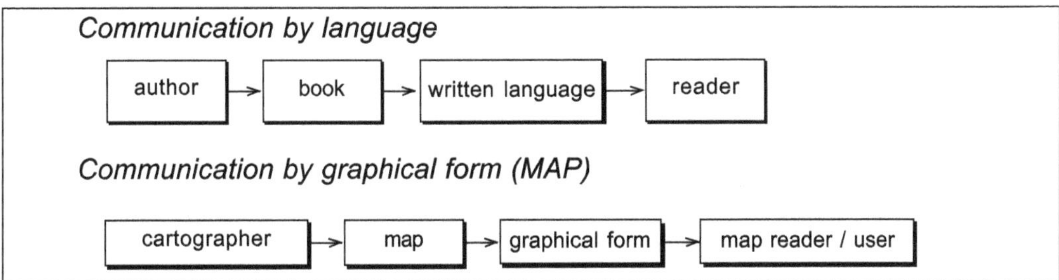

3.5 Cartographic Symbolisation

It is now clear that, the purpose of map is for communication. The map designer (cartographer) designs the map in such a way in the graphical format for communicating his ideas/contents to the map user. In this section we will be discussing about the graphical format used in a map for communication to the map user. Semiology is the study of signs or symbols. In terms of communication symbols have been compared to words in a language. However it is perhaps more accurate to think of the whole map as symbol as well as its constituent parts. A symbol is any drawing or otherwise constructed image used as means of presentation/communication. Fundamental to the understanding of cartographic symbolisation is the concept of the primary visual variables as presented by Jacque Bertin in his influential book, the Semiology of Graphics (Bertin, 1983). To understand symbolisation, one must realise the relationship among the fundamental types of geographical data, the measurement levels, the forms of cartographic symbolisation, the visual variables and graphic elements that makes a specific symbol.

3.5.1 Classification of Symbols

There are great many different kinds of symbols. One has only to think of the numerous different types of maps to realise this. It is possible, however to make a classification of individual symbols. The major broad classification of symbol are : Point symbol, Line symbol and Area symbol.

A brief description of the above three categories are given in this section.

Point symbol used to indicate the location and identify or other characteristics of features of small territorial extent in relation to the map scale. The scale factor is very important in point symbol. The line symbol used wherever features are to be represented has a linear characteristic, for example roads, river, railway etc. Area symbol used to represent features of considerable areal extent, again in relation to map scale. For example : reservoirs, parks, forest coverage etc.

In addition to these symbols, two other types of symbols are qualitative symbols and quantitative symbols. The qualitative symbols deals with quality. Symbols indicate and identify or describe the nature of the features. There are three categories in this type like :

- **Qualitative Point Symbols :** Descriptive or pictorial symbols : Based on actual pictorial representation of the features they are intended to show.
- **Geometric or Purely abstract symbol :** Symbols of Geometric shapes chosen at random. Bears no resemblance to features.
- **Letter or number symbol :** Abbreviations to give identify of specific features. For example, school, post office, etc.

Qualitative line symbols :

Descriptive Symbol : Indicates the description of natural or man made features which exist in the terrain.

Purely abstract or imaginary line Symbol :

e.g., : Grid lines, Political, civil boundaries, latitude, longitude lines etc.

Qualitative area symbols :

These indicate what appears within a specific area on the map. Again the symbol may be descriptive or abstract.

To sum up, qualitative symbols, indicate what a feature is, differentiate it from other feature, and give the location of the features.

All quantitative symbols deal with amounts, they indicate "how much" of a particular feature or phenomenon exists. Quantitative symbols also can be further classified as point, line and Area.

Quantitative point Symbol :

Point symbols can also indicate amounts simply by their presence in the map.

e.g., : dot map, where each dot can represent some fixed number.

Quantitative line symbol :

Isoline are one of the most commonly used quantitative line symbols. Isoline, is a line on which all points have the same value. The most familiar isoline is "isohypse" (hypsoheight), which is more commonly called the "Contour". Isolines are used to indicate a great many different types of information on maps.

e.g., : air pressure (isobars), temperature (isotherms), depth (isobars) etc.

Quantitative area symbols :

Areas enclosed by quantitative area symbols represent variable quantities symbols represent variable quantities

Eg : Population density map, Industry map etc.

Plan and profile symbols :

Depending on the presentation, symbols can also be classified as plan symbols and profile symbols. Point, line, area symbols belong to plan symbols (viewed orthogonally from above.)

Normally point symbols like trees, towers, places of worship (Temple, Church, Mosques etc.) are shown in profile (viewed horizontally).

3.5.2 Variable Governing the Symbology

In the previous paragraphs detailed classification of symbols which are used for cartographic communication are given. All the symbols and their characteristics for visual communication are governed by shape, size, orientation, value, density and colour. Symbols can range from simple geometric shapes (square, circle, triangle etc.) to extremely complex shapes. Size indicates the relevance of a particular symbol in a particular context. Orientation of symbols in the right specified direction is very important in symbolisation. Value denotes value on a grey scale. Density denotes the repetition of lines or dots or any similar feature in a specified area and the colour used judiciously for symbology for communication. Use of colour in cartographic symbolisation is discussed in the next section. The general characteristics of good symbols are given :

Self explanatory : Symbol on a map is to be not only understood by cartographer but also by the MAP USER. The presentation or design of a symbol should be such that, it can be understood without much explanation.

Universality : Symbols should be selected and represented in such a way that it is generally accepted by all types of map users.

3.5.3 Colour in Cartography

A mention has already been made that colour is one of the visual variables used in symbology at the disposal of the cartographer. It is the most complex variable and it is the most important. We, used to think of colours in terms of red, black etc. It should be apparent that these colours are not something that we can feel or touch, but we perceive with our eyes that object around us have a certain colour. If a beam of white light is allowed to pass through a glass prism it is split up into different colours of the light spectrum.

Exactly how the eye does detect colour is still something of a mystery. It seems certain that the eye itself does not analyse colours but this process takes place in the nerve connections to the brain and in the brain itself. It is probable that there are three sensitive systems one for blue, one for red and one for green. These colours of light can be mixed together to give other colours.

We have already learnt that one of the most important variables in cartography is colour and how human eye perceives colour. Based on this properties of colour as visual variable, cartographers have to use the colour in a map to give the map legibility, contrast and pleasing to the eye (map reader). Given below are some of the standard established norms for use of colours in Survey of India maps and other international maps.

Black : Cultural details like non-perenniel streams, river banks, man made features like railway line, power line etc.

Red : Mostly for man made features, buildings, roads etc.

Blue : Water features - all types

Green : Vegetation, forest areas, plantation

Brown : Hill features like rocks, stony waste, boulders and contours.

In addition to the above, number of shades and other forms of colours are used in thematic maps like geological maps, soil maps, land use/land cover and so on.

3.5.4 Standardization of Symbols

The general standardised characteristic features of detail representation are:

- All symbols are standarised and follow certain specifications
- All symbols have proper measurements like height, width etc. Standardised symbols are called "Conventional Symbols" and are shown mostly on the southern margin of the maps. Relief representation was one of the problems faced by the cartographers in early days.

Number of symbols and methodologies were tried for representation of relief. Some are listed below :

Hachuring : Slopes represented by parallel lines. Drawn heavier for steep slopes.

Shading : Slopes represented by shades, by considering as a photograph with illumination coming from fixed direction. Generally light is assumed to come from the N-W direction.

Hypsometric Tints : Each zone between two levels of heights are coloured by graduated colours.

Contouring : Most common and effective method for depicting hill features. Very valuable in Engineering, Military purpose, geomorphology, geology land slope etc.

3.6 Cartographic Generalisation

Generalisation is the selection of and the simplification of representation of features which appear on a map. It is always executed with reference to scale and purpose of the map. All maps are generalised. The degree and nature of generalisation will vary from one map to another. The process of symbolisation can itself be called as a form of generalisation. The two factors which influence generalisation are : scale of the map and purpose of the map.

3.6.1 Scale of the Map

If we examine maps of the same area at scales 1:10,000, 1:100,000, 1:1,000,000 it is immediately obvious that there is a progressive generalisation as the scale decreases. At 1:10,000 scale separate houses and full street pattern would be shown in a town. At 1:100, 000 scale the same town will be shown by out line and with main streets, whereas at 1:1,000,000 scale, the town will be confined to a point symbol by a circle or square. The concept of scale and importance of scale are provided by M. Anji Reddy (2000)

Obviously then the scale of the map has a profound effect on the degree of generalisation. The smaller the scale, the greater the degree of generalisation. Generalisation and the proper approach for generalisation of features are very much essential during "Compilation" of small scale maps from large scale maps.

3.6.2 Purpose of the Map

There are many different types of maps and the features represented on any particular map should always be selected to fulfil the purpose of that map. On a land use map at 1:20,000 scale it would be necessary to show the agricultural field pattern in considerable detail on a standard 1:20,000 topographic map of the same area much less information on the agricultural pattern would be required. Hence the purpose of the map is of prime importance for determining which feature carry more weightage for representation.

3.6.3 Scale and Purpose

The scale and purpose are closely related. The scale must be chosen to suit the intended purpose. If the degree of generalisation has to be in such a way that the information cannot be shown in sufficient detail at the chosen scale, then the scale is too small to suit the purpose of the map and should be enlarged.

3.6.4 Types of Generalisation

Selection : As already mentioned about the purpose of the map is one of the main factor in determining the selection of details which have to be represented. It is absolutely essential that the map contains all the information relevant to its purpose; conversely it is pointless to include detail which is not relevant to the purpose.

Simplification : Also called structural generalisation. Features which are too small or too complex to be shown in full detail must be simplified to avoid loss of legibility or over crowding. Map scale, here, is the main factor involved.

Omittance : In order to preserve the legibility and clarity of the map certain features may be omitted, even though they may in fact be specified for inclusion. Scale and nature of the terrain is the main consideration here.

3.6.5 Broad Directives for Generalisation

Generalisation is a subjective operation. It is difficult, if not altogether impossible, to give a set of rules for generalisation.

Nevertheless there are some broad rules and directives which can be more or less universally applied.

Observe the map purpose at all times : This directive is not confined to generalisation but is one of the basic rules of map design. All maps are intended to fulfil a certain purpose and the cartographic designer must have this purpose in mind. With respect to generalisation, this means that the selection of detail must be consistent with the map purpose and also that any simplification due to scale should meet the requirements of the map user.

Maintain the essential character of the area : The map should aim to give a faithful representation regardless of scale. As we have seen, simplification increases with reduction in scale but the aim should be to preserve the same visual impression of the character of the area.

Consistency in treatment : It is essential to preserve the same degree of generalisation over the map as a whole and over a map series. There is a strong tendency to over simplify areas with a high density of detail and the reverse treatment for areas of scarce details where the temptation is to under generalise. Carried to an extreme this can give an impression that both types of area have similar amount of detail, which gives a misleading impression of the terrain covered by the map.

This is again subjective and has to be over seen by an expert cartographic designer at every stage of generalisation.

3.7 Legibility

In the context of cartography, legibility can be defined as the case with which a map can be read or its contents appreciated by the user. It is essential that the map as a whole is legible and equal that the individual symbols on the map are legible, both in isolation and in map content. Legibility of a map can be viewed in terms of size of symbols, contrast of map features, visual variables and amount of content of the map.

3.7.1 Size of Symbols

Here a distinction has to be made between visibility and legibility. The presence of a very small object may be detected by the eye but this does not necessarily mean that its shape or other characteristics may be recognised. An object may be visible but not legible and the size required to achieve legibility is greater than that required to achieve visibility. It is seen that visibility is not as legibility as well as detection of the form when the symbol is viewed in a map. It has been found that the minimum dimension to achieve legibility in point symbols to be about 1.00 mm. In case of line symbols, the linear dimension increases the case of legibility and visibility. For visibility 0.2 mm, is, however, the safe minimum width for line symbols.

With regard to lettering, 4 point is normally taken as the absolute minimum for the size of lettering. Depending on the back ground, the complexity of the situation and the style or form of lettering 5 or 6 point value may be necessary to achieve legibility.

3.7.2 Contrast

Contrast in an image, the difference it displays with the surroundings and with other similar images. Contrast is vital to map legibility. The points considered for providing proper contrast for legibility are contrast with background and contrast with other symbols. Considering symbols or lettering of small dimension, maximum contrast is afforded by black-on-white or black-on-yellow. These combinations afford maximum case of legibility. But the converse is not true. As a general rule low value colours (dark colours) on a high value (light) back ground should be used for proper legibility for point, lettering and line symbols. It is already discussed that the visual variables, form, size, orientation, value, density and colour are all in fact used to create contrast.

3.7.3 Visual Variables

The form of symbols should not be too complex. Simple shapes are always easier for the eye to distinguish, they tend to be more legible. They are also easier to draw. The size of the symbol must be suited to the scale and the purpose of the map. Symbols which are excessively large can dominate a map, making other details less easy to appreciate. Symbols which vary in orientation will always have a directional element. This is to be used with caution. Otherwise, this can seriously effect legibility. An example is the use of differently oriented bold line screens to indicate areas. The unpleasantness of the visual appearance can influence the legibility. Extreme values tend to dominate the eye, especially black and low value colours. Again this can influence legibility of other details and caution is necessary. The use of high densities, (high number of repetitions per unit length) is the most effective. As the density is reduced, the eye tends to focus on the individual dots or lines making up the symbol and this can effect legibility by drawing the eye away from the main aspect under consideration. The colour is the most useful and the most complex factor in legibility. Even a little colour can greatly increase the legibility, but it can also have a profoundly detrimental effect, if used without due care.

3.7.4 Amount of Details on Map

As already discussed under generalisation, over crowding of details on a map can seriously effect legibility and should be avoided at all costs. Number of tests have been carried out to try and determine the amount of line and point detail appearing on different types of maps. The figures have revealed that one of the most significant elements in overcrowding is the use of two many point symbols.

Legibility and noise are closely related. Good legibility requires that there is an absence or minimum level of noise. Noise is a factor which interferes with the reading of what is shown on the map in the purely visual sense. In other words, a map may be completely legible without being a correct representation of the true facts or ideas which the cartographer wanted to communicate. Hence for proper communication between the cartographer and the map reader, there should be proper level of legibility with minimum level of noise.

3.8 Cartographic Design

Number of factors have already been discussed which are relevant to the cartographic design. As already emphasised, all maps are made with the objective of communicating the required geographical information to a map reader or map user. Hence the complete process of map preparation viz : compilation, symbolization, choice of scale, and other cartographic process like selection of lines, tones and patterns should always be in tune with the requirement of the map user so that the end product (map) will be effective in communicating to the user, the information he wants. The aim of the cartographic design is to present the geographical data in such a fashion that the map, as a whole, appears as an integrated unit and at the same time each item included is clear and legible. The most important elements to be kept in mind, during cartographic design are : clarity and legibility, contrast, figure - ground, colour and balance.

Most of the above components have been discussed in detail in the earlier sections. To sum up, the objective and relevance of each components are given in the following sections.

Once the geographical concepts underlying the purpose and data of the map are clear and correct, then the legibility and clarity in the presentation can be obtained by the proper choice of lines, shapes and colours and by their precise and correct delineation.

No element of cartographic technique is so important as contrast. The degree to which a map appears precise and sharp is dependant on the contrast structure of the map. Contrast is achieved by varying the visual characteristics of the lines and shapes, pattern, value (relative darkness) and the other characteristics of colour.

The most important element in graphic communication that leads to proper identification of the ideas and concepts of the cartographer, by the map reader is called the figure - ground relationship. The cartographer's responsibility is to design his display graphically so that the intended lines, forms or shapes will immediately and without confusion emerge as the dominating figure or figures.

Use of colour in cartographic design has already been discussed in detail. For the various elements of cartographic design, colour is probably the most interesting to the cartographer. The use of colour is also greatly complicated and often leads to contradictions. Colour is so important, that it leads to remarkable subjective effects on the part of the map readers. It is an important aesthetic element and the cartographer must take care to use it with proper care and caution.

Balance in graphic design is the positioning of the various visual components in such a way that their relationship appears logical. In a well balanced design nothing is too light or too dark, too long or too short, too small or too large. Layout is the process of arriving at proper balance. The relative positioning based on the visual importance of the basic parts of a map is called visual balance. The visual balance depends on the relation of each items to the optical centre of the map and to the other items and upon their visual weight. A good example is a thematic map where the cartographer wishes to emphasis one portion of the map or a particular relationship thereupon. In such case, he would like to make them appear above the background base data so that the eye will focus upon these areas and will only incidently look at the locational base material "beneath" the thematic data. In cases like this, the map is made "out-of-balance", intentionally. Here the visual balance is designed to suit the purpose and the specific requirement.

Thus the aim of the cartographer is to balance his visual items so that they look right or appeal natural for the purpose of the map. Many map series, the sheet margins (north and south margin) are for titles, legends, scales, footnotes etc. In such cases balance of boxes & legends are to be carried out appropriately. Survey of India topo-sheet map series is the best examples for such balancing of north and south margin items.

3.9 Thematic Cartography

A thematic map may be defined as a special purpose map showing qualitative and/or quantitative information on specific "features" or "concepts" or "theme" in relation to the necessary topographic details. Land use/ Land cover maps, geomorphology map, geology map, soil map, water quality maps, ground water prospectus maps, road map, industrial maps etc., are some of the examples of thematic maps.

A base map for a thematic map preparation is a map having topographic information and on which the thematic information can be plotted. Topographic maps are most commonly used as base maps not directly as they exist. But the topographic elements are suitably generalised depending upon scale, purpose and theme of the map.

The primary elements of a base map are graticule or grid, drainage pattern, relief, settlements, communication system, administrative unit boundaries, geographical names, human settlements etc.

3.9.1 Representation of Features and Concepts in Thematic Maps

Qualitative mapping concerned with indicating the identify of descriptive representation of features. It describes; what a feature is, differences of it from other features and location of features. Quantitative mapping besides giving qualitative information of the mapped features, it also gives information on amounts (quantities). It indicates how much or how big a feature is, in addition to qualitative information. The examples of qualitative mapping of point, line and area data are

Point : Pictorial point symbols such as temple, tree etc.

Line : Roads, rivers, etc.

Area : Orchard, soil types, geology etc.

The examples of quantitative mapping of point, line and area data are :

Point : Symbol with value (numbers) dot repeating principle

Line : Arrow symbol map, flow line map

Area : Population density, rainfall etc., Isoline mapping, (lines joining points of equal values) choropleth mapping (quantities are shown as they occur within the unit areas)

3.10 Digital Cartography

Computers have taken over the centre stage for storing all types of cartographic data, their manipulation and generation of outputs on screen as well as on hard copy medium. Much of these form essential activities of geographical information system. In such systems suitable combinations of hardware and software are used to process digitized spatial and linked attribute data sets to build up topographic data bases. Maps in digitized format are used for all GIS activities. Maps generated by digital photogrammetry methods are also stored in digital format. Contours, Digital Elevation Models and Orthophotos are also generated from height points by digital photogrammetric systems. Data can be acquired from one or more of the following sources :

3.10.1 Topographical Maps

The most commonly adopted source for collecting cartographic information are topographic maps which are collection of pure data in organized form. The map data can be easily and quickly converted to digital format (Before conversion to digital data, the prevailing rules and regulations regarding conversion of topo-maps to digital format set by Government of India / Survey of India has to be followed).

The digital conversion can be carried out by manual digitization, scanning (conversion to Raster format) and digitisation (conversion from Raster to Vector format).

3.10.2 Photogrammetry Methods

Second major source of data acquisition where the latest aerial photographs of the desired area are used to create a stereo models from which the map is generated.

3.10.3 Satellite Imagery

Digital satellite data can be obtained from National Remote Sensing Agency (NRSA) from the following satellite data.

- RS (INDIA)
- Land Sat (USA)
- SPOT (FRANCE)

The satellite imagery received in digital format can be processed in different stages and further geocoded and prepared in desired formats.

3.10.4 Flow Chart for Preparation of Digital Data Base

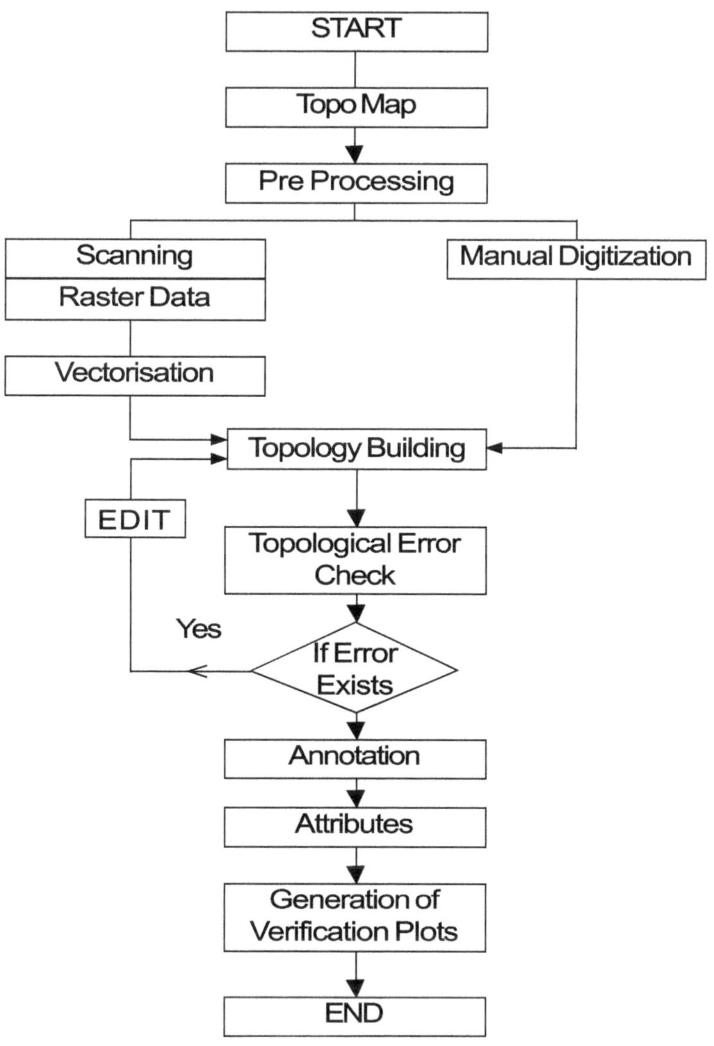

3.11 Conventional Mapping vs. Digital Mapping

Main Characteristics	Conventional	Digital
Inputs	Aerial photo	Aerial photo/satellite imagery
Scale	Well defined	Scale FREE
Resolution	+, –0.15 mm at map scale	Depends on the accuracy of the scale
Identification of objects	By colour, line type symbol, text	By codes and associated attributes
Feature classification/coding	Map legend	Coding table
Symbology	Map Legend	Line/symbol/library
Data collection	2D	2D & 3D
Planimetry/heights	Separately	Combined or separate
Data editing	Manual	Interactive
Data presentation	Manual cartographic processing	Computer assisted cartographic processing
Outputs	Unique standard product (already defined at the stage of the stereo plotting)	Several products : - Line maps (various scales) - Photo maps - Reports - Statistics etc.,

Advantages :

- New graphic products
- Supply of map information in digital form
- Quick response to user needs
- Easy updating of digital topographic database
- Full flexibility; selected area of interest, custom plot, same area plotted at different scales with different presentation and content, etc.,
- New and more interesting job opportunities

Disadvantages :

- High investment for hardware and software
- Highly qualified personnel required
- Benefit at long term

3.11.1 Digital Maps - Advantages

1. Data are maintained in a physically compact format (i.e., the magnetic file).
2. Data can be maintained and extracted at a lower cost per unit of data handled.

3. Data can be retrieved with much greater speed.
4. Various computerized tools allow for a variety of types of manipulation including map measurement, map overlay, transformation, graphic design, and data base manipulations.
5. Graphic and nongraphic (i.e., attribute information) can be merged and manipulated simultaneously in a related manner.
6. Rapid and repeated analytic testing of conceptual models about geography can be performed (i.e., land suitability/capability). This facilities the evaluation of both scientific and policy criteria over large areas in short periods of time.
7. Change analysis can be efficiently performed for two or more different time periods.
8. Interactive graphic design and automated drafting tools can be applied to cartographic design and production.
9. Certain forms of analysis can be performed cost-effectively that simply could not be done efficiently if performed manually (i.e., digital terrain analysis, calculations such as slope aspect, soil intensity, watershed, overlay analysis of multiple sets of complex polygon map sheets etc.,
10. There is a resultant tendency to integrate data collections, spatial analysis, and decision - making processes into a common information flow context. This has great advantages in terms of efficiency and accountability.

3.12 Layout and Numbering of Topographical Maps

Survey of India is the national mapping authority. All the maps produced by the survey of India department are on the geographical layout, i.e., bounding lines are parallels of latitude and meridians of longitude. India and its adjacent countries comprises within the bounding lines 4° to 40° belts of parallels latitude and 44° to 104° of East of longitudes was taken up for mapping by Survey of India under one system.

This large rectangle of geographical lines is divided into 4° belt of latitude and 4° belt of longitude. Each of the 4° square is serially numbered, excluding those squares that do not contain any land.

The layout and numbering to topographical maps or sheets Is based on the sheets of 1/M India and adjacent countries series. Each sheet of this series covers an area of 4° of latitude by 4° of longitude extending over India and adjacent countries. The sheets of this series were published on the scale 1:1M (1:1,000,000) and this series is now obsolete. To find out the number of any sheet the India and adjacent countries index can be referred and it can be seen that the number of the sheet between latitudes 16° and 20° and longitudes 76° and 80° is 56'.

3.12.1 National Numbering System

Three series of topographical maps are maintained by Survey of India

(i) 1:250,000 sheets: (also called degree sheets. Each covers an area of 1° of latitude by 1° of longitude.

So 16^0 sheets cover an area of 1/M sheet. The layout and numbering 1/M (sheet No. 56) is shown in Fig. 3.2.

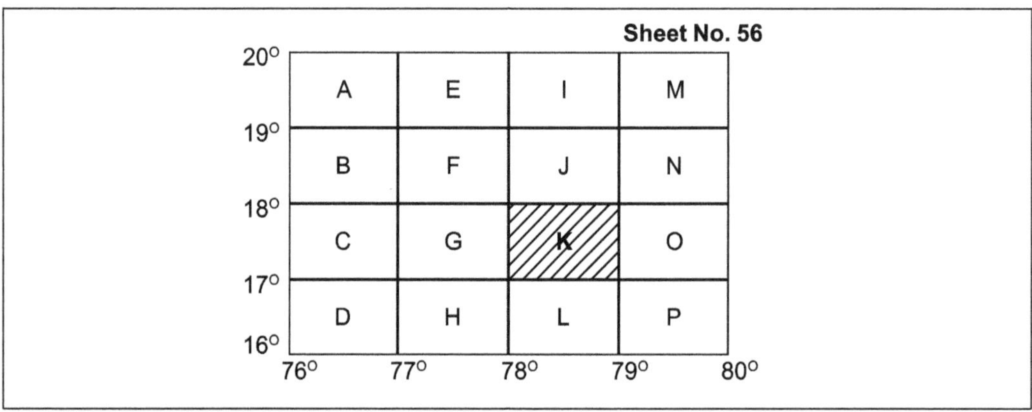

Fig. 3.2 Layout and numbering of sheet 56.

The number of the degree sheet falling between latitude 17° and 18° and longitudes 78° and 79° is 56K.

(ii) 1:50,000 sheets : Each sheet covers an area of 1/4 of latitude by 1/4° of longitude. So one-degree sheet has 16 components 1:50,000 sheets. The layout and numbering of this sheet is shown in the Fig. 3.3.

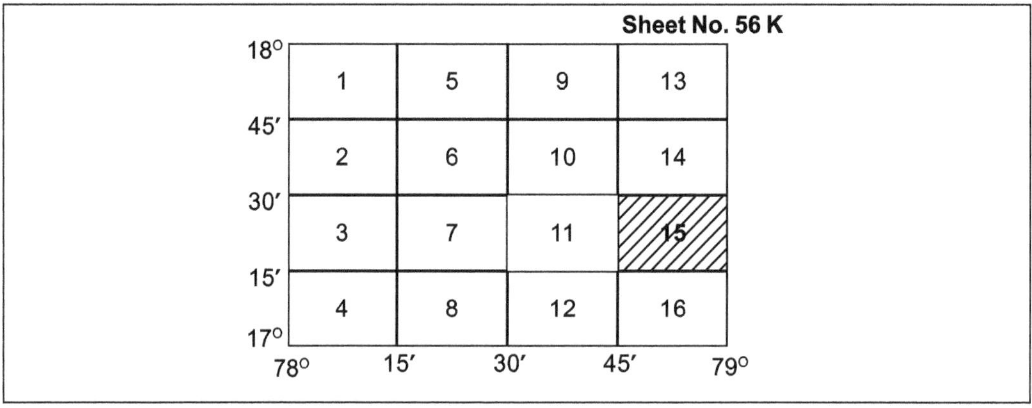

Fig. 3.3 Layout and numbering of sheet 56 K.

The number of sheet falling between latitudes 17°15' & 17°30' and longitude 78°45' and 79°00' is 56K/15.

(iii) 1:25,000 sheets: Each sheet covers an area of 7 ½' of longitude and 7 ½' latitude. So one 1:25,000 sheet has 4-component sheet of the series.

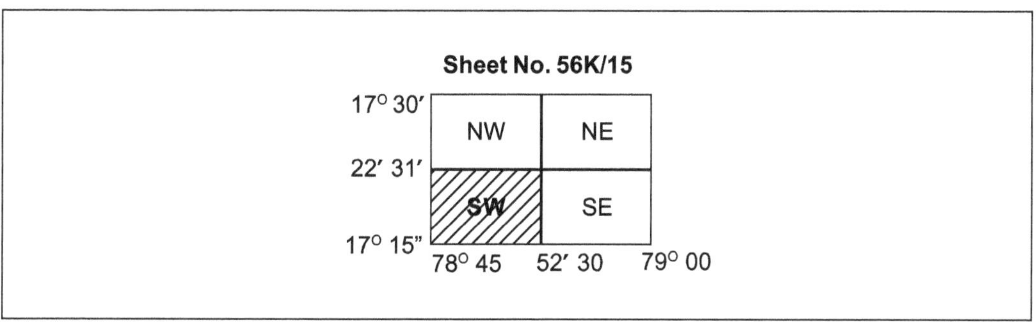

Fig. 3.4 Layout and numbering of sheet 56/K.

The number of sheet falling between latitudes 17°15' and 17° 22' 30" and longitude 78°45' and is 56K/15/SW. The layout of the numbering of this sheet is shown in Fig. 3.4.

3.12.2 International Numbering System

Each sheet of the International Map (Carte International du Monde), which is published on 1/M scale, covers an area of 6° of longitude by 4° of latitude. But pole wards from latitude 60° each sheet covers 12° of longitude by 4° latitude.

In numbering of these sheets belts of 4° of latitude of both North & South of and reckoning from the Equator are designated by the English letters A to V. North and South Belts are distinguished by prefixing N and S respectively in the sheet numbers. The longitudinal belts of 6 degrees reckoning from the date line (180° E or W of Greenwich) and proceeding eastwards are designated by Nos. 1 to 60.

The number of a sheet of the international map series is the letter of its latitudinal belt, followed by the number of its longitudinal belt. For the sheets in the northern hemisphere the letter 'N' is prefixed to the number of a sheet in the southern hemisphere of sheet and letter 'S' to the number of sheet in the southern hemisphere. So by knowing the graticule limits of a sheet we can determine the number of a sheet.

The layout of International numbering system is shown in Fig. 3.5.

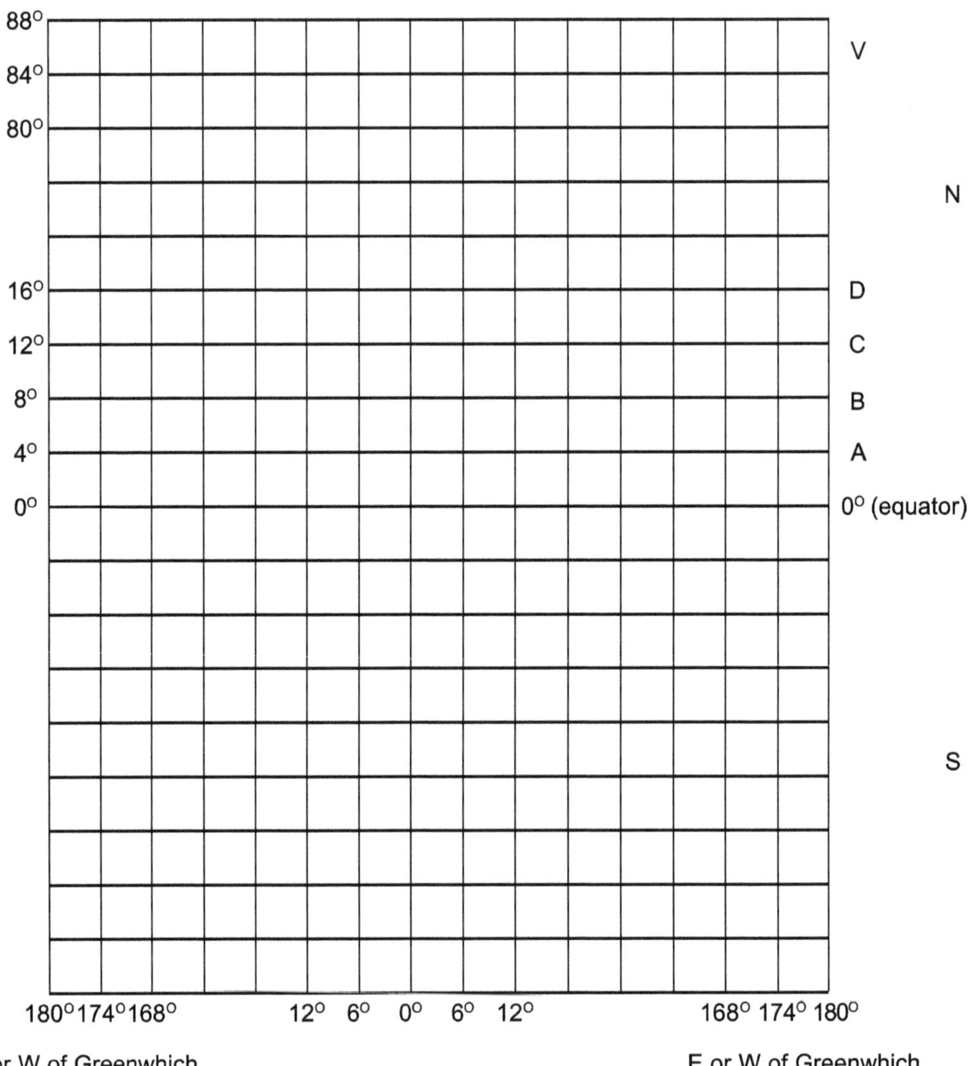

Fig. 3.5 Layout of International numbering system.

Thus the number of international sheet falling between latitudes 16^0 & 20^0 N and longitudes 78^0 & 84^0 E is NE-44

3.13 List of Organisations Engaged in Preparation of Maps

Survey of India (SOI)

Established in the year 1767. Head Quarters at Dehradun. The oldest and premier mapping agency of India.

Geological Survey of India (GSI)

One of the oldest scientific mapping departments in the country. Responsible for carrying out geological and mineral surveys and mapping.

National Atlas and Thematic Mapping Organisation (NATMO)

Established in 1956. Head Quarters Kolkatta. Responsible for preparation of the National Atlas of India and also other thematic maps and Atlases required by the country.

Naval Hydrographic Organisation (NHO)

Controlled by Indian Navy. Head Quarters Dehradun. Responsible for carrying out Hydrographic surveys, of the Indian coasts, and Harbours and to publish nautical charts and connected documents required for navigational and other purposes.

National Remote Sensing Agency (NRSA)

Established in the year 1975. Head Quarters Hyderabad. An autonomous body under Department of Space (DoS), Government of India. Main role is to prepare large scale maps and thematic maps from aerial photographs and satellite imagery. Indian Photo Interpretation Institute, Dehradun is a part of NRSA.

Census Map Office

This office is Under Registrar General of India. Responsible for production of "Demographic maps" and Atlases for the whole of India and its states and union Territories based on the data collected during periodic census.

Other mapping Agencies

Indian Metrological Department

- Climatological Atlas

National Geographical Research Institute

- Several Geophysical Maps of India

Indian Tourist Development Corporation

- Tourist maps of Cities and Towns

Survey and land Records Departments

- State Government Cadastral Maps of respective states.

State Forest Departments

- Forest Maps of respective states.

Director General of Civil Aviation

- Aeronautical charts and Landing charts.

Director, Military Survey

- Maps for Defence Forces

State Remote Sensing Agencies

All states of Indian union have State Government Remote Sensing Agencies under the state government. These are mapping agencies responsible for preparation of state sponsored thematic maps and other large scale maps mostly form satellite imageries.

3.14 Classification of Maps

Based on the published maps by Survey of India (the premier mapping agency for the country) maps can be classified as follows :

- Topographical Maps
- Geographical Maps
- Cadastral Maps
- Thematic Maps
- Large Scale Maps
- Atlas

Topographical Maps

The topographical maps show clearly all the geographical, natural features and man-made features. In India the premier mapping agency, Survey of India is responsible for preparation, publishing and maintenance of the topographical maps for the country. The topographical maps are published in the following scales :

- 1:250,000
- 1:50,000
- 1:25,000

Geographical Maps

Maps covering very large area on small scales are classified as geographical maps. Survey of India is responsible for preparation of the geographical maps listed below and maintenance of these maps, in the following scales.

- Physical Map of India 4.5 M
- Political Map of India 4.5 M
- Rail Map of India 3.5 M
- Road Map of India 3.5 M
- India and adjacent countries series 1:1M

Note : M stands for million

Cadastral Maps

Cadastral maps are maps, which are prepared showing the individual land holdings with their survey numbers. These maps are prepared and maintained by the land survey and settlement departments. These are very important records as they pertain to individual land holdings. All litigations over land dispute are settled based on these cadastral maps. These maps are in one colour and only minimum cartographic information is shown on these maps. The scale of the cadastral maps vary from 1:1000, 1:4000 to 1:8000. Separate reference system is used for cadastral Maps.

Thematic Maps

A Map whose function is to show a particular theme for which it is designed on a topographical base is called a thematic map. The selected theme has more importance than the topography. Examples of thematic maps are : qualitative thematic maps like :

- Land use / Land cover maps
- Geological Maps
- Soil Maps etc.,

and other quantitative maps like

- Population density maps
- Rainfall maps, etc.,

Thematic maps are increasingly used for analysis and study of various data collected for development, planning and geographical studies. The aim of this class of maps is to put life into these inert masses of figures, bring out their hidden meaning and expose the comparative significance of factors in the area; in individually or fused in index combinations.

Atlas

The dictionary meaning of Atlas is "Book of Maps". Example of Atlas Maps are

- ***School Atlas***

  Contains Political and physical maps of India and other different countries, along with map of world and with different types of information.
- National Atlas prepared by NATMO, Kolkatta. National Atlases are prepared mostly on 1:1M scale giving number of statistical information.
- Planning Atlas of States is another example of Atlas maps. Survey of India has prepared A.P. Planning Atlas in collaboration with Osmania University, which is widely used by a number of State Government Agencies.

4
Photogrammetry

4.1 Introduction

Photogrammetry is defined by American Society of Photogrammetric Engineering and Remote Sensing (ASPE & RS) as an art, science and technology of obtaining reliable information about objects, environment through the process of recording, measuring and interpreting photographic images and patterns of recorded electromagnetic radiation through a device. The device is a camera for acquiring a photograph. The science or art of obtaining reliable measurements is by means of photographs. Photogrammetry is frequently decided based on the type of photographs or images used, location of the camera from which they have taken and the purpose of using these photographs. Accordingly, in general, four types of photogrammetry are available today, to the resource scientists, engineers and managers. They are 1. Terrestrial Photogrammetry, 2. Close Range Photogrammetry, 3. Aerial or Space Photogrammetry and (4) Stereo Photogrammetry. The terrestrial photogrammetry deals with the photographs taken from the ground surface as platform for the photographic instruments, like camera. In this photographic system, the distance between the object and the photographer should be 300 mts or more than 300 mts. The terrestrial photogrammetry is called as close range photogrammetry if this distance is less than 300 mts.

Aerial photogrammetry deals with the photographs taken by an aerial camera onboard aircraft. Aerial or space photogrammetry connotes the use of traditional analog photographs (Colwell, 1997) or digital photographs (Hinz 1999) that have been taken from any airborne/space borne platform. These photographs may be either vertical or oblique images based on its geometry and characteristics. The geometry of this type of photographic system is illustrated in Fig. 4.1.

Stereophotogrammetry deals with the overlapping pairs of photographs are obtained, measured and interpreted in a stereoscopic viewing device. This provides a three dimensional view and creates the illusion that the observer or photogrammetrist is viewing a relief model of the terrain. Stereophotogrammetry is essential for creating a digital elevation model of real world by measuring the elevations of 'h - measurements'. This can be performed by using specially designed instruments like stereometer or by computers digitally.

In practice, airborne/space borne photogrammetry is embraced into two broad categories, namely, metric photogrammetry and photo interpretation. In the category of metric photogrammetry, the quantitative measurements are made, from which ground co-ordinates, elevations, distances, areas, volumes and angles can be computed. This is also used to prepare planimetric or topographic maps. This category is often called as quantitative photogrammetry. The most common applications are, (1) The production of topographic maps that contain contours, transportation network and other baseline information of the terrain, (2) The extraction of digital elevation model, (3) The 3-D extraction of planimetric details and contours, (4) Production of orthophoto images.

The second category is photo interpretation in which the photographs are evaluated or interpreted for the extraction of qualitative data like water pollution, classification of soils, geological formations, crops identification and military intelligence etc. (Colwell 1984). This type of photogrammetry is also called as interpretative photogrammetry. Photo-interpretation coupled with fundamental photogrammetric analysis results in the creation of thematic maps of landuse/landcover, hydrogeomorphological and biophysical information maps, which can be used as data sources and data input to GIS.

4.2 Brief history of Photogrammetry

As early as in eighteen century itself the principles of perspective for preparation of maps was suggested by J. H. Lambert. Terrestrial photographs were tried and used for topographic mapping from 1859 to 1909, in France, Canada and Germany. The airplane was first used for obtaining aerial photographs for mapping in 1913.

Instrumentation and techniques contributed during nineth and twentyth century has made photogrammetry the most accurate, efficient and advantageous tool for preparation of topographic maps and also for other applications. With the advances in electronics and computer science, the new concept in photogrammetry called

"Soft copy photogrammetry" or " Digital photogrammetry" has already proven its efficiency in practice. The age old classical analogue and analytical photogrammetric plotters are replaced by computer based "Digital Photogrammetric Work Stations".

4.3 Classification of Aerial Photographs

Broadly, photographs are classified based on the nature of camera and the platform on which the camera is located into two types : terrestrial and aerial photographs.

Terrestrial photographs are the photographs taken with ground based cameras. Camera axis is horizontal photo theodolites are used for taking terrestrial photographs.

Aerial photographs are those photographs taken from an air craft. Camera axis in this case is vertical.

In this chapter, aerial photogrammetric principles are presented as it is more relevant to geoinformatics.

Aerial photographs taken by an aerial camera on board aircraft can be further classified into two categories, namely, vertical and oblique. The vertical photograph for which the camera axis is truly vertical. It can also happen that the camera axis may not remain truly vertical : Normal tilt for preparation of topographical maps can vary from 1^0 to 3^0 such photographs are called nearly vertical. In other words, if the axis of the aerial camera makes an angle 1^0 to 3^0 with the plumbline, such photographs are called as vertical photographs.

In the case of oblique photography, the camera axis is intentionally tilted from the vertical. In this category, there are two types of oblique photographs, namely high oblique and low oblique.

If the camera axis is tilted in such a way that horizon appears on the photograph, such photograph is called as high oblique photograph.

If the amount of tilt is made in such a way that horizon does not appear on photograph, such photograph is called as low oblique photograph. Fig. 4.1 shows the schematic representation of classification of aerial photograph. Fig. 4.2 shows the types of aerial photographs.

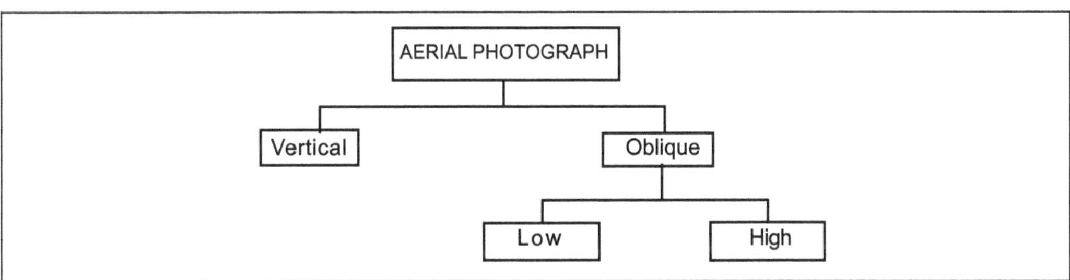

Fig. 4.1 Classification of Aerial Photograph

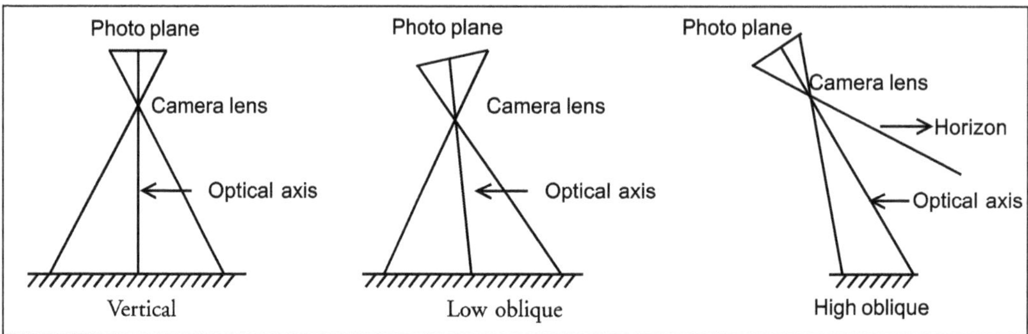

Fig. 4.2 Vertical, Low oblique and High Oblique Photographs

4.4 Organisation of Aerial Photography

Aerial Photography covering a particular area is organised as follows for mapping purpose (Fig. 4.3, Fig. 4.4 and Fig. 4.5).

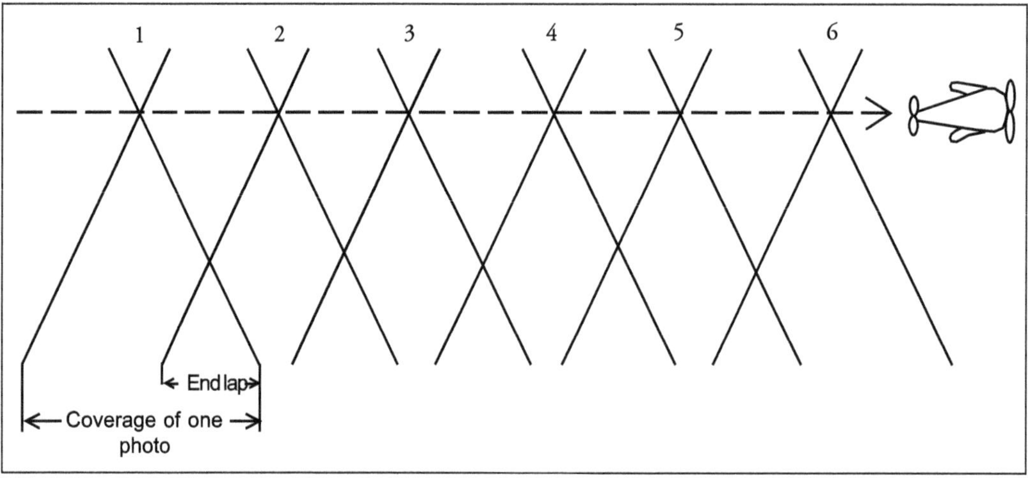

Fig. 4.3 Overlap between two photographs (End Lap)

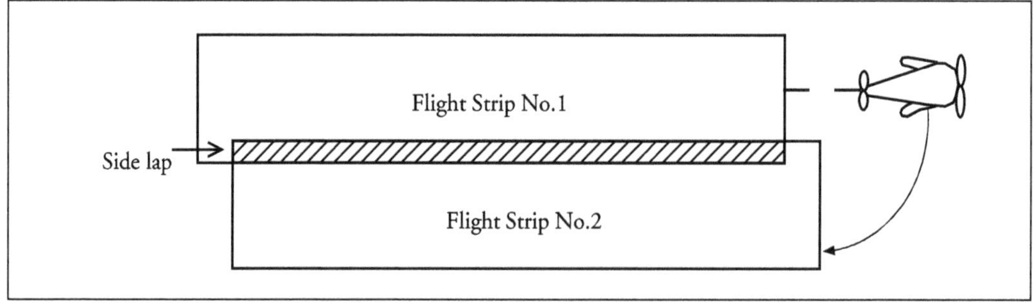

Fig. 4.4 Over lap between two strips - Side Lap

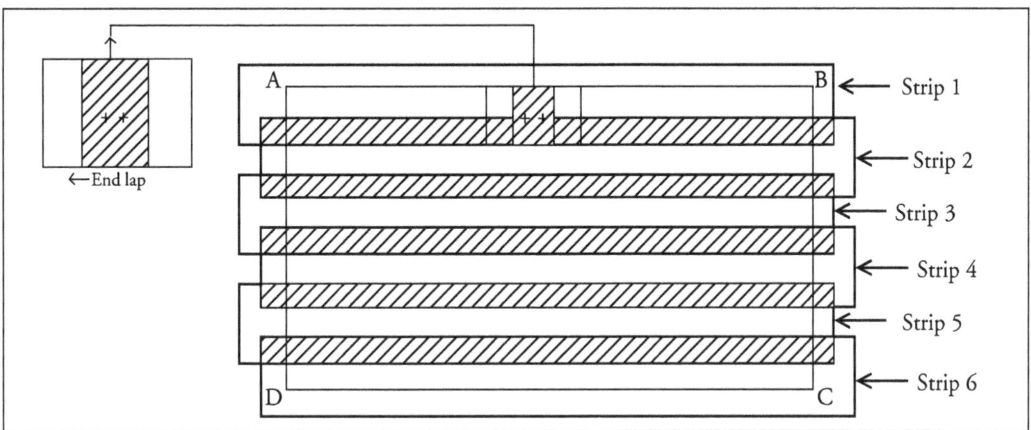

Fig. 4.5 Aerial Photography Block Covering Area ABCD consisting of 6 Strips

Aerial Photograph : One photograph taken from the air at one specified exposure station by camera fitted to air craft.

Aerial Photography STRIP : Number of aerial Photographs along one flight line with specified over laps between the adjoining photographs, normally for the purpose of mapping the area. The over lapping area between the adjoining photographs is kept at 60% or more for mapping. This over lap between adjoining photographs in a strip in called end lap.

Aerial Photography BLOCK : Number of strips covering an area is called block. For the purpose of mapping overlap between each strip is kept at 25% to 30%. This over lap between each strip is called side lap.

4.5 Aerial Cameras

Aerial camera is one of the most important instruments in photogrammetry. Aerial camera is used to obtain aerial photography upon which photogrammetry depends.

Requirements of Aerial Camera are :

- Lens of high geometric quality
- Capable of exposing in rapid succession a great number of photographs to exacting specifications.
- Short cycling time, fast lenses and efficient shutters.
- Capable of functioning under most extreme weather conditions, in spite of air craft vibrations.
- Uses of roll films-capacities ranging from 200 to 400 feet.

Camera mount is a device is used to attach camera to aircraft. The camera mount is designed to carryout the three functions, They are (i) to prevent aircraft vibrations from being transmitted to the camera (ii) to correct the crab and (iii) keeping properly oriented for aerial photos.

4.5.1 Types of Cameras

Broadly aerial cameras are classified in three types.

They are Single lens frame camera, Multi lens frame camera and Convergent camera. Scope of discussions will be single lens frame camera, as they are extensively used for obtaining photographs for mapping purpose.

The features of single lens cameras are related to lens assembly and format of the photographs.

Lens is held fixed to the focal plane to preserve the geometric property and entire format of the film is exposed simultaneously with single click of shutter. The single lens frame cameras are classified into 4 types based on the angular field of view which is an angle subtend by the focal plane of the camera system. If the angular field of view (Fig. 4.6) is less than 60^0, than such cameras are called as narrow angle cameras. The normal angle cameras are those cameras whose angular field of view is upto 75^0. If the angular field of view is from 75^0 to 100^0, then such cameras are wide angle. The last category is super wide angle cameras for which the angular field of view is more than 100^0, some brands of single lens cameras are

RMK 15/23 : f = 150 mm, format 23 × 23cm (f = focal length of the lens).

RC 10 : f = 90 mm, 150 mm and 300 mm format 23 × 23 cm

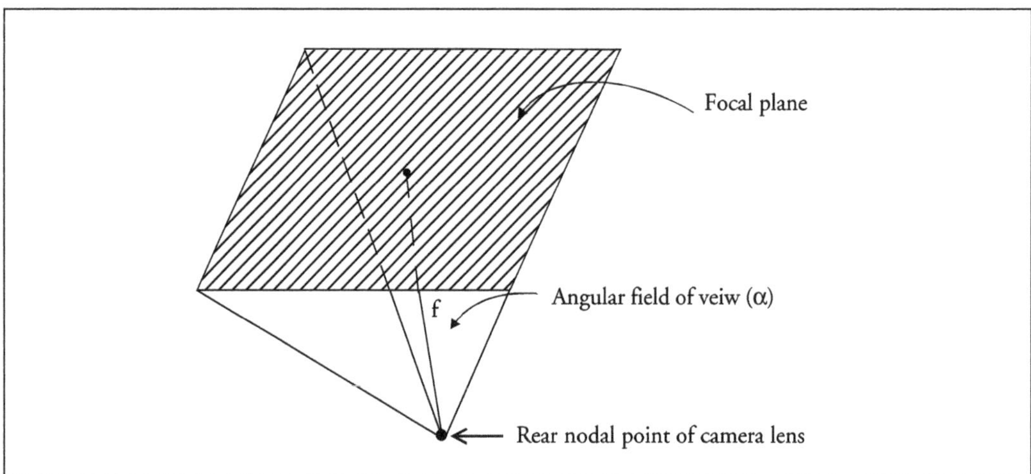

Fig. 4.6 Showing angular field of view (α).

4.6 Components of Aerial Camera

All cameras which are used for aerial photography required for mapping purposes, are the metric cameras. The format size, that is, the actual area covered by photographic emulsion and focal length are known as metric cameras. The basic components or assemblies of a frame metric aerial camera are the magazine, the camera body and the lens cone assembly (Fig. 4.7). The functions of these components in aerial photography are described below.

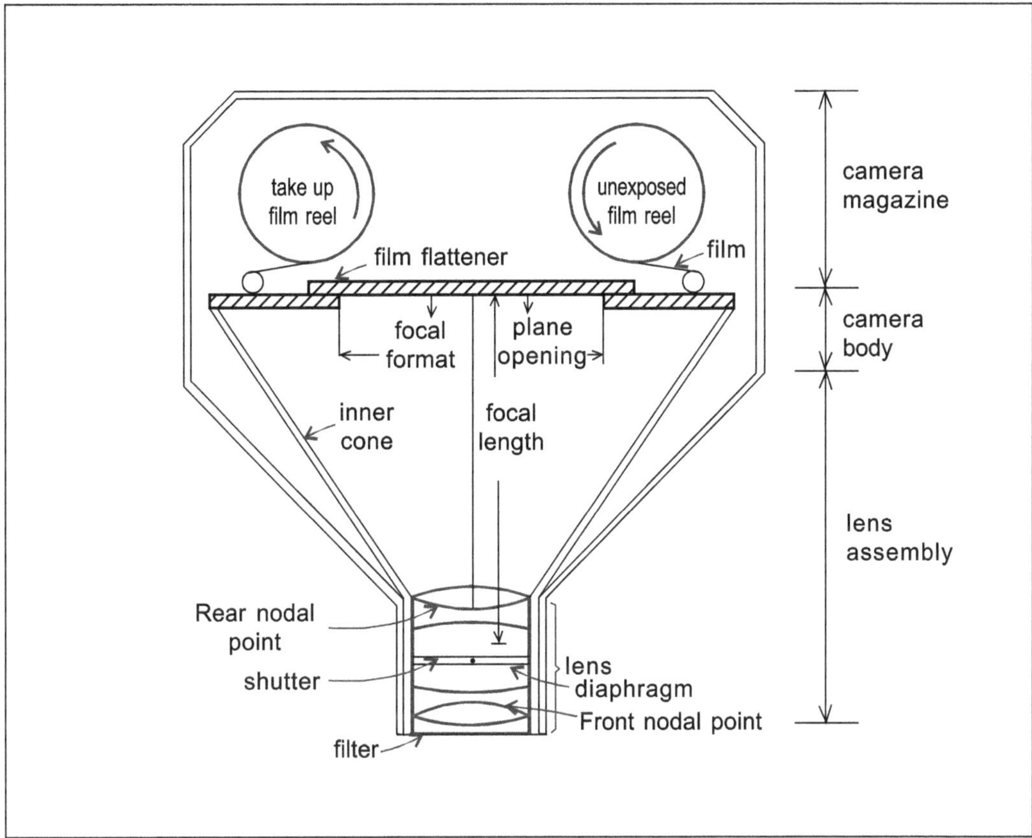

Fig. 4.7 Generalised cross section of a frame aerial mapping camera.

4.6.1 Camera Magazine

The camera magazine houses the reels which hold exposed film and unexposed film contains the film - advancing and film flattening mechanisms. Film flattening may be accomplished in any of the following ways.

(i) By applying tension to the film during exposure.

(ii) By pressing the film firmly against a flat focal - plane glass, which lies in front of the film.

(iii) By applying air pressure into the air tight camera cone, there by film will flat leasing behind the focal plane.

(iv) By drawing the film tightly up against a vaccum plate whose surface lies in the focal plane.

4.6.2 Camera Body

Camera body contains the drive mechanism. Drive mechanism operates the camera through it cycle. The cycle consisting of

(i) Advancing the film
(ii) Flattening the film
(iii) Cocking the shutter
(iv) Tripping the shutter.

Power for the drive mechanism provided by an electric motor.

4.6.3 Lens Cone Assembly

It contains a number of parts and serves several functions. The parts are lens, filter, shutter and diaphragm. It also contains an inner cone of spider.

The camera lens gathers light rays from the object space and brings them to focus in the focal plane behind the lens. The filter serves three purposes, (i) Introduces the effect of atmospheric haze, (ii) It helps to provide light distribution over the entire format, (iii) It protects the lens from damage and dust. The shutter and diaphragm together regulates the length of time, a given amount of light is allowed to pass through the lens to make the exposure. The shutter controls the length of time that light is permitted to pass through the lens. The diaphragm regulates the f - stops of the camera by varying the size of the aperture to control the amount of light passing through the lens.

Camera Mount : The function of camera mount is to absorb the vibration set-up by the aeroplane and maintain vertically of the optic axis at all times during photographic operation. The amount of exposure controls the density of the negative and the quality of details. The recording instrument consists of an assembly of instruments to record relevant data on one of the margins of the negative at the time of exposure.

4.6.4 Resolution of a Lens System

The resolution of a lens is an important characteristic feature of aerial photographic system. The resolution or resolving power of lens is defined as the ability of the lens to show detail. One common method of measuring lens resolution is to count the number of line pairs, i.e., black lines separated by white line spaces of equal thickness which can be distinguished within a width of 1mm in aerial image produced by the lens.

Good resolution is important in aerial cameras because photo images must be sharp and clearly defined for precise measurements and accurate interpretative work. Photographic resolution is not just a function of the lens but also depends on film quality and processing.

Resolution is influenced by a host of parameters, such as the resolving power of the film and camera lens used to obtain an image, any uncompensated image exposure and the atmospheric conditions present at the time of image exposure. An alternative method of determining the film resolution is Modulation Transfer Function (MTF).

The focal length (f) of a lens is the distance from the centre of the lens and the image plane that is the distance from the lens at which parallel light rays are focused to a point.

It is not applicable to a photographic lens. In this, distance is taken from the rear nodal point to the focal plane. For an aerial camera the focal length is defined as the distance between the centre of the lens system and the film plane. It is at this fixed distance away from the camera cone to the focus on the film. For mapping purposes, 152 mm focal length lenses are widely used. Longer focal lengths are used for very high altitude applications. The focal length of a lens, in other words can be defined as the intercept on the optical axis between the rear node of lens and the point at which a sharp image of a distant object is formed. The calibrated focal length of lens is adjusted so as to distribute the effect of lens distortion over the entire field used in aerial camber. This produces an overall mean distribution of radial lens distortion. It is also called as principle distance. The principal distance is the perpendicular distance from the internal perspective centre to the plane of finished negative or print.

4.6.5 Format of a Film

Format of a film is defined as the measurements of a film for one exposure. Various sizes of formats are available such as 70mm by 70mm, 140mm by 140mm or 180 by 180mm. These are used in the earlier cameras but now the format size in most modern cameras has been almost standardised to be as 230mm by 230mm.

Since different companies are manufacturing the aerial cameras differently, it is very important to standardize the format for the production of same size of photographs. The measurements that are in common use are :

Name of Camera	Focal Length (mm)	Format size (mm)	Manufacturer
RMK - A	150 mm	230 *230 mm	Zeiss. W. Germany
WILDRC - 8	150 mm	230 *230 mm	Wild Heebrugh Switzerland
WILDRC - 10	153 mm	230 *230 mm	Wild Heebrugh Switzerland

F - Stop

The diaphragm, in any camera system, regulates the amount of light passing through the camera lens, normally operated automatically and defined by its relative aperture and is equal to focal length / Diameter of diaphragm i.e., relative aperture = f/d = focal length / Diameter of diaphragm. It is also called f-Stop number. The nominal f - Stop setting is 1, 1.4, 2.0, 2.8... and 32. Not all cameras will have all these, but the most explosive cameras will have most of them. In making photographic exposures, the correct amounts of illuminance and time may be regulated by varying f-Stop setting on the camera, while time of exposure is set by varying shutter speed. Variations in f-Stop settings are actually variations in the diameter of the aperture, which can be controlled with a diaphragm a circular shield that enlarges or contracts, changing the diameter of the opening of the lens and this regulating the amount of light that is allowed to pass through it. As the f-Stop increased, the diameter of the lens opening decreases and accordingly the film exposure decreases. Because the area of the lens opening varies as the square of the diameter the change in exposure with f-Stop is proportional to the square root of the f-Stop.

It is not possible to take the photographs by using one camera with fixed focal length and cone angle in order to get all details of the terrain. In aerial photography focal length and angular fields of view play on important role for getting photographs with all required details. Hence it is need to classify the cameras on the basis of these two measurements. For example, for buildup areas, normal angle cameras are used whereas super wide angle cameras are used for the study of hilly terrains. Some time wide - angle cameras are also used for different purpose.

Usually wide-angle camera is used for photographs both types areas, unless the photographs typically required only for a specific purpose.

4.6.6 Photographic Bases

The *photographic bases* are three types, (1) Glass Plates, (2) Film and (3) Paper.

Glassplates are used almost exclusively as base for light - sensitive emulsion. The glass plate can be made almost as plane and constitutes a stable and a non-shrinkable base for the image. This can produce the documentary quality of photographs. But there are disadvantages also. The glass plate is heavily brittle, clumsy to handily and comparatively expensive, changing the negative material becomes complicated which easily courses trouble. And lastly the accommodation problem in the room of the platform.

Now, in aerial photogrammetry, uses film as base for the emulsion. This is non-inflammable and easy to store and handle. It has improved dimensional stability. No accommodation problem. The main draw back of film is, it dimensions variations, of course this change is uniform in all directions. A serious defect is the irregular shrinkage, which occurs in spite of all efforts to improve the film base.

Paper base, it has much lower degree of dimensional stability. For this reason paper prints are not used in photogrammetric work. The distortion in the x - direction is different from distortion in y - direction and this type of differential distortion will be the serious consequences if we neglected. The two types of film are one layer black and white and three layer colour, both of which can be sensitized to either visible only or visible and near infrared wavelengths. The resultant film types are (1) black and white, (2) black and white near infrared, (3) colour film: (a) false colour near IR (b) true colour.

Uses and Merits

1. ***Black and White Film*** : This is orthochromatic emulsion and is sensitive to blue and green. This film is sensitive to broad waveband of visible light is called panchomatic film. The two types of panchromatic black and white film are mapping film which has equal sensitivity of all visible wavelengths. Second one is equal sensitivity to all visible wavelengths. Second one is reconnaissance film which has reduced sensitivity to blue wavelengths to minimise the effect to atmospheric scatter. Mapping film is used in the environmental sciences.

   A Black and White aerial photographs have been used for many applications in the earth sciences, for
      (i) Geological mapping
      (ii) Hydrological investigations and
      (iii) Terrain analysis

   B These are used in agriculture for the identification of crop types, crop disease and soil erosion.

   C Also used is a soil surveys and vegetation survey.

   D The majority of planning authorities through out the world hold these photographs for regional planning, urban planning etc.

2. **Black and white near infrared film :** This film is similar to the above film, but the main difference is its spectral sensitivity which extends beyond visible wavelengths to a wavelength of around 1.0mm, in the near IR region of EMS.

   A Used in studies of vegetation, it has been used for the location of crop diseases, crop type and mapping of forest and semi-national vegetation.

   B It has been used for knowing the soil moisture states of agricultural fields.

   C It has been used for locating the geological boundaries.

4.7 Geometry of Vertical Aerial Photograph

ABCD is the terrain which is photographed. a'b'c'd' is the *negative* obtained during the photography. The negative is a reversal of both tone and geometry of the object space ABCD. The negative is situated at a distance equal to the *focal length* (f) 0'L, of the camera. The *positive* is obtained by contact process. Geometrically it is situated at a distance (OL) equal to the focal length f(O'L) of the camera O'P is the optical axis, which is along the direction of the focal length and perpendicular to both negative and positive planes. The basic geometry of vertical photograph is shown in Fig. 4.8.

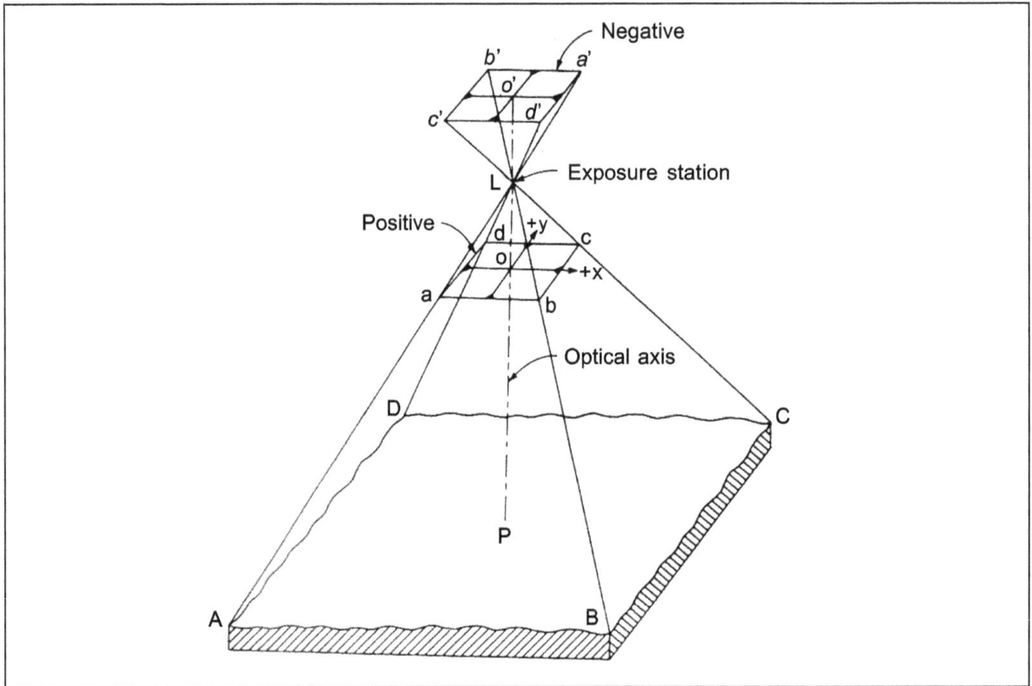

Fig. 4.8 Geometry of vertical photograph.

The first thing that a photogrammetrist/photointerpreter should understand is the geometry of aerial photogrpahy. The aerial photograph is taken by means of an aerial camera on board aircraft / space craft at an altitude based on the scale needed for the given application. The geometry of the photograph is often controlled by the application, scale, size and the condition of the terrain. In general, the geometry of the photography can be known during flight planning. Let us assume that the aircraft is flying at an altitude of 'H' from the Mean Sea Level (MSL). The focal length of the lens used in camera system is 'f'. The focal plane of an aerial camera is the plane in which all incident light rays are brought to focus. In aerial photography the image distances are less than the object distances. Aerial cameras are therefore have their focus fixed for infinite object distances. This is done by setting the focal plane as exactly as possible at a distance equal to the focal length behind the rear nodal point of the camera lens. The focal plane is defined by the upper surface of the focal plane frame. This is the surface upon which the film emulsion rests when an exposure is made. Camera fiducial marks, impressions of holdings, appear on the photograph in the shape of 'V' or 'W' are usually four or eight in number. These fiducial marks are situated in the middle of the sides of the focal plane resulting an impressions on the negative when the picture is taken.The line joining opposite fiducial pints is called fiducial axes. The point of intersection of these fiducial axes is called the principal point. The location of this principal point is based on the position of fiducial points. The principal point may or may not appear at the centre of the photograph. The fudicial axes provide a rectangular coordinate axes which are taken as a photocoordinate system for photogrammetric measurements. The Fig. 4.9 and Fig. 4.10 illustrates the photographic coordinate system.

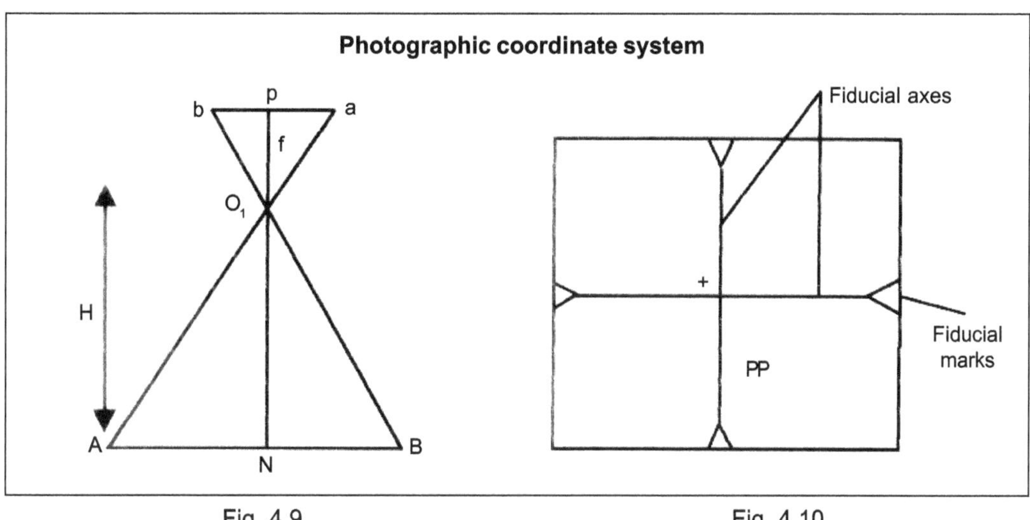

Fig. 4.9 Fig. 4.10

4.8 Scale of a Vertical Aerial Photograph

Based on the geometry of vertical aerial photograph, scale of an aerial photograph can be determined. Scale of a map is different from the scale of an aerial photograph as the map is symbolic representation of earth surface features to some scale and it is uniform throughout the map. The scale of an aerial photograph is different for different points as the photograph is having central projections and the real features are appeared on the photograph. The scale of the photograph can be dealt in two ways based on the nature of the topography. One for flat terrain and the other is for variable or rugged terrain.

4.8.1 Scale of a Vertical Aerial Photograph Over of a Flat Terrain

Let 'A' and 'B' are any two points on the ground and 'a' and 'b' are the corresponding image points on the aerial photograph. 'f' is the focal length of the lens used in aerial camera and 'H' is the flying height of the aircraft by which the photograph be taken. Then the scale of aerial photograph for flat terrain is determined as follows.

From Fig. 4.11.

Δ^{les} aob and AoB are similar triangles.

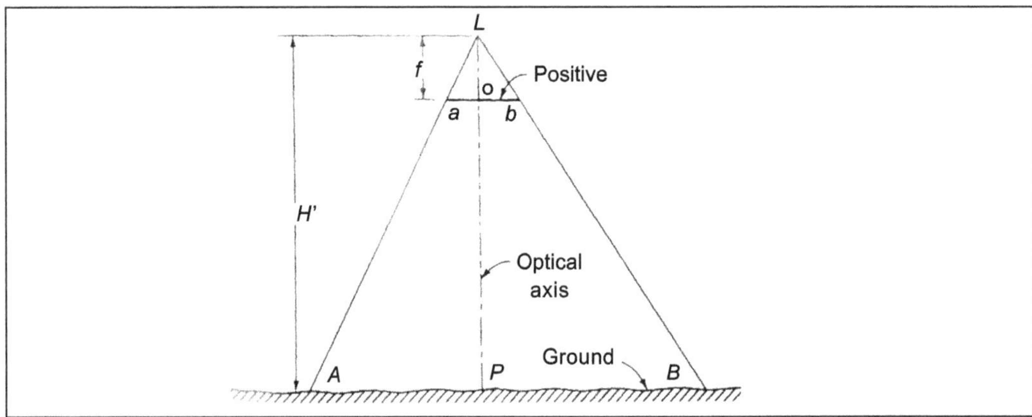

Fig. 4.11. Scale of flat terrain

Scale, $S = \dfrac{\text{Photo Distance}}{\text{Corresponding Ground Distance}} = \dfrac{ab}{AB} = f/H$

i.e., $S = \dfrac{\text{focal length of the lens}}{\text{flying height of aircraft above the terrain}}$

4.8.2 Scale of Vertical Photograph Over Undulating Terrain

Fig. 4.12 scale of vertical photograph-variable terrain camera station-O, flight height H, object point AB image point ab, and elevation of AB from datum h. Ground points AB is imaged ab, on the positive plane; elevation of AB above datum is h AB

The scale at AB = $\dfrac{ab}{AB}$

from similar triangles ab and AB, scale = $\dfrac{f}{H-h}$ considering AB to be infinitesimal, the scale at any point over an undulating terrain with average terrain height h is given by $\dfrac{f}{H-h}$.

Scale of Aerial Photograph for Variable / rugged terrain at the points A, B and C are

$S_A = f / H - h_A$

$S_B = f / H - h_B$ and

$S_C = f / H + h_C$

Where A, B and C are ground points with respect to Mean Sea Level (MSL) (FIg. 4.12) hA, hB are the elevations of A and B where s 'hC' is the depression. Then the scale of points A, B and C can be calculated as given in the above equations.

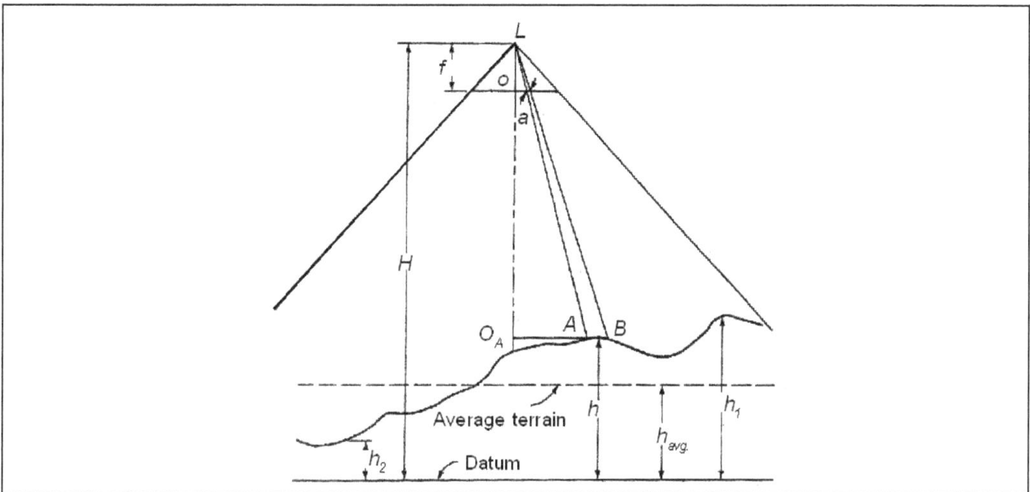

Fig. 4.12 Scale of aerial photograph over a undulating terrain.

4.8.3 Other Methods for Determining Scale of Vertical Aerial Photograph

By measuring ground distance and corresponding distance on photo, between two points, preferably at the same height.

$$\text{Scale} = \frac{\text{Photo Distance}}{\text{Ground Distance}}$$

The exercise can be repeated at 2 to 3 place, for average scale .

By comparison with existing map covering the photographs, measure the distance between two well defined points on photo and corresponding points on maps. Then the scale of aerial photograph is given by

$$\text{Scale} = \frac{\text{Distance on Photo}}{\text{Distance on Map}} \times \text{Scale of map}$$

The exercise can be repeated at two to tree locations and the average scale of the photo can be determined.

4.9 Relief Displacement of Image

In Fig. 4.13 'h' height of the building with A is top point and B is bottom point 'a' and 'b' are corresponding image points on the aerial photos. 'f' and H are focal length of the lens and flying height of the aricarft, respectively. The relief displacement is governed by Vertical exaggeration (V) of the photographic system.

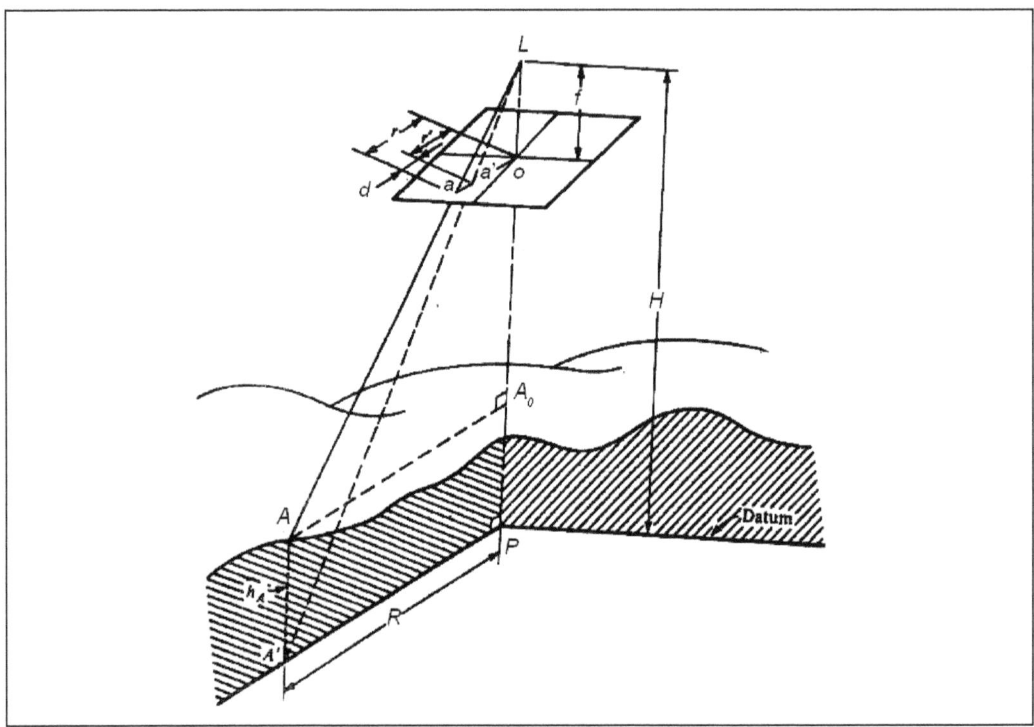

Fig. 4.13 Determination of Relief Displacement.

The displacement of the image due to the height of the object is known as height distortion. The distortion is radial from the principal/plumb point. This means that the distortion or displacement is radial from the plumb point and it increases from the plumb point. Then we can derive the following relationship between h, d (displacement) and 'r' (radial distance) which are depicted in the above figure.

- As the height 'h' of the building increases, the displacement 'd' increases, i.e., $d \ \alpha \ h$
- As the height 'h' of the building increases, the radial distance 'r' increases, i.e., $h \ \alpha \ r$
- As the 'H' increases the displacement 'd' decreases, i.e., $d \ \alpha \ 1/H$

Therefore, we have, $d = r \cdot h/H$

$$\Rightarrow \quad h = d \cdot h/r$$

That is, height of the building is determined by displacement 'd', radial distance 'r' whcih can be measured from the photographs and 'H' can be obtained from the suppliers of photographs. This equation can be used to find the height of any object with the help of a single aerial photograph.

4.10 Stereo Photogrammetry-

4.10.1 Stereoscopic Depth Perception

Methods of judging or measuring depths or distances of objects around us is called as depth perception. Depth perception can be classified as : relative sizes of objects, hidden objects, shadows and differences in eye focussing as per varying distances. Stereoscopic depth perception is much greater degree of accuracy than monoscopic, hence it is of extreme importance in photogrammetry.

Stereoscopic vision is binocular vision. When the left and right eyes are focussed on a certain point, the optical axes of the two eyes converge on that point intersecting at an angle called the parallactic angle. Nearer the object, greater the *parallactic angle* and vice versa. In the Fig. 4.14 L and R are the left and right eyes. The distance between the two eyes is called the *eye base be.* Where the eyes are focussed at point A the optical axes converge, forming parallactic angle ϕ a. Similarly where sighting at point B, the optical axis converge at 'B' forming parallactic angle ϕ b.

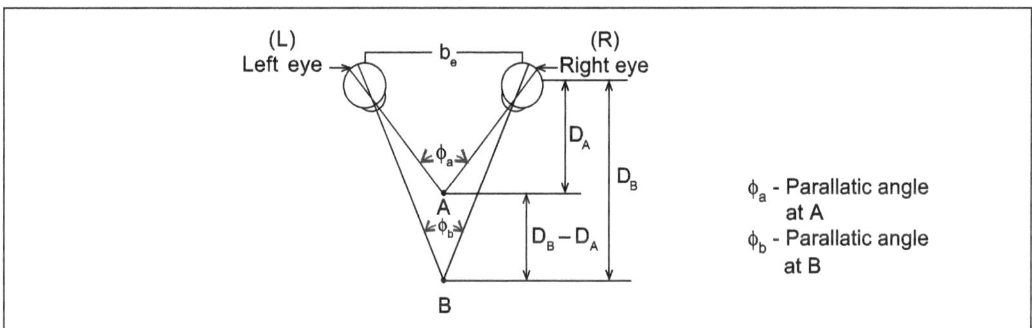

Fig. 4.14 Stereoscopic depth perception as a function of parallatic angle.

The brain associates the distance D_A and D_B with the corresponding parallactic angles ϕ a and ϕ b. The depth between the objects A and B is perceived as the difference between these two parallactic angles. Shortest distance for clear stereoscopic depth perception is about 10 inches for an average adult. Eye base of an average adult is about 6.3 to 6.9 cms or 2.6 inches. Hence the maximum parallactic angle can be calculated by the formula :

$$\phi = 2 \tan^{-1}\left(\frac{1.3}{10}\right)$$

$$= 15^{o} \text{ (app.)}$$

Maximum distance for distinct stereo - for distinct stereoscopic depth perception is not well defined. But for average adult, it is about 2000 ft. Beyond this, parallactic angles are extremely small and changes cannot be differentiated.

4.10.2 Stereoscopic Viewing of Overlapping Aerial Photographs

Fig. 4.15 (a) photographs taken from two exposure stations with building AB in the common overlap area.

Aerial photos are taken from exposure stations L1, L2 so that the building appears on both photos as shown in Fig. 4.15(a). The object point A and B of the top and the ground. This is the situation when aerial photographs of over lapping areas are taken during the flight.

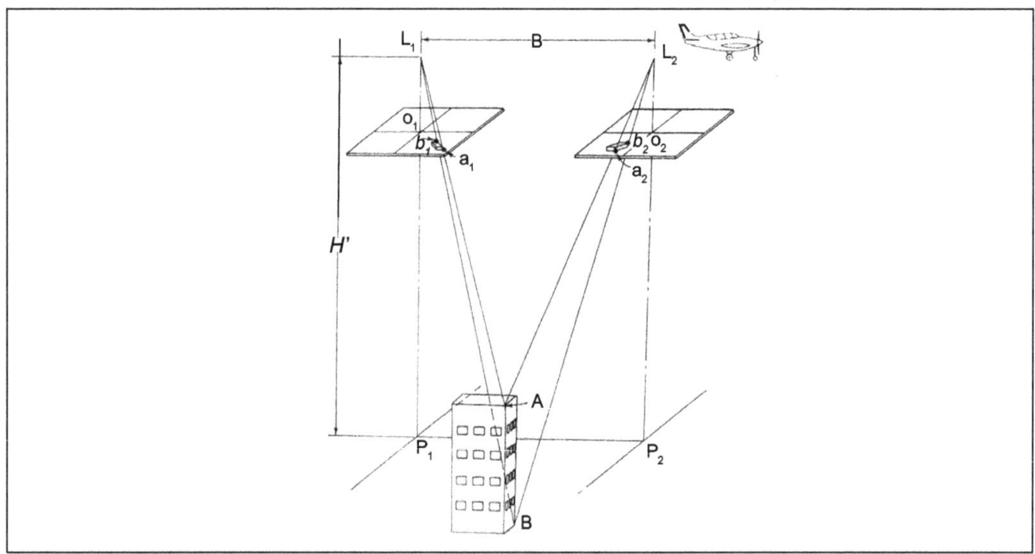

Fig. 4.15 (a) Photographs taken from two exposure stations with builging AB in the common overlap area.

The two photos (left and right) are laid on the table and viewed in such a way that left eye sees only left photo and right eye sees only right photo as shown in Fig. 4.15 (b); the three dimensional impression of the building is obtained. The three dimesional impression appears to be below the table top at a distance 'h' from the eyes. The brain judges the height of the building by associating the depths to point A and B with parallactic angles a and b.

When the eyes view the entire overlap area, the brain receives a continuous three dimesional impression of the terrain. This is achieved by the continuous perception of changing parallactic angles of the infinite number of image points which make up the terrain. The three dimensional model thus formed is called a "*Stereoscopic Model*" and the over lapping pair of photographs is called a "Stereo Pair".

It is difficult to view photographs stereoscopically, without the aid of optical devices. The difficulties can be overcome by an instrument called *Stereoscope.* The most commonly used stereoscopes are Pocket stereoscope and Mirror stereoscope.

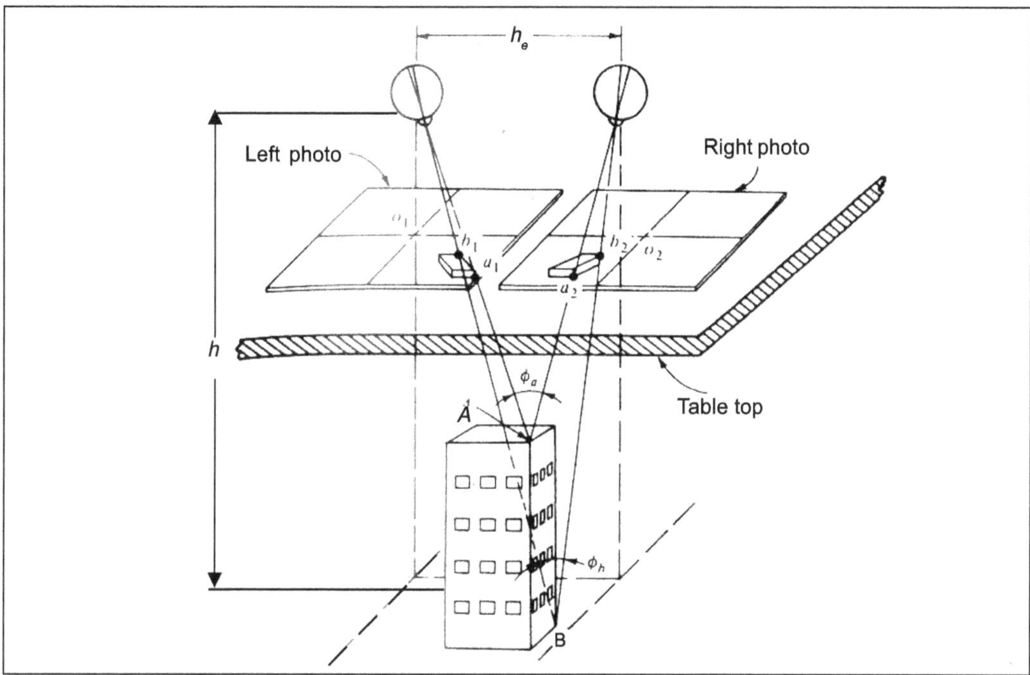

Fig. 4.15 (b) Viewing the building stereoscopically.

Pocket stereoscope are most commonly used and simple construction and consists of two simple convex lenses mounted or a frame. The spacing between the lenses can be varied to accommodate various eye bases.

The Mirror stereoscope has two large wing mirrors and two small eye piece mirrors. Light rays from the photo points a1, a2 are reflected from the mirror surfaces and according to the principles of reflection are received at the eyes, forming the parallactic angle ϕ a. Similarly, for point b1, b2 also forming parallaetic angle ϕ b. The brain automatically associates the depth of the points A and B with their respective parallactic angles ϕ a and ϕ b. This happens for the images of left and right photo, which ultimately generates the 3D stereoscopic viewing, which appear to be below the surface of the photographs.

4.10.3 Procedure for use of Stereoscopes

The following points are to be noted for stereoscopic viewing of aerial photos:

- Place the photographs in such a way that the left photo is seen by the left eye and right by the right eye.
- Reversal will give pseudoscopic view ups and downs and down are reversed in viewing.
- For comfortable (without strain to eye) stereoscopic viewing; the eye base, the line joining the centres of the stereoscopic pairs and flight lines are to be mode parallel.

For this the following procedures are to be carried out :

- Marking of the conjugate principal points of the two over lapping stereopair (The conjugate Principal point for a photograph is the principal pointf of the adjoining photograph transferred to the over lapping photo).

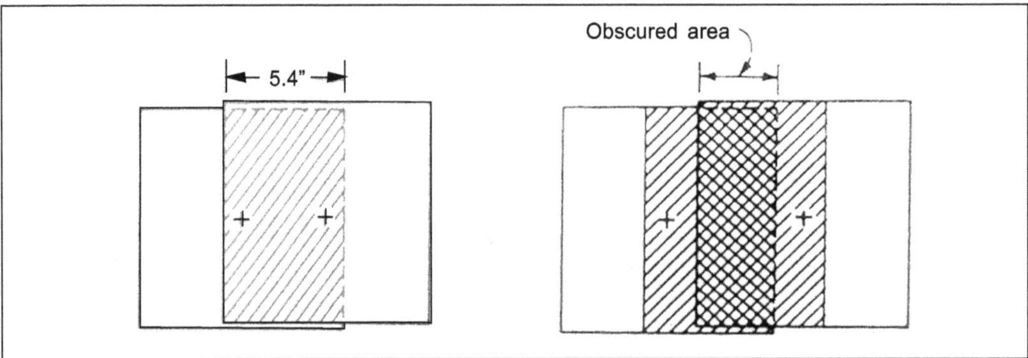

Fig. 4.16 Overlapping area of a stereopair.

4.10.4 Steps for Stereoveiwing by Stereoscope

- Drawing a straight line one a plane (paper fixed) indicating the flight line direction (f f').
- Fix the left photo along the flight line direction (ff') so that line O_1, O_2' direction coincides with the direction ff '.
- Orient the right photo such that line O_1 O_2 O_1' O_2 lie on the straight line ff ';
- Move the right photographs along this direction for getting comfortable stereoscopic view.

Y-parallax

The essential condition for perfect stereoscopic viewing is that the line joining the two corresponding image points must be parallel to the direction of the flight line. If this condition is not satisfied "Y-parallax" exists. Y-parallax causes, strain to eye during stereoscopic veiwing. The more the Y-parallax, the more the strain and vice-versa.

Y-parallax can occur due to

- Improper orientation of photos.
- When flight height H varies at L_1, L_2 of adjoining photographs.
- Hue to lilt in adjoining photograph.

4.11 Stereoscopic Parallax

Apparent displacements in the position of an object, with respect to a frame of reference, by shift in the position of observation. Closer the object, greater the movement. The change in position of an image from one photograph to next caused by the Air Craft's motion is termed as stereoscopic parallax or x-parallax.

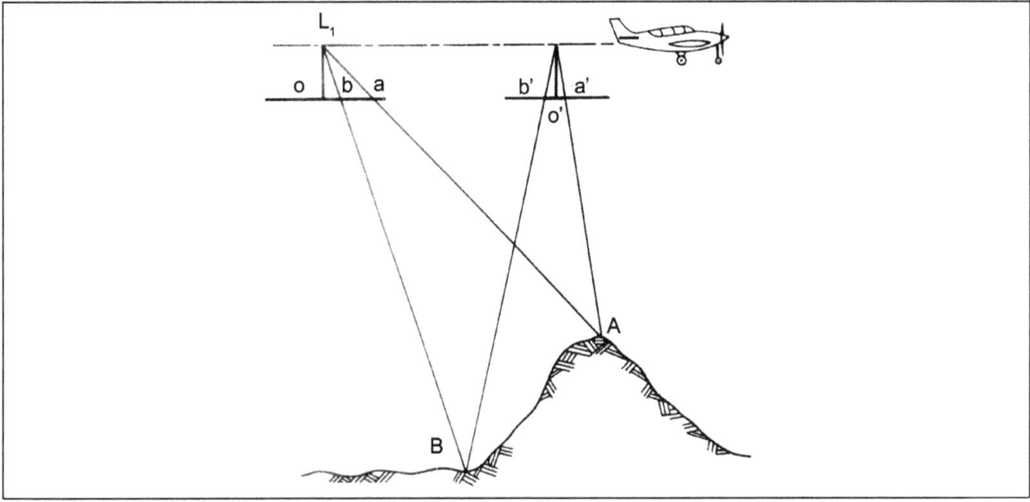

Fig. 4.17 (a) Stereoscopic parallax of vertical over lapping aerial photographs.

A, B ground point are imaged at a, b on left photo and a', b' on right photo parallax of point A is greater than that of B.

Importance of parallax measurements :

- Parallax of any point is directly related to elevation of that point.
- Parallax is greater for higher point and less for lower points.
- Parallax measurements provides basis for determination of elevation of points from photographic measurements.

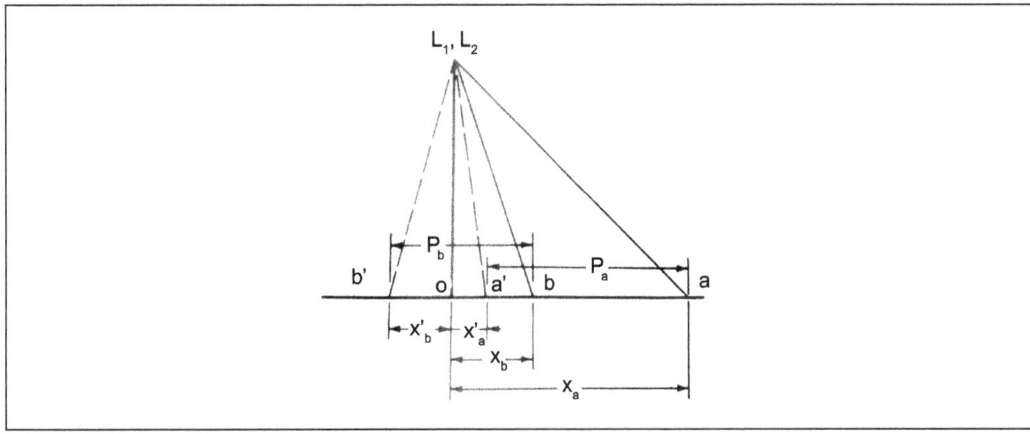

Fig 4.17(b) L, R photographs in superimposition.

Parallax of point A is P_a, parallax of point B is P_b. Parallax expressed in terms of flight line coordinates.

$$P_a = xa - xa'$$

$$P_b = xb - xb'$$

4.11.1 Parallax measurement - Monoscopic

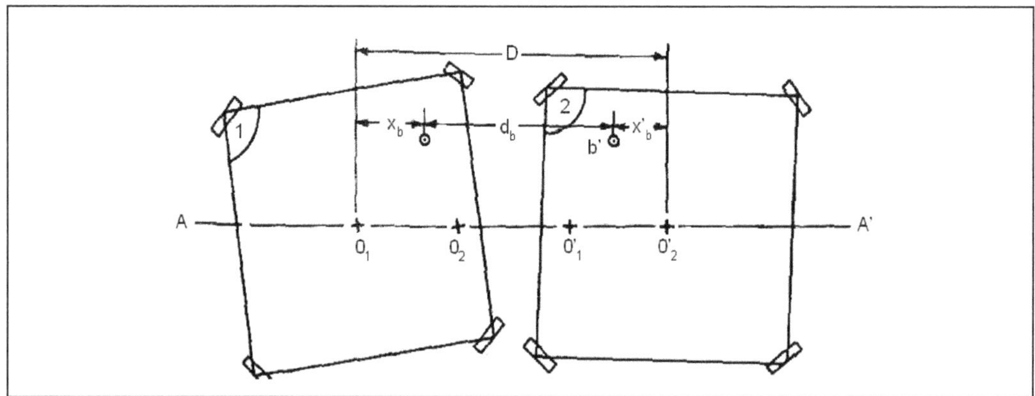

Fig 4.18. Photos 1,2 oriented along flight line axes for stereovision and parallax measurement.

01, 02 - Principal points of photos 1, 2

01', 02' - Conjugate principal points.

xa, x' a - x – Coordinates of point a, a' on photos 1, 2

D - Distance between principal points

da - Distance between image points a, a'

Direct measurement

Parallax of point a = xa-(-x'a)

By measurement of D, da

D, the distance between the principal points O_1, O_2 of the photos is a constant and can be measured. da, measure the distance between the image points a, a'. Parallax Pa is given by **Pa = D - da**

4.11.2 Principal of Floating Mark

Two small identical marks (.) are etched on clear glass. Place these marks on the left and right photographs. Fuse the marks over the stereomodel in such a way that, they fuse into a single mark and appear to rest on the stereomodel at some elevation. If the marks are moved apart and fused, the parallax is increased and the point appears to move down. If the marks are moved closer, the parallax is increased and appears to rise in the stereomodel. This apparent variation in elevation of the point by increasing or decreasing the distances between them is the basis of giving the name “FLOATING MARK” to the point.

The principal of FLOATING MARK in stereo model is extremely important. This is used for point transfer, mapping of terrain and elevation measurement by placing the mark on the terrain at required places and moving along the terrain.

4.11.3 Stereoscopic Methods of Parallax Measurement

By using the principal of floating mark, parallax of stereomodel points are measured. The instrument used along with stereoscope is called "*Parallax bar*".

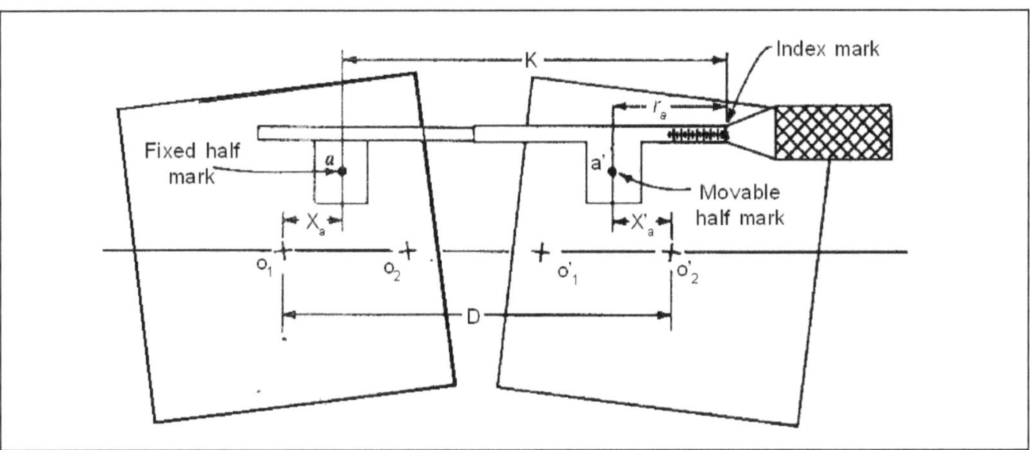

Fig. 4.19

Schematic diagram of parallax bar with measuring principal of parallax of image points a, and a stereo pair of photographs.

Steps

1. Orient the photographs (1) and (2) along flight axis line A A' for comfortable stereovision and fix both photographs.
2. Choose an average elevation point on the terrain, fixed the left hand floating mark of the parallax bar, such that the micrometer drum indicator is at the middle.
3. Left hand mark will be fixed measurements in this particular stereo model.
4. Constant D-Distance between the two principal points.
5. Constant K-Distance between the left hand fixed point of parallax bar and index reading of the micrometer of parallax bar.
6. Adjust the floating mark, such that it rests on the elevation of the fused point a a' in the Stereo model, and take the micro meter reading (ra)

   Parallax of point $Pa = Xa - X'a$

   $= D - (K - ra)$

   $= (D - K) + ra$

   D – K is called the parallax bar Constant (C) for the set up. Substituting,

   $$\mathbf{Pa = C + ra}$$

4.11.4 Determination of parallax bar constant

1. Measure the micrometer reading of any point, by setting the floating mark.
2. Calculate the parallax of the point by monoscopic method.

   Then substituting in this equation.

$$Px = C + rx$$

$$C = Px - rx \qquad \text{..........(1)}$$

3. Similarly take the micrometer of another point by setting the floating mark on the point under Stereofusion, derreading be ry

   than $C = py - ry$ ....(2)

The average value of C from equation (1) and (2) can be taken for final determination of the parallax bar, constant C has to be determined before each observation.

C can also be determined from the parallax reading of the principal point.

Advantage of stereoscope determination by parallax bar are accuracy and speed. Once C is obtained, all parallax can be measured by one measurement.

4.12 Parallax Equations

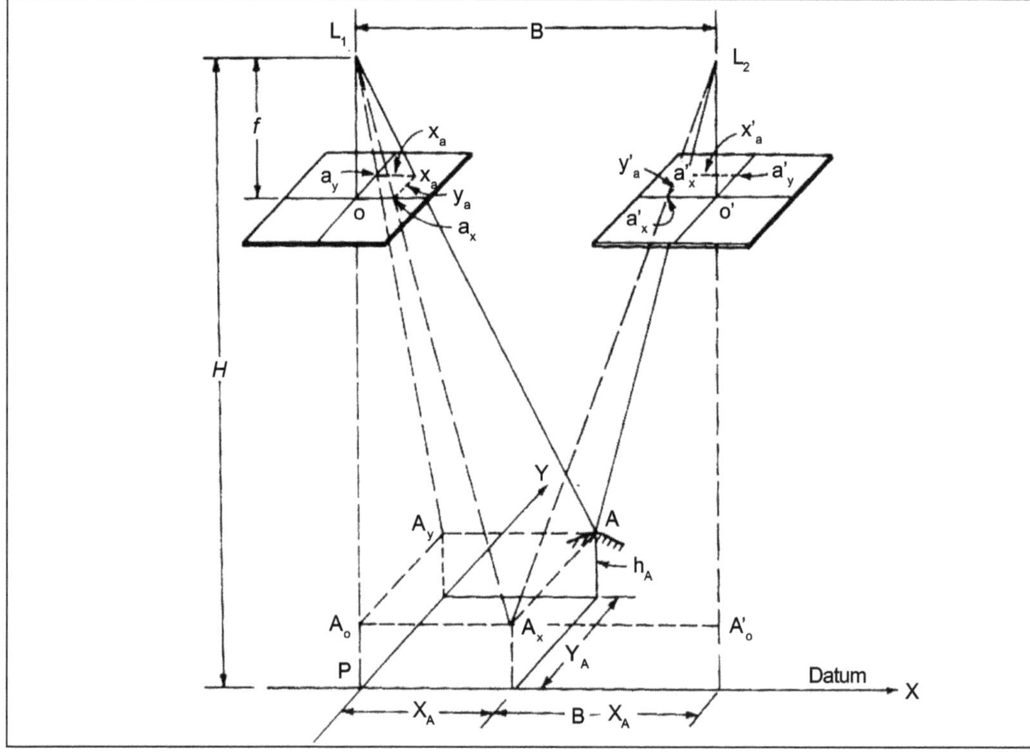

Fig. 4.20 Geometry of overlapping vertical photographs.

B - air base

L_1, L_2 - exposure stations

H - flight height, f - focal length.

A - object on the ground, a, a' images of A or L and R photos.

(XA, YA) ground co-ordinates of point A

(xa, ya), (x'a, y'a) co-ordinates of the image points or left photo and Right photo of ground object A.

Every stereopair of photos have its own unique coordinate system.

Derivation of h_A (ground height of object point A) and (X_A, Y_A)

from lll^r Δ^S L_1O_1 a_y, L_1 A_O A_y, $\frac{Y_A}{H-h_A} = \frac{y_a}{f}$

$$Y_A = \frac{y_a}{f}(H-h_B) \quad ....(1)$$

from lll^r Δ^S L_1O_1 a_x, L A_O A_x, $\frac{X_A}{H-h_a} = \frac{x_a}{f}$

$$X_A = \frac{x_a}{f}(H-h_A) \quad ....(2)$$

from lll^r Δ^S L_2O_2 a'_x, L_2 A'_O A_x, $\frac{B-X_A}{H-h_A} = \frac{-x_a}{f}$

$$X_A = B + \frac{x'_a}{f}(H-h_A) \quad ....(3)$$

Equating (2) and (3)

$$\frac{x_a}{f}(H-h_A) = B + \frac{x'_a}{f}(H-h_A)$$

$$(H-h_A)\left(\frac{x_a}{f} - \frac{x'_a}{f}\right) = B$$

$$H-h_A\left(\frac{x_a - x'_a}{f}\right) = B$$

$$H-h_A = \frac{Bf}{x_a - x'_a}$$

Substituting $P_a = x_a - x'_a$,

$$h_A = H - \frac{Bf}{P_a} \text{ or } P_a = \frac{fB}{H - h_A} \qquad \text{..........(4)}$$

Substituting eqn. (4) in eqns. (2) and (3)

$$X_A = \frac{x_a}{f}\left(H - \left(H - \frac{Bf}{x_a}\right)\right)$$

$$= \frac{x_a}{f}\left(\frac{Bf}{P_a}\right)$$

$$X_A = B\left(\frac{x_a}{P_a}\right) \qquad \text{..........(5)}$$

IIIrly $$Y_A = B\left(\frac{Y_a}{P_a}\right) \qquad \text{..........(6)}$$

Equations (4), (5) and (6) are called *Parallax equations.*

h_A - height of point A above datum.

H - flying height

B - air base, f - focal length of camera.

P_a - parallax of image point A.

(x_A, y_A) ground co ordinates of point A.

(x_a, y_a) photo co ordinates of point a.,

(XA,YA) based or unique co ordinate system with origin at datum principal point of O_1 at P.

Parallax equations are the most useful for a photogrammetrist. The equations enable to complete survey of the stereo overlap area, provided focal length f, and sufficient number of ground controls are available for calculation of B and H.

Conversion of unique arbitrary co-ordinate system of a stereo pair to tree ground co-ordinates system is possible, by two dimensional co-ordinate transformation.

4.12.1 Elevation by Parallax Differences

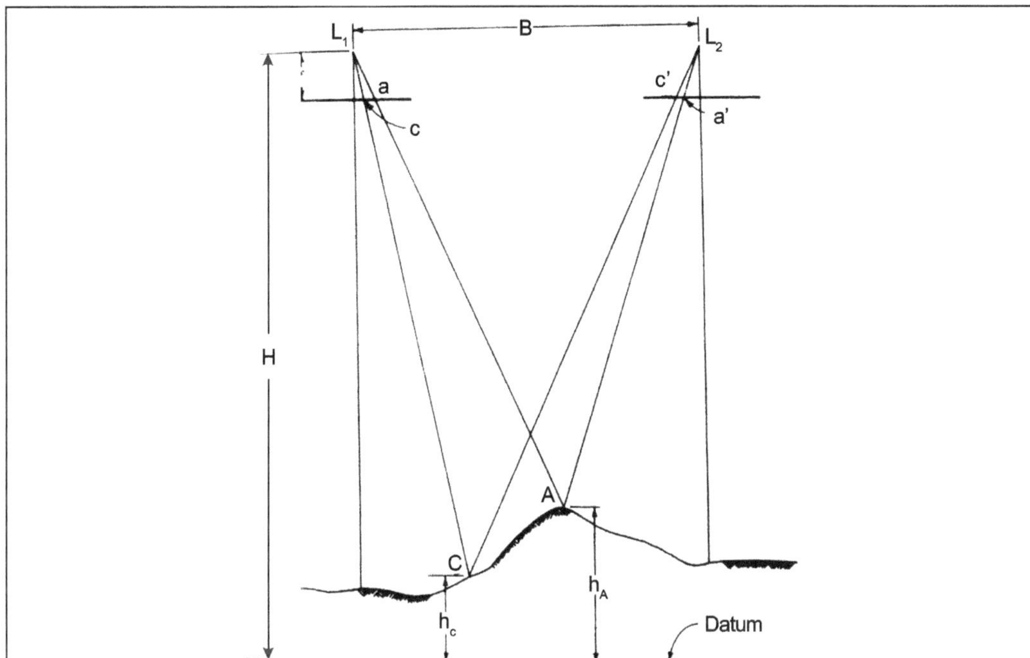

Fig. 4.21 Elevation by parallax differences.

Parallax differences are caused by difference in elevation of the two points. If elevation of one point is known; the parallax differences of the known and unknown point can be applied to determine the elevation of the other point.

In the figure if h_C is known, h_A can be calculated by the difference of parallaxes of A, and C.

$$P_C = \frac{fB}{H - h_C} \qquad \text{..........(7)}$$

$$P_a = \frac{fB}{H - h_A} \qquad \text{..........(8)}$$

$$(8) - (7) = P_a - P_C = \frac{fB}{H - h_A} - \frac{fB}{H - h_C}$$

$$= \frac{fB\{(H - h_C) - (H - h_A)\}}{(H - h_A)(H - h_C)}$$

$$= \frac{fB(h_A - h_C)}{(H - h_A)(H - h_C)}$$

Let $\Delta_P = P_a - P_C$

and substitute $H - h_A = \dfrac{fB}{P_a}$

Then $$\Delta_P = f\,B \frac{(h_A - h_C)}{\frac{fB}{P_a}(H - h_C)}$$

$$\Delta_P = \frac{P_a(h_A - h_C)}{(H - h_C)} \qquad \text{..........(9)}$$

$$\Delta_P = (H - h_C) = P_a\,(h_A - h_C)$$

$$h_A - h_C = \frac{\Delta_P(H - h_C)}{P_a}$$

$$\therefore \qquad h_A = h_C + \frac{\Delta_P(H - h_C)}{P_a} \qquad \text{..........(10)}$$

From equation (10), if the height of a given point is known (say c), height of and other point (A) can be determined, by finding P_a, P_c, Δ_p and knowing the flight height H.

4.13 Stereoscopic Plotting Instruments-Analogue and Analytical

Stereoscopic plotting instruments, commonly called as Stereo plotters are widely used for preparation of topo-graphical maps. The subject of stereo plotters is most important in the study of photogrammetry.

Basic concept : An over lapping pair of aerial photos, (stereo pair) is exposed and transparencies (diapositives) are carefully prepared. After preparation of the diapositives these are placed on the projectors of the stereo plotter in such a way that left photo is put on left project and right photo on right projector. Before plotting of the details and contours, the following orientations are carried out :

4.13.1 Interior Orientation

This stage is for recreating the geometry of the projected rays to duplicate exactly the geometry of the orientation photos. The procedures involved in inner orientation are :

- Preparation of diapositives
- Compensation for image distortions
- Centreing diapositives in the projectors.
- Setting of proper principal distance in the projects.

4.13.2 Relative Orientation

The condition that is fulfilled in relative orientation is that each model point and the two projection centres form a plane in miniature, just like the plane that existed for the corresponding ground point and the two exposure stations. This ensures that the parallactic angle for any point of the stereo model is equal to the original parallactic angle of the corresponding object point at the time of taking photograph. This implies that the projected rays of each the corresponding point on the left and right diapositives must intersect at a point.

Since the X-component Px, of the mismatch of the images can be removed by raising or lowering the plate carriers (Px being a function of elevation of the point), the remaining Y-component Py, called Y-parallax has to be cleared for all points in the stereo model in order to obtain a relatively oriented model.

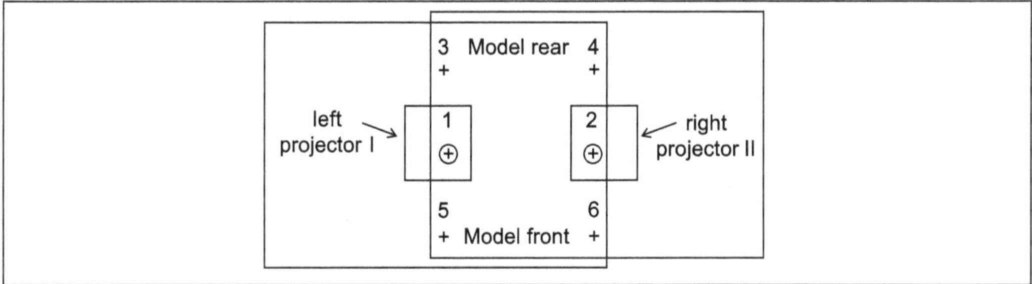

Fig. 4.22(a) Stereo model locations of six points conventionally used for Relative Orientation.

Conventional procedure for clearing Y-parallax at five standard points (plus sixth for check), located in a model, is cleared, then the entire model will be cleared of Y-parallax.

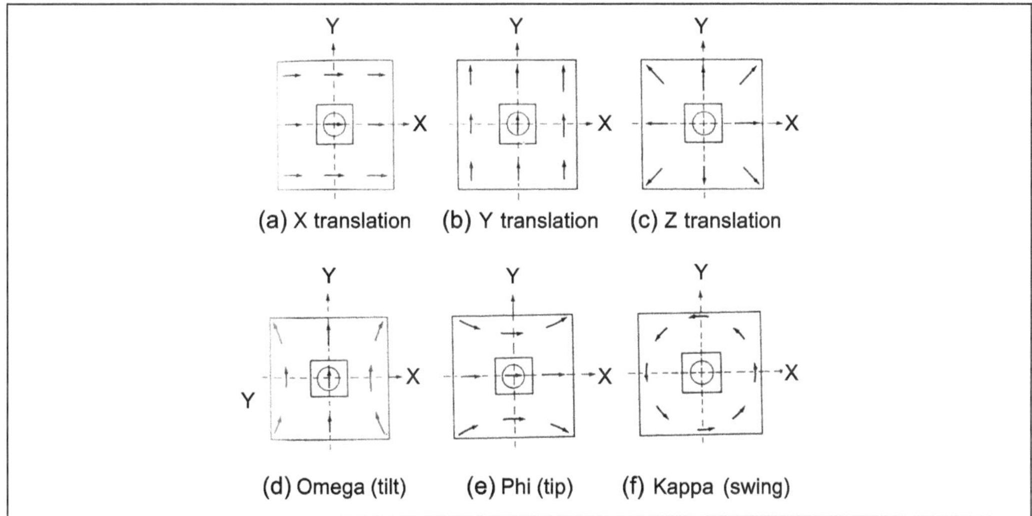

Fig .4.22(b) Stereo model locations of six points conventionally used in relative orientation.

The image movements caused by the six projector motions are given above. The movements are as follows:

X-translation : Only X movement to all images hence no Y-parallax can be removed.

Y- translation : Y movement of all projected images. Hence Y - parallax can be cleared at any point.

Z-translation : Radially outward for Z translation upward and inward forZ translation downword. Y-parallax could be cleared at any point, other than along X-axis.

Omega (Ω) rotation (tilt) : causes Y-movement of all projected images, Hence Y-parallax can be cleared anywhere in the model.

Phi-rotation (tip) : Causes X movement on all projected images, but also imparts Y- Component for points not on X and Y axis.

Kappa - rotation (swing) : Y components to all points everywhere in the model except along Y axis,

Procedure have been detailed after understanding the above image movements for removal of Y-parallax. Relative orientation consists of procedure for clearing Y - parallax for the entire model based on the above movements.

4.13.3 Absolute Orientation

After relative orientation, a true three dimensional model of the terrain in generated. *Model scale fixation* and also *levelling* the model for vertical scale is called Absolute Orientation.

Fixing model scale

Scaling carried out by the formula.

$$Sm = \frac{b}{B} = \frac{h}{H'}$$

also

$$Sp = \frac{f}{H'}$$

∴

$$Sm = \frac{h}{f'} . Sp$$

Model scale can be calculated from photo scale by the above formula where f in the principal distance, h is the optimum projection distance.

Levelling the model: Final step in absolute orientation. Minimum requirement of height points are three which should form a triangle. Four height Control situated at the four corners of the model is preferred along with one at the centre of the model. The model is rotated about X - axis and Y-axis till the heights agree in the triangular set up. After agreement, the fourth and final points are checked and fine corrections to the projectors are carried out.

After the scaling and levelling is carried out, the model is really for plotting of details and contours as per laid out specifications. (Three dimensional data capture).

Map Compilation: These are number of stereo plotters available for map compilation. These can be classified as :

- Optical (analogue)
- Optical Mechanical (analogue)
- Semi-analytical
- Analitical

Different plotters have different methods of procedure for plotting and map compilation.

4.14 Aerial Photography - Planning and Execution

For successful execution of any project based on aerial photography, proper planning and execution of aerial photography mission is required. The basic requirements are the specifications and the preparation of the flight map. The specification for Aerial photography is based on the following item variables:

- Purposed of photography
- Camera and film
- Scale of photography
- End lap and side lap
- Flying height.
- Ground coverage
- Weather conditions
- Season of the year

Purpose of photography : The first and foremost consideration is the purpose for which the photography is taken. For topographic mapping, photos having good metric qualities are needed. For this calibrated cameras, films having fine grained and high resolution emulsion are selected.

Scale of photography : Average photo scale is one of the most important variables that must be selected in Aerial photography planning. For topographic mapping photo scale depends upon the final map scale, contour interval and the Photogrammetric instruments for map preparation. In most of the Photogrammetric instruments the enlargement ratio from photo scale to map scale is five. Hence photo scale for mapping can be fixed at one fifth of the map scale.

End lap and side lap : For topographic mapping and Photogrammetric control extension, the end lap requirement is 60% and the side lap is 30%. The side lap is required for tieing the strips and making the block for the specified area.

Flying height : Once the selection of the camera and scale of photography is finalised, the flying height for the project is determined

Ground coverage : Once the photo scale and camera format dimensions are selected the ground surface area covered by a single photograph can be calculated. In addition, if the end lap and side lap are known the area covered by a *stereoscopic neat model* can be calculated. Neat model is the stereoscopic area between adjacent principal points and up to the middle of the side lap in both directions, extending side ways.

Weather conditions : The weather for aerial photography is an important consideration cloud less days, have sky and atmosphere clear without haze, smoke are chosen for flight.

Season of the year : For topographic mapping the season to be chosen in based on alitude of sun and the trees having leaves fallen or intact. The ground should not be obscured by tree leaves. Normally start of the winter season is chosen in for aerial photo missions.

Flight map : Based on the above specifications and consideration, a flight map is prepared. The flight map gives the project boundaries and flight lines, the pilot must fly to obtain the desired coverage. Flight maps are prepared on existing maps covering the projects Area. (Fig. 4.23).

A transparent templete showing neat model positions based on the scale, end lap, side lap and base is prepared for the project area and super imposed for preparation of the flight map.

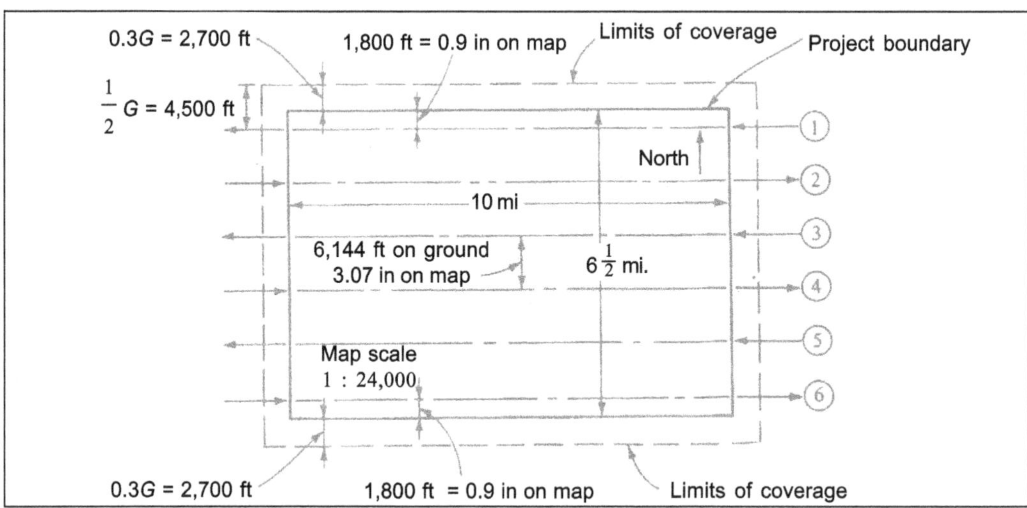

Fig. 4.23 Flight map for project area ABCD

4.15 Control for Aerial Photography

Control points based on some coordinate system are required for any type of mapping project. Similarly, for topographic mapping projects based on Aerial photography also requires ground control. Control selected in this case are sharp well defined points in the object space (ground) so that they are clearly identifiable on the photographs. Ground control for photogrammetry also known as photogrammetric control is classified into two different categories:

- ***Horizontal control :*** Position of point on ground is known with reference to a Horizontal datum or origin.
- ***Vertical control :*** Elevation of a point is known with reference to a vertical datum.

4.15.1 Requirement of Photogrammetric Control

In photogrammetric mapping projects, control are required for orienting stereo models. Further ground surveyed photo-control points are needed for bridging (densifying or supplementing) photo control. The ground surveyed photo-control requirement varies depending on the methods, instruments and personnel. The procedure for densifying control is called Aerial Triangulation. Both horizontal and vertical control are required for Aerial Triangulation.

4.15.2 Methods of Providing Control on Aerial Photographs

Post - pointing : Control points as per requirements are provided on ground on well defined points which are easily identifiable on aerial photograph. The photographs are taken on the ground and the control point locations are transferred on to the photos by identifying them and under stereofusion. This method of transferring, existing fresh control already provided on ground to aerial photographs is called post-pointing.

Pre-pointing : In this methods locations where the points are required are first identified as per planning and targets are mode on the ground at specified locations. The dimension of the target are calculated from the proposed scale of photography. Indicators to the point which are five times bigger than the target mark is laid for quick identification. The back ground contrast is generated by making the target pure white and the back ground colour dark. During aerial photography all such marks are imaged on to the aerial photographs, Although expensive, this methods is much more accurate than post-pointing.

4.16 Aerial Triangulation

For large mapping projects the number of control points needed is extensive and the cost of establishing them by ground methods can be very high. With the help of electronic computers and improved photogrammetric equipment and techniques much of the needed control can be established by photogrammetric control extension often called "Aerial Triangulation".

4.16.1 Aerial Triangulation can be classified into two broad categories

- *Analoguge Method*, where procedure involves, manual relative orientation and measurement of model co-ordinates by using stereoscopic plotting instrument.
- *Analytical Method*, consists of photo coordinate measurements followed by numerical relative orientation and Computation of model coordinates using electronic computer.

4.16.2 Aerial Triangulation - Analogue Method

For analogue method of aerial triangulation, Universal Instruments are used, where "base-in" and "base-out" facility is available. This enables the strip to be formed on the machine itself. The procedure for base-in base-out is as follows.

1. In base-in mode 1-2 photos are kept in L and R projector and observation carried out.
2. After observation photo (1) is removed and 3 is kept at (L) projector and base-out position is obtained. Observation of coordinates is carried out with 3-2 photos at L and R projectors.
3. After this photo 2 in removed and replaced by photo (4). This 3-4 photo is now in L and R projectors and base-in position in model observations are carried out.

The same is Continued next by (3) replaced by (5) and 5-4 Base out position and so-on, till the strip is completed.

After completion, the error curves are plotted on graph paper and all the co-ordinate adjusted and used for photogrammetric plotting.

4.16.3 Independent Model Triangulation

Also referred to as semi-analytical aerial triangulation. In this method each stereopair of a strip is relatively oriented in the plotter, Model coordinates of all control points and pass points are then read and recorded for each stereo model in turn. The coordinate system of each model is independent from others. By means of pass points common to each adjacent models, three dimensional co-ordinate transformation is carried out to tie each successive model to the pervious one. The coordinates of the perspective centres are also determined in each independent model and included as common points for gaining needed geometrical strength (Fig. 4.24). The right exposure station of model 1-2 is the same as the left exposure station of model 2-3.

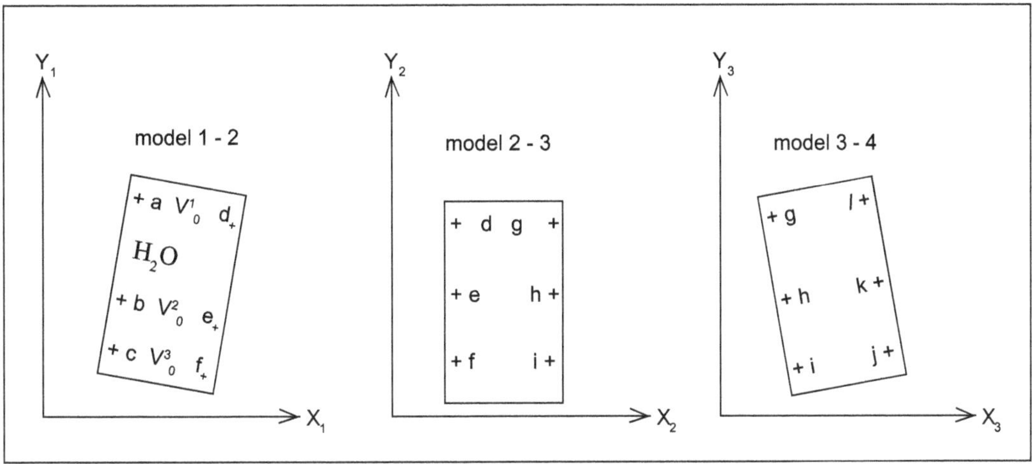

Fig. 4.24. Independent model triangulation observation

To transform model 2-3 into model 1-2 the co-ordinates of common points d, e, f and O_2 of model 2-3 are made to coincide with corresponding model 1-2 coordinates.

By applying coordinate transformations, a continuous strip of stereomodels are formed. This strip may then be adjusted to ground control using polynomial methods to obtain corrected coordinates for all pass points.

For obtaining model coordinates of perspective centres number of methods are available. The grid plate method is most commonly used.

4.16.4 Aerial Triangulation - Analytical Method

Analytical procedure in Aerial Triangulation also involves the basic steps:

- Relative orientation of each model
- Connection of adjacent model to form a continuous strip.
- Adjustment of strip to ground co-ordinates.

In analytical method, the basic input consists of precisely measured photo coordinates of control points and pass points. Relative Orientation is then performed analytically, based upon the measured coordinates and known camera constants.

Basically, analytical method consists of writing condition equations which relate to unknown elements of exterior orientation of each photo to camera constants and measured and refined coordinates. The equations are solved for unknown Orientation parameters and either simultaneously or subsequently the coordinates of pass-points are calculated.

4.16.5 Collinearity or Coplanarity

Collinearity Condition : The collinearity condition specifies that the exposure station, any object point and its photo image all lie along a straight line. (L, a and A) lie along a straight line. Two equations express collinearity conditions of any point. One for x-photo coordinate and other for y-photo coordinate. The basic equations are:

$$x = f\left[\frac{m_{11}(X - X_L) + m_{12}(Y - Y_L) + m_{13}(Z - Z_L)}{m_{31}(X - X_L) + m_{32}(Y - Y_L) + m_{33}(Z - Z_L)}\right]$$

$$y = f\left[\frac{m_{21}(X - X_L) + m_{22}(Y - Y_L) + m_{23}(Z - Z_L)}{m_{31}(X - X_L) + m_{32}(Y - Y_L) + m_{33}(Z - Z_L)}\right]$$

Where x, y are the photo co-ordinates, x, y, z are the ground coordinates, y_L, Y_L, Z_L are the ground co-ordinates of the exposure station, camera focal length and the ms are the functions of rotation angles omega, phi, and kappa.

The collinearity condition equations are non-linear and are linearised by using. Taylor's theorem.

Coplananity Condition : The condition stipulates that the two exposure stations of the stereo pair, any object point and corresponding image points on the two photos, all lie in a common plane. The condition equation is

$$O = B_X (D_1F_2 - D_2F_1) + B_Y (E_2F_1 - E_1E_2) + B_2 (E_1D_2 - E_2D_1)$$

Where :

$$B_X = X_{L2} - X_{L1}$$
$$B_Y = Y_{L2} - Y_{L1}$$
$$B_Z = Z_{L2} - Z_{L1}$$
$$D = (m_{12})\, x + (m_{22})\, y + (m_{32})\, f$$
$$E = (m_{11})\, x + (m_{21})\, y + (m_{31})\, f$$
$$D = (m_{13})\, x + (m_{23})\, y + (m_{33})\, f$$

The coplanarity equation are non-linear and must be linearised using Taylor's theorem.

4.16.6 Blocks of Photos (Block Adjustment)

Control extension can also be made through blocks of photos (two or more side lapping strips). In dealing with blocks number of pass points must be carefully selected so that they appear in the side lap area of adjacent strips.

Using numerical methods, strips are successively joined by matching common tie points and the block is adjusted to available horizontal and vertical field control. Experience has shown that strongest solution is obtained when horizontal control points are located near the periphery of the block. Vertical control must be uniformly distributed throughout the block for block adjustment.

4.16.7 Bundles Adjustment

With the emergence of faster and larger capacity computers, another technique called "*Simultaneous Bundle Block Adjustment*" or "*Bundle Adjustment*" has become practical in analytical photogrammetry. In the procedure all the possible collinearity equations for the entire block of photographs are formulated and solved simultaneously. By solving both control and pass points collinearity equations simultaneously a numerical relative and absolute orientation of all stereo models in the block is achieved.

The symultaneous solutions of all collinearity equations that is highly redundant and is solved by method of least squares, which is mathematically more rigorous than polynomial methods.

4.17 Digital or Soft Copy Photogrammetry

Digital photogrammetry or soft copy photogrammetry has already proven its efficiency in practice. The production of the photos, aerial triangulation, derivation of digital terrain model or photogrammetric plotting are increasingly carried out on "*digital photogrammetric workstation*". The assistance afforded to the operator by image processing tools and the easy interaction with GIS data bases, distinguish digital photogrammetric work stations from the classical analogue analytical plotters.

4.17.1 Scanning Of Aerial Photographs

Digital photogrammetry is now widely used. But the imaging technique have not been changed. Most of the aerial photos are still taken with film cameras. The information content of an aerial photograph requires an average storage space on disc of 1.2 Giga byte, without image compression. The present technology does not permit one to reach such performances with reasonable means under flight conditions. Consequently, film cameras, will remain standard technique for image registration for several years and it will be necessary to convent the analog images to digital format by scanning process. The process in semiconductor field for digital applications is the basis of rapid advances in the image sensor field

4.17.2 Application of GPS in Photogrammetry

Kinematic GPS positioning of air borne platforms coupled with aerial photogrammetry has become a reality. Real time GPS positioning is used for navigation, while post processed carrier phased differencing is employed for exposure station locations. Since it is not possible for exposure station event to control GPS collection, the coordinates are derived by interpolation method. By employing combination of GPS and ground control, the demand for ground control has been reduced. The operational efficiency has increased as the need for expensive and labour intensive ground control has diminished. The technology is rapidly maturing, such that, it will become the most commercially viable approach for aerial mapping.

Airborne GPS offers a flexible and cost-effective alternative to land based methods for obtaining geodetic control needed in aerotriangulation block adjustments.

4.17.3 Integrated GPS Flight Management System

- ***System Details :*** 33 M Hz P. C Wild Re 20 metric mapping camera, GPS receiver and associated peripherals.
- ***Functions :***
  - *GPS receiver Functions :* Operates in continuous kinematic mode (second).
  - *Mapping camera Functions* : For recording of exposure instant location by GPS. The parameters of each exposure event is recorded (all). In Flight Management System the aircraft can be navigated exactly as per the Flight planning Chart.

4.17.4 Aerial Triangulation using GPS Data

Air borne GPS data is used for carrying out Aerial Triangulation Block Adjustments. Application of GPS positioning data in photogrammetry, (GPS photogrammetry) can be carried out in the following ways;

- Determination of ground control points by terrestrial GPS
- GPS flight navigation
- GPS Camera positioning for Aerial Triangulation
- Positioning of Other airborne Sensors.

4.17.5 Storage and Data Compression in Digital Photogrammetry

Storage and retrieving large volumes of high resolution digital imagery in soft copy photogrammetry system require huge amounts of data space and advanced data transfer techniques. The difference in hardware between general purpose systems and photogrammetric work stations is characterized by two features.

- 3D display
- large Capacity short and long term storage subsystems.

Compression is an integral part of any large volume information processing and archiving system. Images usually exhibit quite high degrees of redundancy, they compress well. Handling very large digital images is already a daily practice. Most of the commercial soft copy systems use Industry standard JPEG still, continuous tone lossy image compression scheme. To achieve high performance the compression is typically implemented in hardware.

4.17.6 DEM and DTM in Soft Copy Photogrammetry

The term DEM (Digital Elevation Model) Is some times used generically to mean the digital cartographic representation of earth in any form, rectangular grids, or lattices, triangular net works or irregular spot heights and break lines. In this context, a DEM is a digital cartographic representation of the elevation of land (Z value) at regularly spaced intervals in x and y directions.

The term DTM (Digital Terrain Model) is similar to DEMs, but incorporate the elevation of significant topographic features on land and change points (break lines), which are irregularly placed so as to better characterize the shape of the terrain. Break lines are linear features that describe a change in the smoothness or continuity of the surface.

Advantages of Soft Copy Photogrammetry on DEM/DTM generation : Soft copy photogrammetry has significant benefits over hard copy analytical plotters for the extraction, editing and quality control of DEMs. Soft copy offers the possibility of manual assistance and or automated extraction. In addition, high quality accurate stereo superimposition provides tremendous benefits in effective DTM collection, real time stereo displays and also unmatched capabilities in quality control of DEMS/DTMs.

Soft Copy photogrammetric Work stations : The most essential components of a Digital photogrammetric work station can be identified in the Fig. 4.25.

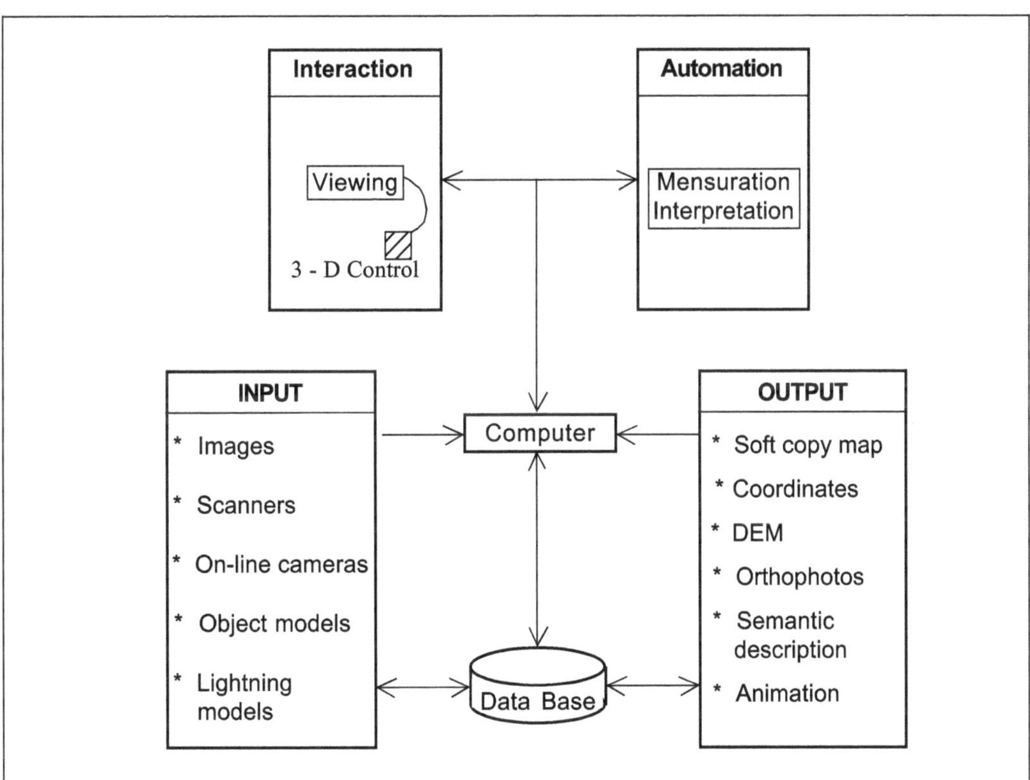

Fig. 4.25 The fundamental components of a soft copy photogrammetric Work station and the related input-output devices, tools and products.

Computer, Data base and Graphics Subsystem : One of the core elements of a soft copy photogrammetry Work Station in the computer.

All basic software and soft copy application software, the user interface, the data transfer and communication are controlled by one or more processing units.

Image and vector data base : In the data base the images, image parameters, preprocessed images, project parameters, lighting models, object models semantic information, resulting vector data, DTM, orthophotos etc. are archived. There are high requirements on the data management, lie, archiving and retrival of the data.

Network : The computer, the data base, peripheral input and output devices are typically connected via very fast bases (eg : VME bus, S bus) or a fast net work like Ethernet.

Interaction : The latest trend in design of soft copy photogrammetric Workstation is to develop the human interface following the principle of uniform interface design. The interaction requires a veiwing system and 2D or 3D control devices.

Veiwing : The display of digital images and super imposed vector information requires high resolution monitor (e.g., 1280 × 1024 pixels or more). In 8 bit colour or 24 bit true colour with a non-destructive overlay. Images are displayed in monomodel or in stereo mode using special techniques. Use of an additional control monitor with input menus and process information is very common.

Stereo Veiwing : For many applications stereo veiwing is essential. The stereo veiwing requires the separation of the two images of a stereo pair. The separation is based on temporal separation (alternate display of two images) or radiometric separation (polarization or anaglyph principle) or on spatial separation (split screen and stereoscope) or on combination of these principles (Fig. 2.26(a) & (b)).

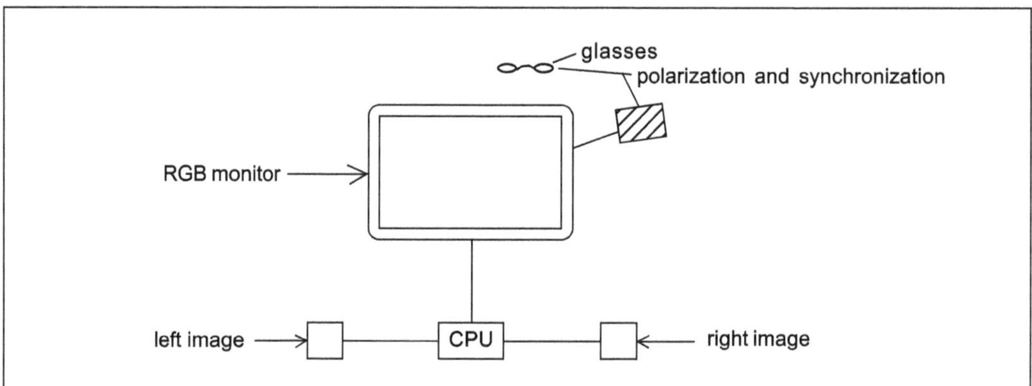

Fig. 4.26(a) Stereoveiwing by polarization. Left and right image at 120 Hz and synchronized polarization for veiwing by polarized glance.

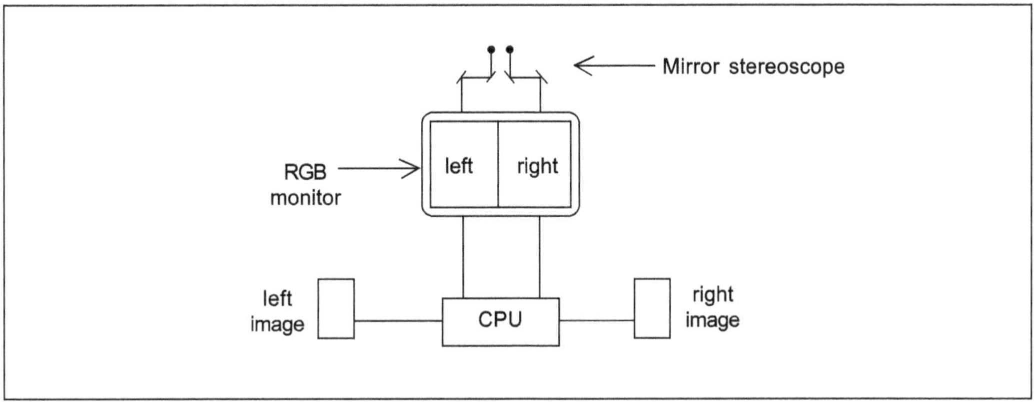

Fig. 4.26(b) Stereoveiwing by optical separation. Left and righ image are displayed on split screen veiwing in performed through a mirror stereoscope.

Super-Imposition : The graphical display of object data on the image is called superimposition. In the case of stereo it is called stereographic superimposition. For 3D superimposition in required for high quality overlay.

Control : Subpixel positioning is an essential prerequisite of a soft copy photogrammetric Workstation. The cursor (floating mark) can be easily defined in any desired shape and size. These are solutions specially for 2D and 3D control. A mouse is most often used for the positioning. Some systems use hand wheels and foot and wheel to control movement of floating mark. Advanced 3D cursors allow the control of 3D movement and after programmable function buttons for special tasks.

2 D	3D
- Mouse - Track ball - Key board keys - Hand wheels	- 3D Mouse - Track ball - key board keys - 3 D cursor with function buttons - Hand wheels & foot wheel / foot disc

5
Remote Sensing

5.1 Introduction

Remote sensing is the science and art of obtaining information about an object, area or phenomenon through an analysis of the data acquired by a device which is not in contact with the object, area or phenomenon under investigation. In the present context, the definition of remote sensing is restricted to mean the process of acquiring information about any object without physically contacting it in anyway regardless of whether the observer is immediately adjacent to the object or millions of miles away. It is further required that such sensing may be achieved in the absence of any matter in the intervening space between the object and the observer. Consequently, the information about the object, area or any phenomenon must be available in a form that can be impressed on a carrier vacuum. The information carrier, or communication link, is electromagnetic energy. Remote sensing data basically consists of wavelength intensity information acquired by collecting the electromagnetic radiation leaving the object at specific wavelength and measuring its intensity.

Remote sensing of earth's environment comprises measuring and recording of electromagnetic energy reflected from or emitted by the planet's surface and atmosphere from a vantage point above the surface, and relating to such measurements to the nature and distribution of surface materials and atmospheric conditions. Sensors mounted on aircraft or satellite platforms measure the amounts of energy reflected from or emitted by the earth's surface. These measurements are made at a large number of points distributed either along a one-dimensional profile on the ground below the platform or over a two-dimensional area on either side of the ground track of the platform. The sensors scan the ground below the satellite or aircraft platform and as the platform moves forward, an image of the earth's surface is built up. Fig. 5.1 shows how a sensor on board satellite scans along line AB. Each scan line of a remotely sensed image is a digital or numerical record of radiance measurements made at regular intervals along the line. A set of consecutive scan lines forms an image (Mather, 1987). Two-dimensional image data can be collected by means of two types of imaging sensors, namely, nadir looking or side looking sensor.

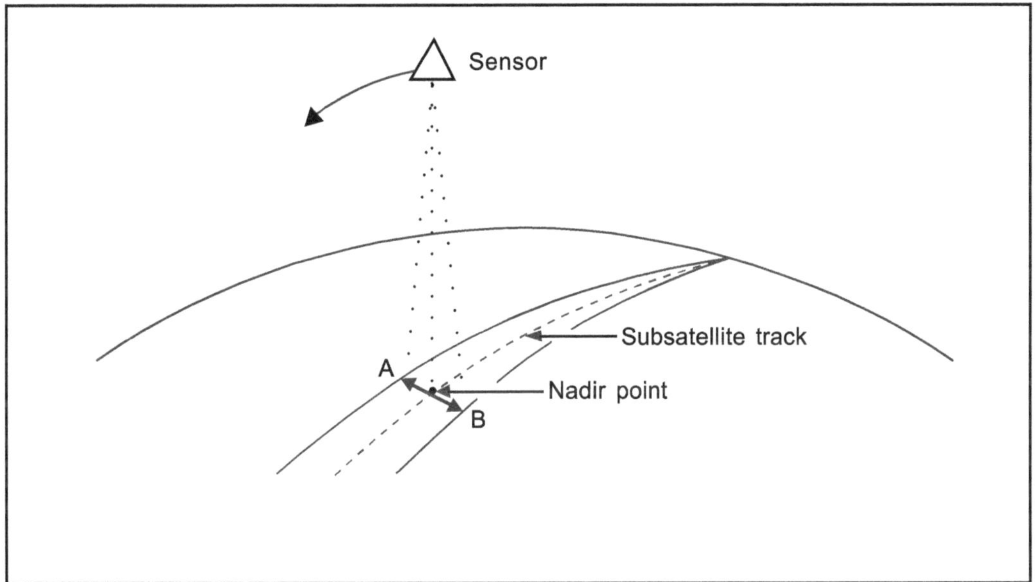

Fig. 5.1 Sensor on-board satellite scans along line AB. As the platform moves forward, an image of the swath region is built up.

In the case of nadir looking, the ground area to either side of the satellite or aircraft platform is imaged, whereas an area of the earth's surface lying to one side of satellite track is imaged by means of side looking sensor. Spatial patterns evident in remotely sensed images are interpreted in terms of geographical variation in the nature of material forming the surface of the earth. Such materials may be vegetation, exposed soil and rock, or water. These materials are not themselves detected directly by remote sensing, and their nature inferred from the measurements made. The characteristic of digital image data is that they can be adjusted so as to provide an estimate of physical measurements of properties of the targets, such as, radiance or reflectivity (Mather, 1987).

Broadly, there are two types of sensing systems to record the information about any target. They are active sensing system and passive sensing system. An active sensing system generates and uses its own energy to illuminate the target and records the reflected energy which carries the information content or entropy. Synthetic aperture radar (SAR) is one of the best examples of active sensing systems. These sensing systems operate in the microwave region of electromagnetic spectrum and include radiation with wavelengths longer than 1 mm. These systems do not rely on the detection of solar or terrestrial emissions as the solar irradiance in microwave region is negligible. The active remote sensing operation principles and the general details of latest imaging radar systems are described in the following chapter. The second type of remote sensing systems are passive systems mainly depending on the solar radiation operates in visible and infrared region of electromagnetic spectrum. The nature and properties of the target materials can be inferred from incident electromagnetic energy that is reflected, scattered or emitted by these materials on the earth's surface and recorded by the passive sensor (for example, a camera without flash). The remote sensing system that uses electromagnetic energy can be termed as electromagnetic remote sensing.

5.2 Electromagnetic Remote Sensing Process

The generalised processes involved in electromagnetic remote sensing system or passive remote sensing system, namely, data acquisition and data analysis are outlined below and a schematic diagram of electro-magnetic remote sensing process is shown in Fig. 5.2. The data acquisition process comprises distinct elements, namely, (i) energy sources, (ii) propagation of energy through the atmosphere, (iii) energy interactions with earth's surface features (iv) airborne/spaceborne sensors to record the reflected energy and (v) generation of sensor data in the form of pictures or digital information. These elements are described in detail further in this chapter.

The data analysis process involves examining the data using various viewing instruments to analyse pictorial data which is called the 'visual image interpretation techniques and computers to analyse digital data through a process known as digital image processing techniques. The analysis of a data utilising visual image interpretation involves use of the fundamental picture elements, namely tone, texture, pattern, size and shape in order to detect and identify various objects. Aerial or satellite imagery are seen through stereoscopic instruments to obtain three-dimensional images. There are many photogrammetric instruments today for visual-interpretation and for transferring the details on to base maps. If the data is available in digital form, it can be analysed on interactive computer systems for extracting statistical data or classified to obtain thematic information about resources. The scene is interactively analysed using computers by comparing with the actual "signature" of the object collected through field visits. This system of classification of objects is quite accurate and depends on the dispersion of training data sets over the area of the scene.

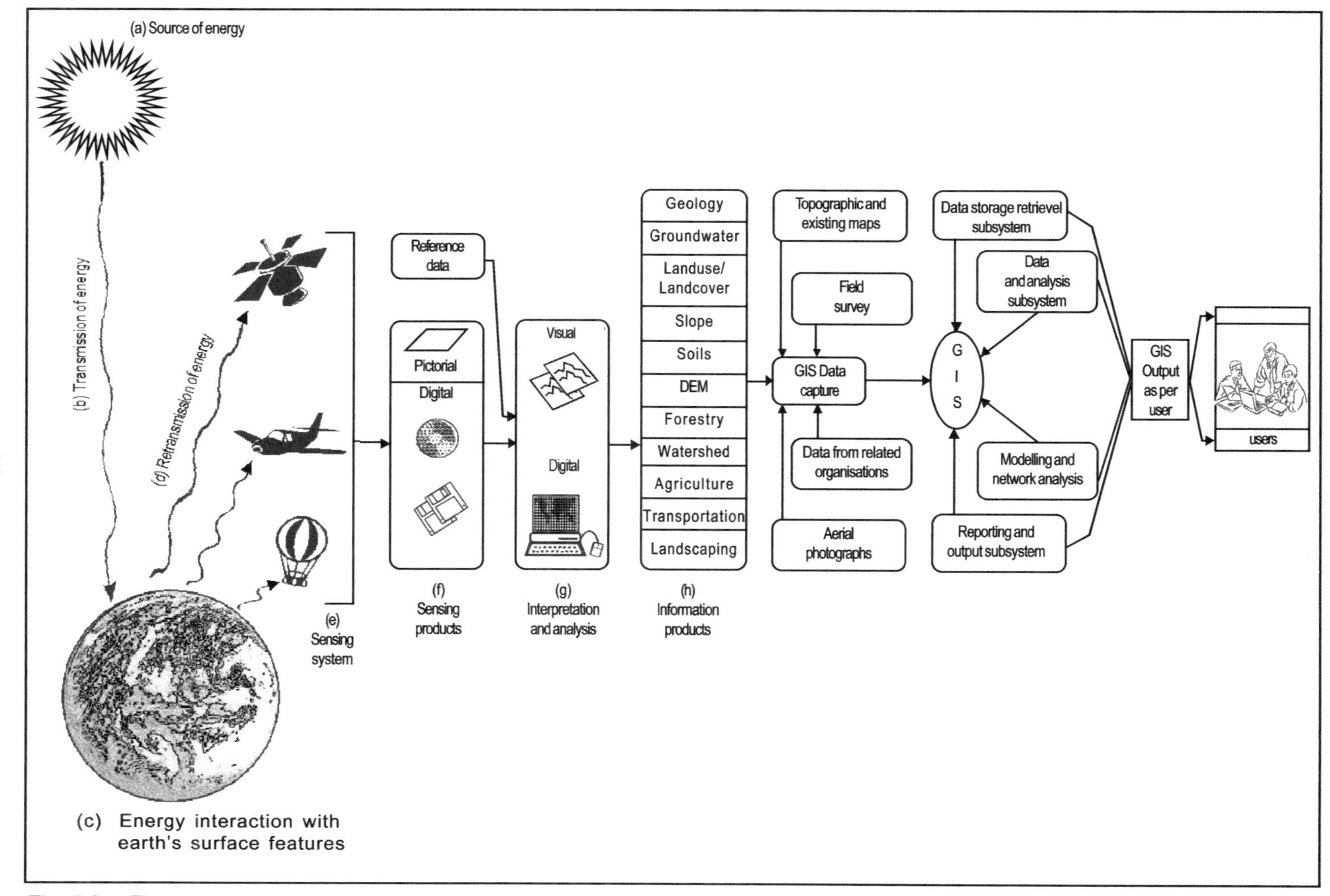

Fig. 5.2 Electromagnetic Remote Sensing Process with overview on GIS.

Digital data can also be used for certain functions like edge detection and height extraction, using specialised image processing techniques. Computer analysis of remotely sensed images offers several advantages like quick processing of large volumes of data and special image processing possibilities, such as geometrical and other types of corrections, scale changing, band ratioing, contrast enhancement, edge enhancement, and feature enhancement. Besides the standard peripherals, the digital system consists of high resolution display terminal for interaction with the resource scientist and output devices like plotters. Sample reference ground data, called training data, is collected and used conjunctively in order to obtain a more accurate thematic information by means of image classification. This thematic information layers are then used as input data for a GIS.

Whether it is the pictorial data analysis or digital data analysis, reference data should be generated as supporting data for the entire remote sensing data analysis. Reference data also called as ground truth or field check, is an essential part of remote sensing data processing. Reference data is used to serve any or all of the following purposes: (i) to analyse and interpret remotely sensed data, (ii) to calibrate a sensor, and (iii) to verify information extracted from remote sensing data. (Lillesand, T.M. and keifer, R.W, 1994). Collection of reference data consists of either time-critical or time-stable measurements. Time critical measurements are those made in cases where ground conditions, such as, vegetation conditions or water pollutants change rapidly with time. Time stable measurements like the geology of the area of interest are those involving the materials under observation, which do not change with time.

5.3 Physics of Radiant Energy

Radiant energy, which like all other energies expressed in joules, is the energy associated with electromagnetic radiation. The rate of transfer of radiant energy is called the flux and has watts as the units of power. Density implies distribution over the surface on which the radiant energy falls. If radiant energy falls upon a surface then the term irradiance (E) is used in place of radiant flux density. If the flow of energy is away from the surface, as in the case of thermal energy emitted by the earth or incoming solar energy which is reflected by the earth, then the term radiant exitance or radiant emittance as measured in units or Wm^{-2} is used (Mather, 1987).

Radiance (L) is defined as the radiant flux density transmitted from a small area on the earth's surface and viewed through a unit solid angle. It is measured in watts per square meter per steradian (Wm^{-2} Sr^{-1}). The concepts of the radian and steradian are illustrated in Fig. 5.3. The other important terms we come across remote sensing technology is 'reflectance' denoted by e. It is defined as the ratio between the irradiance and the radiant emittance of an object. When remotely sensed images collected over a time period are to be compared, it is most appropriate to convert the radiance values recorded by the sensor into reflectance in order to eliminate the effects of variable irradiance over the seasons of the year.

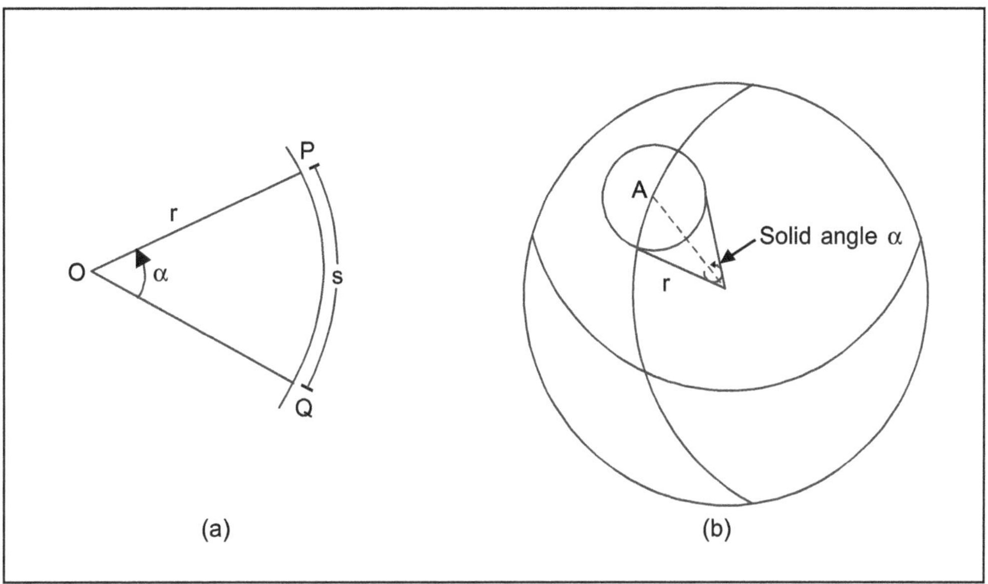

Fig. 5.3 (a) The angle formed when arc length s equals r, the radius of the circle, is equal to 1 radian. Thus, angle α = s/r radians. There are 2p radians (360 degrees) in a circle.

(b) A steradian is the solid (three-dimensional) angle formed when the area A delimited on the surface of the sphere is equal to the square of the radius of the sphere. A need not refer to a uniform shape. Thus angle a = A/r^2 steradians (sr). There are 4π radians in a sphere. (Mather, 1987)

The reflectance characteristic of earth's surface features may be quantified by measuring the portion of incident energy that is reflected. It is a dimensionless quantity. The quantities described above are very often used to refer to particular narrow wavebands rather than to the whole spectrum. The terms are then preceded by the word 'spectral', as in 'spectral radiance for a given waveband is the radiant flux density in the waveband per unit solid angle per unit wavelength' (Curran, 1988). The sun's light is the form of electromagnetic radiation most familiar to human beings. The light as reflected by physical objects travels in a straight line to the observer's eye. On reaching the retina, it generates electrical signals which are transmitted to the brain by the optic nerve. These signals are used by the brain to construct an image of the viewer's surroundings. This is the process of vision and it is closely analogous to the process of remote sensing; indeed, vision itself is a form of remote sensing. The set of all electromagnetic waves is called the electromagnetic spectrum, which includes the range from the long radio waves, through the microwave and infrared wavelengths to visible light waves and beyond them to the ultraviolet and to the short wave X-and gamma rays (Fig. 5.4).

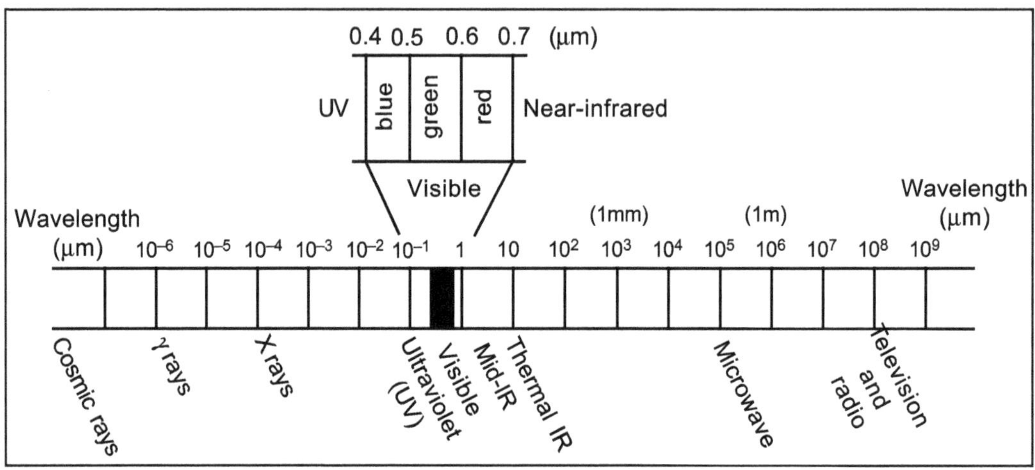

Fig.5.4 Electromagnetic Spectrum.

5.3.1 Nature of Electromagnetic Radiation

A basic difficulty encountered in analysing electromagnetic energy and interaction with matter derives from the dual nature of the behavior of electromagnetic energy. To be precise, in some situations electromagnetic energy behaves like waves, while in others it displays the properties of particles. This has been a very controversial point for the past 250 years concerned the nature of electromagnetic energy. This controversy has been sufficient to explain the nature of visible light, even though originally it was not realised that light is a form of electromagnetic energy.

Electromagnetic wave theory formulated by Maxwell in 1962, succeeded in characterising the electric and magnetic fields and their relation to charges and current, and expressing these relationships in a set of partial differential equations now known generally, as Maxwell's equations. Maxwell demonstrated that it was possible to have wave-like configurations of electric and magnetic fields. Maxwell's equations explain a great variety of phenomena relating to propagation, dispersion, reflection, refraction, and interference of electromagnetic waves; but they do not explain the interaction of electromagnetic energy with matter on an atomic and molecular level. In 1990 Planck found that, in order to calculate the correct distribution of energy emitted by a black body, he could not assume that the constituent oscillators gain and lose energy continuously. He was rather forced to assume that a particular oscillator of frequency 'V' is able to exist only in discrete states whose energies are separated by the interval hv, where 'h' is known as the Planck's constant. Planck's ideas were applied and extended shortly afterwards.

The decisive step was provided in 1925 by De Broglie who proposed that waves accompanied particles, or particles had a wave nature such that l = h/mv, where l is the wavelength and mv (mass x velocity) is the momentum. However, it was Schrodinger in 1926 who formulated wave mechanics in terms of a wave equation. The Schrodinger wave equation for atomic-molecular scale problems is not really derivable, and should be regraded as the counterpart of Newton's laws of motion for macroscopic bodies. It is used and accepted, not because of its derivation showing validity, but because when properly applied it yields correct results consistent with observation and experiment. The schrodinger wave equation directly yields the allowed energy levels of an atomic or molecular systems.

Based on the historical development of understanding the nature of electromagnetic energy, it is presently possible to furnish a consistent and unambiguous theoretical explanation for all optical phenomena using a combination of Maxwell's electromagnetic wave theory and the modern quantum theory. Maxwell's theory deals primarily with the propagation and macroscopic optical effects of electromagnetic energy, while quantum theory is concerned with the atomic molecular absorption and emission aspects of radiation.

Maxwells Theory

The four differential equations that form the basis of electromagnetic theory are generally referred to as "Maxwells's equations," and they are expressed in mathematical terms. The electric and magnetic fields may exist in regions where no electric charges are present. When the fields at one point in space vary with time, then some variation of the fields must occur at every other point in space at some other time, and consequently, changes in the fields propagate throughout space. The propagation of such a disturbance is called an electromagnetic wave. According to Maxwell the electromagnetic state at a point in a vacuum can be specified by two vectors : E, the electric field in volts per meter and H, the magnetic field in ampere turns per meter. These vector quantities are completely independent of each other in the static case, and are determined by the distribution of all charges and currents in space. In the dynamic case, however, the fields are not independent, but rather their space and time derivatives are interrelated as expressed by the curl (∇) equations.

$$\nabla \times \mathbf{E} = \mu_0 \frac{\partial \mathbf{H}}{\partial t} \quad \text{and} \quad \nabla \times \mathbf{H} = \varepsilon_0 \frac{\partial \mathbf{E}}{\partial t} \qquad \text{..........(5.1)}$$

where $\mu_0 \cong$ permeability of the vacuum = 4 $\lambda \times 10^{-7}$ h/m and $\varepsilon_0 \cong$ permitivity of the vacuum = 8.85×10^{-12} farads/m. and

$$\nabla . \mathbf{E} = 0 \quad \text{and} \quad \nabla . \mathbf{H} = 0 \qquad \text{..........(5.2)}$$

Eq. (5.2) indicate that there is no charge at the point in question, and this is true in both the static and the dynamic cases. There four equations are "Maxwell's equations" for a vacuum. The four equations or both the fields satisfy the same formal partial differential equation,

$$\nabla^2(X) = \frac{1}{C^2} \times \frac{\partial^2(X)}{\partial t^2} \qquad \text{..........(5.3)}$$

where X = E or H, and $c = \frac{1}{\mu_0 \varepsilon_0}$ and this is called the wave equation, which occurs in connection with many different kinds of physical phenomena. The major implication of the equation is that changes in the fields E and H propagate through space at a speed equal to the constant value c, which is known as the speed of light, with a measured value of 2.9979 × 10^8 m/s. The Maxwell curl equations are precisely the same for isotropic nonconduction media as they are for vacuum, except that the vacuum constants μ_0 and ε_0 are replaced by corresponding constants for the medium, denoted μ and ε. It can be shown, for the case where the spatial variation occurs in the z direction, that the function $E_Z t = E_0 \cos(k_z - wt)$ is a solution to the wave equation. This is the fundamental solution to the wave equation, representing a plane harmonic wave for which the solution is the same as that for the magnetic field. It can be shown that the magnetic and electric components are perpendicular to each other and that these plane waves are both perpendicular to the direction of propagation (Fig. 5.5). In summary, it can be seen that all electromagnetic radiation is energy in

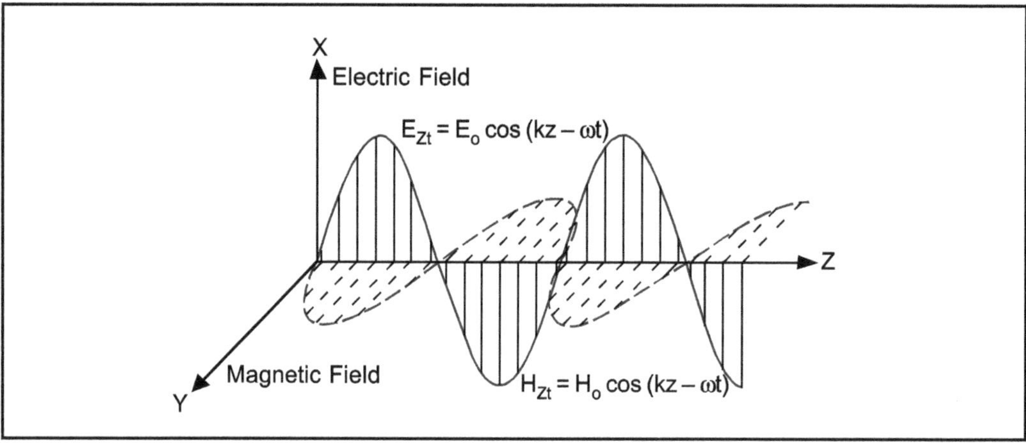

Fig. 5.5 Electromagnetic wave : the electric (vertical) and magnetic (horizontal) components are indicated.

transit and can be regarded as a wave motion. It consists of inseparable oscillating electric and magnetic fields that are always mutually perpendicular to each other and to the direction of propagation, the rate of propagation being constant in a vacuum.

Quantum Theory

The basic idea of quantum theory is that radiant energy is transmitted in indivisible packets whose energy is given in integral parts, of size hν, where **h** is Planck's constant = 6.6252 × 10^{-34} j - s, and ν is the frequency of the radiation. These are called quanta or photons.

The dilemma of the simultaneous wave and particulate views of electromagnetic energy may be conceptually resolved by considering that energy is not supplied continuously throughout a wave, but rather that it is carried by photons. The classical wave theory does not give the intensity of energy at a point in space, but gives the probability of finding a photon at that point. Thus the classical concept of a wave yields to the idea that a wave simply describes the probability path for the motion of the individual photons.

The particular importance of the quantum approach for remote sensing is that it provides the concept of discrete energy levels in materials.The values and arrangement of these levels are different for different materials. Information about a given material is thus available in electromagnetic radiation as a consequence of transitions between these energy levels A transition to a higher energy level is caused by the absorption of energy, or from a higher to a lower energy level is caused by the emission of energy. The amounts of energy either absorbed or emitted correspond precisely to the energy difference between the two levels involved in the transition. Because the energy levels are different for each material, the amount of energy a particular substance can absorb or emit is different for that material from any other materials. Consequently, the position and intensities of the bands in the spectrum of a given material are characteristic to that material.

5.3.2 Electromagnetic Spectrum

The electromagnetic spectrum may be defined as the ordering of the radiation according to wavelength, frequency, or energy. The wavelength, denoted by λ, is the distance between adjacent intensity maximum (for example) of the electromagnetic wave, and consequently, it may be expressed in any unit of length. Most commonly wavelength is expressed in meters (m) or centimeters (cm); microns or micrometers (μ or μm = 10^{-4} cm); nanometers (nm = 10^{-7} cm); or Angstrom units (Å = 10^{-8} cm). The frequency denoted by ν, is the number of maxima of the electromagnetic wave that passes a fixed point in a given time. Its relationship to wavelength is simply, $\nu = C/\lambda$

where, 'c' is the speed of light. Frequency is commonly expressed in reciprocal centimeters, also called wave numbers (cm^{-1}) or cycles per second (cps) which are also called Hertz (Hz). The wavelengths may assume any value, although for most practical purposes the spectrum is usually presented between 10^{-16} and 10^7 m, or from the cosmic ray to the audio range. However, wavelengths as long as 10^{11}m have been detected by sensitive magnetometers.

No matter what the wavelength of the electromagnetic radiation, it is all generated by electrically charged matter. However, there is no universal radiation generator that provides a useful intensity of radiation at all wavelengths for practical purposes, and there is no universal wavelength resolving instrument or universal detector. Consequently, the spectrum has been divided into regions that bear names related to the sources that produce it, such as, the "ray" regions, or as extensions from the visible range such as, the ultraviolet and the infrared regions, or according to the way in which wavelengths in a range are used such as, radio and television. The extent of the wavelength ranges corresponding to these names were made arbitrarily, and the decision as to where the divisions should be was made mostly on the basis of the limits imposed by the range of the human eye (visible), the properties of optical materials, and the response limits of various sources and detectors.

In brief, the electromagnetic spectrum is the continuum of energy that ranges from meters to nano-meters in wave length, travels at the speed of light, and propagates through a vacuum like the outer space (Sabins 1987). All matter radiates a range of electromagnetic energy, with the peak intensity shifting toward progressively shorter wave lengths at an increasing temperature of the matter. In general, the wavelengths and frequencies vary from shorter wave length high frequency cosmic waves to long wavelength low frequency radio waves. The wave lengths of greatest interest in remote sensing are visible and near-infrared radiation in the wave band 0.4 - 3 micrometers, infrared radiation in the wave band 3-14 micrometers and microwave radiation in the wave band 5-500mm.

Spectral Wave Bands

Visible light is electromagnetic radiation with wavelengths between 0.4 μm and 0.7 μm. The eye is not uniformly sensitive to light within this range and has its peak sensitivity at 0.55 μm (Fig. 5.6). This peak in the response function of the human eye corresponds closely to the peak in the sun's radiation emittance distribution. Electromagnetic radiation with wavelengths shorter than those of visible light (0.4μm) is divided into three spectral regions, namely, the gamma rays, X-rays, and ultraviolet rays. Because of the effect of scattering and absorption, none of these bands is used in satellite remote sensing.

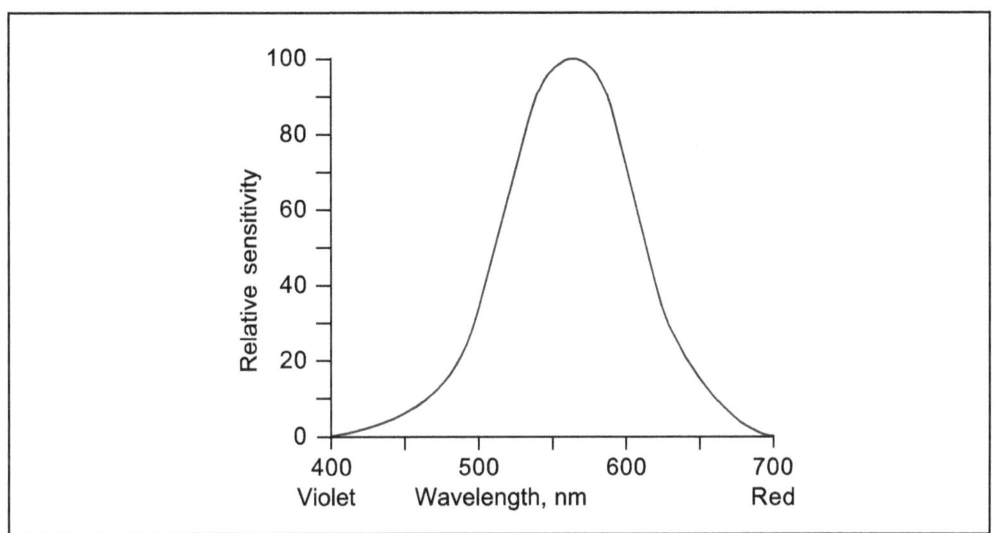

Fig. 5.6 Response Function of Human Eye.

Wavelengths longer than the visible red are sub-divided into the infrared (IR), microwave and radio frequency wavebands. The infrared waveband, extending from 0.7 μm to 1 μm is not a uniform region. Short wavelength or near - IR between 0.7 μm and 0.9 μm behaves like visible light and can be detected by special photographic film. Infrared radiation with a wavelength upto 3 μm is reflected by the surface of the earth. Beyond a wavelength of 3 μm, IR radiation emitted by the earth's surface can be sensed in the form of heat. The region of the spectrum composed of electromagnetic radiation with wavelengths between 1 mm and 300 cm is called the microwave band and radiation at these wavelengths can penetrate the clouds. The microwave band is thus a valuable region for remote sensing. Beyond the microwave region is the radioband of very long wavelengths used in certain radar applications. The electromagnetic wavebands with their utility in remote sensing are described in Table 5.1.

5.4 Energy Source and ITS Characteristics

All objects whose temperature is greater than an absolute zero (273°k), emit radiation. All stars and planets emit radiation. Our chief star, the sun is almost a spherical body with a diameter of 1.39×10^6 km at a mean distance from the earth equal to 1.5 x 10 km. The continuous conversion of hydrogen to helium which is the main constituents of the sun, generates the energy that is radiated from the outer layers. If the energy received at the edge of earth's atmosphere were distributed evenly over the earth, it would give an average incident flux density of 1367 w/m^2. This is known

TABLE 5.1 Electromagnetic Spectral Regions (Sabines, 1987)

Region	Wavelength	Remarks
Gamma ray	<0.03nm	Incoming radiation is completely absorbed by the upper atmosphere and is not available for remote sensing.
X-ray	0.03 to 3.0nm	Completely absorbed by atmosphere. Not employed in remote sensing.
Ultraviolet	0.3 to 0.4μm	Incoming wavelengths less than0.3μm are completely 0.4μm absorbed by ozone in the upper atmosphere.
Photographic UV band	0.3 to 0.4μm	Transmitted through atmosphere. Detectable with film and photodetectors,but atmospheric scattering is severe
Visible	0.4 to 0.7μm	Imaged with film and photodetectors.Includes reflected energy peak of earth at 0.5μm.
Infrared	0.7 to 1.00μm	Interaction with matter varies with wavelentgth. Atmospheric transmission windows are separated.
Reflected IR band	0.7 to 3.0μm	Reflected solar radiation that contains information about thermal properties of materials. The band from 0.7 to 0.9μm is detectable with film and is called the photographic IR band.
Thermal IR	3 to 5μm band	Principal atmospheric windows in the 8 to 14μm thermal region. Images at these wavelengths are acquired by optical mechanical scanners and special vidicon systems but not by film. Microwave 0.1 to 30 cm longer wavelengths can penetrate clouds, fog, and rain. Images may be acquired in the active or passive mode.
Radar	0.1 to 30 cm	Active form of microwave remote sensing. Radar images are acquired at various wavelength bands.
Radio	>30 cm	Longest wavelenth portion of electromagnetic spectrum. Some classified radars with very long wavelengths operate in this region.

as the solar constant. 35 percent of the incident radiant flux is reflected back by the earth. This includes the energy reflected by clouds and atmosphere. 17 percent of it is absorbed by atmosphere and 48 percent is absorbed by the earth's surface materials (Mather, 1987).

If the sun were a perfect emitter, it would be an example of an ideal black body. A black body transforms heat energy into radiant energy at the possible maximum rate consistent with Planck's law which defines the spectral exitance of a black body as follows (Henderson, 1970):

$$M_l = \frac{C_1}{{}_\lambda 5\left[\exp(C_2/\lambda T) - I\right]} \qquad ....(5.4)$$

Where, $C_1 = 3.742 \times 1O^{-16}\ Wm^{-2}$, $C_2 = 1.4388 \times 10^{-2}\ m^0k$

λ = Wavelength (μm)

T = Temperature (in ^{0}k)

M_λ = Spectral exitance per unit wavelength.

Curves showing the variation of spectral exitance for black bodies at different temperatures are shown in Fig. 5.7. The wavelength at which the maximum spectral exitance is achieved is reduced as the temperature increases. The dashed line in Fig. 5.7 which joins the peaks of the spectral exitance curves, is described by

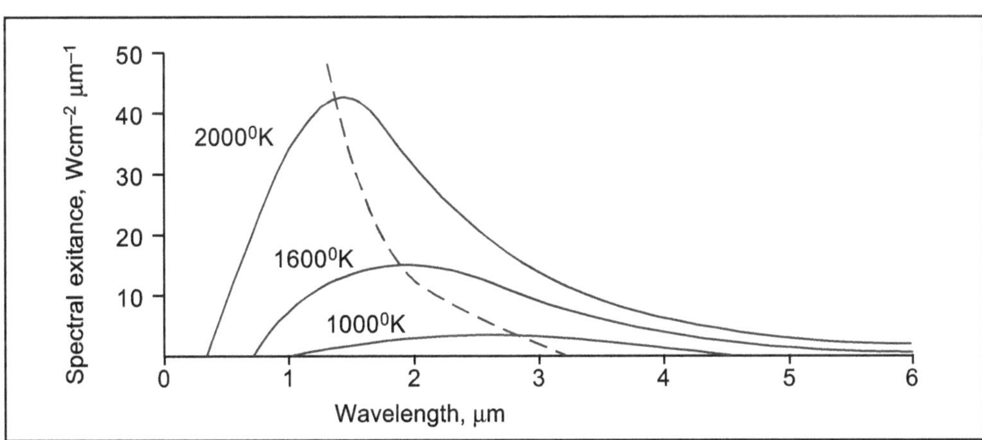

Fig. 5.7 Spectral Exitance curves for black bodies.

Wien's displacement law. The law gives the wavelength of maximum spectral exitance (λ_m) in the following form :

$$\lambda_m = C_3 / T \qquad ....(5.5)$$

where, $C_3 = 2.898 \times 10^{-3}$ mk, T = temperature of body

The total spectral exitance of a black body is given by the following Stefan Boltzman law:

$$M = \sigma T^4 \qquad \text{..........(5.6)}$$

in which, $\sigma = 5.6697 \times 10^{-8}\ Wm^{-2}\ k^{-4}$ (Stefan - Boltzman constant)

The distribution of the spectral exitance for a black body at 5900^0k closely approximates the sun's spectral exitance curve (Mather, 1987), while the earth can be considered to act like a black body with a temperature of 290^0k (Fig. 5.8.) The solar radiation, maximum of which occurs at 0.47mm, is within the visible spectrum. Wavelength dependent mechanisms of atmospheric absorption alter the solar irradiance that actually reaches the surface of the earth. Fig. 5.9 shows the spectral irradiance from the sun at the edge of the atmosphere (solid curve) and at the earth's surface (dashed line).

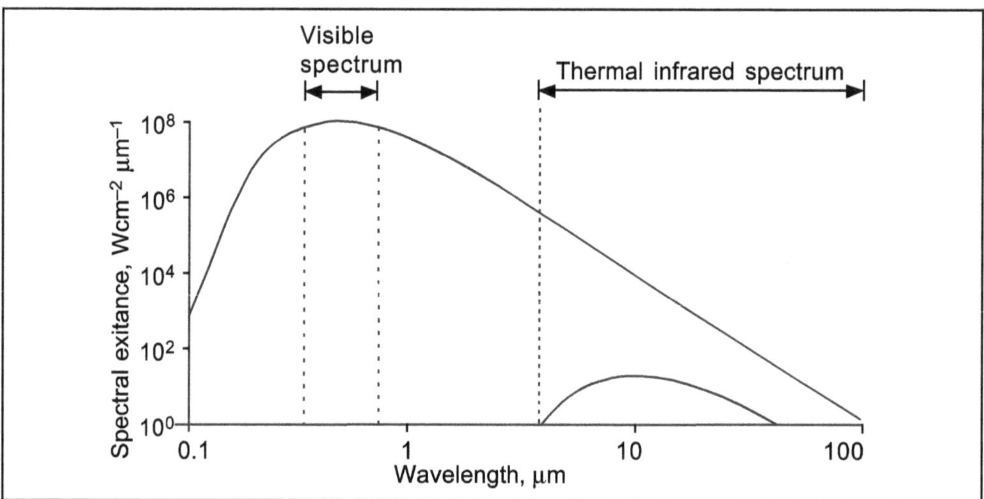

Fig. 5.8 Spectral exitance curves at temperatures of Sun and Earth.

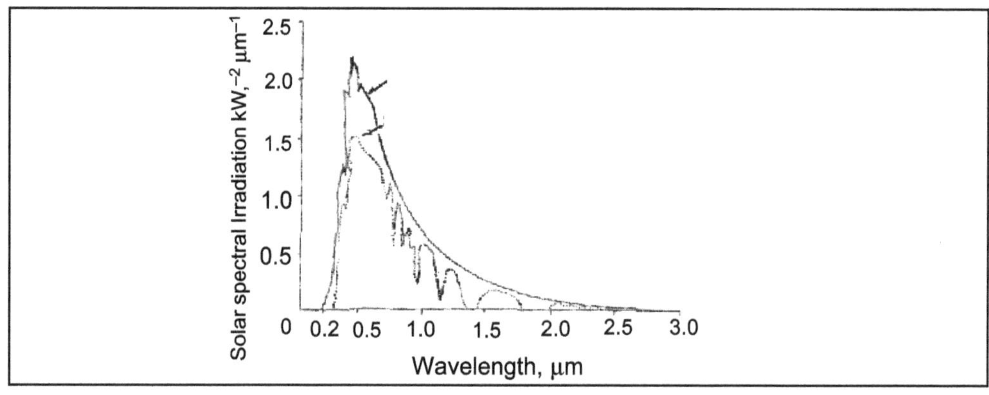

Fig. 5.9 Solar irradiance curves.

The characteristics of radiation sources impose some limitations on the range of wavebands that can be utilized in remote sensing. Generally, the selection of wavebands for use depends on (a) the characteristics of the radiation source, (b) the effects of atmospheric absorption and scattering, and (c) the nature of the target.

5.5 Atmospheric Interactions with Electromagnetic Radiation

All electromagnetic radiation detected by a remote sensor has to pass through the atmosphere twice, before and after its interaction with earth's atmosphere. This passage will alter the speed, frequency, intensity, spectral distribution, and direction of the radiation. As a result atmospheric scattering and absorption occur (Curran, 1988). These effects are most severe in visible and infrared wavelengths, the range very crucial in remote sensing.

During the transmission of energy through the atmosphere, light interacts with gases and particulate matter in a process called *atmospheric scattering*. The two major processes in scattering are selective scattering and non-selective scattering. Rayleigh, Mie and Raman scattering are of selective type. Non selective scattering is independent of wavelength. It is produced by particles whose radii exceed 10mm, such as, water droplets and ice fragments present the clouds. This type of scattering reduces the contrast of the image. While passing through the atmosphere, electromagnetic radiation is scattered and absorbed by gasses and particulates. Besides the major gaseous components like molecular nitrogen and oxygen, other constituents like water vapour, methane, hydrogen, helium and nitrogen compounds play an important role in modifying the incident radiation and reflected radiation. This causes a reduction in the image contrast and introduces radiometric errors. Regions of the electromagnetic spectrum in which the atmosphere is transparent are called *atmospheric windows.* The atmosphere is practically transparent in the visible region of the electromagnetic spectrum and therefore many of the satellite based remote sensing sensors are designed to collect data in this region. Some of the commonly used atmospheric windows are 0.38 – 0.72 μm (visible), 0.72 –3.00 μm (near infrared and middle infrared) and 8.00 –14.00 μm (thermal infrared). Fig. 5.10. Shows relative scatter as

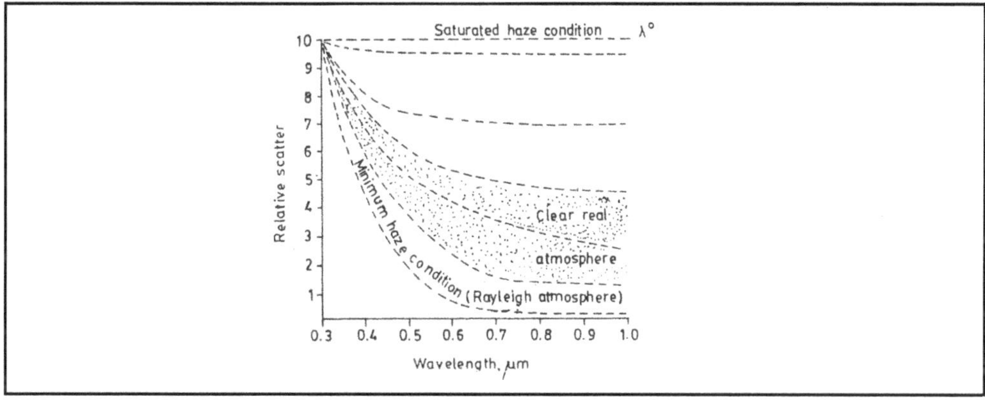

Fig. 5.10 Relative scatter for various levels of atmospheric haze.

a function of wavelength from 0.3 to 1 μm of the spectrum for various levels of atmospheric haze. The characteristics of all the four types of scattering in the order of their importance in remote sensing are given in Table 5.2.

Table 5.2 Types of Atmospheric Scatter in Order of Importance (Curran. 1988)

Type of Scatter	Size of effective atmospheric	Type of effective atmospheric particles	Scatter particles	Effect of scatter on visible and near visible wavelength
Rayleigh	Smaller than the wavelength of radiation.	Gas molecules	Molecule absorbs high energy radiation and re-emits. skylight scatter is inversely proportional to fourth power of wavelength.	Affects short visible wavelengths, resulting in haze in photography, and blue skies.
Mie	Same size as the wavelength of radiation.	Spherical particles fumes and dust	Physical scattering under overcast skies.	Affects all visible wavelengths
Non-selective	Larger than the wavelength of radiation.	Water droplets and dust.	Physical scattering by fog and clouds.	After all visible wavelengths equally, resulting white fog and clouds
Raman	Any	Any	Photon has elastic collision with molecule resulting in a loss or a gain in energy; this can decrease or increase wavelength.	Variable

5.5.1 Atmospheric Properties

The main part of the radiance measured from high flying aircraft or satellite stems form multiple scattering in the atmosphere. Therefore, the remaining signal can be interpreted in terms of suspensions only after a careful correction for the atmospheric contribution. For this reason the varying optical parameters of atmosphere must enter the radiative transfer calculations (Fischer J, 1989). Before we study the effects of solar radiation and atmospheric properties, we shall consider the mass quantities which determine the spectral upward radiance. The source of the shortwave radiation field in atmosphere is the Sun emitting in a broad spectral range. The extra terrestrial irradiance at the top of the atmosphere, the solar constant, depends on the black body emission of the Sun's photosphere and on the scattering and absorption process in the Sun's chromosphere. Important Fraunhofer lines caused by the strong absorption in the Sun's chromosphere show some prominent drops in the spectral distribution of the solar radiation. Fig. 5.11 shows the solar irradiance at the top of the earth's atmosphere to be between 0.4 and 0.8 μm as determined by Necked and Labs.

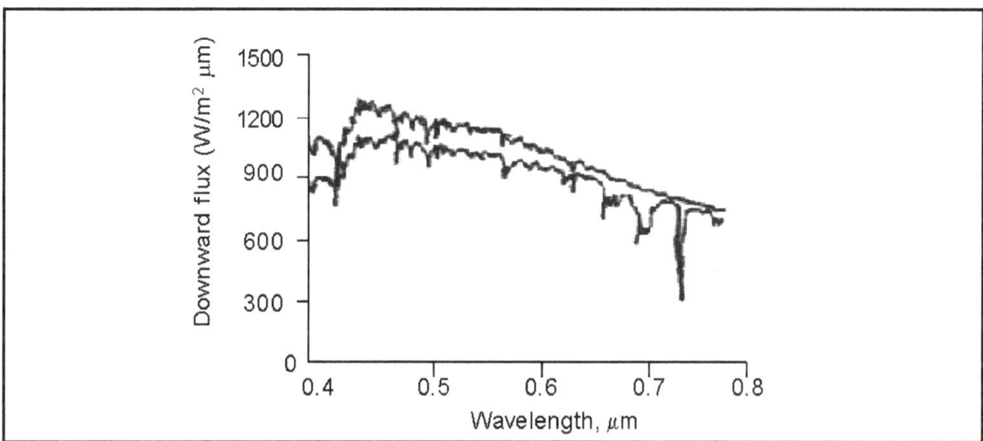

Fig. 5.11 Solar irradiance at the top of the atmosphere illuminating the Earth between 0.4 mm – 0.8 mm.

5.5.2 Absorption of Ozone

Ozone is a trace gas in the atmosphere mainly confined to stratospheric heights between 20 and 40 km with a maximum concentration near 25 km. At these levels, ozone dominates the short wave radiation budget, while at other heights its influence is nearly negligible. The Chappuis band of ozone in the visible is the only ozone band used to detect the oceanic constituents from space.

The transmission of the chlorophyll fluorescence to the top of the atmosphere is hindered through the absorption by water vapour and molecular oxygen in their vibration action bands. In order to study the selective gaseous absorption in the radiative transfer calculations the transmission functions of O_2 and H_2O are computed from absorption line parameters by explored through areas of Lorentz's theory of collision broadening. The contribution from resonance broadening is negligible in the spectral region considered. Also the Doppler line broadening, which is small when compared with Lorentz line widths, is neglected since the area absorption takes place in the atmosphere below 40km (Barrow 1962). The transmission functions are averaged over 1nm wavelength intervals. The reduction in the solar flux due to absorbtion and scattering by a clear mid-latitude summer atmosphere. Response studies for the temperature and pressure dependence of the transmission function have been performed and show only a weak influence for the temperature effect. The pressure impact is not negligible and has to be accounted for. Air molecules are small compared to the wavelength of the incoming sunlight. Hence, the extinction through molecular scattering can be determined with Rayleigh theory. The necessary property for the determination of the scattering coefficient of the vertical profile of the atmospheric pressure has been here estimated after (Hunt, et. al, 1973). Since molecular scattering within the atmosphere depends mainly on pressure, the scattering coefficient can be estimated by climatological measurement.

Atmospheric spectral turbidity variations are caused by variations in aerosol concentration, composition and size distribution. The vertical distribution of the aerosols is taken from (Adler and Ken, 1963). The phase functions of aerosols are nearly wavelength independent within the visible and near infrared. For the radiative transfer calculations the scattering functions are estimated by Mie theory. The range of atmospheric turbidity values used to study the effects of aerosol scattering on the measured spectral radiances correspond to horizontal visibilities at the surface between 6 and 88 km.

Atmospheric constituents like water vapour, carbon dioxide, and ozone are the primary absorbers of electromagnetic energy. The regions of the electromagnetic spectrum in which atmospheric absorption is low are called atmospheric ‘windows’ and it is through these ‘windows’ that remote sensing of the earth’s surface is made. The atmospheric windows are not totally free from atmospheric absorption as gases and suspended particles in the atmosphere absorb certain amount of radiation from ground objects, resulting in a decrease in radiation at the sensor. These effects are variable in space and time and often necessitate the atmospheric correction of remotely sensed data.

5.5.3 Atmospheric Effects on Spectral Response Patterns

The energy recorded by a sensor is always modified by the atmosphere between the sensor and the ground. As shown in Fig. 5.12, the atmosphere influences the radiance recorded by a sensor in two ways, namely, (a) it attenuates or reduces the energy illuminating a ground object and (b) the atmosphere acts as a reflector itself adding the path radiance to the signal detected by the sensor. These two atmospheric effects are expressed mathematically as follows (Lillesand, and Kiefer 1979):

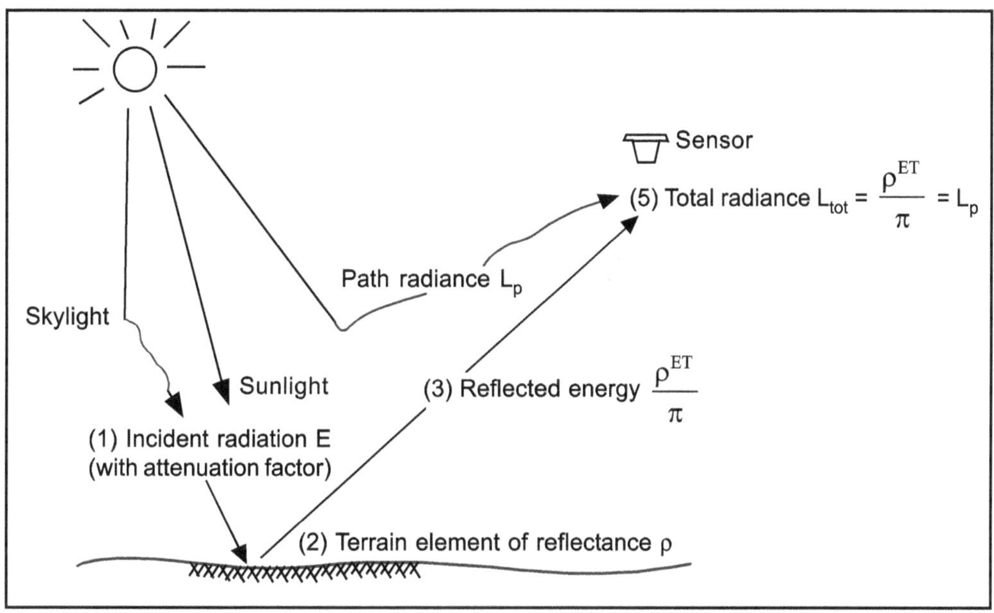

Fig. 5.12 Atmospheric Effects influencing the Spectral Radiance.

$$L_{tot} = \frac{\rho ET}{\pi} + L_P \qquad \text{..........(5.7)}$$

Where, L_{tot} = Total spectral radiance measured by the sensor

L_P = Path radiance from the atmosphere

ρ = Reflectance of object

E = Irradiance on object or incoming energy.

T = Transmitted energy

The irradiance (E) is caused by directly reflected 'sunlight' and diffuse 'skylight', which is the sunlight scattered by the atmosphere. The amount of irradiance depends on seasonal changes, solar elevation angle, and distance between the earth and sun.

5.6 Energy Interactions with Earth's Surface Materials

When electromagnetic energy is incident on any feature of earth's surface, such as a water body, various fractions of energy get reflected, absorbed, and transmitted as shown in Fig. 5.13. Applying the principle of conservation of energy, the relationship can be expressed as:

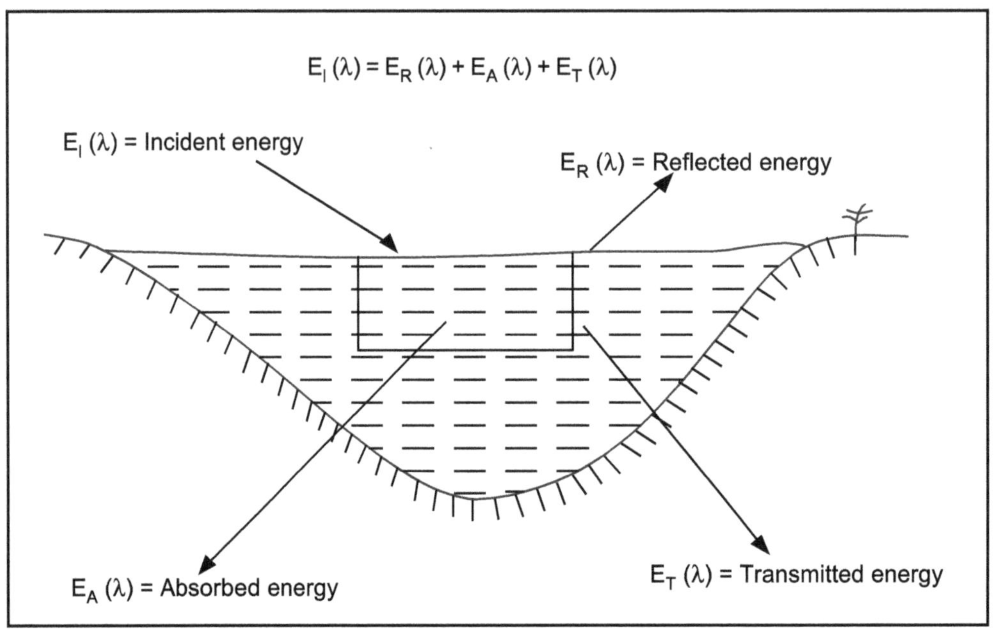

Fig. 5.13 Basic Interactions Between Electromagnetic Energy and a water body.

$$E_I(\lambda) = E_R(\lambda) + E_A(\lambda) + E_T(\lambda) \quad ....(5.8)$$

Where, E_I = Incident energy

E_R = Reflected energy

E_A = Absorbed energy

and, E_T = Transmitted energy

All energy components are functions of wavelength, (l). In remote sensing, the amount of reflected energy $E_R(\lambda)$ is more important than the absorbed and transmitted energies. Therefore, it is more convenient to rearrange these terms like

$$E_R(\lambda) = E_I(\lambda) - [E_A(\lambda) + E_T(\lambda)] \quad ....(5.9)$$

Eq. (5.9) is called balance equation. From this mathematical equation, two important points can be drawn. Firstly,

$$\frac{E_R(\lambda)}{E_I(\lambda)} = \frac{E_I(\lambda)}{E_I(\lambda)} - \left[\frac{E_A(\lambda)}{E_I(\lambda)} + \frac{E_T(\lambda)}{E_I(\lambda)}\right] \quad ....(5.10)$$

According to principles of physics, it is known that

$\frac{E_R(\lambda)}{E_I(\lambda)}$; $\frac{E_A(\lambda)}{E_I(\lambda)}$ and $\frac{E_T(\lambda)}{E_I(\lambda)}$ are called reflectance, absorbance and transmittance and can be denoted as $\rho(\lambda)$, $\alpha(\lambda)$ and $\gamma(\lambda)$. Simply, it can be understood that, the measure of how much electromagnetic radiation is reflected off a surface is called its reflectance. The reflectance range lies between 0 and 1. A measure of 1.0 means that 100% of the incident radiation is reflected off the surface, and a measure '0' means that 0% is reflected. The reflectance characteristics are quantified by "spectral reflectance, ρ_λ which is expressed as the following ratio :

$$\rho_\lambda = \frac{E_R(\lambda)}{E_I(\lambda)}$$

$$= \frac{\text{energy of wavelength '}\lambda\text{' reflected from the object}}{\text{energy of wavelength '}\lambda\text{' incident upon the object}} \qquad \text{..........(5.11)}$$

Eq. (5.10) can be written as

$$\rho(\lambda) = 1 - [\alpha(\lambda) + \gamma(\lambda)] \qquad \text{..........(5.12)}$$

Since, almost all earth surface features are very opaque in nature, the transmitance $\gamma(\lambda)$ can be neglected. According to Kirchoff's law of physics, the absorbance is taken as emissivity (ξ). Therefore Eq. (5.12) becomes :

$$\rho(\lambda) = 1 - \xi(\lambda) \qquad \text{..........(5.13)}$$

Eq. (5.13) is the fundamental equation by which the conceptual design of remote sensing technology is built. If $\xi(\lambda)$ is a zero, then $\rho(\lambda)$, that is, the reflectance is one, which means, the total energy incident on the object is reflected and recorded by sensing systems. The classical example of this type of object is snow (white object). If $\xi(\lambda)$ is one, then (λ) is a zero indicating that whatever the energy incident on the object, is completely absorbed by that object. Black body such as lamp smoke is an example of this type of object. Therefore it can be seen that the reflectance varies from 0 (black body) to 1 (white body). When we divide the incident energy on both sides of the balance equation, we get the proportions of energy reflected, absorbed and transmitted which vary for different features of the earth depending on the material type. These differences provide a clue to differentiate between features of an image. Secondly, from the wavelength dependency of the energy components, it is evident that even within a given feature type, the proportion of reflected, absorbed, and transmitted energies may vary at different wavelengths. Thus two features which are indistinguishable in one spectral range, may exhibit a marked contrast in another wavelength band. Because many remote sensing systems operate in the wavelength regions in which reflected energy predominates, the reflectance properties of terrestrial features are very important.

5.6.1 Spectral Reflectance Curves

A basic assumption made in remote sensing is that a specific target has an individual and characteristic manner of interacting with incident radiation. The manner of interaction is described by the spectral response of the target. The spectral reflectance curves describe the spectral response of a target in a particular wavelength region of electromagnetic spectrum, which, in turn depends upon certain factors, namely, orientation of the sun (solar azimuth), the height of the Sun in the sky (solar elevation angle), the direction in which the sensor is pointing relative to nadir (the look angle) (Fig. 5.14), and nature of the target, that is, state of health of vegetation.

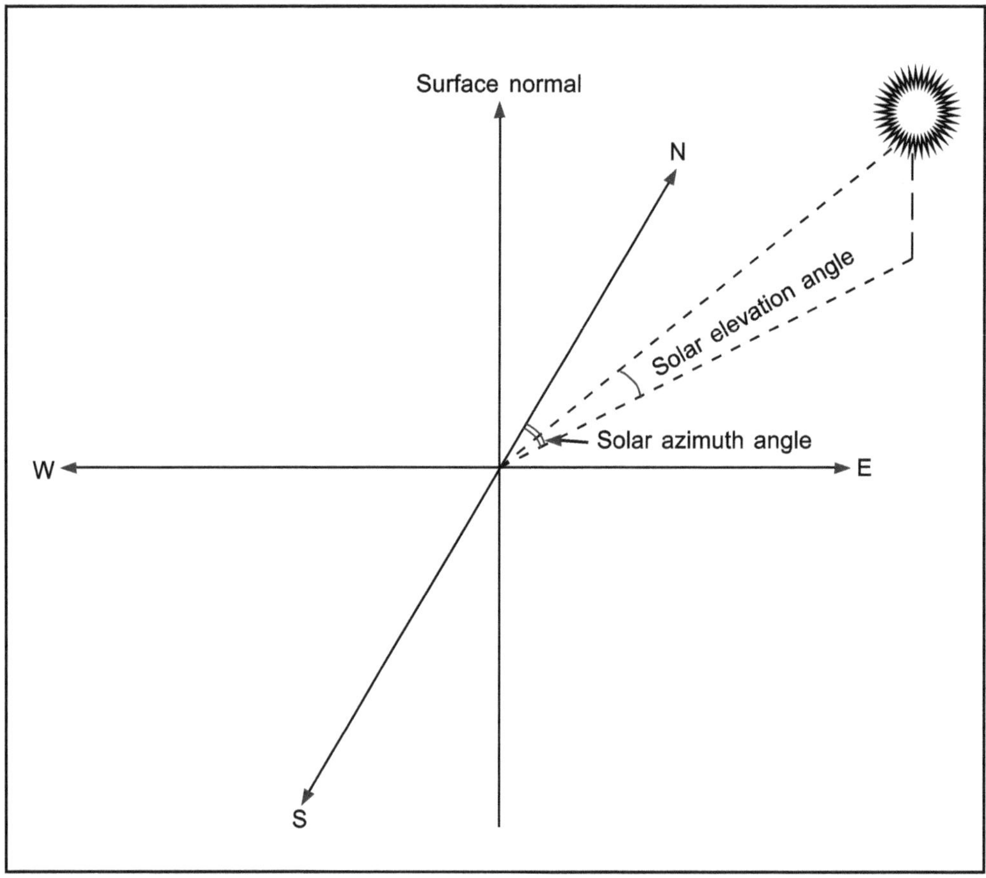

Fig 5.14 Solar elevation and azimuth angles. Elevation is measured upwards from the horizontal plane. Azimuth is measured clockwise from north. The zenith angle is measured from the surface angle, and equals 90 minus elevation angle, in degrees.

Every object on the surface of the earth has its unique spectral reflectance. Fig. 5.15 shows the average spectral reflectance curves for three typical earth's features : vegetation, soil and water. The spectral reflectance curves for vigorous vegetation manifests the "Peak-and-valley" configuration. The valleys in the visible portion of the spectrum are indicative of pigments in plant leaves. Dips in reflectance (Fig. 5.15) that can be seen at wavelengths of 0.65mm, 1.4mm and 1.9mm, are attributable to absorption of water by leaves. The soil curve shows a more regular variation of reflectance. Factors that evidently affect soil reflectance are moisture content, soil texture, surface roughness, and presence of organic matter. The term spectral signature can also be used for spectral reflectance curves. Spectral signature is a set of characteristics by which a material or an object may be identified on any satellite image or photograph within the given range of wavelengths. Sometimes, spectral signatures are used to denote the spectral response of a target.

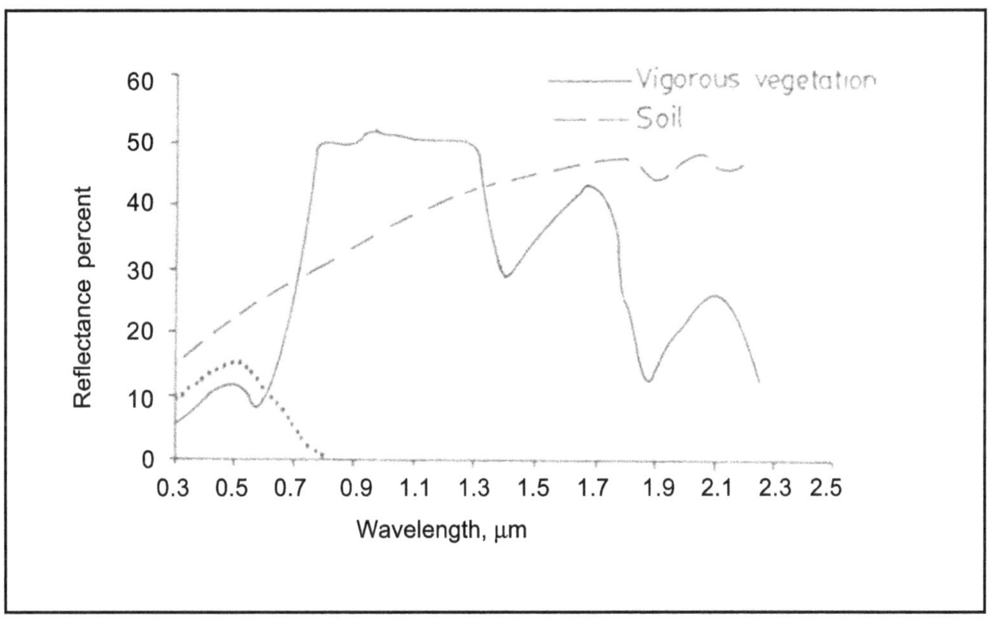

Fig. 5.15 Spectral Reglectance Curves or Spectral Signatures of Typical Features of earth's Surface.

The characteristic spectral reflectance curve Fig. 5.15 for water shows that from about 0.5µm, a reduction in reflectance with increasing wavelength, so that in the near infrared range, the reflectance of deep, clear water is virtually a zero (Mather, 1987). However, the spectral reflectance of water is significantly affected by the presence of dissolved and suspended organic and inorganic material and by the depth of the water body. Fig. 5.16 shows the spectral reflectance curves for visible and near-infrared wavelengths at the surface and at 20 m depth. Suspended solids in water scatter the down welling radiation, the degree of scatter being proportional to

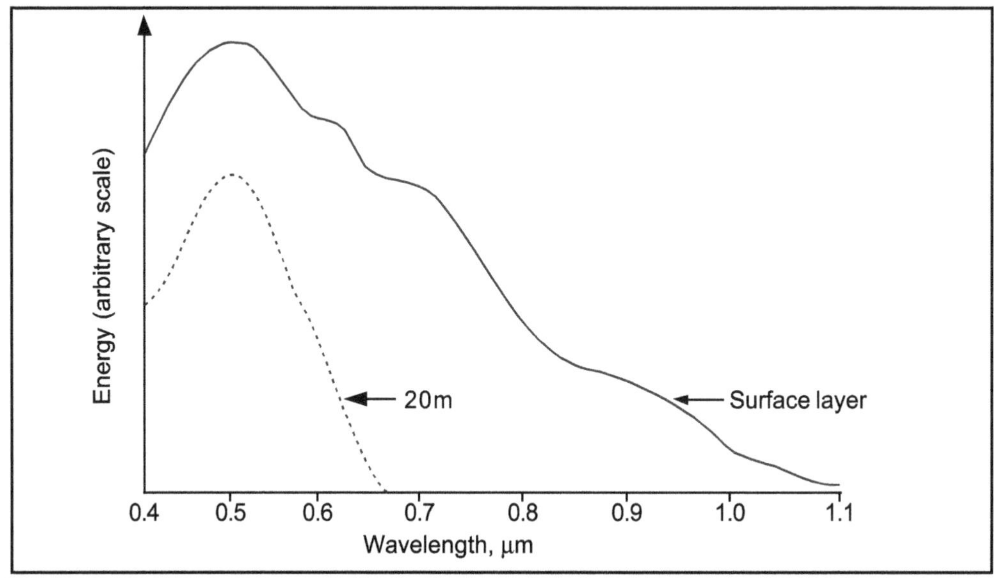

Fig. 5.16 Spectral reflectance curves for water at different depths.

the concentration and the colour of the sediment. Experimental studies in the field and in the laboratory as well as experience with multispectral remote sensing have shown that the specific targets are characterised by an individual spectral response. Indeed the successful development of remote sensing of environment over the past decade bears witness to its validity. In the remaining part of this section, typical and representative spectral reflectance curves for characteristic types of the surface materials are considerd. Imagine a beach on a beautiful tropical island. Fig. 5.17 shows interaction

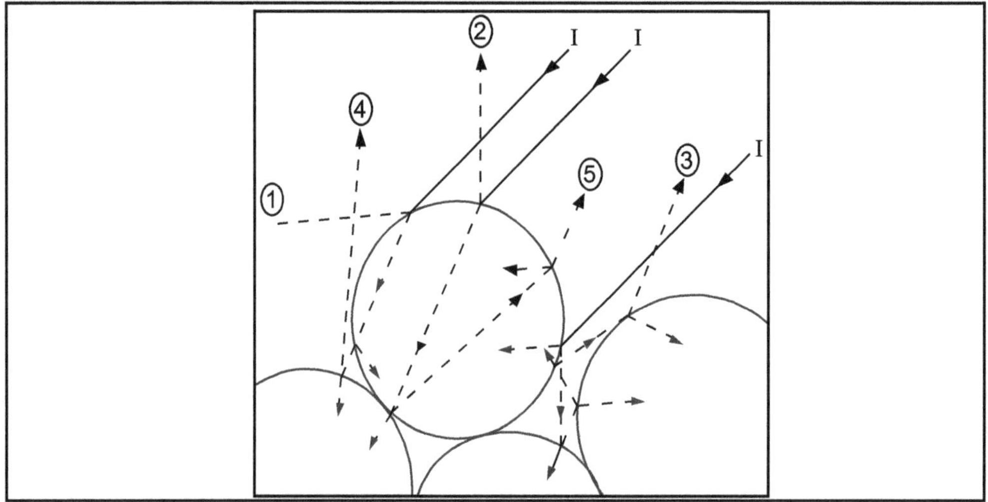

Fig. 5.17 Interaction between electromagnetic radiation and the top layer of particles comprising a mat surface. Solid lines (I) represent incident rays, lines 4 and 5 are volume rays. (After Vincent and Hunt 1968.) (Courtesy of "Applied Optics," Optical Society of America.)

of electromagnetic radiation with the top layer of sand grains on the beach. When an incident ray of electromagnetic radiation strikes an air/grain interface, part of the ray is reflected and part of it is transmitted into the sand grain. The solid lines in the figure represent the incident rays, and dashed lines 1, 2, and 3 represent rays reflected from the surface but have never penetrated a sand grain. The latter are called specular rays by Vincent and Hunt (1968), and surface-scattered rays by Salisbury and Wald (1992); these rays result from first-surface reflection from all grains encountered. For a given reflecting surface, all specular rays reflected in the same direction, such that the angle of reflection (the angle between the reflected rays and the normal, or perpendicular to the reflecting surface) equals the angle of incidence (the angle between the incident rays and the surface normal).

The measure of how much electromagnetic radiation is reflected off a surface is called its reflectance, which is a number between 0 and 1.0. A measure of 1.0 means the 100% of the incident radiation is reflected off the surface, and a measure of 0 means that 0% is reflected. In the case of first-surface reflection, this measure is called the specular reflectance, which will be designated here as $r^s(\lambda)$. The λ in parentheses indicates that specular reflectance is a function of a wavelength. The reason that $r^s(\lambda)$ is a function of a wavelength is that the complex index of refraction of the reflecting surface material is dependent on a wavelength. The term *complex* means that there is a real and imaginary part to the index of refraction. Every material has a complex index of refraction, though for some materials at some wavelengths, only the real part of the complex index of refraction may be nonzero.

For a sand grain with complex index of refraction $N(\lambda) = n(\lambda)[1 - ik(\lambda)]$, the specular reflectance is expressed by Fresnel's equation (Jenkins and White 1957), as follows :

$$r^s(\lambda) = \frac{[n(\lambda)-1]^2 + n^2(\lambda)k^2(\lambda)}{[n(\lambda)+1]^2 + n^2(\lambda)k^2(\lambda)} g(\theta, \phi) \qquad \text{..........(5.12)}$$

where,

$r^s(\lambda)$ = specular reflectance of one reflecting grain $(0 < r^s(\lambda) < 1)$

$m(\lambda)$ = intrinsic spectral index of refraction of the grain

$k(\lambda)$ = intrinsic spectral index of absorption of the grain

$g(\theta, \phi)$ = generally nonzero function of the angle of incidence (ϕ) and angle of observation (θ) with respect to the macroscopic surface

The final reflectance of a specular ray bouncing off multiple grains of sand is simply the multiplicative product of specular reflectance from all of the encountered air/grain interface. For instance, if the specular reflectance of three grains for a particular wavelength of electromagnetic radiation were 0.9, 0.8 and 0.7, respectively, the final reflectance of a specular ray bouncing off all three grains would be (0.9)(0.8)(0.7) = 0.504. The specular reflectance of the beach surface, $R^s(\lambda)$, is the average of all the individual specular ray reflectance.

Rays of electromagnetic radiation that have been transmitted through some portion of one or more grains are called volume rays. These are shown as dashed lines 4 and 5 in Fig.5.17. The equation for the volume reflectance, $r^s(\lambda)$, of a sand grain is complicated because it depends on both the transmittance of the grain and the interface reflectance of the top of that grain and the underlying grain (s).

The average $r^s(\lambda)$ for all the grains in the beach from which electromagnetic radiation is reflected is defined as the volume reflectance of the beach, $R^s(\lambda)$. The total reflectance of the beach, $R^T(\lambda)$, is the averaged sum of the specular and volume reflectance, as follows :

$$R^T(\lambda) = \frac{[R^S(\lambda) + R^V(\lambda)]}{2} \qquad \text{..........(5.13)}$$

The dependence of $R^s\lambda$ and $R^v(\lambda)$ are markedly different, as demonstrated in Fig.5.18 for the case of a uniform grain size and varying wavelength. Three important observations can be summarised from the above discussion on the beautiful beach island.

(i) The spectral locations of absorption bands depend on chemical composition of the material; for example, quartz and calcite absorption bands in the thermal infrared wavelength region have different spectral locations because Si and O ions in quartz are connected by a "spring" with a different bond strength than that of the " spring" connecting Ca and O ions in calcite.

(ii) The brightness, or magnitude, of the spectral reflectance depends primarily on the size of the reflecting grains.

(iii) Absorption bands appear as reflectance minima in transparent materials (such as quartz and calcite in the visible wavelength region), whereas absorption bands appear as reflectance maxima in opaque materials.

Note that when we use the terms transparent or opaque to explain optical behavior, we must designate both a wavelength region and the material because the complex index of refraction of any material is generally non-constant over a large range of wavelength.

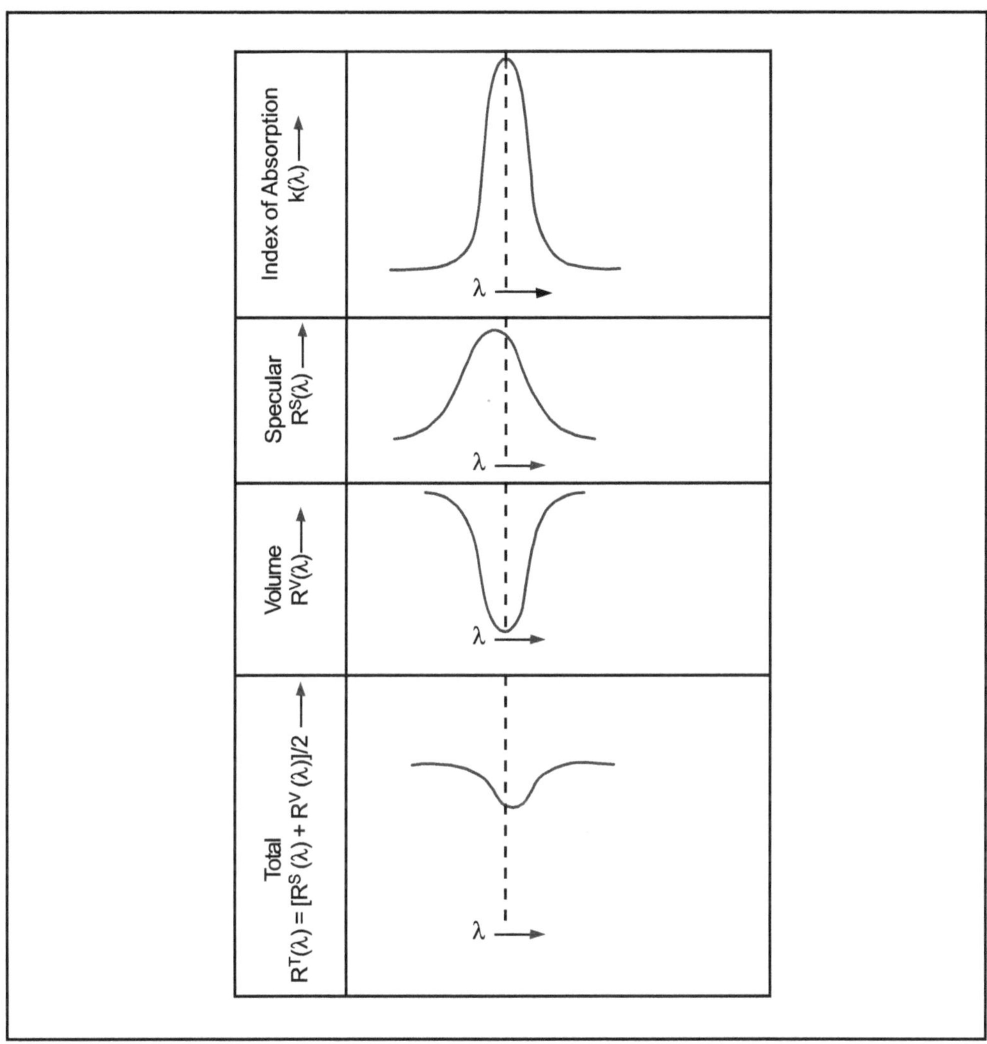

Fig. 5.18 Absorption coefficient, specular reflectance, volume reflectance, and total reflectance vs. wavelength for a spectral feature. (After Vincent and Hunt 1968.) (Courtesy of "Applied Optics," Optical Society of America.)

To consider the effect on reflectance of mixing several minerals together, let us take the simpler case of a particulate medium consisting of several mineral constituents, with air filling the interstices between particles. It is possible for us to estimate the spectral reflectance of a mixed-mineral particulate sample by using a linear combination of the reflectance spectra of its mineral constituents, weighed by the percentage of area on the sample's surface that is covered by each mineral constituent. The following equation demonstrates this estimation for the total spectral reflectance of a mixed particulate sample at wavelength λ (Robert K. Vincent., 1997) (Fig. 5.19(a)).

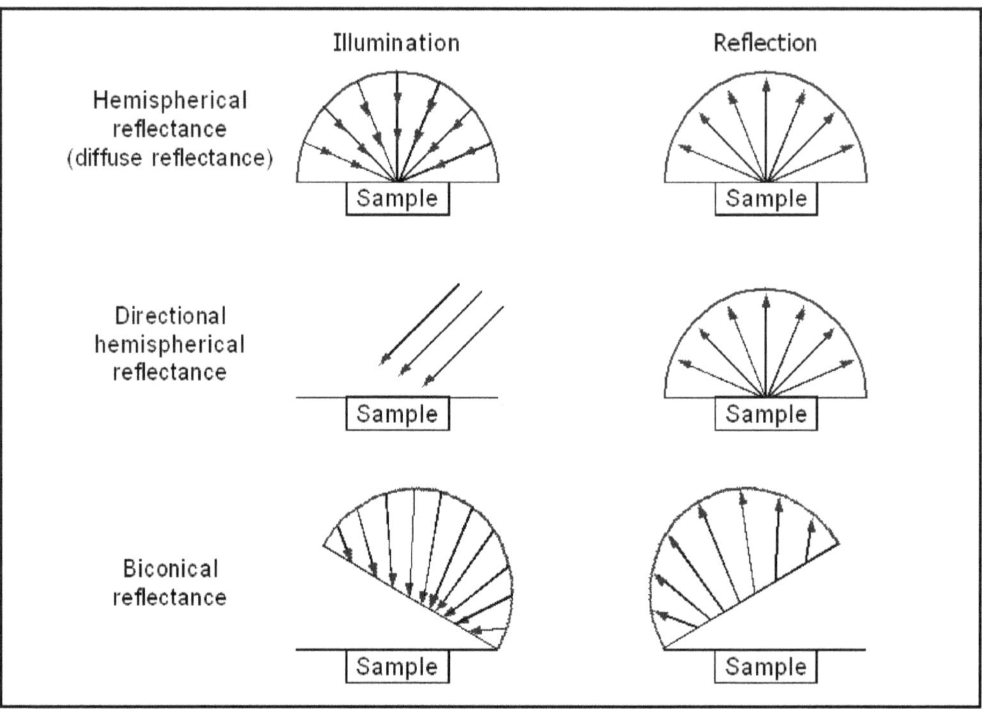

Fig. 5.19(a) Diagrams of the illumination and reflection geometries of (a) hemispherical reflectance. (b) directional hemispherical reflectance, and (c) bioconical reflectance. (Courtesy of R. K. Vincent.)

$$R^{T}(\lambda)=\sum_{i=1}^{n} f_i R_i^{T}(\lambda) \qquad \text{..........(5.14)}$$

where

f_1 = fraction or percentage of the ith mineral constituent covering the sample surface, where $\sum_{i=1}^{n} f_i$ = 1.0

n = total number of mineral constituents in the particulate sample

$R_i^{T}(\lambda)$ = spectral reflectance (the total of specular and volume reflectance) at wavelength l for the ith mineral constituent alone.

Thus, so far we have talked about volume reflectance and specular reflectance on the basis of whether electromagnetic rays did or did not penetrate one or more grains in a soil or rock surface. Now we need to define some reflectance terms that relate to the manner in which the soil or rock surface is illuminated, as well as, how the reflected energy from its surface is measured. The most fundamental term for reflectance used in this book is defined as spectral hemispherical reflectance or diffuse reflectance.

5.7 Cossine Law

Assume that this is a photographic system (Fig. 5.19(b)). Let

f = Focal length of the lens used

θ = Angle made by the normal with the axis of ray from the object to the image

d_{AS} = Object area

d_{AI} = Image area

MK = R, distance from image area to lens system

KN = r, distance from object area to lens system.

We know that,

$$\text{the solid angle subtended by object area} = \frac{d_{As} \cos \alpha}{R^2}$$

$$\text{the solid angle subtended by image area} = \frac{d_{Ai} \cos \alpha}{R^2}$$

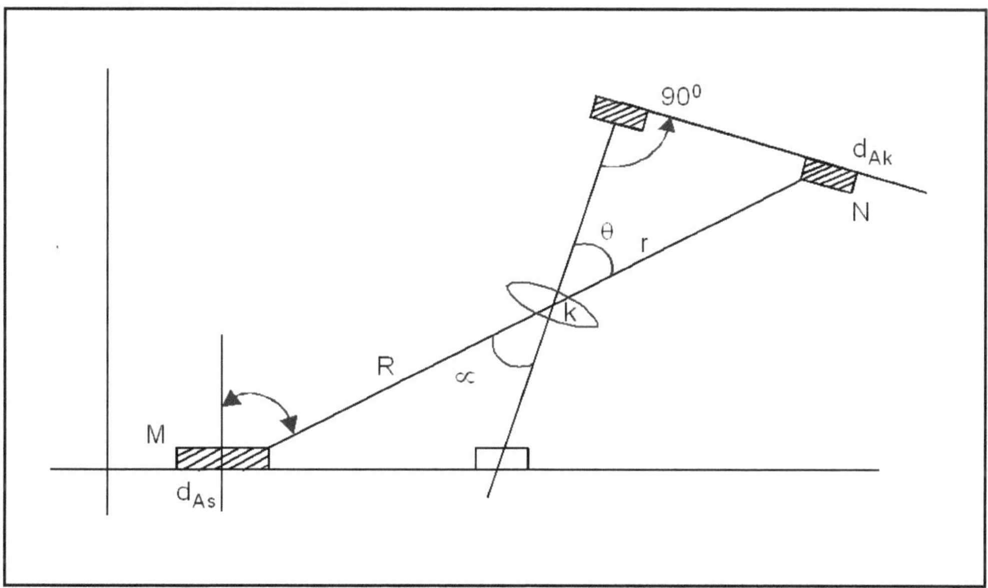

Fig. 5.19(b) Photographic system that explains the Cossine law.

At the centre of lens the solid angles are equal

That is, $$\frac{d_{As}\cos\theta}{R_2} = \frac{d_{Ai}\cos\alpha}{r^2}$$

$$d_{AS} = \frac{R^2 d_{Ai}\cos\gamma}{r^2\cos\alpha} \quad \text{..........(1)}$$

Now, $$dw = \frac{\text{Normal area}}{R^2}$$

We know that, Normal area = $\pi\ (D/2)^2 \cos\alpha$

But we also know that,

$$L\ (\text{Radiance}) = \frac{d^2\theta}{d(A\cos\theta)dw}$$

$$df = Ld_{AS}\ \text{Cosq}\ dw \quad \text{..........(2)}$$

But, $$dw = \frac{\text{Normal area}}{R^2}$$

$$= \frac{\pi(D/2)^2\cos\alpha}{R^2} = \frac{\pi D^2\cos\alpha}{4R^2}$$

From 2 $$d^2f = L\ d_{AS}\ \text{Cosfd}\ \frac{\pi D^2\cos\alpha}{4R^2}$$

$$= \frac{\pi D^2 L\cos 0\cos\alpha d_{As}}{4R^2}$$

Finally, $$dE = \frac{\pi L\cos^4\gamma}{4N^2}$$

This relation is called Cossine Law. As "a" decreases the intensity of energy increases, and vice versa. This common effect in photographic images is called vignetting.

5.8 Earth Resources Satellites

There are three distinct groups of earth resources satellites. The first group of satellites record visible and near visible wavelengths. The five satellites of Landsat series which are the first generation earth resources satellites are a classic example of this group. The four IRS satellites and the more improved SPOT series of these satellites may be considered the second generation earth resources satellites of the same group. Group two satellites carry sensors that record thermal infrared wavelengths and include the Heat Capacity Mapping Mission satellites, namely, Explorer series. Group three satellites are deployed with sensors that record microwavelengths. The seasat series and the ERS are examples of this group.

5.8.1 Landsat Satellite Programme

National Aeronautics and Space Administration (NASA) of USA with the co-operation of the U.S. Department of Interior planned the launching of a series of Earth Resources Technology Satellites (ERTS). ERTS-1 was launched by a Thor-Delta rocket on July 23, 1972 and it operated until January 6,1978. It represented the first unmanned satellite designed to acquire data about the earth resources on a systematic, repetitive, medium resolution, multispectral basis. Subsequently, NASA renamed the ERTS programme as "Landsat" programme to distinguish it from the series of meteorological and oceanographic satellites that the USA launched later. ERTS-1 was retrospectively named Landsat-1. Five Landsat satellites have been launched so far and this experimental programme has evolved into an operational global resource monitoring programme. Three different types of sensors have been flown in various combinations on the five missions. These are Return Beam Vidicon (RBV) camera system, the Multispectral Scanner (MSS) system and the Thematic Mapper (TM). Table 5.3 summarizes the characteristics of Landsat-1, through 5, and the sensors used on these satellites and orbital characteristics. Landsat images have found a large number of applications, such as, agriculture, botany, cartography, civil engineering, environmental monitoring, forestry, geography, geology, land resources analysis, landuse planning, oceanography, and water quality analysis.

5.8.2 SPOT Satellite Programme

France, Sweden and Belgium joined together and pooled up their resources to develop the System Pourl' Observation dela Terre (SPOT), an earth observation satellite programme. The first satellite of the series, SPOT-1 was launched from Kourou Launch Range in French Guiana on February 21,1986 aboard an Ariance Launch vehicle (AIV). This is the first earth resource satellite system to include a linear array sensor employing the pushbroom scanning technique. This enables side-to-side off-nadir viewing capabilities and affords a full scene stereoscopic imaging from two different viewing points of the same area. The high resolution data obtained from SPOT sensors, namely, Thematic Mapper (TM) and High Resolution Visible (HRV), have been extensively used for urban planning, urban growth assessment, transportation planning, besides the conventional applications related to natural resources. The characteristics of SPOT satellite and HRV sensor are given in Table 5.4.

5.8.3 Indian Remote Sensing Satellite (IRS)

The IRS mission envisages the planning and implementation of a satellite based remote sensing system for evaluating the natural resources. The principal components of the mission are: a three axis stabilised polar sunsynchronous satellite with multispectral sensors, a ground based data reception, recording and processing systems for the multispectral data, ground systems for the in-orbit satellite control including the tracking network with the associated supporting systems, and hardware and software elements for the generation of user oriented data products, data analysis and archival. The principal aim of the IRS mission is to use the satellite data in conjunction with supplementary/complementary information from other sources for survey and management of natural resources in important areas, such as, agriculture, geology and hydrology in association with the user agencies. IRS series of satellites are IRS IA, IRS IB, IRS IC, IRS ID and IRS P4 apart from other satellites which were launched by the Government of India. The orbital and sensor characteristics of IRS IA and IB are the same and IRS IC and IRS ID have almost similar characteristics. IRS P4 is an oceanographic satellite, and this will be discussed in the next section. IRS has application potential in a wide range of disciplines such as management of agricultural resources, inventory of forest resources, geological mapping, estimation of water resources, study of coastal hydrodynamics, and water quality surveying.

The sensor payload system consists of two pushbroom cameras (LISS-II) of 36.25 m resolution and one camera (LISS-I) of 72.5 m resolution employing linear Charge Coupled Device (CCD) arrays as detectors. Each camera system images in four spectral bands in the visible and near IR region. The camera system consists of collecting optics, imaging detectors, inflight calibration equipment, and processing devices. The orbital characteristics of the IRS-1A, 1B satellites and the sensor capabilities are given in Table 5.5. As IRS-1D satellite is the latest satellite of the series and hence the system overview of IRS - 1D is provided.

Table 5.3 Characteristics of Landsat Satellites and Their Sensors

Satellite Capabilities :

Particulars	Landsat - 1 to 3	Landsat - 4 & 5
Altitude	919Km	705Km
Orbit	Near-Polar Sun-Synchronous	Near-Polar Sun-Synchronous
Inclination	99.09 Degree	98.2 Degrees
Period	103 minutes	99 minutes
Equatorial crossing time	0930 Hours	0945 Hours
Repeat Cycle	18 Days	16 Days
Swath Width	185 Km	185 Km
Datarate	15.O6MBPS	84.9MBPS

Sensor Capabilities :

Sensor	Mission	Channel	Spectral Spatial Resolution (Microns)	Spatial Resolution	Radiometric Resolution
RBV	Landsat - 1 to 3	1	0.475-0.575	80m	6 bits (127 levels)
		2	0.580-0.680	80m	
		3	0.690-0.830	80m	
		4	0.505-0.750	80m	
MSS	Landsat - 1 to 5	1	0.5-0.6	79/82m*	6 bits (127 levels)
		2	0.6-0.7	79/82m*	
		3	0.7-0.8	79/82 m*	
		4	0.8-1.1	79/82 m*	
		5	10.4-12.6	240 m	
TM	Landsat - 4 & 5	1	0.45-0.52	30m	8 bits (255 levels)
		2	0.52-0.60	30m	
		3	0.63-0.69	30m	
		4	0.76-0.90	30m	
		5	1.55-1.75	30m	
		6	2.08-2.35	30m	
		7	10.4-12.5	120m	

* The Spatial Resolution is 79m for Landsat-12 & 3. It is 82m for Landsat 4 & 5.

Table 5.4 Characteristics of Spot Satellite and HRV Sensor

Spot Satellite		
Orbit	:	Near-polar Sun-synchronous
Altitude	:	832 km
Inclination	:	98.7 Degrees
Equatorial Crossing Time	:	10.30 Hours
Repeat Cycle	:	26 Days
HRV Sensor		
Channel		**Waveband (Microns) Multispectral**
1		0.50-0.59
2		0.61-0.68
3		0.79-0.89
Panchromatic		
1		0.51-0.73
Spatial resolution	:	20-m (Multispectral) (at nadir) 10 m (panchromatic)
Radiometric resolution	:	8 bits (Multispectral) 6 bits (Panchromatic)
Swath Width	:	117 Km (60 km per HRV, 3Km overlap)
Angularfield of view	:	4.13 Degrees.
Off-nadir viewing	:	+ 27^0 in 45 steps of 0.6^0 (= + Km from nadir)

The IRS-1D is a three-axes body stabilized satellite, similar to IRS-1C. Since IRS-1C and 1D are similar in orbital characteristics and sensor capabilities, the details of IRS-1D are discussed as it is a very recent satellite. It will have an operational life of three years in a near polar sunsynchronous orbit at a mean altitude of 780 Km. The payload consists of three sensors, namely, Panchromatic camera (PAN), linear imaging and self-scanning sensor (LISS-III) and wide Field sensor (WiFs). The satellite is equipped with an On-Board Tape Recorder (OBTR) capable of recording limited amount of specified sensor data. Operation of each of the sensors can be programmed. The payload operation sequence for the whole day can be loaded daily on to the on-board command memory when the satellite is within the visibility range. The ground segment consists of a Telemetry Tracking and Command (TTC) segment comprising a TTC network, and an Image segment comprising data acquisition, data processing and product generation system along with data dissemination centre. The over view of IRS-1D mission is to provide optimum satellite operation and a mission control centre for mission management, spacecraft operations and scheduling. The three sensors on board IRS-1D and IRS-1C are described in the following paragraph.

The panchromatic camera provides data with a spatial resolution of 5.2-5.8m (at nadir) and a ground swath between 63 Km –70 Km (at nadir). It operates in the 0.50 – 0.75 microns spectral band. This camera can be steered upto $\pm 26^0$. storable upto ±398 Km across the track from nadir, which in turn increases the revisit capability to 3 days for most part of the cycle and 7 days in some extreme cases. Table 5.6 shows the specifications and resolution of PAN. Fig. 5.20 shows the PAN image showing parts of Mumbai city and Fig. 5.21 shows the IRS-1D PAN camera by which this image is obtained. The LISS-III sensor provides multispectral data collected in four bands of the visible, near infra-red (V,NIR) and short wave infra-red (SWIR) regions. While the spatial resolution and swath in the case of visible (two bands) and NIR (one band) regions are between 21.2m to 23.5m and 127 Km-141 Km. respectively, they are between 63.6m to 70.5m and 133 Km to 148 Km for the data collected in SWIR region. (Table 5.7). Plate 3 shows IRS-1D LISS III FCC image (band 2, 3, 4) and corresponding black and white images of band 2, band 3, and band 4 of path 108, row 56 showing Culcutta and surrounding areas whereas Fig. 5.22 shows the LISS III camera.

The Wide Field Sensor (WiFS) sensor collects data in two spectral bands and has a ground swath between 728 Km to 812 Km with a spatial resolution of 169m to 188m. (Table 5.8). The camera by which the WiFS image is formed is shown in Fig. 5.23.

Table 5.5 Particulars of Indian Remote Sensing Satellites (IRS Series)

Characteristics of Satellite		
Orbit	:	Near-polar, Sun-synchronous
Altitude	:	904 Km
Inclination	:	99.03 Degrees
Equatorial Crossing Time	:	1000 Hours
Repeat Cycle	:	22 days
Eccentricity	:	0.002
Period	:	103 minutes

Sensor Capabilities

Linear Image Scanning System : LISS

No. of LISS Cameras	LRC (One)*	MRC (two)**
No. of Spectral Bands	4	4
IFOV (Microrad)	80	40
Geometric Resolution	72.5	36.25
Swath Width	148 Km	74 Km
Radiometric Resolution	7 bits	7 bits
Band-to-Band	0.5	0.5

* Low Resolution Camera
** Medium Resolution Camera

Table 5.6 Details of PAN of IRS ID

S.No.	Parameter	Specification
1.	Spatial resolution (m)(at Nadir)	5.2 to 5.8
2.	A. Swath (Km) at Nadir B. Swath Steering Range (Deg) C. Step size (Deg) D. Repeatability (Deg)	63 to 70 ± 26 ± 0.09 ± 0.1
3.	Spectral band (micron)	0.50 – 0.75
4.	Camera Square Wave Response (SWR) (at Nyquist frequency)	> 0.20
5.	Quantisation (Bits)	6
6.	Signal to Noise R atio (SNR) (at saturation radiance)	> 64
7.	Saturation radiance (Gain 1) (mw/cm^2-str-micron)	35 – 40
8.	Integration time (ms)	0.873
9.	Data rate (Mbps)	84.903

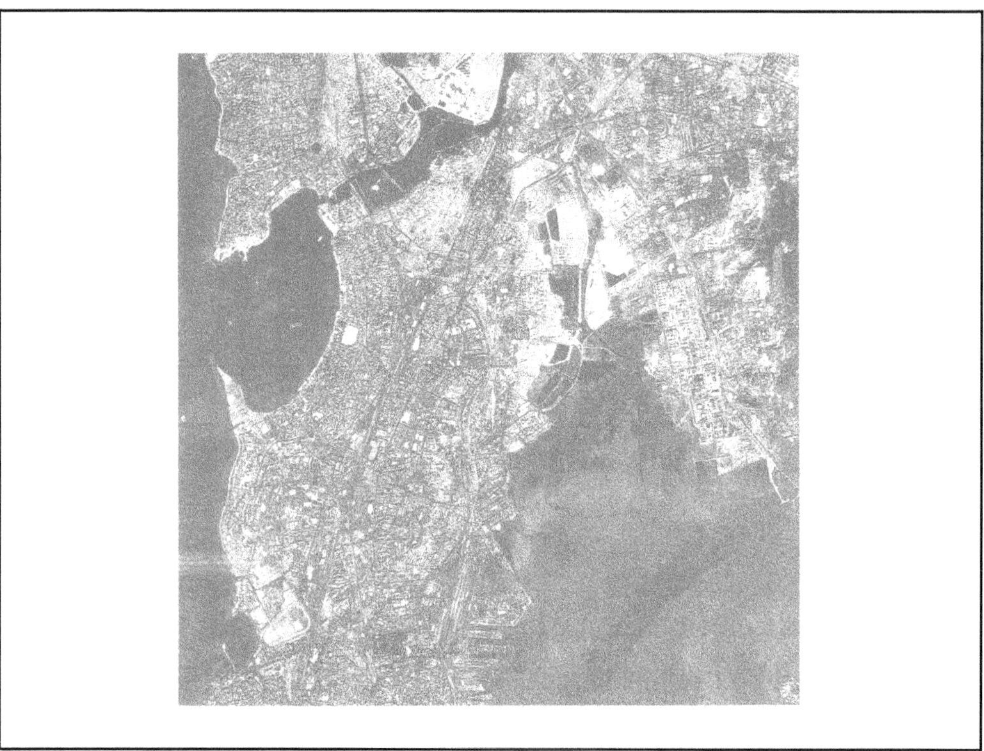

Fig. 5.20 PAN image showing parts of Mumbai city

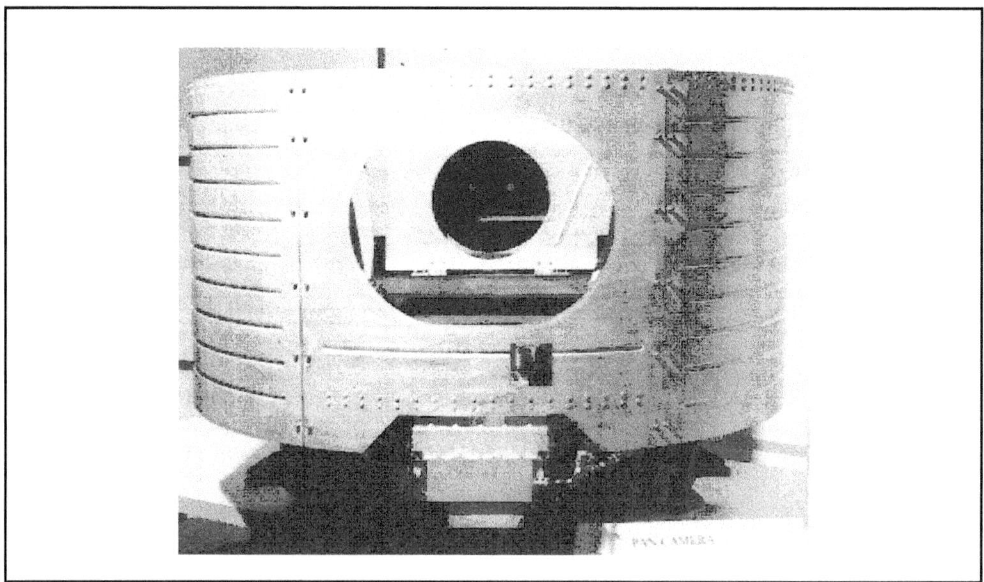

Fig. 5.21 IRS-1D PAN Camera

Table 5.7 Details of LISS III of IRS-1D

Sl.No.	Parameter	Specification	
1.	Spatial resolution (m)	B2,B3,B4	21.2 to 23.5
		B5	63.6 to 70.5
2.	Swath (Km)	B2,B3,B4	127 to 141
		B5	133 to 148
3.	Spectral band (microns)	B2	0.52 - .059
		B3	0.62 – 0.68
		B4	0.77 – 0.86
		B5	1.55 – 1.70
4.	Camera Square Wave	B2	>40
	Response (SWR)	B3	>40
		B4	>35
		B5	>30
5.	Quantisation (bits)	7	
6.	Signal to Noise Ratio	>128	
7.	Saturation Radiance (Gain 1)		
	(mw/cm^2-str-micron)	B2	29 ± 1.5
		B3	28 ± 1.5
		B4	28 ± 1.5
		B5	3.25 ±25
8.	Integration time (ms)	B2, 3 and 4	3.6
		B5	10.8
9.	Data rate (Mbps)	B2, 3 and 4	35.7904
		B5	1.3906
10.	Band to Band registration (pixels)		

Fig. 5.22 IRS-1D LISS-III Camera

Fig. 5.23 IRS-1D WiFS Camera

Table 5.8 Details of WiFS of IRS ID

Sl.No.	PARAMETER	SPECIFICATION	
1.	Spatial resolution (m)	169 to 188	
2.	Swath (Km)	728 to 812	
3.	Spectral band (micron)	B3 B4	0.62 – 0.68 0.77 - 0.86
4.	Square Wave Response	B3 B4	> 0.34 > 0.20
5.	Quantisation (bits)	7	
6.	Signal to Noise Ratio (at saturation)		>128
7.	Saturation radiance (mw/cm^2-str-micron)	B3 B4	28 ± 1.5 31 ± 1.5
8.	Integration time (ms)	28.8	
9.	Data rate (Mbps)	2.0616	

5.8.4 AEM Satellites

The (Heat capacity Mapping Mission) HCMM satellite is the first of a small and relatively inexpensive series of NASA's Applications Explorer Mission (AEM) satellites. Launched in April 1978, it lasted till September 1980. Table 5.9 summarises the details of AEM satellite and HCM sensor characteristics. The orbits of the satellite are arranged to ensure that images of each scene are obtained during the periods of maximum and minimum surface temperature for the determination of thermal inertia. The data from HCMM are intended primarily for conversion to thermal inertia maps for geological mapping. However, the images have found wider applications, such as, vegetation mapping, vegetation stress detection, microclimatology, soil moisture mapping, snowmelt prediction, and monitoring industrial thermal pollution.

5.9 Meteorological Satellites

Meteorological satellites designed specifically to assist in weather prediction and monitoring, generally incorporate sensors that have very coarse spatial resolution compared to land-oriented systems. These satellites, however, afford a high frequency global coverage. USA has launched a multiple series of meteorological satellites with a wide range of orbit and sensing system designs. The first of these series is called the NOAA, an acronym for National Oceanic and Atmospheric Administration. These satellites are in near-polar, sunsynchronous orbits similar to those of 'Landsat and IRS'. In contrast, another series of satellites which are of essentially meteorological type, called Geostationary Operational Environmental Satellite (GOES) series and Meteosat operated by European Space Agency, are geostationary, remaining in a constant relative position over the equator.

5.9.1 NOAA Satellites

Several generations of satellites in the NOAA series have been placed in orbit. The satellites NOAA-6 through NOAA-10 contained Advanced Very High Resolution Radiometer (AVHRR). The even-numbered missions have daylight (7.30 A.M.) north-to-south equatorial crossing and the odd-numbered missions have night time (2.30 A.M.) north-to-south equatorial crossing. The basic characteristics of these missions and the AVHRR instrument are listed in Table 5.10. Apart from routine climatological analyses, the AVHRR data have been used extensively in studies of vegetation dynamics, flood monitoring, regional soil moisture analysis, dust and sandstorm monitoring, forest wild fire mapping, sea surface temperature mapping, and various geological applications, including observation of volcanic eruptions, and mapping of regional drainage and physiographic features.

Table 5.9 Details of AEM Satellite and HCMM Sensor

Characteristics of Satellite		
Orbit	:	Near-polar Sun-synchronous
Altitude	:	620 Km
Inclination	:	97.6 Degrees
Equatorial CrossingTime	:	1400 Hours (Ascending) Q200 Hours (Decending)
Repeat Cycle	:	16 days

HCMM Sensor Capabilities	
Channel	**Waveband (Microns)**
1	0.5-1.1
2	10.5-12.5
Spatial Resolution	500 m (channel -1) 600m(Channel-2)
Swath width	716 Km

Table 5.10 Particulars of NOAA Satellite and AVHRR Sensor

NOAA Satellite		
Orbit	:	ear-polar, Sun-synchronous
Altitude	:	833-870 Km
Inclination	:	98.7 Degrees
Equatorial crossingTime	:	0730 and 1930 Hours
		1400 and 0200 Hours
Repeat Cycle	:	12 Hours
Period	:	102 Minutes
AVHRR Sensor Capabilities		
Channel		**Waveband (Microns)**
1		0.58 - 0.68
2		0.725 - 1.10
3		3.55 - 3.93
4		10.3 - 11.3
5		11.5 - 12.5
Spatial resolution	:	1.1 Km (at nadir)
Radiometric resolution	:	1 0 bits (1024 levels)
Swath Width	:	3000 Km.

5.9.2 GOES Satellites

The GOES programme is a cooperative venture between NOAA and NASA. The Geostationary Operational Environmental Satellites (GOES) are part of a global network of meteorological satellites spaced about 70^0 longitude apart around the world. The GOES images are distributed in near real-time for use in local weather forecasting. They have also been used in certain large area analyses such as regional snow cover mapping. The details of GOES satellites and the sensors they are carrying are shown in Table 5.11.

5.9.3 NIMBUS Satellites

This is one of the ocean monitoring satellites launched in October 1978. This satellite carries the Coastal Zone Colour Scanner (CZCS) designed specifically to measure ocean parameters. The details of the six bands in which the CZCS operates and the characteristics of NIMBUS-7 satellite are presented in Table 5.12. The CZCS has been used to measure sea surface temperatures, detection of chlorophyll and suspended solids of near-shore and coastal waters.

5.9.4 Meteosat Series

Meteosat-1 of this series launched in November 1977 failed in November 1979. Meteosat-2 was launched June 1981, and like GOES its sensors recorded images every half an hour in three wavebands (Table 5.13). Data from Meteosat sensors have been used for many meteorological applications like studies of synoptic climatology, sea surface temperature, land surface temperature, windvector determinations and monitoring all types of natural disasters.

5.10 Satellites Carrying Microwave Sensors

Microwave imaging is gaining increasing importance with a better understanding of the relation between image tone and the earth's surface characteristics. The clear advantage of microwave sensor is its capacity to penetrate cloud cover. Satellites that carry microwave sensors are seasat with Synthetic Aperture Radar (SAR), European Remote Sensing Satellite (ERS)-1and Radarsat.

5.10.1 Seasat

It is an experimental satellite designed by NASA to establish the utility of microwave sensors for remote sensing of oceans. Images of the land are also obtained giving environmental scientists their first synoptic view of the earth in microwavelengths.

Table 5.11 Particulars of Goes Satellites and their Sensors

GOES Satellite		
Orbit	:	Geostationary
Altitude	:	33,367-48,390 Km
Inclination	:	1.9 - 0.2 Degrees
Repeat Cycle	:	Twice per hour
Period	:	1430 - 1436 Minutes
Visible Infrared Spin Scan Radiometer (VISSR)		
Channel	**Waveband (Microns)**	**Ground Resolution**
1	0.55-0.70	14Km
2	10.5-12.6	8Km
VISSR Atmospheric Sounder (VAS)		
1	0.55-0.70	14Km
2	4.496-14.81	16Km

Table 5.12 Characteristics of Nimbus Satellites and CZCS Sensor

NIMBUS Satellites		
Orbit	:	Near-polar1 Sun-synchronous
Altitude	:	955 Km
Inclination	:	99.3 Degrees
Repeat Cycle	:	6 Days)
Period	:	104 Minutes
Overpass	:	1200 Hours

CZCS Sensor Capabilities	
Channel	**Waveband (Microns)**
1	0.433 - 0.453
2	0.510 - 0.530
3	0.540 - 0.560
4	0.660 - 0.680
5	0.700 - 0.800
6	10.50 - 12.50
Spatial resolution: 825 m	
Swath Width:1 566 Km.	

Table 5.13 Details of Meteosat and Sensor Wavebands

Meteosat Satellites		
Orbit	:	Geostationary
Altitude	:	35900 Km
Position	:	Equator, O° Longitude
Sensor Wavebands		
Channel		**Wavelength (Microns)**
1		0.4 - 1.1
2		5.7 - 7.1
3		10.5 - 12.5
Spatial resolution	:	2.4 Km (Channel - 1)
	:	5 KM (Channel-2&3)
Radiometric Resolution	:	8 bits (256 levels)
Provides coverage of entire hemisphere.		

The satellite carries five sensors, two of which are of potential interest to environmental scientists, these being a radiometer and synthetic aperture radar. The characteristics of the satellite and these two sensors are given in Table 5.14. The data obtained from seasat are used for ocean studies, mapping sea ice, urban patterns landcover, geology and hydrological features.

5.10.2 European Remote Sensing Satellite-1

ERS-1 launched in May 1991 has a synthetic aperture radar capable of penetrating clouds (Table 5.15). The objectives of this mission are study of coastal zones, dynamics of ice cover, and monitoring of global weather and landuse.

5.10.3 Radarsat

Radarsat - 1 launched on November 28, 1995 is the first satellite of Canadian Space Agency. The orbit for Radarsat is sunsynchronous at an altitude of 798 km and inclination of 98.6^0. The orbit period is 10.7 min. and the repeat cycle is 24 days. Radarsat is a right -looking sensor facing east during the ascending orbit and west during the descending orbit. The Radarsat-SAR is a C-band (5.6 cm) system with HH polarisation. The system can be operated in a variety of beam selection modes providing various swath widths, resolutions and look angles. Table 5.16 shows modes in which the system operates. Fig. 5.24 shows the radarsat image of the Zhao Qing, China . The primary applications of Radarsat are land cover mapping, agricultural and forestry monitoring, disaster monitoring, landslide identification, flood monitoring, and snow distribution mapping.

5.11 OCEANSAT-1 (IRS-P4)

The successful launch of IRS-P4 satellite at Sriharikota, India on May 26,1999, was a fully indigenous effort which will help India to know more about its, oceans, their wealth and the costal ecosystems. This satellite has onboard two main sensors, namely, the Ocean Color Monitor (OCM) and Multi Frequency Scanning Microwave Radiometer (MSMR). The potential scientific investigations, application areas, and calibration and validation of OCM and MCMR are listed below as outlined by ISRO (1999). The satellite and sensor characteristics of IRSP4 are given in Table 5.17 and 5.18.

Ocean colour Monitor (OCM)

Scientific Investigations

Atmospheric correction algorithms
Aerosol characterization
Ocean primary productivity
Models for fish stock assessment
Inverse modeling techniques for retrievel of oceanic parameters
Space-time variability of plankton growth due to eco-physiological

Table 5.14 Seasat Satellite and Its Sensors

Seasat Satellite		
Orbit	:	Near-polar Circular
Altitude	:	790
Km Inclination	:	108 Degrees
Repeat Cycle	:	152 Days
Period	:	101 Minutes
Synthetic Aperture Radar (SAR) Sensor		
Frequency	:	1274.8 GHz
Wavelength	:	23.5 Cm (L-Band)
Polaraization	:	H H (Horizontal & Horizontal)
Depression Angle	:	20^0
Spatial Resolution	:	25 m
Swath width	:	100 Km
Radiometer Sensor		
Visible band	:	0.47-0.94 microns
Spatial Resolution	:	2 Km.
Thermal infrared band	:	10.5 - 12.5 microns
Spatial resolution	:	4 Km.

Table 5.16 SAR Imaging Mod Characteristics

ERS - 1 Satellite		
Orbit	:	Sun - Synchronous
Altitude	:	785 Km
Repeat cycle	:	3 days
Equator crossing time	:	1030A.M.
ERS-1 SAR		
Frequency	:	C - band (5.3 0Hz + 0.2 MHz)
Swath width	:	100 Km
Incident angle	:	23 Degrees mid swath (350 in roll-tilt)
Spatial resolution	:	30 m x 30 m
Radiometric resolution	:	2.5dB for 0o = -18 db
Chirp band width	:	15,55 = 0.1 MHz
Chirp length	:	37.1+0.05 US
Peak power	:	4.8 Kw
Antenna size	:	10 mxl. m
A/D sampling rate	:	18.96 M samples
Quantisalion	:	51, SQ, OGRC (Nominal 61, 6Q OBRC (experimental)

Table 5.17 Specifications of IRS-P4 OCM

1.	Ground resolution (m)		:	
	Real-Time Mode (Full resolution Mode)		:	360 (across-track)
			:	236 (along-track)
2.	Swath (km)		:	1420
3.	Instrument Spectral Bands	Wavelength Range (nm)	SNR at Ocean Radiance's Radiance*	Expected SNR
	C-1	402-422	9.1	307.6
	C-2	433-453	8.4	440.7
	C-3	480-500	6.6	415.0
	C-4	500-520	5.6	408.8
	C-5	545-565	4.6	412.2
	C-6	660-680	2.5	345.6
	C-7	745-785	1.6	393.7
	C-8	845-885	1.1	253.6
	*In mw/(cm^2-Sr-mm)			
4.	MTF (at Nyquist)		> 0.2	
5.	Quantisation		12 bits	
6.	Data Rate (Mbps)		20.8 Mbps	
7.	Transmission Frequency		X-band	
8.	Along track steering (To avoid sunlight)		$\pm 20^0$	

Source : ISRO, 1999

Fig. 5.24 Radarsat image.

Table : 5.18 Specifications of IRS – P4 MSMR

1.	Frequencies (GHz)	6.6 10.65 18.0 21.0
2.	Polarisation	V & H for all frequencies
3.	Spatial Resolution (km x km)	6.6 (V & H) : 105 X 68 10.65 (V & H) : 66 X 43 18.0 (V & H) : 40 X 26 21.0 (V & H) : 34 X 22
4.	Swath	1360 km for all frequencies
5.	Nominal Incidence Angle (deg.)	49.7
6.	Radiometric Sensitivity	Better than 1° K
7.	Dynamic Temp. Range	$10^{o} - 330^{o}$ K
8.	Data Rate (kbps)	6.4

parameters
Models for quantitative estimation of sediments
Optical properties of water
Regional and global climate and environmental changes
Bio-geochemical cycles
Modeling Carbon fluxes
Synergistic use of data from multiple missions
Development of new techniques for image analysis and classification

Application Demonstration

Synergistic use of SST-Chlorophyll in fishery forecast
Identification and monitoring of phytoplankton bloom
Coastal processes, Sediment dynamics and Coastal pollution
Vegetation dynamics / agriculture applications
Snow-ice studies

Calibration and Validation

Characterization of sensor performance
Algorithm development
Geophysical parameter validation
Inter-comparison with similar ocean colour sensor from different missions.

Multifrequency Scanning Microwave Radiometer

Application Demonstration

Sea state predictions
Oceanic process studies
Weather and climate studies
Monsoon and atmospheric predictions
Cyclone genesis and development
Air – sea interaction studies
Polar ice studies
Circulation and wave studies
Mixed layer depth studies
Rainfall estimation
Latent heat flux estimation
Soil moisture estimation

Calibration and Validation

Characterisation of sensors
Algorithm development
Geophysical parameter validation
Data assimilation in numerical models
Inter-comparison with similar sensors from different missions.

5.12 IKONOS Satellite Series

IKONOS satellite is one of the high resolution satellites launched on September 24th, 1999 by USA. This is a very recent satellite with a resolution higher than that of other existing satellites. The revisit time will be more frequent for higher latitudes and less frequent for latitude closer to the equator. This satellite is designed to occupy a 681 KM sunsynchronous orbit having an equatorial crossing time of 10:30 AM. The ground track of the systems repeats every 11 days, but the revisit time for imaging is less than 11 days, based on latitude and the tilt of the system. The system collects the data at angles upto 45^0 from vertical in both along-track and across-track directions. This provides an opportunity for frequently covering an area, and also enables the use of the data for stereoscopic or 3-D study of the terrain.

IKONOS employs linear array technology and collects data in a Panchromatic band (0.45-0.90 μm) with spatial resolution of 1m. It also covers the terrain in Multispectral bands at a nominal spatial resolution. These spectral bands include 0.45 to 0.52 μm (blue), 0.52 to 0.60 μm (green) 0.63 to 0.69 μm (red) and 0.76 to 0.90 μm (near infrared). The Panchromatic and multispectral bands can be combined to produce pan-sharpened multispectral imagery, with an effective resolution of 1m data collection over 2048 gray levels (11 bits). The ground processing software has the capability to rapidly process on mosaic imagery so as to create seamless image products with a consistent fixed ground sample distance. The IKONOS data has been extensively used for urban growth assessment studies, municipal planning, utility management, facility management and os on. It is very much useful in application areas like the creation of Cadastral Information System (CIS) where large scale mapping is necessary. Table 5.19 shows the orbital characteristics of IKONOS. Fig. 5.25 is an IKONOS image of a part of Beijing, China with 1m resolution.

Table 5.19 Orbital Information of IKONOS Satellite

Altitude	681 km.
Inclination	98.1 degrees
Speed	7 km/sec.
Descending nodel	10:30 a.m.
Orbit time	98 minutes
Orbit type	Sun-synchronous
Swath	11 km at nadir.
Image area	11 km × 11 km.

Fig. 5.25 IKONOS Image

6

Geodesy

6.1 Definition of Geodesy

Geodesy, is that branch of applied mathematics which, determines, by observation and measurement, the size and shape of earth, the coordinates of points, the lengths and directions of lines on the earth's surface, and the variation of terrestrial gravity. Geodesy is generally considered to have two separate parts. These are Geometrical Geodesy and Physical Geodesy. Geometrical geodesy concerns the size and shape of the earth, the inter-continental ties among the land masses of the earth, determination of positions, length of lines, and azimuths. As the name implies it has to do with the Geometry of the earth.

Physical Geodesy concerns the gravity field of the earth of the direction and magnitude of the physical force which links the earth to objects on its surface and in surrounding space. A study of the gravity is useful in determining the shape but not the size of the earth. Although these two parts of Geodesy may be studied separately, in practice they are closely related.

6.2 Problem of Geodesy

The problem of Geodesy is to determine the figure and the external gravity field of the earth, as well as to determine the mean earth ellipsoid from parameters observed on and exterior to earth's surface. By the figure of earth it is meant as the "The physical and mathematical surface of the earth". The physical surface of the earth and reference surfaces, viz. the Geoid and the Ellipsoid are discussed in the following paras.

6.2.1 Physical Surface of the Earth

The physical surface of the earth is the border between the solid or fluid masses and the atmosphere. The ocean floor also has been included in the formulation of geodetic problem being the founding surface between the solid terrestrial body and the oceanic water masses. The ocean surface (70% of the earth's surface of constant gravity potential) of earth gravity field. While it is necessary to make observations and measurements on or near the physical surface of the earth, it would be quite impossible to perform detailed and extensive computations on a surface whose definition requires an infinite number of parameters.

6.2.2 The Geoid

A possible surface upon which to compute is mean sea level, or its extension into the land areas of earth if channels were cut through the continents so that the sea could enter and come to an undisturbed and free state of equilibrium. We must assume these channels to be every narrow so that the quantity of water removed from the oceans would be negligible. This undisturbed water surface is known as the bounding equipotential surface of the earth or, more commonly, the Geoid. The surface of the geoid is every where normal to the direction of gravity. Another possible definition of the Geoid is "The surface of the earth in the Geometrical sense is nothing more than that surface which intersects every where the direction of gravity at right angles, and a part of which coincides with the surface of the oceans. Like the physical surface, the Geoid is unsuitable as a mathematical model for computations be cause it is again defined by an infinite number of parameters.

6.2.3 The Ellipsoid

If the earth were only slightly flattened, a sphere would be an ideal mathematical surface. It is a simple figure defined only by one parameter, its radius. As it is, the earths departure from a true spherical shape is too great and so the thought of a sphere must discarded.

But some mathematical figure must be adopted. It must be simple enough so that computations are not overly difficult, but must be nowhere depart from the true figure of the earth by an amount, which will give intolerable errors in the results. This figure is the ellipsoid of revolution, generally referred to simply as the ellipsoid. It is produced by rotating an ellipse about its minor axis, with major axis generating the equatorial plane. This ellipsoid of revolution, approximates an oblate spheroid and the terms "ellipsoid by revolution" and spheroid are used interchangeably.

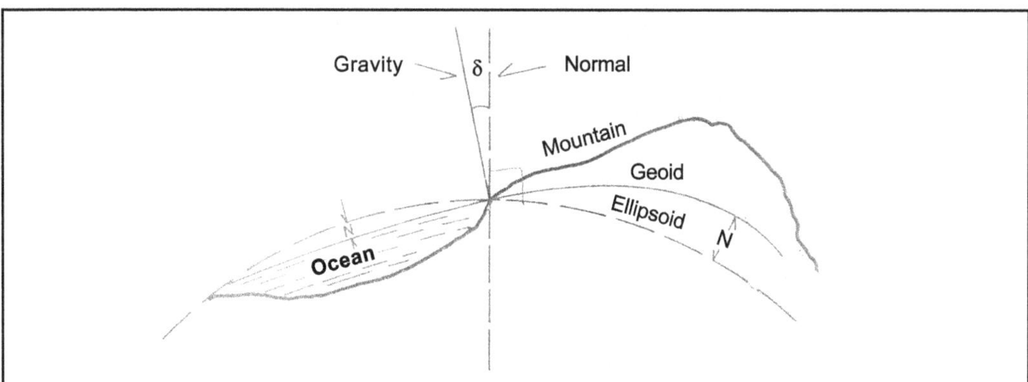

Fig. 6.1 Relationship among the physical surface the geoid and the ellipsoid.

The Fig. 6.1 shows the relationship among the physical surface, the geoid and the ellipsoid of revolution at a small section of earths surface. The following points are to be noted to understand the relationship between the geoid and the ellipsoid.

- At sea the geoid coincides with the free unbounded water surface.
- Under the continents it rises above the ellipsoid.
- The vertical separation between the Geoid from the ellipsoid is known as the geoid height and designated as N, The geoid height may be positive or negative, depending on whether it is above or below the ellipsoid.
- The angle between the ellipsoid and geoid at any point is called the deflection of the vertical and designated as 'δ'. It is also the angle between the direction of gravity and the normal to the ellipsoid at any specific point.

6.3 The Ellipsoid of Revolution

The geoid is undulating or warped, and the ellipsoid of revolution most nearly approximates the geoid with least departure there from. It is analogous to the work of a civil engineer who builds a road through hilly country and seeks to find the route where the cuts and fills most nearly balance. The ellipsoid of revolution is a sensible mathematical model upon which to compute and is the 'corner stone' of geometrical geodesy.

6.3.1 Mathematical Properties of Ellipsoid of Revolution

Since the ellipsoid of revolution has been adopted as the closest reasonable approximation to the figure of the earth, it is necessary to have a fundamental understanding of its mathematical properties. The ellipsoid is generated by rotating an ellipse about its minor axis.

The Fig. 6.2 is an ellipse PP is the minor axis, has length 2b, and is the axis of rotation for generating the ellipsoid. EE' is the major axis, has length 2a and generates equatorial plane when the ellipse is rotated about its minor axis.

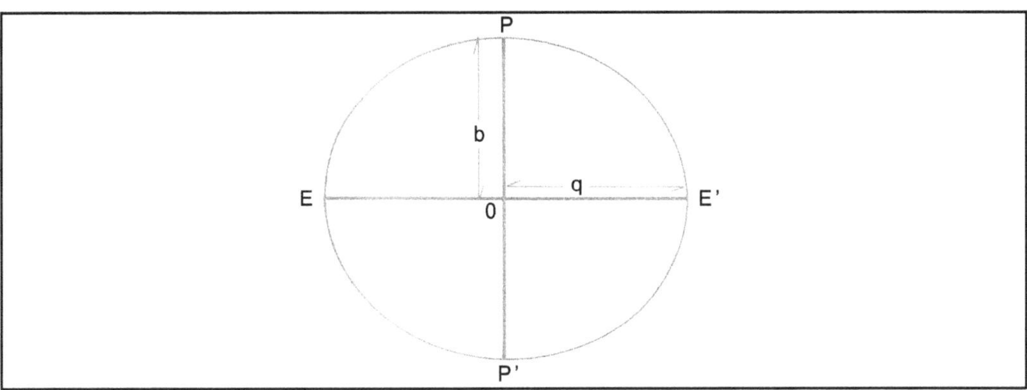

Fig. 6.2 The ellipse

A basic principle of the ellipse is that a circle of radius equal in length to the semimajor axis and with P (or P') as the centre cuts the major axis EE' at F and F[1].

The points F and F' are called the "face of the ellipse". The distance

$$OF = OF' = \sqrt{a^2 - b^2} \text{ or } \left(a^2 - b^2\right)^{\frac{1}{2}}$$

The flattening of the ellipse is designated as f, and f is defined as

$$f = \frac{(a - b)}{a} = \left(1 - \frac{b}{a}\right)$$

Occasionally, the reciprocal of the flattening is used to avoid fractions.

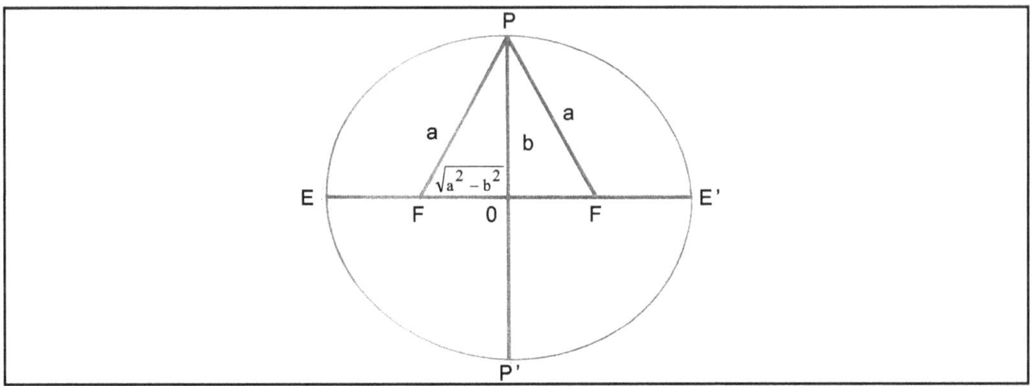

Fig 6.3 Basic Properties of Ellipse

The first eccentricity of the ellipse is known as 'e'. It is the distance between the centre of the ellipse and a focus divided by the semimajor axis.

$$e = \frac{DF}{OE} = \frac{\left(a^2 - b^2\right)^{1/2}}{a}$$

Squaring both sides,

$$e^2 = \left(a^2 - b^2\right)/a^2 = 1 - \frac{b^2}{a^2}$$

The second eccentricity of the ellipse is known as e'. If is the distance between the centre of the ellipse and a focus divided by semi minor axis

$$e' = OF/OP = \left(a^2 - b^2\right)^{\frac{1}{2}}/b$$

Squaring both sides $e'^2 = \frac{\left(a^2 - b^2\right)}{b} = \left(\frac{a^2}{b^2} - 1\right)$

The eccentricity and flattening have a definite mathematical relationship.

$$e^2 = 1 - \frac{b^2}{a^2} \text{ or } \frac{b^2}{a^2} = 1 - e^2$$

or $\left(\frac{b}{a}\right) = \left(1 - e^2\right)^{\frac{1}{2}}$

Substituting in $f = 1 - \frac{b}{a} \Rightarrow f = 1 - \left(1 - e^2\right)^{\frac{1}{2}}$

$$1 - f = \left(1 - e^2\right)^{\frac{1}{2}}$$

Squaring both sides

$$1 - e^2 = 1 - 2f + f^2$$

From this $e^2 = 2f - f^2$

f^2 being small quantity, we have

$$e^2 = 2f$$

6.3.2 Principal Parameters of Ellipsoid

The parameters a,b,f, e and e' are principal parameters of ellipsoid of revolution. Only two of these parameters need to define the ellipsoid, provided one is linear, e' is seldom used. Universally a and f are used to define the ellipsoid.

Ideally an ellipsoid approximates the shape of the entire earth. During the past few hundred years, there have been dozens of determinations of earth's form and dimensions. The following table gives the most frequently used ellipsoidal parameters. The names of the principal investigators along with year of findings and with the equatorial radius and flattening are shown in the table.

Table. 6.1 Parameters of Commonly Used Ellipsoids

Reference ellipsoid	a(meters)	*f*
Everest 1830	6, 377, 304	1/300.8
Bessel 1841	6, 377, 397	1/299.2
Clarke 1866	6, 378, 206	1/295.0
Clarke 1880	6, 378, 249	1/293.5
Hayford 1910	6, 378, 388	1/297.0
Krasovskii 1938	6, 378, 245	1/298.3
Hough 1956	6, 378, 270	1/297.0
Fischer 1960	6, 378, 166	1/298.3
Kaula 1961	6, 378, 165	1/298.3
Fischer 1968	6, 378, 150	1/298.3

A regional datum uses an ellipsoid which best approximates the shape of the earth in that region. This ellipsoid is fixed to the surface at some point in that region. This is called the surface origin At this surface origin, coordinates and orientation are assumed to be free from error and the survey for the entire region is adjusted thereto. Different ellipsoids are used as datum in different parts of the world. Examples are :

The Everest ellipsoid is used in India and Burma, The Clarke ellipsoid is used in Africa, The krasovskii ellipsoid is used by Modern Russia. The U.S Department of Defense has its own World Geodetic system (WGS), which is now being adopted by other countries.

6.4 Coordinate System of Rotational Ellipsoid

In the ellipsoid shown in the Fig. 6.4 above a spatial x,y,z, Cartesian co ordinate system is introduced. The origin of the system is situated at the centre O of the figure, Z - axis coincides with the minor axis of the ellipsoid. The equation of the surface of the ellipsoid is given by.

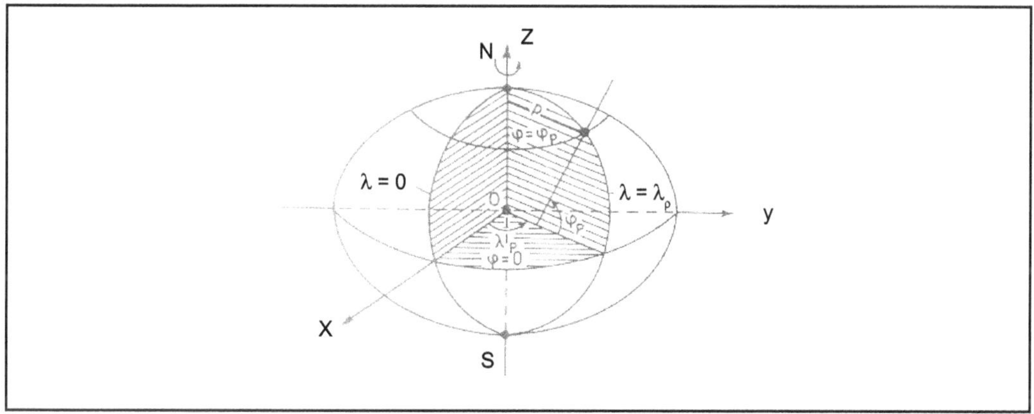

Fig 6.4 Ellipsoidal geographic coordinates.

$$\frac{x^2+y^2}{a^2}+\frac{Z^2}{b^2}-1=0$$

Consider a point 'P' on the ellipsoid. Then the system of ellipsoidal and georaphic co ordinates of P is defined by geographic latitude 'ϕ' and the geographic longitude λ, (also called geodetic latitude and longitude). ϕ is the angle measured in the meridian plane between the equitorial plane (x, y plane) of the ellipsoid and the surface normal at P. λ is the angle measured in the equitorial plane between the zero meridian (x - axis) and the meridian plane of P. The ellipsoidal meridian plane is formed by the surface normal through P and Z - axis. ϕ and λ are defined to have angular values, but they may also be considered as curvilinear surface co ordinates. The co ordinate lines of the orthogonal system are meridians and parallels.

6.5 Spatial Ellipsoid Coordinate system

For the spatial determination of points on the physical surface of the earth with respect to the rotational ellipsoid the height h, above the ellipsoid is introduced in addition to the geographic coordinates ϕ, λ. Height h, is measured along the surface normal.

The spatial ellipsoidal coordinates ϕ,λ,h are designated as geodetic coordinates. The point on the ellipsoid is obtained by projecting the surface point P along the ellipsoidal normal.

The co ordinate surface (ϕ, λ and w) of this system are orthogonal. The co ordinate lines (ϕ - line = geodetic meridian, λ - line = geodetic parallel, h - line = ellipsoidal normal) represent planar curves.

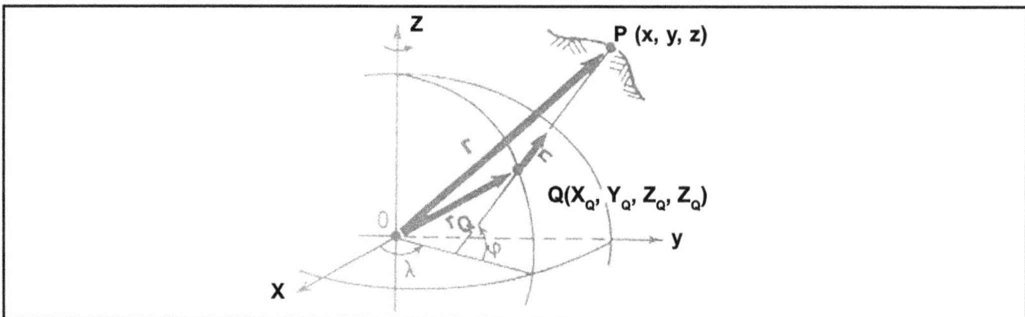

Fig. 6.5. Spatial ellipsoidal coordinates.

The point Q on the ellipsoid is deduced by,

$$h=\frac{\sqrt{x^2+y^2}}{\cos\phi}-N,$$

$$\phi=\text{arc tan}\ \frac{Z}{\sqrt{x^2+y^2}}\left(1-e^2\frac{N}{-N+h}\right)^{-1}$$

$$\lambda=\text{arc tan}\ \frac{y}{x}$$

6.6 Computations on the Ellipsoid

6.6.1 Need for a mathematical surface

The geodesist must use the reference ellipsoid for a particular country or region because the initial necessity is a mathematical surface upon which the computations are to be carried out. Procedures for making computations on the ellipsoid for the purpose of obtaining meaningful information as length of lines, azimuth and co ordinates are discussed in this section.

6.6.2 Reduction of Base Line to Reference Ellipsoid

Suppose a geodetic surveyor has measured a base line for subsequent use in triangulation. If he has carried out his measurements by use of an invar tape, the corrections for grade, alignments, temperature, for grade, alignment, temperature, sag, standarisation of tape and other corrections, if any are carried out. If the measurements were made by electronic methods, such as Geodimeter or Tellurometer the customary corrections of reduction to horizontal would have been performed. The measured and corrected base line is still not usable because it still lies on the physical surface of the earth. It must be reduced to the surface of reference ellipsoid and all subsequent computations performed on that mathamatical surface.

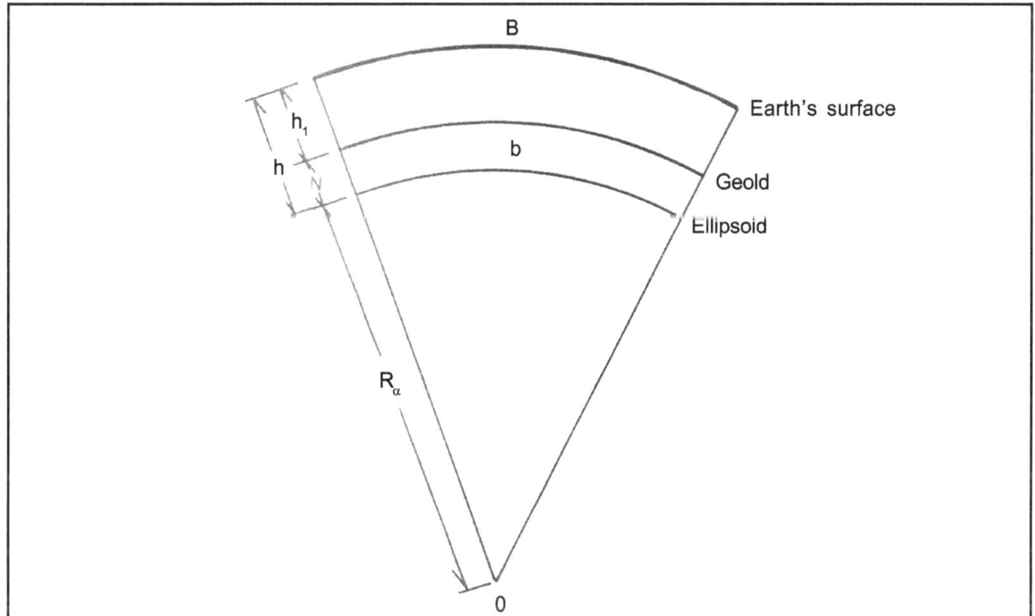

Fig 6.6. Reduction of a base line to the reference elliposoid.

In the figure 6.6 B is the base line as measured and corrected on the surface of the earth. If the two ends of the base line are at different elevations, the mean value should be used in the reduction to the surface ellipsoid. The base line required is 'b' lying on the surface of the reference ellipsoid. If Geoid, which is the sea level surface projected inland is defined then geoid height is known. Let the geoid height be N, above the ellipsoid. By means of levelling, the height h_1, of the earth surface above sea level has been determined. Let $N+h_1 = h$. h is the distance of the earth surface from the reference ellipsoid. The notation R_α is the radius of curvature of the earth in the plane of the base line.

Then by simple propotion

$$\frac{b}{B} = \frac{R_\alpha}{R_\alpha + h}$$

$$\Rightarrow \qquad b = B\left(\frac{R_\alpha}{R_\alpha + h}\right) = B\left(\frac{1}{1 + h/R_\alpha}\right)$$

By use of the series $\frac{1}{1+x} = 1 - x + x^2 - x^3 + \ldots..$

$$b = B\left(1 - \frac{h}{R_\alpha} + \frac{h^2}{R^2{}_\alpha} - \ldots..\right)$$

Therefore the amount of correction needed is

$$B - b = \frac{Bh}{R_\alpha} - \frac{Bh^2}{R_\alpha{}^2} + \ldots..$$

The Geoid height N may be positive or negative depending upon whether the geoid is above or below the ellipsoid. A sketch has to be drawn in every case in order to place the geoid, the ellipsoid and the physical surface of the earth in their correct relative locations.

6.6.3 Effect of height of point observed on the azimuth of a line

Because of the flattening of the reference ellipsoid, the normals projected vertically upward from two points on the ellipsoid generally do not lie in the same plane. One of the normals will be inclined with respect to the other. For this reason there will be an error in the observed azimuth of a station depending on the height above the surface of the ellipsoid.

Unless the point sighted upon is at a high elevation (in excess of 1000 meters), the error is smaller than the probable error of an observed direction and therefore, is negligible. In mountainous terrain where observation may be made upon stations located at extremely high altitudes, a correction must be applied to the observed direction.

The Correction to the azimuth is

$$\Delta_\alpha = \frac{\rho he^2 \cos^2\phi \, \text{Sin}2\alpha}{2N(1-e^2)}$$

where Δ_α the correction in seconds of arc, p is 206, 264.8 the number of seconds in a radian, h is the height of the observed station above the ellipsoid in metres, e is the flattening of the ellipsoid, ϕ is the latitude of the station observed $\angle_\alpha$ is the azimuth to the station observed and N is the radius of curvature in the prime vertical at latitude.ϕ

6.6.4 Convergence of meridians

At the equator, all meridians are parallel, because they intersect the equator at right angles. At poles all the meridians have converged to a common point. Between the equator and the pole the amount of convergence can be determined mathematically.

The Figure 6.7 is drawn as a sphere for convenience, than a ellipsoid. The figure represents a wedge shaped slice of widths λ, bounded by the equator and the two meridional planes in the northern hemisphere. The arc AB is the segment of a parallel of latitude lying at latitude ϕ. The radius of the sphere is r. The radius of the parallel of latitude is r cos ϕ. The tangents to the sphere at the points A and B, two line are extended in each plane of its respective meridian. The two lines will intersect along the extension of the polar axis and form angle c, which is the amount of convergence of the two meridians. The length of the two lines is r $\angle$cot ϕ.

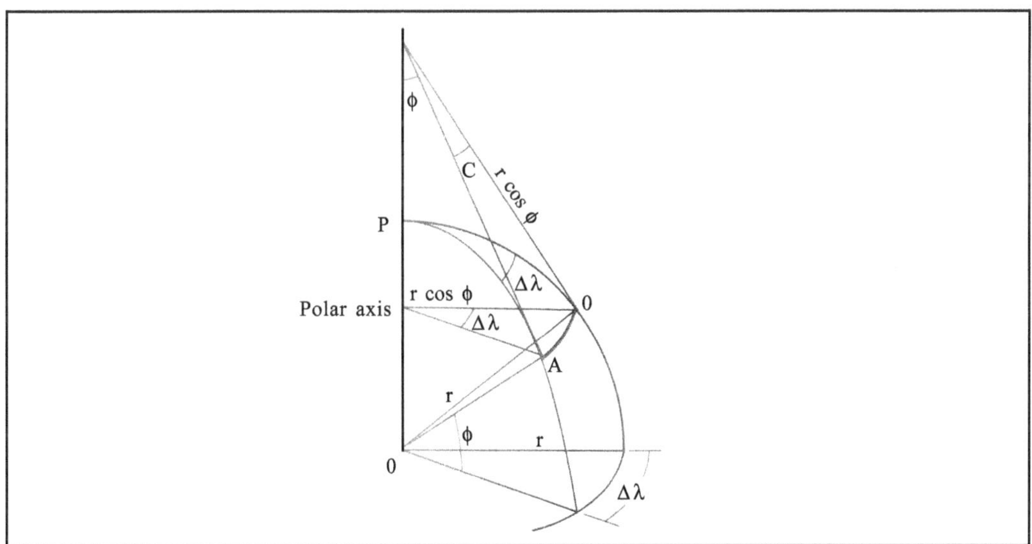

Fig. 6.7 Convergence of the meridians

From the figure it is found that arc AB is common to two triangles Applying the formula l = rϕ,

$$\text{arc AB} = r \,\text{cost}\, \phi \,\Delta\lambda \quad .....(1)$$

$$\text{also arc AB} = r \cot \phi . c \quad .....(2)$$

equating (1) and (2)

$r \cos \phi.\ \Delta\lambda = r \cot \phi.\ c$

$$c = \Delta \lambda \frac{\text{Cos}\,\phi}{\text{Cot}\,\phi} = \Delta \lambda \text{ Sin } \phi.$$

This is the amount of convergence at the latitude ϕ on a sphere. This quantity of convergence is used for surveys of limited extent or for checking the azimuths of lines in a traverse.

The equation for convergence of meridians on an ellipsoid is more complicated. The value is given by

$$C = \Delta \lambda \text{ Sin } \phi_m \text{ Sec } \frac{\Delta\phi}{2} + (\Delta \lambda)^3.F$$

where C : convergence in seconds of arc

$\Delta\lambda$: difference in longitude between A and B

ϕm : average latitude in between A and B

$\Delta\phi$: difference latitude between A and B

$$F = 1/12 \sin \phi m \cos^2 \phi_m \sin^2 1''$$

From the above equation, the value of convergence at the equator and at the pole can be deduced.

At the equator convergence should be zero

$$C \text{ (at equator)} = \Delta\lambda.0.1 + (\Delta\lambda)^3.0 = 0$$

At the pole convergence should be equal to the difference in longitude

$$C \text{ (at pole)} = \Delta\lambda.\ 1.1 + (\Delta\lambda)^3\ .0 = \Delta\lambda.$$

6.6.4.1 Forward and back Azimuths

The practical use of convergence in geodesy lies in the fact that the forward and backward azimuths of a line do not differ by 180^0 as in the case of plane surveying, but by 180^0 plus the convergence.

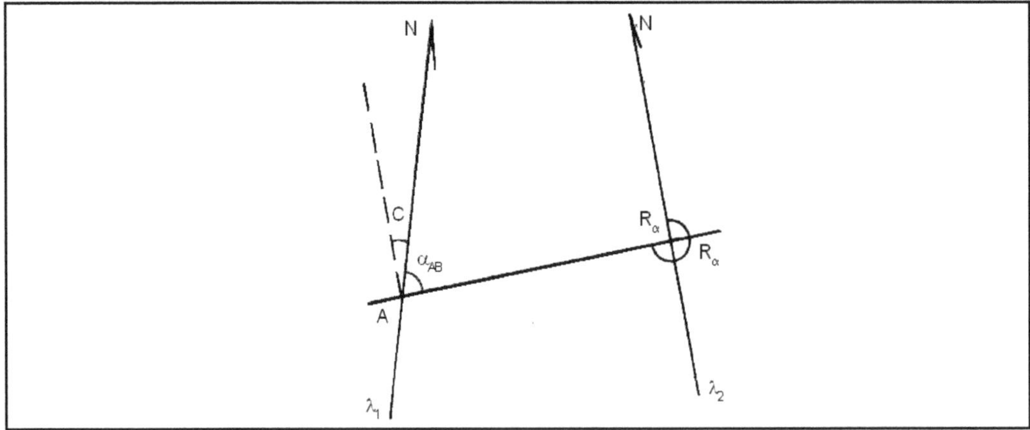

Fig. 6.8 Forward and back Azimuths

The figure 6.8 shows two meridians λ and λ_2 converging towards the pole. Each lines in a true north south line but they are not parallel. Point A lies on the meridion of λ, and B on meridian of λ_2. The azimuth of the line AB measured with respect to meridian λ, is α_{AB} The azimuth of the line BA with reference to its meridian λ_2 is α_{BA}.Through the point A a line is drawn parallel to meridian λ_2. The angle this line makes with the meridian λ_1 is the convergence C. Hence

$$\alpha_{BA} = \alpha_{AB} + C + 180^0$$

which shows that the forward and backward azimutt is of line differ by 180° plus convergence.

6.6.5 The plane curves and Geodesic line

6.6.5.1 Plane Curves on a Sphere

On a sphere, the shortest distance between two points A and B is the arc of the great circle connecting these points. If a theodolite is set up at point A, levelled and sighted at point B, the sight plane would trace out the arc of a great circle. If the instrument ware moved to point B levelled and sighted at point A the sight plane would again trace out the arc of a great circle. These two arc would be identical because of the normals at these two points intersect at a common point, the centre of the sphere.

6.6.5.2 Plane Curves on an Ellipsoid

On an ellipsoid the above case is not true. Because of flattening, the normals at different latitudes never intersect. The higher the latitude the lower the point where the normal intersects the polar axis. If the theodolite is setup at a point A (figure 6.9) and levelled its vertical axis will coincide with the normal to the ellipsoid at that point. (In this discussion the gravity anomalies are not considered and assumed that deflection of the vertical is zero).

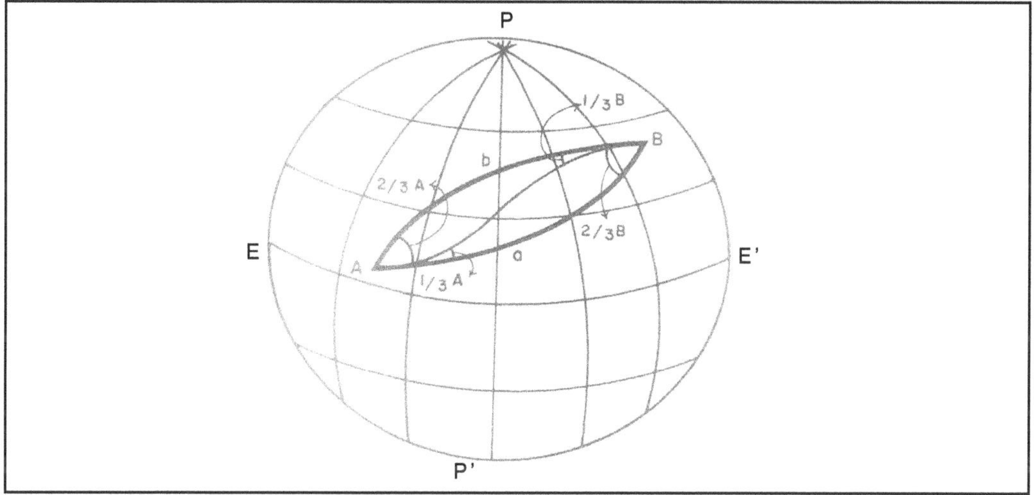

Fig. 6.9 The Plane curves and geodesic line.

If the theodolite is sighted at point B the plane of sight will trace out the line A a B. If the theodolite is moved to point B and levelled its vertical axis will now coincide with the normal at that point. Since B is at a higher altitude the normal will intersect the polar axis at a lower point and so the plane of sight from B to A will trace out the line BbA which is north of the other line. Both planes contain the chord AB, but the two normals are titled with respect to each other, so the trace on the surface must be different in each case. These two curves Aa B and Bb A, are called plane curves.

6.6.5.3 The Geodesic Line

On an ellipsoid, the shortest line that can be drawn between two points is called geodesic line. It is also known as geodetic line. Unlike the plane curves the geodesic line has double curvature. If we were to imagine that the, theodolite is moved between A and B of Fig. 6.9 so that the vertical plane of the instrument always contained the normal to the ellipsoid, the path of the theodolite would be the geodesic line. When traced completely around the ellipsoid, the geodesic line will not generally return to the starting point but will pass through that point on the equator on a sightly different longitude and then from another loop around the ellipsoid. Normally the geodesic line lies between the plane curves and divides the included angle in the ratio approximately 2 to 1 as indicated in Fig. 6.9.

6.6.5.4 Numerical Relationship - Two plane curves and geodesic line

Figure 6.10 explains the numerical relationship between the geodesic line and the two plane curves.

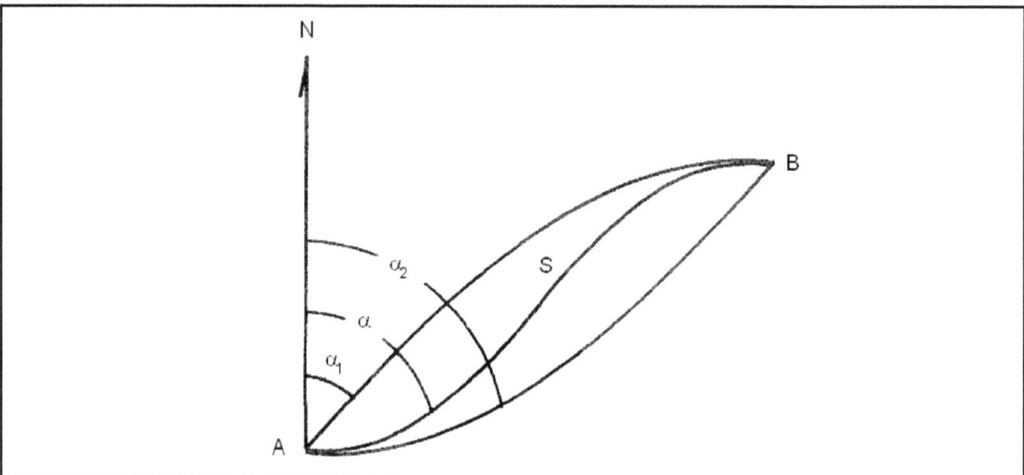

Fig. 6.10 Comparison of azimuths of the two plane curves and the geodesic line.

The azimuth of the geodesic line is the azimuth of the northern plane curve is α_1 and of the lower plane is α_2. The following relation ship exists:

$$\alpha_2 - \alpha_1 = \frac{\rho}{4r^2} e^2 S^2 \cos^2 \theta \sin 2\alpha$$

$$\alpha_2 - \alpha = \frac{\rho}{12r^2}\, e^2\, S^2 \cos^2\theta \sin 2\alpha$$

$$\alpha - \alpha_1 = \frac{\rho}{6r^2}\, e^2\, S^2 \cos^2\theta \sin 2\alpha.$$

where ρ - number of seconds in a radian

S - distance AB

and θ is the latitute of the geodesic line corresponding the intersection of the point where the azimuth is measured.

6.6.5.5 Equaton to the Geodesic Line

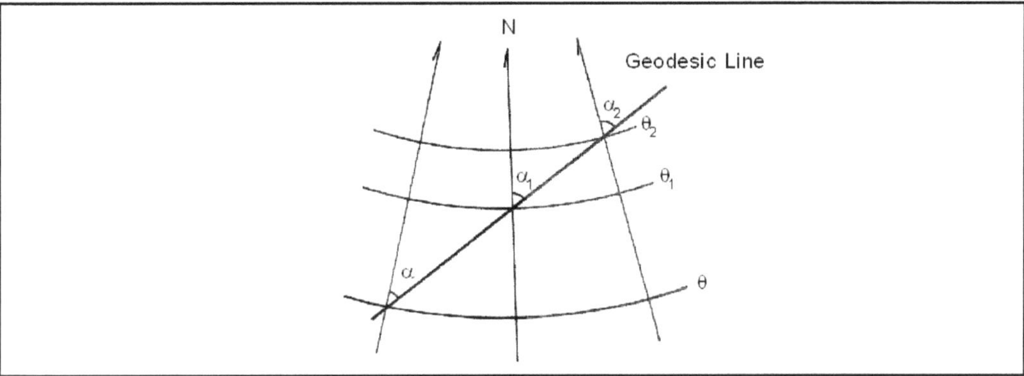

Fig 6.11 The geodesic line

In the figure 6.11 the geodsic line is shown cutting three meidians Because of convergence of meridians, the geodesic line intersets each successive meridian at different angles. Let $^{1}\angle\alpha$, α_1, α_2 be the azimuths and $^{2}\angle\theta$, θ_1, θ_2 the reduced latitude at each of the three meridians where the geodesic line intersects. The equation of the geodesic line is $\cos\theta \sin\alpha = \cos\theta_1 \sin\alpha_1 = \cos\theta_2 \sin\alpha_2 =$ constant. At the point where the geodesic line crosses the equator, $\theta = 0$ and $\sin\alpha$ equals the constant. At the point where $\alpha = 90$, the constant become the cosine of the latitude.

6.6.6 Calculation of Coordinates

The final products of geodesy are geodetic coordinates, distances and directions on the reference ellipsoid. The equation are complex and difficult to derive and are beyond the scope of this chapter.

6.6.6.1 Data for Computation

The station No. 1 shown is Fig. 6.12 is known to us. The co ordinates of station No.2 have to be computed with respect to station No. 1.

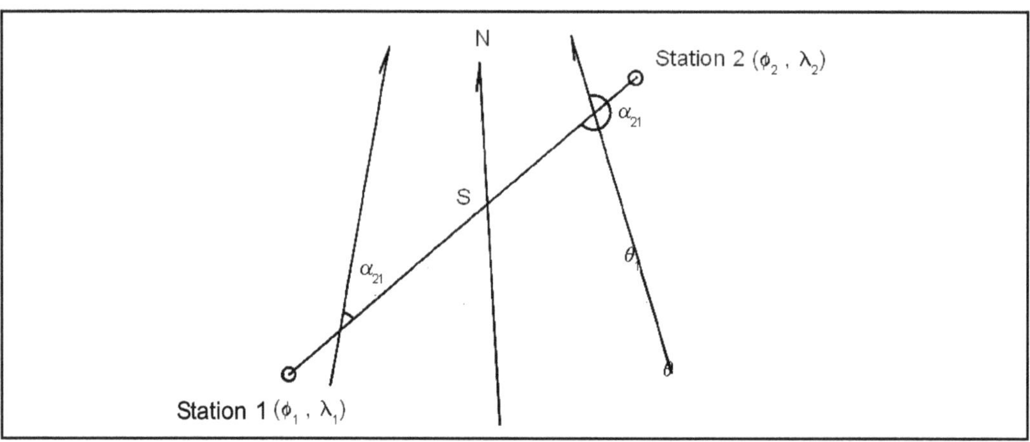

Fig. 6.12 Elements for calculation of coordinates

The first requirement is the reference ellipsoid data on which the computations are carried our. The data are

a : semimajor axis of reference ellipsoid

f : flattening of the reference ellipsoid.

Data of the known point (station No.1)

ϕ_1 - latitude of station No.1

λ_1 - longitude station No.1

S - distance between the two points 1 and 2

α_{12} - Azimuth of the line measured at the known point.

The elements to be computed for the unknown point 2 are

ϕ_2 - latitude of the station No.2

λ_2 - longitude of station No.2

and α_{21} - the back azimuth of line S measured at station 2.

Summary of the conditions :

Given	To Calculate
a	ϕ_2
f	λ_2
ϕ_1	α_{21}
λ_1	
S	
α_{12}	

The equation for the solutions are given

6.6.6.2 Difference in latitude

$$\Delta\,\phi = SB\cos\alpha\ - S^2 C\sin^2\alpha,$$
$$- D\,(\Delta\,\phi)^2 - hS^2 E\sin^2\alpha \qquad .....(1)$$

where,

$$\hbar = SB\cos\alpha$$

$$B = 1/R\sin 1''$$
$$C = \tan\phi\,/\,RN\sin 1$$

$$D = \frac{3e^2\,\text{Sin}\phi\,\text{Cos}\phi\,\text{Sin}1''}{2\left(1-e^2\,\text{Sin}^2\phi\right)}$$

$$E = \frac{\left(1+3\tan^2\phi\right)\left(1-e^2\,\text{Sin}^2\phi\right)}{6a^2}$$

6.6.6.3 Difference in longitude

$$\Delta\lambda = SA\sin\alpha\sec\phi_2 \qquad .....(2)$$

6.6.6.4 Difference in Azimuth

$$\Delta\ \alpha = \Delta\lambda\sin\phi_m\ \frac{\text{Sec}\,\Delta\phi}{2} + (\Delta\lambda)^3 F \qquad .....(3)$$

The coordinates of the station No.2 and back azimuth are:

$$\phi_2 = \phi_1 + \Delta\phi \qquad .....(4)$$
$$\lambda_2 = \lambda_1 + \Delta\lambda \qquad .....(5)$$

and
$$\alpha_{21} = \alpha_{12} + \Delta\,\alpha + 180^0 \qquad .....(6)$$

After determining the differences of latitude, longitude and azimuth between station No(1) (known) and station No(2) (unknown), the coordinates ϕ_2, λ_2, and azimuth α_{21} can be determined by applying the differences as shown in equation (4),(5) and (6).

As in triangulation the coordinates of the unknown point should always be computed from two known points.

6.7 Gravity

The size and the shape of the earth interms of its departure from the mathematically defined surface (the ellipsoid) must be specified. Thorough study of earth's gravity and its potential, the departures of the earth from the reference ellipsoid may be defined. The centre of the ellipsoid may be placed at the centre of the earth's gravity and the shape (flattening) of the ellipsoid may be determined. The application of gravity for the above purpose falls in the realm of physical geodesy To explain this aspect Physical geodesy, the following are to be discussed :

* Expressions for gravity and potential of a rotating sphere
* Expression for Geodetic deviation of the earth from the ellipsoid as functions of gravity variations.
* Measuring gravity on earth
* Reducing gravity values to those on a surface defined by on a surface defined by mean sea level

Only brief discussions on the above topics are given in the following sections.

6.7.1 Expressions for Gravity and Potential

Expression for potential 'V' is (sphere)

$$V = \frac{GM}{\rho}$$

Expression on for gravitation of sphere is

$$F = \frac{GM}{\rho^2}$$

When G is the Gravitational constant,

M - mass of the sphere and ρ the radius of curvature.

6.7.2 Geoid Undulations and Deflections of Vertical

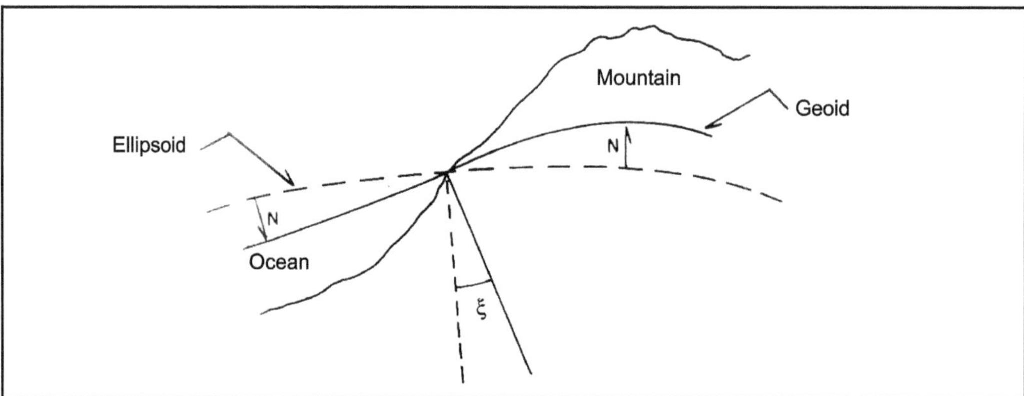

Fig. 6.13 Result of mass anomalies. Mass excess corresponds to positive gravity anomaly and positive geoid height.

The Fig. 6.13 indicates the result of deviation from a uniform mass. The actual mass expenses and deficiencies (mass anomalies) cause geoid surface to vary from that of the ellipsoid. In the case of mass excess the geoid is above the ellipsoid. The plumbing of an instrument which defines the verticals to the geoid is diverted from the ellipsoid normal and towards mass excess. In the case of mass deficiency, the geoid is below the ellipsoid. The basis for gravimetric theory is that the defections of the vertical, geoid warp (geoid undulation) and gravity anomalies all result from the same cause mass anomalies. The gravity anomalies can be observed and then employed to compute the geometric deviation of the geoid from the ellipsoid.

Symbols used :

U - Potential of the ellipsoid

W - Potential of Geoid

γ - Gravity on the ellipsoid (normal gravity)

g - Gravity on the Geoid.

Δg - G - γ = gravity anomaly.

The condition that the potentials for of the ellipsoid and geoid are equal is

$$U = W = UW + T = \text{constant.}$$

Where UW is the normal potential of the geoid (potential if these were no mass anomalies) and T, the disturbing potential.

Even if U = W, the two equipotential surface do not necessarily coincide. They may be separated by geoid undulation or height.

6.7.3 Measurement of Gravity on Earth

The purpose of gravity measurements and their benefits have been discussed. The methods of acquiring values for gravity is discussed below.

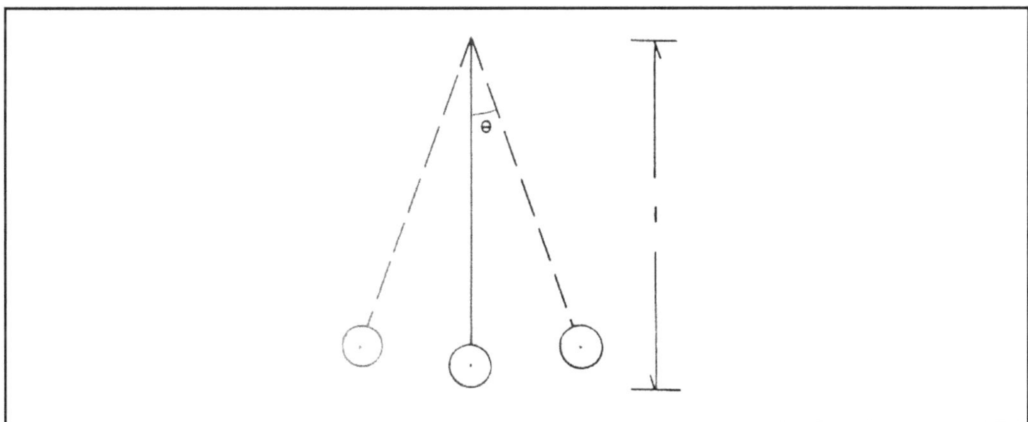

Fig. 6.14 Simple pendulum and differentiation of gravity.

The gravity can be determined with a simple pendulum. The period of a simple pendulum is

$$P = \pi \left(\frac{1}{g}\right)^{\frac{1}{2}} \left(1 + \frac{\theta^2}{16} + \ldots\right)$$

Where P is the half period, l is the length, g, gravity and θ the angle of swing. Since is small the last portion can be ignored. Then,

$$P = \pi \left(\frac{l}{g}\right)^{\frac{1}{2}}$$

or

$$g = \frac{\pi^2}{p^2}.l$$

6.7.4 Reducing Gravity Values to Those on a Surface Defined by a Mean Sea Level

Gravity measurements made on earth's surface must be reduced to the Geoid for computation of Gravity anomalies.

The reductions must attempt to produce the value of 'g' that would have occurred if it were possible to observe on the Geoid surfaces. A few of the reduction methods are listed below. Detailed discussions on the reduction methods is beyond the scope of this chapter.

The methods are :

6.7.4.1 Free - Air Reduction

This reduction merely takes the gravity observed at elevation H and reduces it to Geoid.

6.7.4.2 Bouguer Reduction

This reduction moves the mass between earth's surface and the Geoid to infinity and reduces the point to the Geoid.

6.7.4.3 Isostatic Reduction

After number of discussion two working theories are proposed

- Depth of compensation (equal Pressure) be measured from the topography rather than from Geoid.
- Airy theory - concluded that normal thickness of earth is crust (no roots no antiroots) is close to 30 Km.

6.8 Satellite Geodesy

6.8.1 Introduction

In the field Geodesy, satellites (artificial or otherwise) have made global surveys a realistic objective. The Geodetic positions may be determined directly on a global datum with origin at the earth's center of gravity.

6.8.2 Artificial Satellite

Natural satellite (the moon) has serious observational limitation. Artificial satellites are observed electronically or optically to measure range, range rate or direction from the observer to the satellite to determine Geodetic position. The measurements are made after determining the orbit of the artificial satellite.

6.8.3 Satellite Orbit

An orbiting satellite travels in an elliplical path with the earth's centre at one of the foci. This ellipse is defined by its semimajor axis (a) and the square of its eccentricity e^2. If e = o, the ellipse is a circle, the satellites position is simply a function of time and polar coordinates. When the path is a true ellipse, the calculation becomes difficult. The distance from the centre of earth to the satellite varies and also the angular velocity, which is not constant.

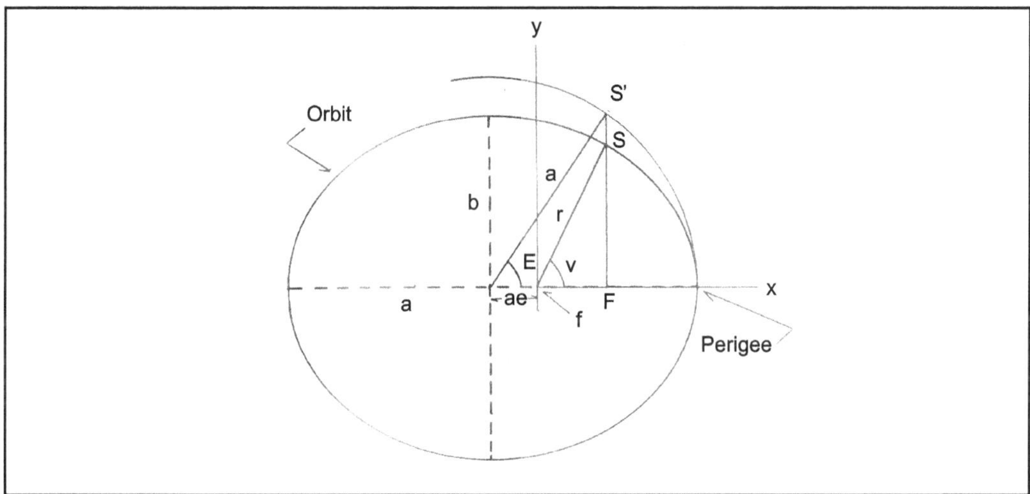

Fig. 6.15 Orbital plane of a satellite.

The figure 6.15 shows the orbital plane based on which the calculations are made. The following data are given in the Figure.

angle V - true anomaly

angle E - eccentric anomally M is called the mean anomaly and is calculated as follows :

$$M = n\,(t - T) \qquad \text{.....(1)}$$

Where

$$n = \frac{2\pi}{P} \qquad \text{.....(2)}$$

Where n is the angular rate, P the period of satellite orbit, T the time of perigee passage (closest approach to orbit foci) and t is the time of satellite position after T.

The intermediary M in related to the Geometric E by Kepers equation

$$M = E - e \sin E \qquad .....(3)$$

If the satellite position is to be determined as a function of time, this equation is solved by iteration. The procedure is given below :

1. Compute M for a time t from equation (1) above
2. Compute approximate E1 with $E1 = M + e \sin M + 1/2\, e^2 \sin 2M$
3. Substitute this value in equation (3) and compute M1

   $$M_1 = E_1 - e \sin E_1$$

   now $\Delta M = M1 - M$
4. Determine the correction for E_1 with

   $$\Delta E = \frac{\Delta M}{1 - e \operatorname{Cos} E}$$

5. Compute E_2

   $$E_2 = E1 + \Delta E$$

6. With E_2, repeat this sequence until $\Delta M = 0$ (or nearly zero). With value of E, referring to the axillary x and y coordinates (Fig. 6.15).

   $x = r \cos v = a \cos E - ae = a\,(\cos E - e)$

   $y = r \sin v = b \sin E = a\,(1 - e2)^{1/2} \sin E$

   r and v the polar coordinates of the position of satellite S can be calculated as follows

   $\gamma = (x^2 + y^2)\ 1/2 = a\,(1 - e \cos E)$

   $$\tan v = \frac{y}{x} = \frac{(1 - e^2)^{\frac{1}{2}} \operatorname{Sin} E}{\operatorname{Cos} E - e}$$

The above set of equation defines the satellite motion and its position in the plane of the ellipse.

6.8.4 Celestial Coordinate System

The celestial coordinate system is defined by the right ascension (α) and declination (δ). The position of the satellite 'S' can be determined by solving the right spherical triangle NDS as shown in the Fig. 6.16.

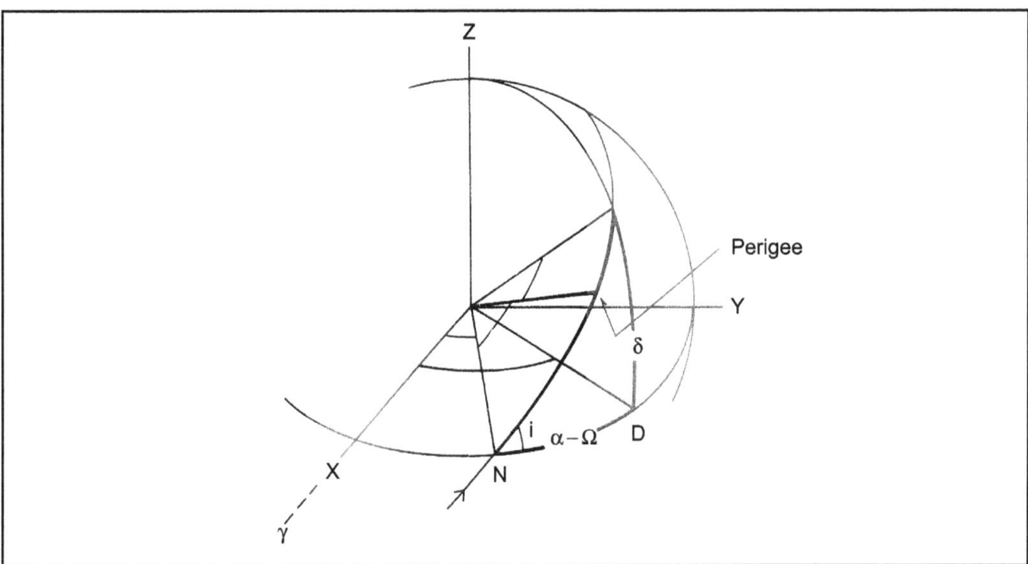

Fig. 6.16 Orbital elements.

The solutions are given by

$$\sin \delta = \sin (\varpi + \upsilon) \sin i$$
$$\tan (\alpha - \Omega) = \cos i \tan (\varpi + \upsilon)$$

The above equations are true only if the satellite were orbiting a uniform spherical earth in a vacuum. But this is not the case. Due to disturbing influence of drag, attraction of the moon and sun, radian pressure, tides and relativity causes perturbations of the orbit of the satellite. Because of this perturbations the element of the elliptical orbit (a,e, P and T) and angles of the celestial sphere (Ω, ϖ and i) are not constant. A detailed discussion on orbit perturbations are beyond the scope of this section.

In practice, the problem of computing satellite position with varying elements of orbit can be solved by assuming the variable ellipse to be tangent to actual orbit at the satellite point. Such a changing ellipse is said to osculate with the true orbit. If all the disturbing forces (perturbations) were removed, the satellite would travel the path defined by the ellipse at the moment of osculation.

6.8.5 Geodetic Position From known orbit

As discussed in the previous section, the coordinates of an orbiting satellite are referred to a system whose origin coincides with the earth's centre of gravity (CG). By observing a satellite of a known position, the observer may be located in the CG centered system. The geodetic position of the observer can be calculated from direction, range and range rate observations.

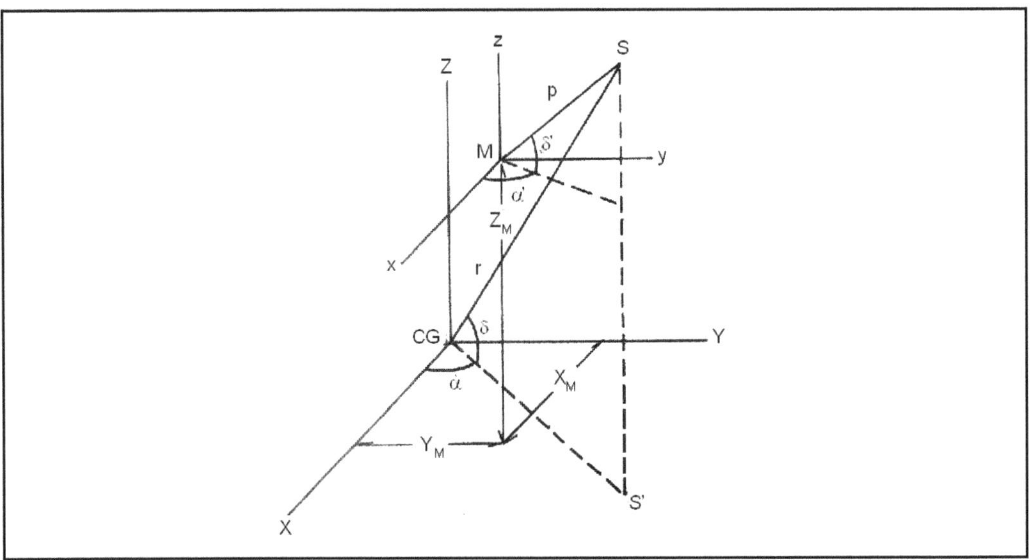

Fig. 6.17 Celestial and topocentric coordinate systems with directions observations.

Two coordinate systems are shown in Fig. 6.17 (1) with earth's CG (X, Y, Z system) and (2) at the observer (x,y,z topocentric system). The coordinate axes of these systems are parallel. The following coordinates are explained. α, (right ascension) and (declination) of the satellite coordinates viewed by an observer at the earth's center. α^1, δ^1 celestial coordinates of satellite actually observed (topocentric).

P - range from observer to satellite

γ - range from satellite to

(x_M,y_M,z_M) rectangular coordinate system, at the moment of observation.

In X, Y, Z system,

$$\text{Cos}\ \alpha = \frac{X}{r \,\text{Cos}\, \delta}$$

$$\text{Sin}\ \alpha = \frac{Y}{r \,\text{Cos}\, \delta}$$

$$\text{Sin}\ \delta = \frac{Z}{r}$$

The satellite coordinates can be expressed as

$$X = r\ \text{Cos}\ \delta\ \text{Cos}\ \alpha$$

$$Y = r\ \text{Cos}\ \delta\ \text{Sin}\ \alpha$$

$$Z = r\ \text{Sin}\ \delta$$

The satellite coordinates, in topocentric system

$$\text{Sin } \delta' = \frac{Z}{P'} \text{ and } P = \frac{Z}{\text{Sin}\delta'}$$

Therefore $x = \rho \cos \delta^1 \cos \alpha^1 = z \cot \delta^1 \cos \alpha^1$

$$y = \rho \cos \delta^1 \sin \alpha^1$$

$$= z = \rho \text{ Sin } \delta' \text{ Sin } \alpha'$$

$$z = \rho \sin \delta'$$

6.8.6 Coordinate Transformation in Equatorial Plane

The earth rotates in celestial X, Y, Z, system and therefore X_m Y_m and Z_m will vary with time. To overcome this, the XYZ system will be expressed in terms of the fixed geodetic rectangular system "UVW". This transformations accomplished by rotating the xyz system around the z-axis (assumed to coincide with W- axis) by the hour angle H between the vernal equinox and the Greenwich meridian. H corresponds to the time of satellite flash.

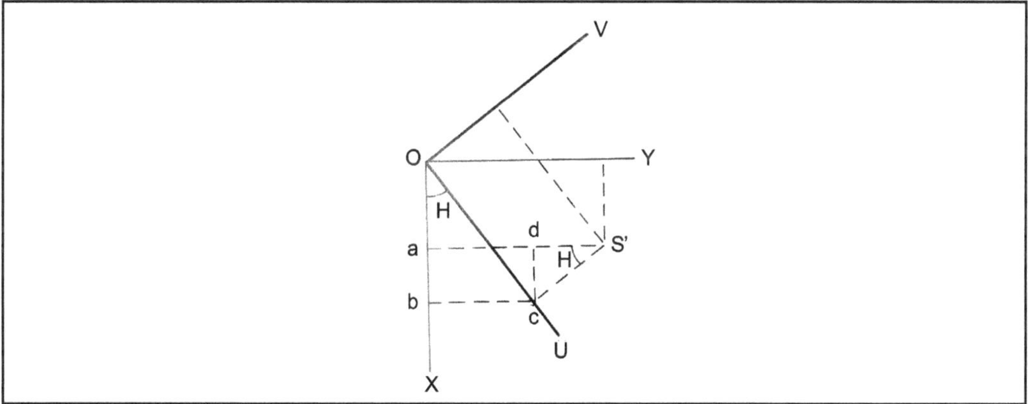

Fig. 6.18 Coordinate transformation in equatorial plane.

The relationship of transformation is explained by Fig. 6.18.

X = Oa = Ob - ab = Ob - dc = Oc CosH - S'c Sin H = U Cos H - V Sin H

Y = S'a = S'd + ad = S'd + bc = S'c Cos H + oc Sin H = V Cos H + U Sin H

Z = W

6.8.7 Range Observations on three Satellite Points

In this study, the proposition is considered when range is observed rather than direction of Satellites of known orbit.

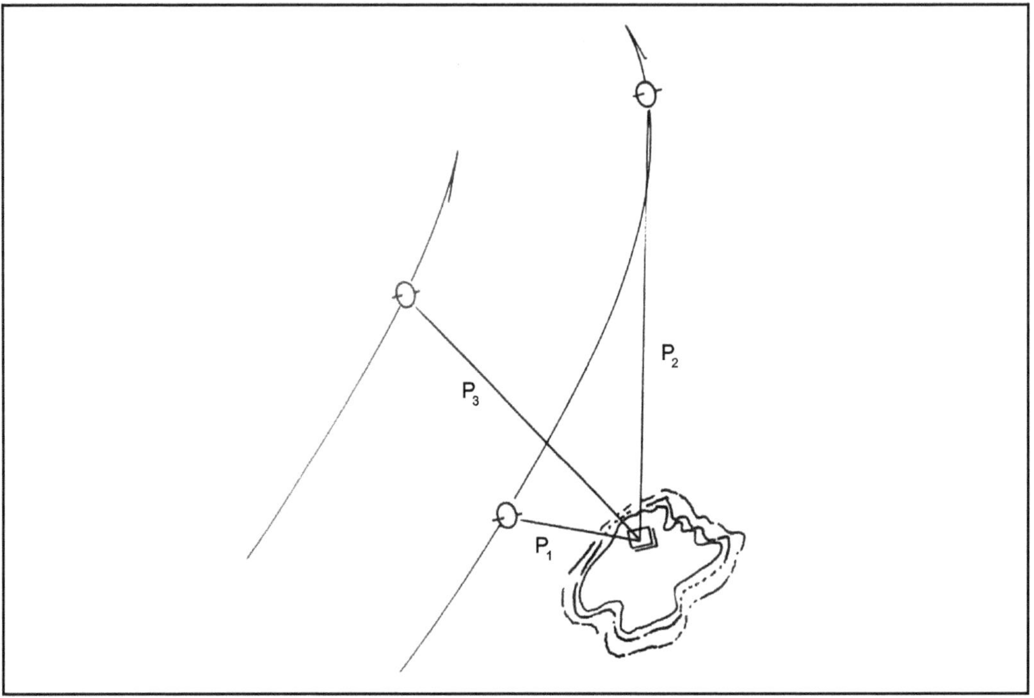

Fig. 6.19 Range observation on three Satellite points.

The Fig. 6.19 depicts range observations P on three Satellite points. These three ranges can be written in terms of Satellite and observer coordinates.

$$\rho_1{}^2 = (U_1 - U_M)^2 + (V_1 - V_M)^2 + (W_1 - W_M)^2$$

$$\rho_2{}^2 = (U_2 - U_M)^2 + (V_2 - V_M)^2 + (W_2 - W_M)^2 \quad \text{and}$$

$$\rho_3{}^2 = (U_3 - U_M)^2 + (V_3 - V_M)^2 + (W_3 - W_M)^2$$

The above are three equations with three unknowns, the observer's coordinates - U_M, V_M & W_M. These equations (2nd degree) can be solved and the unknown coordinates (U_M, V_M & W_M) can be determined.

From geometric point of veiw, a single pass containing all three Satellite points would produce a very weak solution. For this reason at least two Satellite passes are required to provide a realistically unique solution.

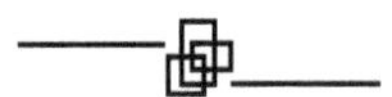

7

Global Positioning System (GPS)

7.1 Introduction

The NAVSTAR Global Positioning System (GPS) is a satellite-based radio-positioning and time-transfer system designed, financed, deployed, and operated by the U.S. Department of Defence. GPS has also demonstrated a significant benefit to the civilian community who are applying GPS to a rapidly expanding number of applications. The major attractions of GPS are :

- The relatively high positioning accuracies, from tens of metres down to the millimetre level.
- The capability of determining velocity and time, to an accuracy commensurate with position.
- The signals are available to users anywhere on the globe: in the air, on the ground, or at sea.
- It is a positioning system with no user charges, that simply requires the use of relatively low cost hardware.
- It is an all-weather system, available 24 hours a day.
- The position information is in three dimensions, that is, vertical as well as horizontal information is provided.

The number of civilian users is already significantly greater than that of the military users. However, for the time being the U.S. military still operates several "levers" with which they control the performance of GPS (Bomford, 1980).Nevertheless, despite the handicap of GPS being a military system there continues to be tremendous product innovation within the civilian sector, and it is ironic that this innovative drive is partly directed to developing technology and procedures to overcome some of the constraints to GPS performance which have been applied by the system's military operators.

7.2 System Design Considerations

Development work on GPS commenced within the U.S. Department of Defence in 1973, the motivation being to develop an all-weather, 24-hour, global positioning system to support the positioning requirements for the armed forces of the U.S. and its allies. (Parkinson, B.W, 1994). The system was therefore designed to replace the large variety of navigational systems already in use, and great emphasis was placed on the system's reliability and survivability. In short, a number of stringent conditions had to be met :

- suitable for all classes of platform: aircraft (jet to helicopter), ship, land (vehicle-mounted to handheld) and space (missiles and satellites),
- able to handle a wide variety of dynamics,
- real-time positioning, velocity and time determination capability to an appropriate accuracy,
- the positioning results were to be available on a single global geodetic datum,
- highest accuracy to be restricted to a certain class of user,
- resistant to jamming (intentional and unintentional),
- redundancy provisions to ensure the survivability of the system,
- passive positioning system that does not require the transmission of signals from the user to the satellite(s),
- able to provide the service to an unlimited number of users,
- low cost, low power, therefore as much complexity as possible should be built into the satellite segment, and
- total replacement of the Transit satellite and other terrestrial navaid systems.

This led to a design based on the following essential concepts:

- A one-way ranging system, in which the satellites transmit signals, but are unaware of who is using the signal (no receiving function). As a result the user (or listener) cannot easily be: (a) detected by the enemy (military context), or (b) charged for using the system (civilian context).
- Use of the latest atomic clock and microwave transmission technology, including spread-spectrum techniques.

- A system that makes range-like measurements with the aid of pseudo-random binary codes modulated on carrier signals.
- Satellite signals that are unaffected by cloud and rain.
- A multiple satellite system which ensures there is always a sufficient number of satellites visible simultaneously anywhere on the globe, and at any time.
- Positioning accuracy degradation that is graceful.

What was perhaps unforeseen by the system designers was the power of product innovation, which has added significantly to the versatility of the GPS as a system for precise positioning and navigation. For example, GPS is able to support a number of positioning and measurement modes in order to satisfy simultaneously a variety of users, from those satisfied with general navigational accuracies (tens of metres) to those demanding very high (sub-centimetre) relative positioning accuracies. It has now so penetrated certain applications areas that it is difficult for us to imagine life without GPS!

7.2.1 Technological Advances

Rarely have so many seemingly unrelated technological advances been required to make a complex system such as GPS work. Briefly they are :

Space System Reliability : The U.S. space program had by 1973 demonstrated the reliability of space hardware. In particular, the Transit system had offered important lessons. The Transit satellites were originally designed to last 2-3 years in orbit, yet some of the satellites have operated well beyond their design life. In fact Transit continued to perform reliably for over 25 years.

Atomic Clock Technology : With the development of atomic clocks a new era of precise timekeeping had commenced. However, before the GPS program was launched these precise clocks had never been tested in space. The development of reliable, stable, compact, space-qualified atomic frequency oscillators (rubidium, and then cesium) was therefore a significant technological breakthrough. The advanced clocks now being used on the GPS satellites routinely achieve longterm frequency stability in the range of a few parts in 10^{14} per day (about 1 sec in 3,000,000 years!). This long-term stability is one of the keys to GPS, as it allows for the autonomous, synchronised generation and transmission of accurate timing signals by each of the GPS satellites without continuous monitoring from the ground.

Quartz Crystal Oscillator Technology : In order to keep the cost of user equipment down, quartz crystal oscillators were proposed (similar to those used in modern digital watches), rather than using atomic clocks as in the GPS satellites. Besides their low cost, quartz oscillators have excellent short-term stability. However, their long-term drift must be accounted for as part of the user position determination process.

Precise Satellite Tracking and Orbit Determination : Successful operation of GPS, as well as the Transit system, depends on the precise knowledge and prediction of a satellite's position with respect to an earth-fixed reference system. Tracking data collected by ground monitor stations is analysed to determine the satellite orbit over the period of tracking (typically one week). This reference ephemeris is extrapolated into the future and the data is then up-loaded to the satellites. Prediction accuracies of the satellite coordinates, for one day, at the few metre level have been demonstrated.

Spread-Spectrum Technology : The ability to track and obtain any selected GPS satellite signal (a receiver will be required to track a number of satellites at the same time), in the presence of considerable ambient noise is a critical technology. This is now possible using spread-spectrum and pseudo-random-noise coding techniques.

Large-Scale Integrated Circuit Technology : To realise the desired low cost, low power and small size necessary for much of the user equipment, the GPS program relies heavily on the successful application of VLSI circuits, and powerful computing capabilities built onto them.

7.3 GPS System Eelements

The GPS system consists of three segments (Fig. 7.1).

- The Space Segment: comprising the satellites and the transmitted signals.
- The Control Segment: the ground facilities carrying out the task of satellite tracking, orbit computations, telemetry and supervision necessary for the daily control of the space segment.
- The User Segment: the entire spectrum of applications equipment and computational techniques that are available to the users.

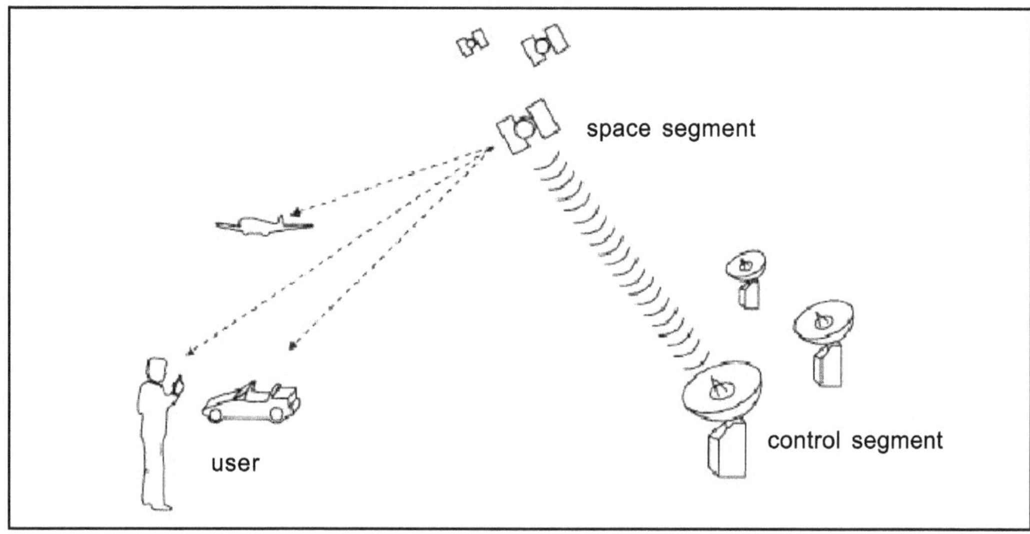

Fig. 7.1 GPS System Elements.

7.3.1 The Space Segment

The Space Segment consists of the constellation of spacecraft and the signals broadcast by them which allow users to determine position, velocity and time. The basic functions of the satellites are to :

- Receive and store data transmitted by the Control Segment stations.
- Maintain accurate time by means of several onboard atomic clocks.
- Transmit information and signals to users on two L-band frequencies.
- Provide a stable platform and orbit for the L-band transmitters.

Several constellations of GPS satellites have been deployed, and several more are planned. The experimental satellites, the so-called "Block I" satellites, were built by the Rockwell Corporation. The first was launched in February 1978, and the last of the eleven satellite series (one exploded on the launchpad) was launched in 1985. The operational series of GPS satellites, the "Block II" and "Block IIA" satellites, were also built by the Rockwell Corporation. The 20 replacement "Block IIR" series of satellites, first launched in 1997, are built by the General Electric Corporation (now the Lockheed Martin Corporation). The "Block IIF" series are still in the design phase and may, for example, incorporate an additional civilian transmission frequency. They are planned for launch from 2005 onwards. The operational satellite I.D.s are separated into three space vehicle numbering series: SVN 13 through 21 for the Block II, SVN 22 through 40 for Block IIA, and SVN 41 and above for the Block IIR satellites.

The current status of the GPS constellation and such details as the launch and official commissioning date, the orbital plane and position within the plane, the satellite I.D. number(s), etc., can be obtained from several electronic GPS information sources on the Internet. The following sections describes the general satellite orbit characteristics.

Each GPS satellite transmits a unique navigational signal centred on two L-band frequencies of the electromagnetic spectrum, permitting the ionospheric propagation effect on the signals to be eliminated. At these frequencies the signals are highly directional and so are easily reflected or blocked by solid objects. Clouds are easily penetrated, but the signals may be blocked by foliage (the extent of blockage is dependent on the type and density of the leaves and branches, and whether they are wet or dry). The signal is transmitted with enough power to ensure a minimum signal power level of -160dBw at the earth's surface (the maximum it is likely to reach is about -153dBw, (Spilker, 1995). The satellite signal consists of the following components (Fig. 7.2) :

- The two L-band carrier waves.
- The ranging codes modulated on the carrier waves.
- The so-called "navigation message".

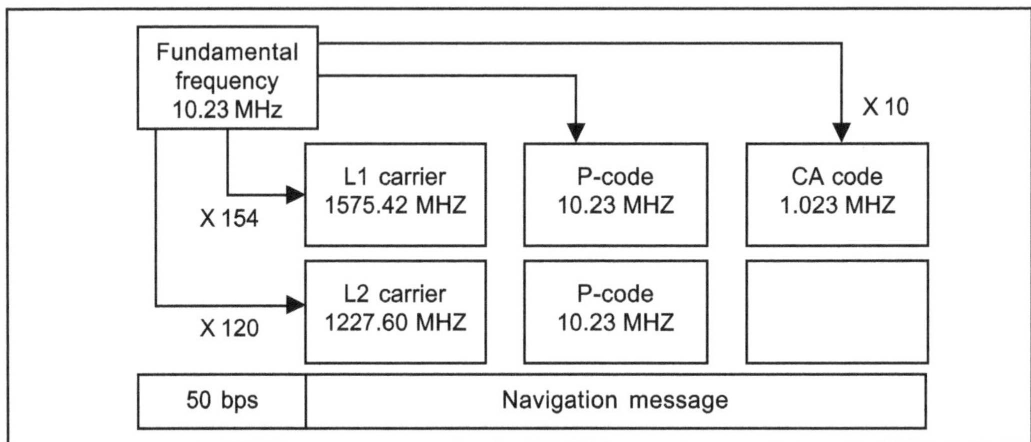

Fig. 7.2 GPS Satellite Signal Components.

Modulated onto the carrier waves are the PRN ranging codes and navigation message for the user. The primary function of the ranging codes is to permit the signal transit time (from satellite to receiver) to be determined. The transit time when multiplied by the velocity of light then gives a measure of the receiver satellite "range" (in reality the measurement process is considerably more complex). The navigation message contains the satellite orbit information, satellite clock parameters, and pertinent general system information necessary for real-time navigation to be performed. All signal components are derived from the output of a highly stable atomic clock. Each GPS satellite is equipped with several cesium and rubidium atomic clocks.

7.3.2 The Control Segment

The Control Segment consists of facilities necessary for satellite health monitoring, telemetry, tracking, command and control, satellite orbit and clock data computations, and data uplinking. There are five ground facility stations: Hawaii, Colorado Springs, Ascension Island, Diego Garcia and Kwajalein. All are owned and operated by the U.S. Department of Defense and perform the following functions:

- All five stations are Monitor Stations, equipped with GPS receivers to track the satellites. The resultant tracking data is sent to the Master Control Station.
- Colorado Springs is the Master Control Station (MCS), where the tracking data are processed in order to compute the satellite ephemerides and satellite clock corrections. It is also the station that initiates all operations of the space segment, such as spacecraft manoeuvring, signal encryption, satellite clock-keeping, etc.
- Three of the stations (Ascension Is., Diego Garcia, and Kwajalein) are Upload Stations allowing for the uplink of data to the satellites. The data includes the orbit and clock correction information transmitted within the navigation message, as well as command telemetry from the MCS.

Overall operation of the Control and Space Segments is the responsibility of the U.S. Air Force Space Command, Second Space Wing, Satellite Control Squadron at the Falcon Air Force Base, Colorado.

Each of the upload stations can view all the satellites once a day. All satellites are therefore in contact with an upload station three times a day, and new navigation messages as well as command telemetry can be transmitted to the GPS satellites every eight hours if necessary. The computation of : (a) the satellite orbits or "ephemerides", and (b) the determination of the satellite clock errors, are the most important function of the Control Segment. The first is necessary because the GPS satellites function as "orbiting control stations" and their coordinates must be known to a relatively high accuracy, while the latter permit a significant measurement bias to be reduced.

The GPS satellites travel at high velocity (of the order of 4 km/sec), but within a more or less regular orbit pattern. After a satellite has separated from its launch rocket and it begins orbiting the earth, it's orbit is defined by its initial position and velocity, and the various force fields acting on the satellite. In the case of the gravitational field for a spherically symmetric body (a reasonable approximation of the earth at the level of about 1 part in 103) this produces an elliptical orbit which is fixed in space - the Keplerian ellipse. Due to the effects of the other, non-spherical gravitational components of the earth's gravity field, and non-gravitational forces, which perturb the orbit, the actual trajectory of the satellite departs from the ideal Keplerian ellipse (Fig. 7.3).

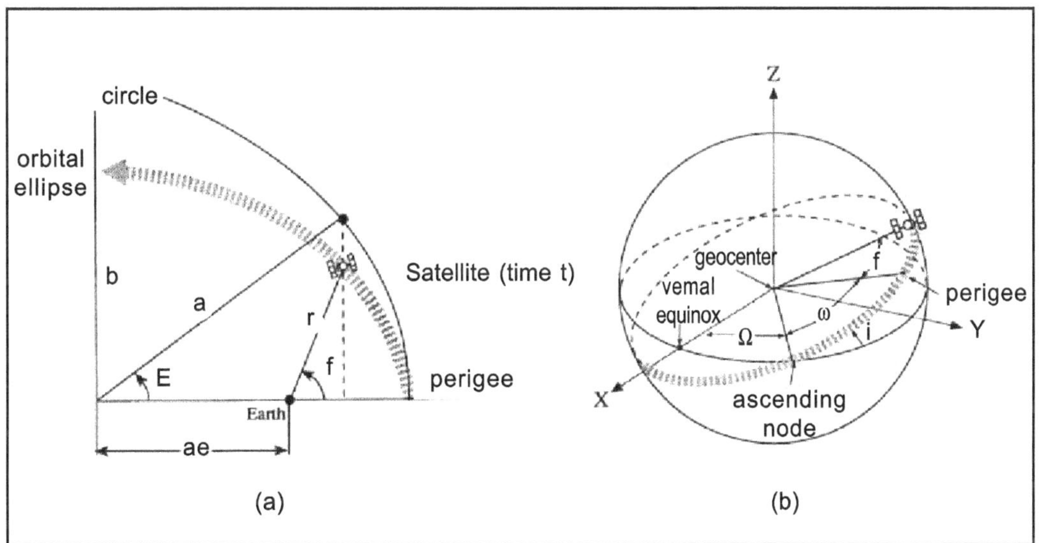

Fig. 7.3 (a) The Keplerian Ellipse, and (b) Keplerian Orbital Elements.

The most significant forces that influence satellite motion are:

- the spherical and non-spherical gravitational attraction of the earth,
- the gravitational attractions of the sun, moon, and planets (the "third body" effects),
- atmospheric drag effects,
- solar radiation pressure (both direct and albedo effects), and
- the variable part of the earth's gravitational field arising from the solid earth and ocean tides.

To determine the motion of a satellite to a high precision these perturbing forces need to be modelled accurately. If these forces were known perfectly and the initial position and velocity of the satellite were given, then the integration of the Equations of Motion would give the satellite's position and velocity at any time in the future.

However, the perturbing forces are not known to sufficient precision. An "orbit computation" process is therefore performed, in which satellite observations obtained at tracking sites of known position (in the case of GPS, the monitor stations of the Control Segment) are analysed in order to produce an orbit that is a "best fit" to the available observations. This involves the adjustment of the appropriate parameters of the orbit, possibly together with several additional force model parameters (Seebera, 1993). Determining the orbit is a complex procedure, and in the case of GPS satellites this process occurs on a continuous, automatic basis.

The product of the orbit computation process at the MCS is the satellite ephemeris (or trajectory). A satellite ephemeris may be expressed in a number of forms :

- a list of 3-D coordinates, and velocities, at regular intervals of time,
- the Keplerian elements at some reference epoch, plus their rate-of-change with time,
- a polynomial representation of the trajectory in a suitable reference system, such as along-track, cross-track, or radial components, or
- satellite position and velocity at some reference time epoch, and requiring these values to be derived for subsequent times by integrating the Equations of Motion.

The GPS broadcast ephemeris, as represented in the navigation message, is actually a combination of all of the above orbit representations (Langley, R. B, 1991b.). The orbital ephemerides are expressed in the reference system most appropriate for positioning, which is an earth-fixed reference system such as WGS84. Hence the Control Segment has the function of propagating the satellite datum, which users connect to via the transmitted satellite ephemerides.

The behaviour of each GPS satellite clock is monitored against GPS Time, as maintained by an ensemble of atomic clocks at the GPS Master Control Station. The satellite clock bias, drift and drift-rate relative to GPS Time are explicitly determined in the same procedure as the estimation of the satellite ephemeris. The clock behaviour so determined is made available to all GPS users via clock error coefficients (defining the mis-synchronization with GPS Time) in a polynomial form broadcast in the navigation message. However, what is available to users is really a prediction of the clock behaviour for some future time interval. Due to random deviations — even cesium and rubidium oscillators are not entirely predictable — the deterministic models of satellite clock error are only accurate to about 20 nanoseconds. This is not precise enough for accurate range measurement.

As the GPS system matures we can expect that the satellites will operate with greater independence from the ground-based Control Segment, without significant degradation in performance. For example the "Block IIR" and "Block IIF" satellites will have a crosslink capability enabling between-satellite communication and ranging. The satellites will talk to each other! This data will be processed to produce the ephemeris information within the space segment, with relatively little operator control having to be exercised.

7.3.3 The User Segment (The Applications)

This is the part of the GPS system with which we are most concerned — the space and control segments being largely transparent to the operations of the navigation function. Of interest is the range of GPS :

- Applications,
- Equipment, and
- Positioning strategies.

The "engine" of commercial GPS product development is, without doubt, the user applications. Each day new applications are being identified, each with its unique requirements with regards to: accuracy of the results, reliability, operational constraints, user hardware, data processing algorithms, latency of the GPS results, etc. To make sense of the bewildering range of GPS applications it may be useful to classify them according to the following :

1. Land, Sea and Air Navigation and Tracking, including enroute as well as precise navigation, collision avoidance, cargo monitoring, vehicle tracking, search and rescue operations, etc. While the accuracy requirement may be modest and the user hardware is generally comparatively low cost, the reliability, integrity and speed with which the results are needed is generally high.
2. Surveying and Mapping, on land, at sea and from the air. Includes geophysical and resource surveys, GIS data capture surveys, etc. The applications are of relatively high accuracy, for positioning in both the static and moving receiver mode, and generally require specialised hardware and data processing software.
3. Military Applications: Although these are largely mirrored by civilian applications, the military GPS systems are generally developed to "military specifications" and a greater emphasis is placed on system reliability.

4. Recreational Uses, on land, at sea and in the air. The primary requirement is for low cost instruments which are very easy to use.
5. Other specialised uses, such as time transfer, attitude determination, spacecraft operations, atmospheric studies, etc. Obviously such applications require specially developed, high cost systems, often with additional demanding requirements such as real-time operation, etc.

GPS user equipment has undergone an extensive program of development, both in the military and civilian area. In this context, GPS "equipment" refers to the combination of : hardware, software, and operational procedures or requirements.

While the military R&D programs have concentrated on achieving a high degree of miniaturisation, modularisation and reliability, the civilian user equipment manufacturers have, in addition, sought to bring down costs and to develop features that enhance the capabilities of the positioning system. The following general remarks can be made. Civilian users have, from the earliest days of GPS availability, demanded ever increasing levels of performance, in particular higher accuracy and improved reliability. This is particularly true of the survey user seeking levels of accuracy several orders of magnitude higher than that of the navigation user. In some respects the GPS user technology is being driven by the precise positioning market-in much the same way that automotive technology often benefits from car racing. Yet another major influence on the development of GPS equipment has been the increasing variety of civilian applications. For although there may exist a similar positioning accuracy requirement across many user applications, to address a particular application in the most satisfactory manner, a specific combination of hardware and software features is often required.

It is expected that the worldwide market for GPS receiver equipment will grow from about US$1 billion at the present time, to over US$8 billion by the year 2010. Market surveys suggest that the greatest growth is expected to be in the commercial and consumer markets such as ITS applications, integration of GPS and cellular phones, and portable GPS for outdoor recreation and similar activities. These could account for more than 60% of the GPS market by the turn of the century. There are at present over 100 manufacturers of GPS instruments of varying kinds.

7.3.4 The User Segment (Positioning Principles)

The basic concept of GPS positioning is that of positioning-by-ranges. The geometrical principles of positioning can be demonstrated in terms of the intersection of locii. In the two-dimensional case, a measured range to a known point constrains the position to lie on circle with the measured range as radius. In three dimensions a measured range to a known point constrains the position in 3-D space to lie on the surface of a sphere centred at the known point, with radius being the measured distance (Fig. 7.4). In the case of GPS, the distance measurement is made to a satellite with known position (coordinates are obtained from the satellite ephemeris data transmitted within the navigation message), however the principle applies to any range measuring positioning system, terrestrial or satellite-based.

In two dimensions, position can be defined as the intersection of two circles, involving distances d1 and d2 to two known points. Note that there are two possible solutions, only one of which is correct. In general one solution can be discarded rather easily through apriori knowledge of approximate position and velocity. Another possibility is to measure another range to a third point and if all ranges are measured without error the intersection of three LOPs is a single uniquely defined point.

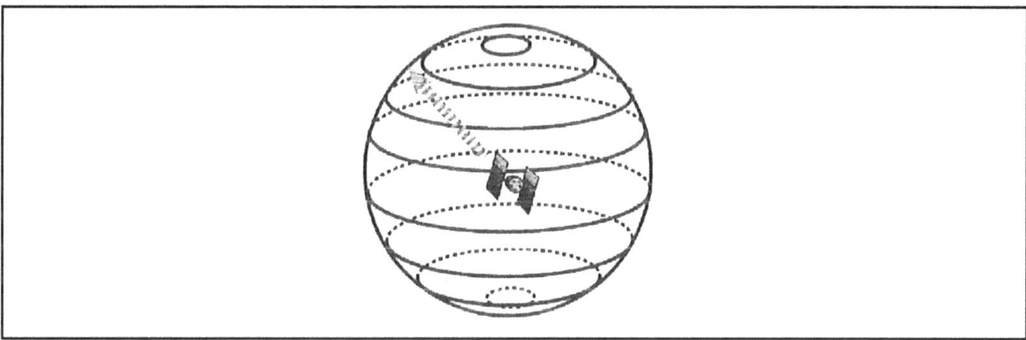

Fig. 7.4 Surfaces of Position for Range Measurements.

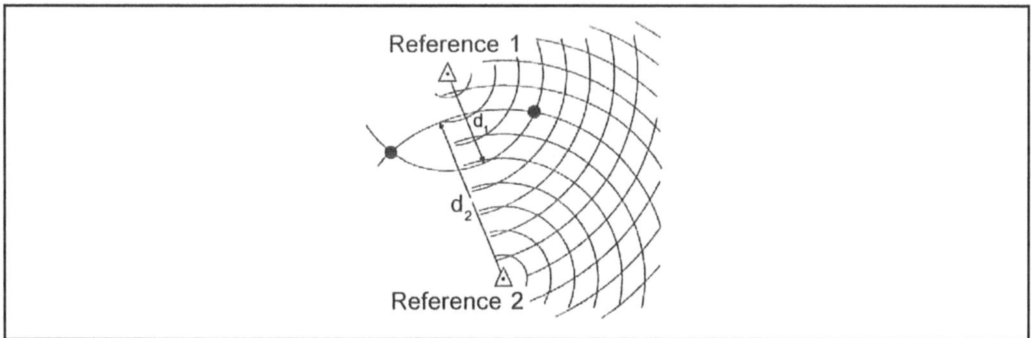

Fig. 7.5 The Intersection of Circular Lines of Position for 2-D Positioning.

In the three-dimensional case, the intersection of three spheres describes two points in space, only one of which is correct (Fig. 7.6). Hence, a minimum of three ranges are required, to three separated known points, in order to solve the 3-D position problem. The quality of the positioning solution is dependent, amongst other things, on the accuracy with which the ranges can be measured and the geometry of the intersection.

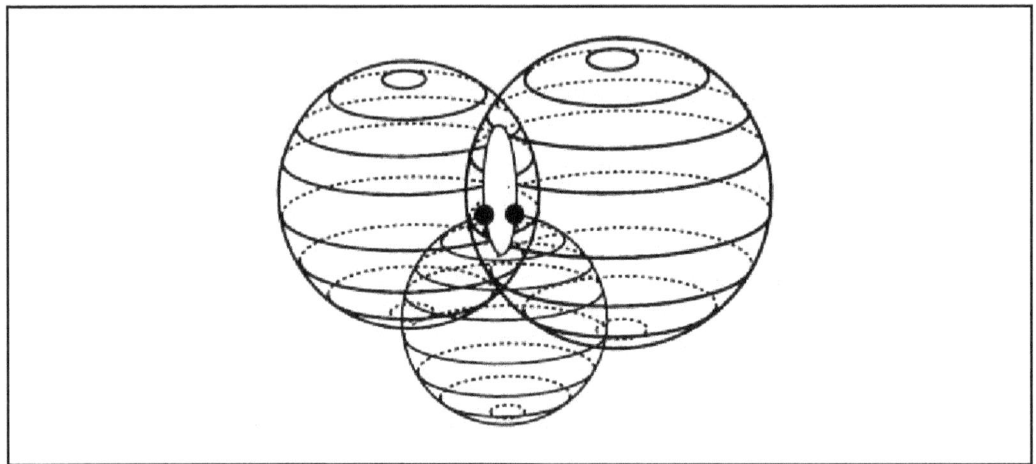

Fig. 7.6 Intersection of Surfaces of Position Based on Range Measurements.

If the point being positioned is stationary, the 2 (or 3) ranges do not need to be measured simultaneously. If the point is moving however, all ranges must be measured simultaneously (or over an interval of time during which the point has not moved by an amount greater than the uncertainty of the "fix"). Because the GPS constellation was designed to ensure at least four satellites are always visible anywhere on the earth, satellite positioning using simultaneously measured ranges is the basic positioning strategy for most navigation applications.

However, there is still the issue of how to account for measurement biases, as the technology used for making GPS range measurements does not give calibrated distance from the receiver to the satellite. Disturbing influences and errors in fact contaminate the range measurements to an unacceptable degree, and hence the basic positioning principle is modified in several ways to satisfy the varying levels of accuracies required by different applications.

7.4 GPS Satellite Constellation and Signals

7.4.1 GPS Constellation Design

The operational Block II/IIA satellite constellation was to be fully deployed by the late 1980's. However, a number of factors, the main one being the Space Shuttle Challenger disaster (28 January 1986), has meant that the GPS system only became operational in the 1990's as far as most users were concerned. Full Operational Capability was declared on 17 July 1995 — 24 Block II/IIA satellites operating satisfactorily. At an altitude of approximately 20,200km, a constellation of 24 functioning GPS satellites is sufficient to ensure that there will always be at least four satellites visible, at all unobstructed sites on the globe. Typically there are 6 to 10 satellites visible most of the day. The U.S. Department of Defense has undertaken to guarantee 24 satellite coverage 70% of the time, and 21 satellite coverage 98% of the time.

As the GPS satellites are in nearly circular orbits, at an altitude of approximately 20,200km above the earth (Fig. 7.7), this has a number of consequences :

- Their orbital period is approximately 11hrs 58mins, so that each satellite makes two revolutions in one sidereal day (the period taken for the earth to complete one rotation about its axis with respect to the stars).
- At the end of a sidereal day (23hrs 56mins in length) the satellites are again over the same position on earth.
- Reckoned in terms of a solar day (24hrs in length), the satellites are in the same position in the sky about four minutes earlier each day.
- The orbit groundtrack approximately repeats each day, except that there is a very small drift of the orbital plane to the west which is arrested by periodic manoeuvres.

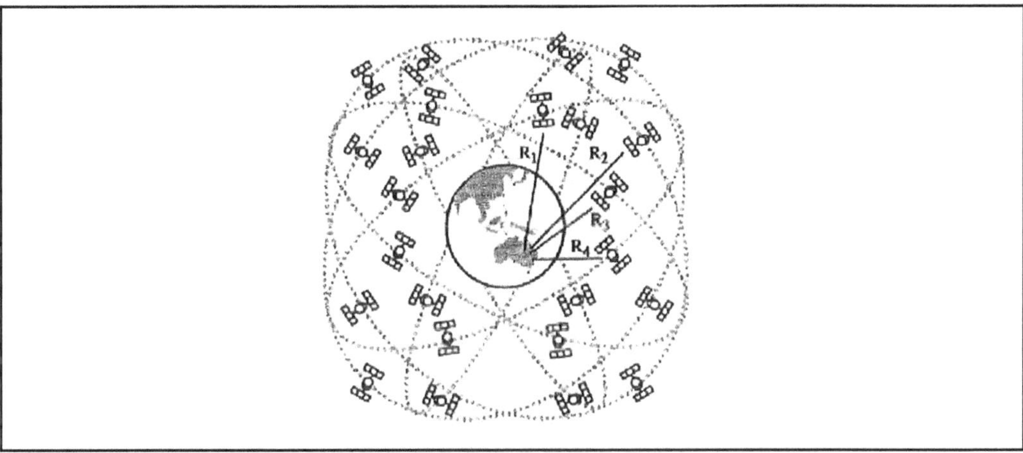

Fig. 7.7 The GPS Constellation "Birdcage".

The following general remarks can be made with regard to satellite constellation design for navigation purposes:

- The higher a satellite, the longer it is visible above the horizon (the extreme case is the geostationary satellites).
- The higher a satellite, the better the coverage due to longer fly-over passes and extended visibility of the satellite across large areas of the earth.
- The higher a satellite, the less the rate-of-change of distance, and the lower the Doppler frequency of a transmitted signal.
- The greater the angle of inclination, the more northerly the track of the sub-satellite point across the surface of the earth.
- No satellite can be seen simultaneously from all locations on the earth.

- Depending on the positioning principles being employed, there may be a requirement for observations to be made to more than one satellite simultaneously from more than one ground station.

The Block II satellites are deployed in six orbital planes at 60o intervals about the equator, with each containing four satellites. The satellites can be moved round their orbits if it becomes necessary to "cover" for a failed satellite. The orbital planes are inclined at an angle of 55^0 to the equatorial plane.

As the satellites are at an altitude of more than three times the earth's radius, a particular satellite may be above an observer's horizon for many hours, perhaps 6-7 hours or more in the one pass. At various times of the day, and at various locations on the surface of the earth, the number of satellites and the length of time they are above an observer's horizon will vary. Although at certain times of the day there may be up to 12 satellites visible simultaneously, there are nevertheless occasional periods of degraded satellite coverage (though naturally their frequency and duration will increase if satellites fail). "Degraded satellite coverage" is generally defined in terms of the magnitude of the Dilution of Precision (DOP) factor, a measure of the quality of satellite geometry. The highest the DOP value, the poorer the satellite geometry. For example, if all the visible satellites are located in the same part of the sky, the intersections of the SOPs will be very obtuse.

7.4.2 GPS Signal Components

The basis of the GPS signal are the two L-band carrier signals. These are generated by multiplying the fundamental frequency f_o (10.23MHz) by 154 and 120, yielding the two microwave L-band carrier waves L1 and L2 respectively. The frequency of the two waves are: $f_{L1} = f_o \times 154 = 1575.42$ MHz, and $f_{L2} = f_o \times 120 = 1227.6$ MHz. These are radio frequency waves capable of transmission through the atmosphere over great distances, but which cannot penetrate solid objects. Note that all GPS satellites transmit carrier waves at the same two L-band frequencies (unlike the GLONASS system, where a different frequency is assigned to each satellite. However, the L-band carrier waves themselves carry no information, and must be modified (or modulated) in some way. In the Global Positioning System the L-band carrier waves are modulated by two ranging codes, and the navigation message. The two distinct GPS ranging codes are :

- The C/A code (sometimes referred to as the "clear/access" or "coarse/ acquisition" code), sometimes also referred to as the "S code".
- The P code (the "private" or "precise" code) was designed for use only by the military, and other authorised users.

The L1 carrier was designed to be modulated with both the P and C/A codes, whereas the L2 carrier would be modulated only with the P code. Under the policy of Anti-Spoofing the P code is encrypted through modulation by a further secret code (the "W code") to produce a new "Y code". Both carrier signals contain the navigation message.

The C/A and P (or Y) codes provide the means by which a GPS receiver can measure one-way ranges to the satellites. These codes have the characteristics of random noise, but are in fact binary codes generated by mathematical algorithms and are therefore referred to as "pseudo-random-noise" or PRN codes. Both the C/A and P code generating algorithms are known, and are based on a simple Tapped Feedback Shift Register scheme. One C/A code is assigned to each GPS satellite (the PRN code number is often used as the satellite I.D.). Each C/A code is a 1023 "chip" long binary sequence, generated at a rate of 1.023 million chips per second (that is, 1.023 MHz), thus the entire C/A code sequence repeats every millisecond. The P code is a far more complex binary sequence of 0's and 1's, being some 267 days long with a chipping rate at the fundamental frequency f_0 (10.23 MHz). The resolution of this code (length of the P code chip) is ten times the resolution of the C/A code. Instead of assigning each satellite a unique code, as is the case with the C/A code, the P code is allocated such that each satellite transmits a one week portion of the 267 day long PRN sequence, restarting the code sequence at the end of each week.

To measure one-way range (from satellite to receiver), a knowledge of the ranging code(s) is required by the user's receiver. Knowing which PRN code is being transmitted by a satellite means that a receiver can generate a local replica of the same code sequence. These PRN codes possess a very important attribute : a given C/A (or P or Y) code will fully correlate with an exact replica of itself only when the two codes are aligned, and has a low degree of correlation with other alignments.

As the same P (or Y) code is synchronously modulated on both carrier waves, any difference in signal transit time of the same PRN sequence is due to the retardation of the two L-band signals by a different amount as they travel through the ionosphere. The effect of the ionosphere on signal propagation is essentially a function of signal frequency, hence measurement on both frequencies is a very effective way of overcoming the ionospheric signal delay.

7.4.3 The Civilian - Military Relationship

Although GPS is a military navigation system the civilian sector represents an important (and rapidly growing) user group that has increasingly lobbied the U.S. Government in order to influence: (a) the direction of GPS system development; (b) official GPS policy concerning enhancement and control; and (c) the design of follow-on systems to GPS for the 21st century. Several policy decisions have already been made which impact on GPS performance. Some of these actions were agreed to during the early system design phase, while others were made only after much of the present system had been deployed :

- Two PRN ranging codes are implemented. The C/A code is intended for general, civilian use, while the P code was reserved for military and other authorised users. Because of the P code's higher measurement resolution it was expected that the accuracy of positioning using the P code would be much better than that possible using the C/A code. However, it was found that the performance of C/A code positioning was often no worse than that of P code positioning by a factor of two. (The latest C/A code tracking technology permits ranging quality almost as good as P code ranging and hence, all other things being equal, positioning accuracy close to that expected from the processing of P code ranges should be possible.)

- These two levels of positioning performance were designed into the GPS system from the very beginning. The positioning service based on using C/A code ranging data is referred to as the "Standard Positioning Service" (SPS), while the service based on P code ranging data is known as the "Precise Positioning Service" (PPS).
- As a result of the demonstration of a surprisingly good level of SPS accuracy, the policy of "Selective Availability" (SA) was endorsed in order to artificially widen the gap between the two levels of positioning. SA is an intentional degradation of the accuracy of GPS horizontal positioning to 100m and vertical positioning at the 160m (at the 95% confidence level) for SPS users. SA is implemented through an encryption of the navigation message whereby a part of the transmitted ephemeris and satellite clock data is falsified (the so-called "epsilon" effect) and the satellite clock is "dithered" (the so-called "delta" effect). SA therefore affects the precision of all measurements, code and carrier phase. SA does not affect PPS users who have user equipment able to decipher the correct ephemeris and clock error data.
- Any GPS hardware manufacturer is able to construct a P code ranging receiver (the P code PRN generation algorithm is published). The non-military market for P code receivers was always assumed to be very small, and one that could be controlled by the U.S. Government through the issuance of "export licences", etc. However, a significant demand by the land surveying market for dual-frequency phase tracking GPS receivers (code-correlating phase tracking receivers need a knowledge of the P code PRN in order to make a carrier phase measurement) led to an expansion in the production of P code capable positioning equipment.
- Under another policy known as "Anti-Spoofing" (AS), access is denied to the P code modulated on both L-band frequencies. AS was implemented on 31 January 1994, through the encryption of a secret "W code" onto the P code. The rationale behind this decision was that by keeping the military PRN code secret, an enemy of the U.S. could not jam the signal using a ground-based transmitter, nor "spoof" military GPS receivers by transmitting a false P code signal from a satellite. However, several GPS receiver manufacturers have developed proprietary techniques for making dualfrequency measurements even in the presence of AS.

- Dual-frequency observations will lead to more accurate positioning results than single frequency observations, because the ionospheric bias can be eliminated from the code range measurements. The fact that the C/A code is only modulated on the L1 carrier is therefore an intentional design decision to ensure that the SPS service cannot deliver the accuracy that the PPS service can, even with SA off.

It must be emphasised that the situation as far as the policies on SA and AS is under almost continuous review, particularly since the Presidential Decision Directive on the "U.S. National GPS Policy" was released in March 1996.

7.4.4 Why is the GPS Signal so Complicated?

GPS was designed as a complex military navigation system, which would also be used by civilians, hence we can summarise the reasons why the signal is so complicated as:

(a) The requirement for GPS to be multi-user system is most easily satisfied if it designed as a listenonly, or self-positioning system. An unlimited number of users could be supported because no return signal was necessary, however, the measurements and positioning procedures must then be based on one-way (satellite-to-receiver) signals.

(b) The very important requirement for real-time positioning influenced a number of design criteria related to the satellite signals:

- Simultaneous measurements from many satellites, but all signals are at the same frequency (though Doppler-shifted) — need to identify signals by use of different codes
- Unambiguous range measurements were required — need to determine signal delay
- Satellite positions are needed — broadcast ephemerides

Hence the PRN codes were used to distinguish the different satellite signals. The measurement of a "range" to a particular satellite required that a comparison be made of the PRN code generated within the receiver (corresponding to the satellite in question) to the PRN code sequence contained within the incoming satellite signal.

(c) The requirement for GPS to support high accuracy positioning also had implications for satellite signal design:

- High frequency modulation — P code at 10 MHz
- Dual-frequency signal — permits ionospheric delay estimation
- Microwave carrier frequency — 1.2 to 1.6 GHz

(d) The anti-jamming requirement could be partly met through the use of spread-spectrum codes. This also fulfilled the requirement for a low power signal that could still be detected even though it was below the ambient signal noise level. By ensuring that some of the PRN codes are secret (known only to military and a few authorised users), jamming or "spoofing" of signals was made much more difficult. (However, it is possible to jam a GPS receiver if the jamming source is within several metres of the receiver.)

(e) As the system was intended as a "dual-use" technology, satisfying both military and civilian users, this had several implications the most important of which was that the civilians were not to have the same level of accuracy as the military users :

- The use of two PRN codes — P and C/A code
- Restriction on the dual-frequency signal for civilians — C/A code only on the L1 frequency
- Microwave carrier frequency — 1.2 to 1.6 GHz

7.4.5 GPS Satellite Ephemerides

How are the GPS satellite ephemerides computed? As the forces of gravitational and non-gravitational origin perturb the motion of the GPS satellites, the coordinates of the satellites in relation to the WGS84 reference system must be continually determined through the analysis of tracking data. In the case of the GPS broadcast ephemeris, this procedure is a three-step process (Langley, R. B, 1991a)

- An off-line orbit determination is performed through the analysis of tracking to generate a "reference orbit". This is an initial estimate of the satellite trajectories computed from about one week's of tracking from the five Control Segment monitor stations.
- An on-line daily updating of the reference orbit using a Kalman filter as new data are added. This provides the current estimates of the satellite orbit which is used to predict the future orbit.
- The ephemeris is estimated for 1 to 14 days into the future. To obtain the necessary broadcast information, curve fits are made to 4 to 6 hour portions of the extrapolated ephemeris, and hourly orbit parameters determined.

Note that these parameters are not true Keplerian elements as they only describe the ephemeris over the interval of applicability and not for the whole orbit. (Although only intended for use during the transmission period, they do, however, adequately describe the orbit over intervals of 1.5 to 5 or more hours, with a graceful degradation in accuracy.). The user can derive the earth-centred, earth-fixed WGS84 Cartesian coordinates of the GPS satellite from the broadcast orbital parameters, using an algorithm described in (Vandierendonck, A. J., et.al, 1980) and implemented in every GPS receiver.

7.5 GPS Measurements

7.5.1 The Transmitted Signal

The signal that actually leaves a GPS satellite antenna is a combination of the three components: carrier wave, ranging codes and navigation message. The generation of the signal to be transmitted is carried out in a number of steps, and relies on the fact that all the components are derived by multiplying or dividing the fundamental frequency (Fig. 7.2). There are two distinct procedures for the combination of signal components: (a) BPSK modulation of binary sequences onto the carrier waves, and (b) modulo-2 addition of binary sequences.

Biphase Shift Key Modulation (BPSK) is the technique used to added a binary signal to a sine wave carrier. This amounts to causing a 180° phase shift in the carrier each time the binary sequence undergoes a transition from "0" to "1", or "1" to "0". The P code plus navigation message is modulated on both the L1 and L2 carriers, while the C/A code plus navigation message is only modulated on the L1 carrier.

An example of binary-to-binary modification of codes is the modulo-2 addition of the navigation message data to the C/A code. Because of the difference in frequency between the C/A code and the message stream this has the effect of inverting 20 C/A code binary "states" whenever the data bit of the navigation message is equal to "1". Conversely, when the data bit is "0" the C/A code sequence remains unaffected. The same satellite message is also modulated onto the P code sequence using this modulo-2 addition procedure. There are two main types of measurements that can be made on the GPS signals (Langley, R. B, 1993) :

- range observations based on the PRN codes, sometimes referred to as "code range" or "code phase", and
- carrier phase observations, which are more precise range-type measurements, but which have a much higher degree of "ambiguity" than the code ranges.

7.5.2 The GPS Range Measurements

The PRN codes are accurate time marks that permit the receiver's navigation computer to determine the time of transmission for any portion of the satellite signal. Before considering how ranging is carried out we need to describe, in general terms, how the incoming satellite signal is processed within a GPS receiver. (Parkinson, 1994 and Kaplen 1996)

Neglecting any discussion on extraneous noise (in relation to which PRN codes have special properties), and assuming that the only satellite signals received at the antenna originate from one GPS satellite, the following simplistic procedure is carried out within a receiver "channel". The L1 carrier modulated by the C/A code is converted to a signal of lower frequency and then mixed with a locally on a different time scale (due to nonsynchronization of receiver clock to GPS Time and the travel time of the signal from the satellite to antenna) to that of the incoming C/A code. As soon as the incoming signal and the receiver C/A code sequences are aligned within the receiver the ones and zeros of the two codes cancel, leaving the received carrier modulated only by the binary navigation message.

The extraction of the code range, or more precisely the determination of the amount by which the receiver generated PRN code must be shifted to align it with the incoming signal, is carried out with the aid of a PRN code correlator in some form "delay-lock loop" scheme. How accurately is this carried out? The C/A code has a sequence rate of 1.023 Mbps, corresponding to a resolution of about 300m (speed of light divided by the frequency). The P (or Y) code, on the other hand, has a socalled "chip" rate of 10.23 Mbps, and hence an effective resolution of about 30m. As a rule-of-thumb, alignment of the incoming and receiver generated codes is generally possible to within about 1-2% of the chipping rate, hence the measurement precision of C/A code ranging is of the order of 3-5m, and for P code ranging it is of the order of 0.3-0.5m. However, under the policy of "Anti-Spoofing", the Y code is encrypted and is therefore not available to civilian applications.

It should be noted, however, that modern "narrow correlator" C/A code measurement technology as implemented in several "top-end" GPS receivers has demonstrated ten times better correlation performance for the C/A code correlation than that quoted above.

7.5.3 The GPS Carrier Phase Measurements

The wavelengths of the carrier waves are very short (approximately 19cm for L1 and 24cm for L2) compared to the C/A and P code wavelengths. Assuming a measurement resolution of 1-2% of the wavelength, this means that carrier phase can be measured to millimetre precision compared with a few metres for C/A code measurements. Unfortunately a phase measurement is ambiguous as it cannot discriminate one (L1 or L2) wavelength from another. In other words, time of transmission information for the L-band signal cannot be imprinted onto the carrier wave as is done using PRN codes (this would be possible only if the PRN code frequency was the same as the carrier wave, rather than 154 or 120 times lower in the case of the P code, and 1540 or 1200 times lower for the C/A code). The basic phase measurement is therefore an angle in the range 0o to 360o. It is nevertheless the basis for high precision GPS positioning.

7.5.4 Ranging Using PRN Codes

Consider for a moment a perfect system, where all satellite clocks are synchronized to the same time system: the GPS Time (GPST). Furthermore, the ground receiver's clock also maintains the same synchronization, and none of the clocks drift from this GPST scale. Now suppose the satellite starts transmitting its L1 carrier (modulated with the combined C/A PRN code and navigation data). At the same instant, the receiver begins generating the C/A PRN code corresponding to that particular satellite. Under these circumstances the satellite and receiver generated C/A codes would be output in unison. When the satellite signal is received, however, it will be lagging the receiver generated code due to the signal transit time. Multiplying the time offset required to align the two codes (in effect determining the signal transit time) by the speed of light yields the satellite to receiver distance.

Measuring ranges simultaneously in this fashion to three satellites would fix one's position at the intersection of three spheres of known radii (the satellite ranges), centred at each satellite whose coordinates can be calculated from the navigation message. In reality the situation is more complex :

- GPS receivers are equipped with crystal clocks that do not keep the same time as the more stable satellite clocks (the satellite clocks can be nearly synchronised to GPST using the clock correction model transmitted in the navigation message). Consequently each range is contaminated by the receiver clock error. This range quantity is therefore referred to as pseudo-range, and in order for the user to derive position from pseudo-range data, the receiver equipment is required to track (a minimum of) four satellites, and solve for four unknown quantities: the three-dimensional position components and the receiver-clock offset (from GPST). This is the basis of GPS real-time navigation, and why GPS could be considered an example of a time-differenceof- arrival system.
- There is in fact a 300 km "ambiguity" in the C/A code pseudo-range measurements (300 km is the approximate length of the C/A code sequence). That is, all measured "distances" appear to have a range of 0 to 300 km. This ambiguity is resolved in a number of ways, but the easiest to assume that if the approximate receiver position is known to within say 100 km, the "missing" component of the distance can be determined, and hence the raw pseudo-range measurement can be corrected for this ambiguity to obtain the full satellite-receiver distance.
- Ranging (and hence receiver position determination) can be carried out using the C/A code or the P code. P code ranging can be done on the combination of the two frequencies, hence eliminating the bias due to ionospheric refraction. Furthermore, the C/A code is "coarser", and hence the C/A derived ranges are subject to greater measurement "noise". The absence of a C/A code on L2 is intentional, as one of the accuracy limitations of the GPS system for the general class of civilian users.
- As previously mentioned, this differentiation between ranging codes, and the formulation of policies for their use (in peacetime and in times of global emergencies), is responsible for the provision of two GPS services: The Precise Positioning Service based on P or Y code (dual-frequency) ranging, and the Standard Positioning Service based on single frequency C/A code ranging.

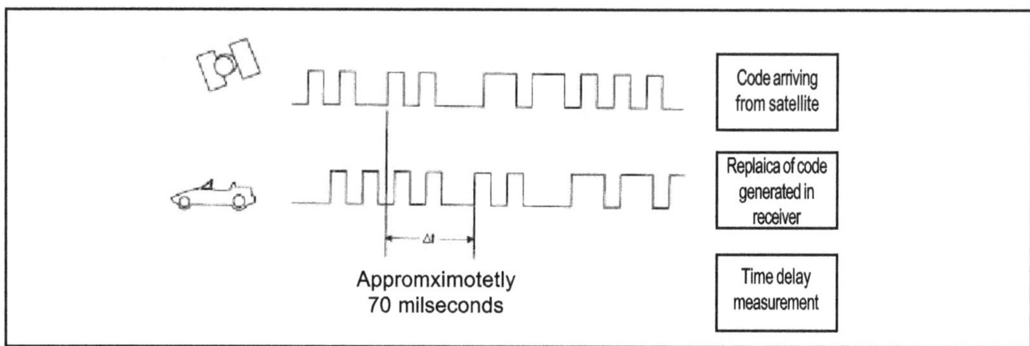

Fig. 7.8 One-Way Ranging Using PRN Codes.

7.5.5 An Observation Model of the Pseudo-Range

The observation equation for a receiver-clock-biased range (Langley, R. B. 1991d)is

$$p = \rho + \varepsilon_{rc}(t_r) . c \qquad \text{..........(7.1)}$$

where c is the velocity of electromagnetic radiation in a vacuum (or simply the "velocity of light"), erc is the receiver clock error caused by the receiver oscillator (assume satellite clock time is "true" time) at time of reception t_r, p is the measured range and ρ is the true "geometric" range. Each observation made by the receiver can be parameterised as follows :

$$(x^s - x)^2 + (y^s - y)^2 + (z^s - z)^2 = (\rho - \varepsilon_{rc} . c)^2 \qquad \text{..........(7.2)}$$

where x^s, y^s, z^s is the coordinate of the satellite and x, y, z is the coordinate of the receiver. Note that the time argument has been discarded (satellite coordinates are time-dependent, and so are the receiver coordinates if the receiver is moving).

As the 3-D coordinate of the satellite is known, then each measurement p contains four parameters which may be considered unknown: the 3-D coordinate of the receiver (x, y, z) and the receiver clock error (ε_{rc}). By making four measurements, to four different satellites, the following system of equations is obtained:

$$(x^{s1} - x)^2 + (y^{s1} - y)^2 + (z^{s1} - z)^2 = (p^1 - \varepsilon_{rc} . c)^2$$
$$(x^{s2} - x)^2 + (y^{s2} - y)^2 + (z^{s2} - z)^2 = (p^2 - \varepsilon_{rc} . c)^2$$
$$(x^{s3} - x)^2 + (y^{s3} - y)^2 + (z^{s3} - z)^2 = (p^3 - \varepsilon_{rc} . c)^2$$
$$(x^{s4} - x)^2 + (y^{s4} - y)^2 + (z^{s4} - z)^2 = (p^4 - \varepsilon_{rc} . c)^2 \qquad \text{..........(7.3)}$$

which has a unique solution **(see, for example [15]).** If more than four measurements are made the method of Least Squares, or other estimation techniques, can be used to determine the optimum solution. Does the receiver clock error have to be estimated at each epoch? That depends upon such factors as :

- How well the clock error is estimated.
- How often the position solution is carried out.
- The stability of the clock.

Under the assumptions that (a) there is a range measurement uncertainty of about 1 metre attributable to receiver "noise", and (b) the receiver is equipped with a quartz crystal clock (stability of 0.1 nanoseconds/second), then the uncertainty of the clock time after 30 seconds is as great as the range measurement error. Hence, once the clock error were determined, it would have to be independently reestimated at least every 30 seconds, otherwise it would dominate the range error budget. This is one of the reasons why this procedure could not be used if pseudo-range measurements to different satellites were not made "simultaneously" (here, within 30 seconds of each other), so that the clock error could be assumed to be a constant.

This is the conventional strategy used for pseudo-range-based GPS positioning: the receiver clock error is simply treated as an additional unknown and the navigation problem can be considered as 4-D estimation. However, the GPS satellite clock error is assumed to be a known quantity, and parameters defining this clock error are transmitted in the navigation message. The ionospheric delay is also modelled using transmitted parameters. An estimate of the tropospheric delay may be obtained from a simple tropospheric model such as Hopfield. All other biases are assumed to be insignificant compared to the measurement noise.

7.5.6 Coping with the Satellite Clock Bias

As the satellite clock error is the largest source of GPS measurement bias it deserves closer study. Under the assumption that the satellite clock error is an unknown quantity, the observation equation for such a satellite-biased range :

$$p = \rho + \varepsilon^{sc}(T^s) \,.\, c \qquad \text{..........(7.4)}$$

ε^{sc} is the satellite clock error caused by the satellite oscillator not being synchronized to "true" time (GPST). p is the measured range, ρ is the true range and T^s is the time of transmission. Each observation made by the receiver can be parameterised as in equation (1.1), except for the replacement of the term ε_{sc} for ε_{rc} :

$$(x^s - x)^2 + (y^s - y)^2 + (z^s - z)^2 = (p - \varepsilon_{sc} \,.\, c)^2 \qquad \text{..........(7.5)}$$

The 3-D coordinate of the satellite signal transmitter (x^s, y^s, z^s) is known, hence in the case of three range observations there are six unknowns in the system: the 3-D coordinate of the receiver (x_{r1}, y_{r1}, z_{r1}) and the three satellite clock error terms (ε^{sc1}, ε^{sc2}, ε^{sc3}). Hence, at first glance, six satellite-biased range observations are required to solve this positioning problem. It is not, however, possible simply to make observations to more satellites as each new observation introduces a new satellite clock parameter. There are two options for overcoming this dilemma.

It is possible to take advantage of the fact that all observations made to a particular satellite are biased by the same amount (if made at the same time, or close enough together so that the satellite clock error can be assumed to have not changed by a significant amount). If three range observations are made from another station, whose coordinate is known (x_{r2}, y_{r2}, z_{r2}), then it is possible to obtain a system of six equations in six unknowns:

$$(x^{s1} - x_{r1})^2 + (y^{s1} - y_{r1})^2 + (z^{s1} - z_{r1})^2 = (p_{r1}{}^{s1} - \varepsilon^{sc1} . c)^2$$
$$(x^{s2} - x_{r1})^2 + (y^{s2} - y_{r1})^2 + (z^{s2} - z_{r1})^2 = (p_{r1}{}^{s2} - \varepsilon^{sc2} . c)^2$$
$$(x^{s3} - x_{r1})^2 + (y^{s3} - y_{r1})^2 + (z^{s3} - z^{r1})^2 = (p_{r1}{}^{s3} - \varepsilon^{sc3} . c)^2 \quad \text{..........(7.6)}$$
$$(x^{s1} - x_{r2})^2 + (y^{s1} - y_{r2})^2 + (z^{s1} - z^{r2})^2 = (p_{r2}{}^{s1} - \varepsilon^{sc1} . c)^2$$
$$(x^{s2} - x_{r2})^2 + (y^{s2} - y_{r2})^2 + (z^{s2} - z^{r2})^2 = (p_{r2}{}^{s2} - \varepsilon^{sc2} . c)^2$$
$$(x^{s3} - x_{r2})^2 + (y^{s3} - y_{r2})^2 + (z^{s3} - z^{r2})^2 = (p_{r2}{}^{s3} - \varepsilon^{sc3} . c)^2$$

for which a unique solution can be obtained. In a conceptual sense this is the basis of differential GPS positioning, although in reality there are several possible implementations.

The other strategy for accounting for satellite clock error is for the GPS operators to periodically determine the clock error. As the satellite clocks have significantly better long-term drift characteristics than the receiver clocks, a suitable clock error model could be a time polynomial :

$$\varepsilon^{sc} = a_0 + a_1 (t - t_{oc}) + a_2 (t - t_{oc})^2 \quad \text{.........(7.7)}$$

where :

- a_0 is the clock bias term
- a_1 is the clock drift term
- a_2 is the clock drift-rate
- t is satellite clock time
- t_{oc} is some reference epoch for the definition of the coefficients

What is actually available to users via the navigation message is a prediction of the satellite clock error behaviour for some time into the future (24 hours or more). Such deterministic models of satellite clock error are accurate to about 20 nanoseconds, or about six metres in equivalent range, depending upon the time since last navigation message update. Hence the measured pseudo-ranges may be corrected for satellite clock error, and then the observation model in equation (7.1) can be used.

Selective Availability complicates matters because it is a further artificial "dithering" of the satellite clock causing an additional several tens of metres error in the pseudo-range measurement.

7.6 GPS Instrumentation

The following components of a generic GPS receiver can be identified (Fig. 7.9):

- ***Antenna and Preamplifier :*** Antennas used for GPS receivers have broadbeam characteristics, thus they do not have to be pointed to the signal source like satellite TV receiving dishes. The antennas are compact and a variety of designs are possible. There is a trend to integrating the antenna assembly with the receiver electronics.

- ***Radio Frequency Section and Computer Processor :*** The RF section contains the signal processing electronics. Different receiver types use somewhat different techniques to process the signal. There is a powerful processor onboard not only to carry out computations such as extracting the ephemerides and determining the elevation/azimuth of the satellites, etc., but also to control the tracking and measurement function within modern digital circuits, and in some cases to carry out digital signal processing.
- ***Control Unit Interface :*** The control unit enables the operator to interact with the microprocessor. Its size and type varies greatly for different receivers, ranging from a handheld unit to soft keys surrounding an LCD screen fixed to the receiver "box".
- ***Recording Device :*** in the case of GPS receivers intended for specialised uses such as the surveying the measured data must be stored in some way for later data processing. In the case of ITS applications such as the logging of vehicle movement, only the GPS-derived coordinates and velocity may be recorded. A variety of storage devices were utilised in the past, including cassette and tape recorders, floppy disks and computer tapes, etc., but these days almost all receivers utilise solid state (RAM) memory or removable memory "cards".
- ***Power Supply :*** Transportable GPS receivers these days need low voltage DC power. The trend towards more energy efficient instrumentation is a strong one and most GPS receivers operate from a number of power sources, including internal NiCad or Lithium batteries, external batteries such as wet cell car batteries, or from mains power.

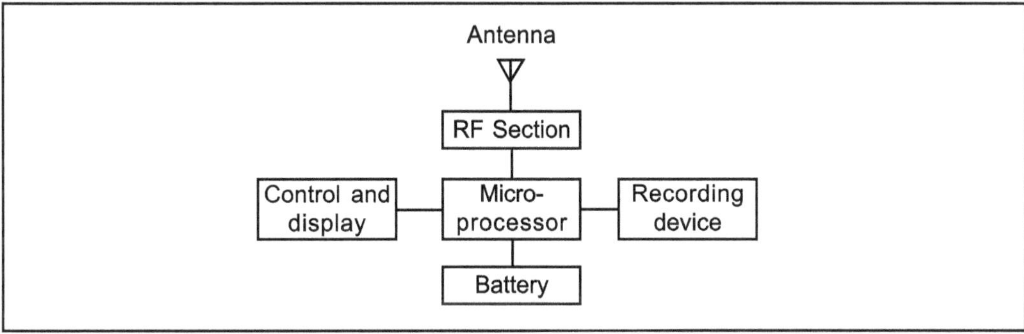

Fig. 7.9 The Generic GPS Receiver.

The antenna and RF technology components are briefly discussed below. For further details the reader is referred to (Wells, D. E, et.al. 1987).

7.6.1 Antennas

The task of the antenna is to convert the energy of the arriving electromagnetic waves into an electric current that can be processed by the receiver electronics. There are a number of special considerations as far as antenna design is concerned:

- the antenna must be able to pick and discriminate very weak signals,
- the antenna may need to operate at just the L1 frequency, or at both the L1 and L2 frequencies,
- as the signals are right-hand circularly polarised, the GPS antenna must also be right-hand circularly polarised,
- antenna gain pattern that enhances the ability of the RF section to discriminate against multipath signals (such as left-hand circularly polarised signals),
- a stable electrical centre,
- low cost, and
- reliable.

There have been several types of GPS antennas used:

- monopole or dipole configurations,
- quadrifilar helices,
- spiral helices,
- microstrip,
- choke ring, and other multipath resistant designs

In short, the antennas are required to be rugged, simple in construction, have stable phase centres, be resistant to multipath, have good gain and pattern coverage characteristics. The microstrip antenna is almost universally used for navigation applications.

7.6.2 Signal Processing Principles for Code-Correlating Receivers

The RF section of a GPS receiver converts the incoming (preamplified) signal to a signal of a more manageable frequency. This intermediate frequency is obtained by mixing the incoming signal with a pure sinusoidal signal generated by the local oscillator (the quartz "clock"). The frequency of this beat frequency is the difference between the original (Doppler-shifted) received carrier frequency and the local oscillator. The intermediate or beat frequency is then processed by the signal tracking circuitry.

There are several classes of signal processing techniques that can be employed to make observations, as well as several proprietary implementations of tracking technologies. This is particularly the case for instruments which track carrier phase. It is beyond the scope of this chapter to discuss in detail the electronic circuitry and the reader is referred to Kaplen 1996, and Spilkar 1995. As our interest is the pseudo-range measuring instruments, we need only consider the issues of the "code-correlating" signal processing technique, and some brief details of the processing architecture within the receiver, and in particular the form of the tracking "channel(s)". Code-correlating receivers employ tracking loops to extract the useful measurements from the beat signal. A typical GPS receiver contains two types of tracking loops :

- the delay-lock, or code-tracking, loop, and
- the phase-lock, or carrier tracking, loop.

The delay-lock loop is used to make the alignment of the PRN code sequence (C/A or P code) that is contained in the signal from a satellite with an identical PRN code generated within the receiver. A correlator in the delay-lock loop continuously cross-correlates the two code streams, time shifting the receiver generated stream until alignment is obtained. The time shift is then converted to a pseudo-range measurement. Once the code-tracking loop is aligned, the PRN code can be stripped from the satellite signal. The resulting signal then passes to the phase-lock loop where the satellite message is extracted. Once the local oscillator is locked onto the satellite signal it will continue to follow the variations in the phase of the carrier as the satellite-receiver distance changes.

A GPS receiver may be described as continuous or switching depending on the type of channel(s) it has. A continuous-tracking receiver has dedicated hardware channels, and each channel tracks a single satellite, maintaining continuous code and/or phase lock on the signal. Each channel is controlled and sampled by the receiver's microprocessor with input/output operations being performed fast enough so that tracking is not disturbed. Continuous-tracking receivers may enjoy a signal-to-noise advantage over switching receivers in that the satellite signal is continuously available and may be more frequently sampled. A further advantage is a potential redundancy capability. Should one of the hardware channels fail, it may still be possible to obtain sufficient data to determine a position. One disadvantage of a multi-channel receiver is that the differences in signal path delay in the channels, the so-called inter-channel biases, must be well calibrated. This is the most common architecture used in GPS receivers today.

A switching receiver has hardware channels which sequentially sample the incoming signal from more than one satellite. There may be a single channel for all satellites, or each channel may track, say, two satellites. Code and/or carrier phase tracking for the individual signals is controlled by software (or, more usually, the "firmware") within the microprocessor. As a result, greater demands are placed on the microprocessor in a switching receiver and its programming is necessarily more complex — in effect hardware complexity is exchanged for software complexity. If the cost of hardware components is a significant factor in determining the selling price of a receiver then in principle, the cost of a switching receiver should be cheaper than a continuous receiver (this is certainly the case for the low cost GPS navigation units).

There are three basic kinds of switching receivers, distinguished by the time required to sequence through the signals tracked by a particular channel. A multiplexing channel is one for which the sequencing time to sample all satellites assigned to the channel is equal to 20 milliseconds, the period of one bit in the satellite navigation message. The sampling can be arranged so that no message bit boundary is spanned by any tracking interval. In this way, the messages from all satellites phase tracked by the channel can be read simultaneously. A multiplexing channel can be used to obtain both L1 and L2 data by alternating between the frequencies every 20 milliseconds. If a channel switches between signals at a rate which is asynchronous with the message bit rate, the channel is referred to as a sequencing channel. A fast sequencing channel is one which takes the order of a second or so to sequence through the signals. A slow sequencing channel may take several seconds or even minutes. A single sequencing channel would lose bits in a particular satellite message during those intervals spent sampling the signals from other satellites. Consequently, sequencing receivers may have an extra hardware channel just for message decoding. Alternatively, the navigation message must be decoded before the receiver starts the tracking cycle for real-time positioning (note that the message only changes once an hour).

The configuration of channels is selected so that, for example in the case of a navigation receiver, a minimum of four satellites can be tracked at the same time. This may call for a single switching channel, or two switching (between two satellite signals) channels, or even five channels (one used for calibrating the other four channels, and decoding the navigation message).

7.6.3 Trends In GPS Instrumentation

It is impossible to precisely predict trends in GPS instrumentation. The task is no easier if we focus our attention only to a specific market segment, such as positioning hardware for ITS applications. Nevertheless, speculation based on R&D activities being undertaken at present gives us some clues :

- The third generation of GPS receivers presently available already exhibit significant gains in miniaturisation, reduction in power consumption, and portability, over the earlier models. We can expect this trend to continue.
- The choice confronting GPS manufacturers is whether to maintain a broad-based development program, and market a product that is versatile enough to satisfy many applications (including the provision of interchangeable components such as antennas, RF units, memory, etc., to accomplish this), or to focus on specialist applications (military, navigation, differential navigation, surveying, kinematic, etc.). To date, most manufacturers are split between those with products for the high precision surveying market, and those focussing on the low cost, high volume navigation market. Only a few address all markets.

- There have been many predictions of low cost GPS receivers. However there is an enormous range of prices from the "top-of-the-line" surveying receivers to the "bare-bones" GPS "engines" implemented on one or more chips. GPS boardsets with basic navigation functionality are available at less than one hundred U.S. dollars each (when purchased in high volumes).
- There is a trend towards product differentiation, with many different configurations of tracking channels, data recording options, and, in particular, software options. Although some of these trends may be due to manufacturers wanting to give their product "an edge" in the marketplace, it is equally valid to suggest that this is in response to the different demands (some very specialist) of the market. However, even in the case of complex ITS systems, it is possible to identify the "basic GPS component", and distinguish it from the "add-ons" .
- An exciting marriage is possible between satellite navigation and satellite communication, combining real-time positioning with instantaneous transmission of position. GPS receivers may be fitted to many different platforms (some unmanned, for example rail rolling stock and ship cargo containers), and their locations remotely monitored at a central site via a satcom link. In addition, for land navigation these could include electronic map displays to aid the driver. What is clear is that for many applications the navigation data provided by a GPS receiver will merely be the first link in a complex information system.
- New tracking electronics, such as "narrow-correlator" technology, would improve pseudo-range measurement precisions, as would a combination of phase and pseudo-range measurement. Both are, however, still relatively expensive technologies and are not found on "bare-bones" GPS boardsets. Further electronic tracking refinements will lead to more multipath-resistant receivers.

8
Map Projections

8.1 Introduction

The earth is round; Maps are flat. If a particular map is to show only a very small portion of the earth, the roundness of the earth will be insignificant. If on the other hand the map is to show a large portion of the earth on a small scale, the roundness of earth will present a major problem. Some kind of deformation will be necessary.

8.2 Evolution of Map Projection

Figure 8.1 shows the evolution of map projection. If all points within some large portion of the earth are to be represented on a flat map, two transformations of the sphere or spheroid (ellipsoid) must take place. First there must be a scale reduction to make such a huge area fit into the limits of the paper (plane surface); Second, there must be a systematic way of deforming the rounded surface of the sphere or spheroid to make it flat. Figure 8.1 shows graphically the two transformations. There are infinite number of ways, literally, to accomplish the second step.

8.3 Reference Lines on Earth

Certain reference lines are points have been established on the earth. The equator and the two poles are known to all. Lines running north - south from pole to pole, are *meridians*. One of them passing through Greenwich, England, has been chosen

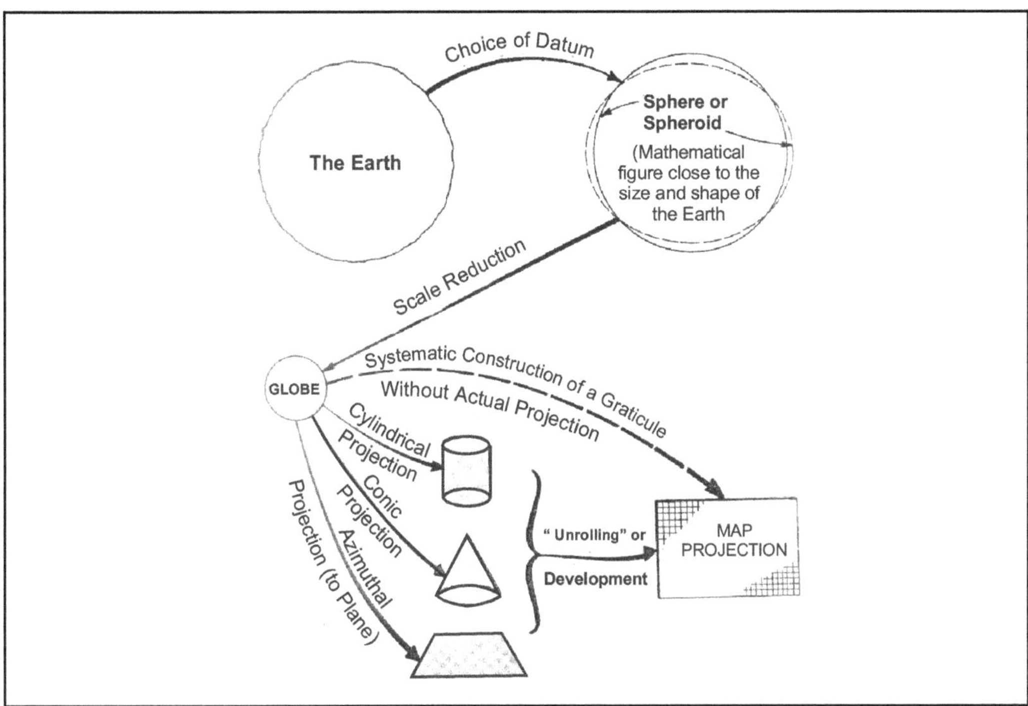

Fig. 8.1 Evolution of map projection. Dashed arrow represents developable surfaces other than cylindrical, conical or azimuthal.

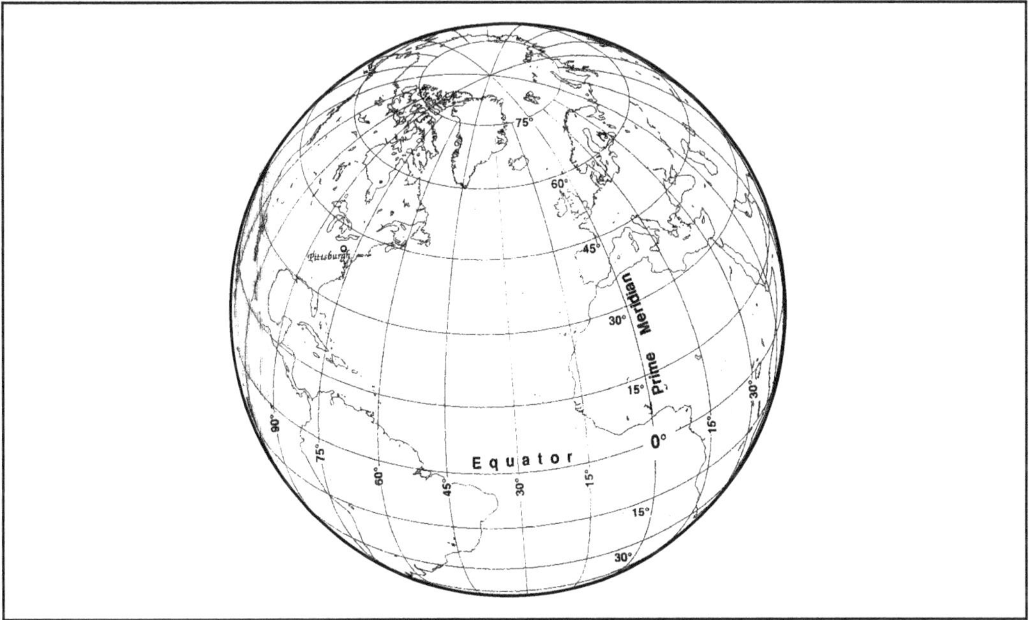

Fig. 8.2 The network of meridians and parallels called "the graticule".

arbitrarily to be the *Prime Meridian*. It is assigned the angular value 0°. The other meridians are identified by their distance, in angular units, east or west of the Prime Meridian. The line crossing all the meridians at right angles and running parallel to the equator are called parallels. They are identified by their angular distance north or south of equator. The co ordinates of a place on the earth is denoted by the meridian of longitude (denoted by the Greek letter (λ - lambda) and parallels of lattitude (denoted by the Greek letter (ϕ - phi). Fig. 8.2 shows network of meridians and parallels on the earth. The co-ordinates of Pittsburgh is denoted by = λ 40°N and ϕ = 80°W.

8.4 Developable Projection Surfaces

The network of meridians and parallels is called graticule. The graticule becomes reference network for all points on the globe just as it is on the earth itself. A point on the globle having a certain ϕ and λ can be transferred to the surrounding surface, which then may be "un-rolled" or developed to form a plane map. The functions used for transformation are purely mathematical and bear some resemblence to the geometrically projected cases as shown in Fig. 8.3(a) to (c). Fig. 8.3(a) : Cylinder, tangent to globe at the equator and developed or "unrolled" to map. (Cylindrical Projection) Fig. 8.3(b) : Cone, tangent to globe along a parallel and developed or "unrolled" to map (conical projection) Fig. 8.3(c) Plane tangent to globe at the north pole and part of the resulting map. (Azimuthal Projection)

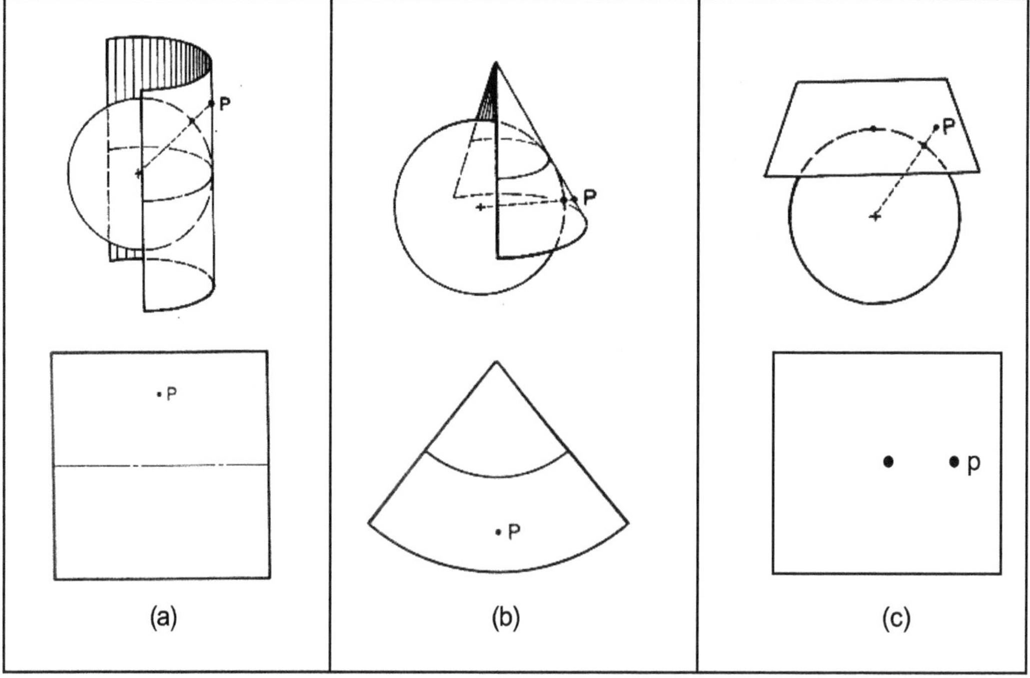

Fig. 8.3 Developable projection surfaces. In all cases point 'P' on the globle is projected to the surface.

8.5 Characteristics of Map Projections

It is now clear that an infinite number of Map Projections are possible. About 15 or 20 of them are in common use. It is more common to classify the projections as cylindrical, conic and azimuthal as discussed in previous section important characteristics that some projections have in common are discussed below.

8.5.1 Standard Lines

Standard lines are defined as lines which do not change length when projected from a generating globe to a map. Some projections have only one standard line but other have more than one. Several of the conic projections share a characteristic of having two parallels.

8.5.2 Equidistant Projections

No map projection can offer a uniform scale. But some of them have it in one direction. A projection can have a scale factor in the north south direction (for example). All meridians are standard. This kind of projection is classified as equidistant projections.

8.5.3 Azimuthal Projections

The characteristic which all azimuthal projections share; namely; that directions to all points with respect to a central point are not deformed during projection from globe to map.

8.5.4 Equal Area Projections

If the relative size of all features on a generating globe is maintained during the process of projection to a map, the projection is said to be equal - area (or equivalent). If area is to be preserved, then scale cannot be. In such a case, given a feature on the globe, the plotting will have to be done with scale factor greater than 1.000 in one direction and less than 1.000 in another direction. Such "Compensatory Scale Factors" on an equal area projection will occur in perpendicular directions called "Principal directions" Example : If a circle with radius of 1cm is drawn upon the surface of a large generating globe, it will appear as an ellipse when projected to an equal area map, as shown in Fig. 8.4, 8.5 and 8.6. One semi axis is reduced to 0.5 cm, and the other will be increased to 2.0 cm in order that the ellipse may contain the same area as the circle.

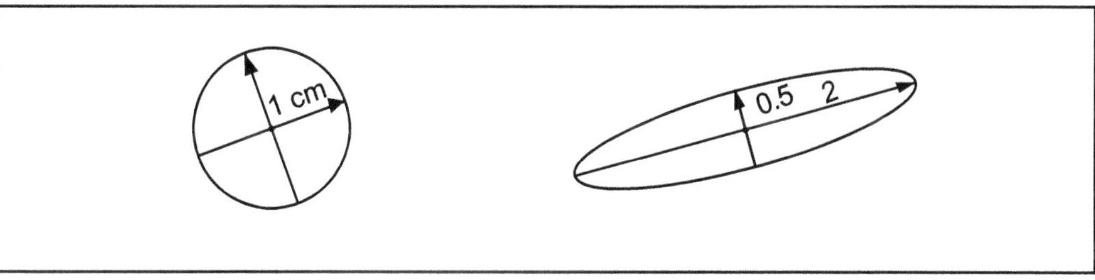

Fig. 8.4 Equal area projection. A circle on globe will project as an ellipse on map

Fig. 8.5 Conformal projection. A circle on a globe will project as another circle with different scale.

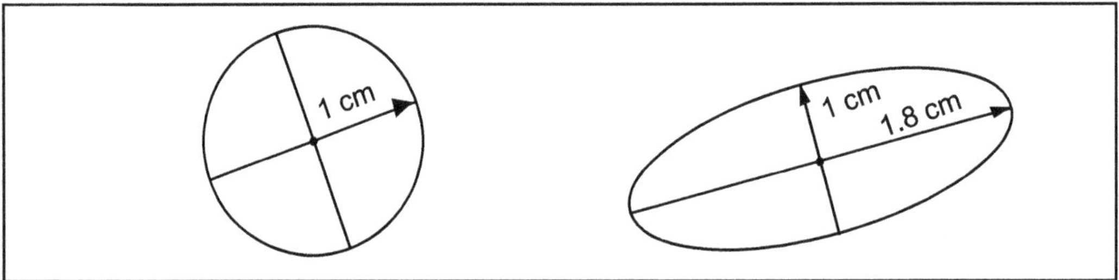

Fig 8.6 Neither equal area nor conformal. Circle gets distorted.

8.5.5 Conformal Projections

While equal area projections have compensatory scale factors, to avoid change in its area, conformal projections equal scale factors in all directions at any one point. A tiny circle is not distorted, but it becomes larger or smaller circle depending on its location on its map. Conformal projections preserve *shape* instead of size. They are also called orthomorphic projections. Conformal projection are ideal for setting up plane co-ordinate system where area of survey is small compared to portion of the earth on a particular system. The angles and scale factor will be same for the entire area.

8.5.6 Treatment of both Properties

A map cannot be both equal area and conformal. It cannot have minimum and maximum scale factors at a point which are compensatory and yet equal, some projections are neither equal area nor conformal. The circle referred in earlier sections is distored in both size and shape. (Fig 8.6).

8.6 Equidistant Projections with one Standard Parallel

It is possible to generate very simple equidistant projections on three basic projection surfaces the cylinder, cone and plane. Each case there is one standard parallel, all meridians are standard (scale factor is 1.000 in north - south directions). The projections discussed in this chapter are cylindrical, conical and polar azimuthal.

8.6.1 Cylindrical Equidistant Projection with one Standard Parallel

Cylinder is wrapped all the way round the generating globe. The standard parallel is the equator. The length is $2\pi R$. The meridians are standard also and drawn as straight vertical lines of length πR. Fig. 8.7 shows the projection of the whole world in this projection.

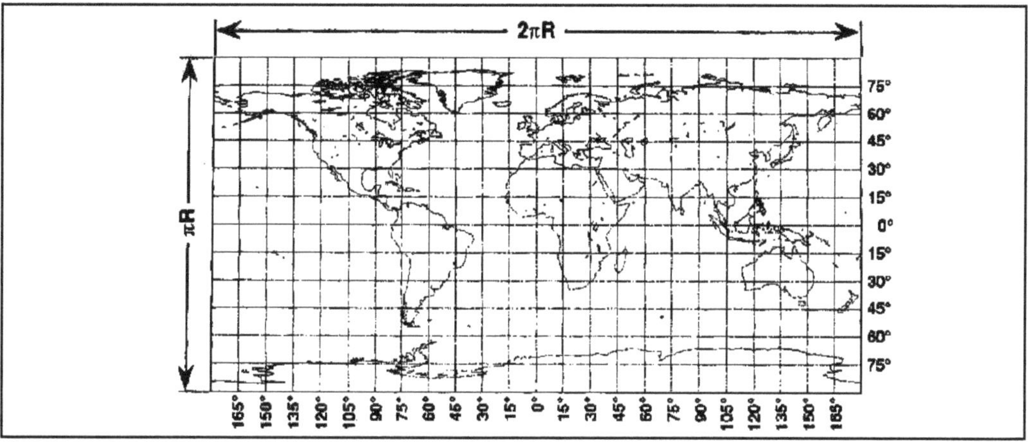

Fig. 8.7 Cylindrical equidistant Projection with one standard parallel

8.6.2 Conical Equidistant Projection

The vertex of the cone (A) to the point of tangency (T) is calculated as follows : Refer Fig. 8.8(a).

$$AT = \frac{R}{\tan\phi} = R\cot\phi$$

The distance AT is used for drawing the standard parallel on the map. The radius of the standard parallel on the globe is r cos ϕ and hence its length is 2π R cos ϕ. After "unrolling" the cone on the map parallel will have the same length but the radius used in it will be AT or R cot ϕ as shown in Fig. 8.8(b).

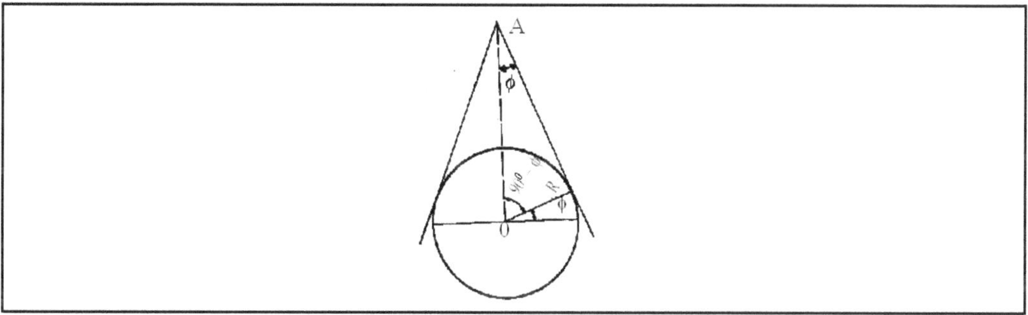

Fig. 8.8(a) Cross - Sectional view of globe and tangent cone.

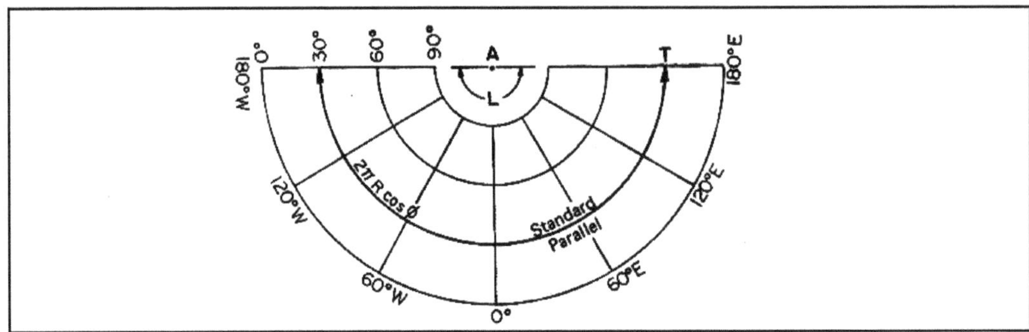

Fig. 8.8(b) Developed cone for Northern Hemisphere conical equidistant projection with standard parallel at 30.

The central angle at A called L, in radians is equal to

$$L = \frac{2\pi R \cos\phi}{R \cot\phi} = 2\pi \cos\phi \tan\phi = 2\pi \sin\phi^{c}$$

$$L \text{ in degrees} = 2\pi \sin\phi \,.\, \frac{360^{0}}{2\pi} = (360 \sin\phi)^{o}$$

In case of conical equidistant projections sin ϕ is the "Constant of the Cone" (K). The standard parallel is divided into equal parts and the meridians are drawn as straight radial lines of standard length. As in case of cylindrical equidistant North pole will be a line, instead of a point.

8.6.3 Polar Azimuthal Equidistant Projection

A plane which is tangent to the pole can be treated as a special cone with altitude zero and standard parallel at 90^{0} N or S. The distance AT is R cot 90^{0} = 0. The constant of the cone K = Sin 90^{0} = 1.000, Thus the central angle L at A, (pole) will be 360^{0} of longitude 360^{0}. Fig. 8.9 gives the diagramatic polar azimuthal equidistant projections.

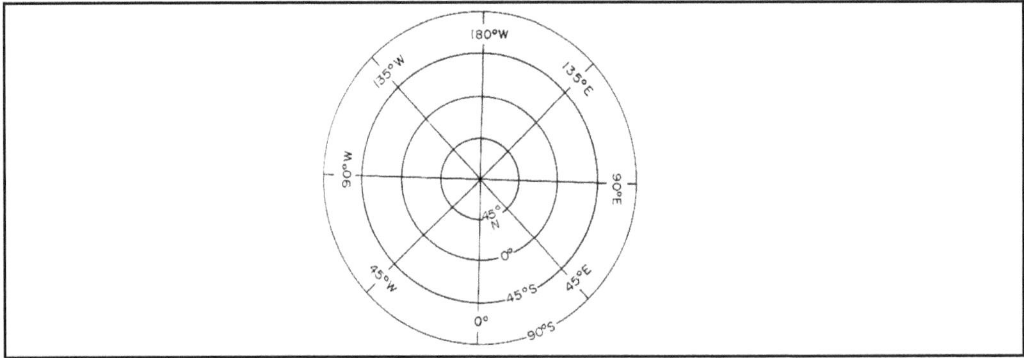

Fig. 8.9 Polar azimuthal equidistant projection.

The graticule has a fan shaped appearance like regular conics. Azimuthal projection show all directions from the centre without distortions. In polar case meridians will be shown with their actual difference in longitude. The parallels will be equally spaced concentric circles.

8.7 Equidistant Projections with two Standard Parallels

8.7.1 Cylindrical Equidistant with two Standard Parallels

This projection is constructed in the same way as equidistant projection with one standard parallel with the exception, namely, that the equator is drawn with scale factor less than 1.000. The net work of perfect squares shown in figure is compressed to a smaller east - west dimension. As a result of East - West compression, there will be two parallels which were too long earlier, but now corrected. Ref. Fig. 8.10.

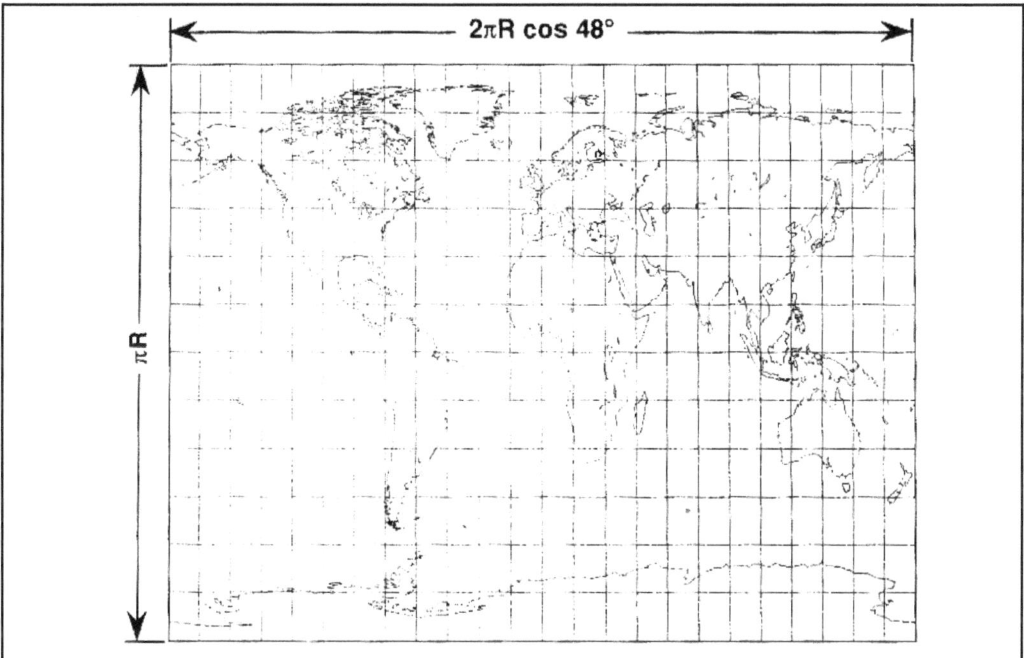

Fig. 8.10 Cylindrical equidistant projection with two standard parallel.

8.7.2 Conical Equidistant with two Standard Parallels

In the conical equidistant with one standard parallel, other parallel is added mathematically, with out truly projecting them. In the conical equidistant with two standard parallels there is no real cone at all. Radius for drawing each standard parallel comes directly from the decision that two parallels and all meridians should be standard.

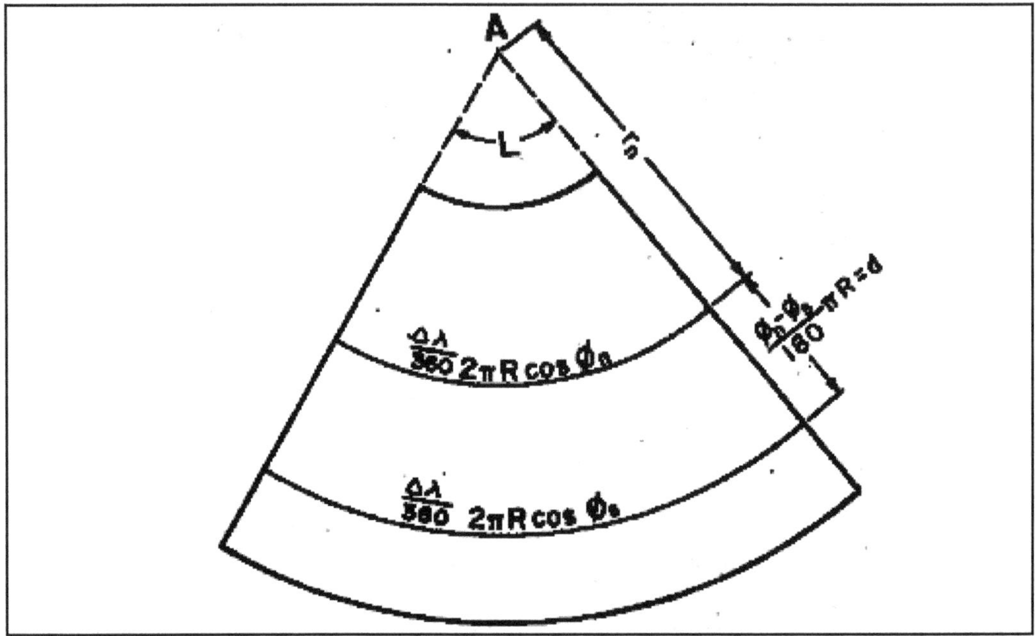

Fig. 8.11 Conical equidistant projection with two standard parallels.

In Fig. 8.11 the latitude of the northern standard parallel is ϕ_n and southern is ϕ_s. The width of the region to be mapped in degrees of longitude is called $\Delta\lambda$. The radius to draw ϕ_n is r_n and the central angle at A is L d is the parallel spacing.

From the Fig. 8.11,

$$L = \frac{(\Delta\lambda/360)\,2\pi R \,\text{Cos}\,\phi_S}{r_n + d}$$

$$= \frac{(\Delta\lambda/360)\,2\pi R \,\text{Cos}\,\phi_n}{r_n}$$

equating the two expressions

$$\frac{\text{Cos}\,\phi_S}{r_n + d} = \frac{\text{Cos}\,\phi_n}{r_n}$$

$$\text{Cos}\,\phi_S = \frac{(r_n + d)\,\text{Cos}\,\phi_n}{r_n}$$

$$= \text{Cos}\,\phi_n + \frac{d\,\text{Cos}\,\phi_n}{r_n}$$

$$r_n = \frac{d \cos \phi_n}{\cos \phi_S - \cos \phi_n}$$

and

$$d = \frac{\phi_n - \phi_S}{180^0} \pi R$$

The constant of cone K is

$$K = \frac{R \cos \phi_n}{r_n}$$

or

$$K = \frac{180}{\pi_n} \left(\frac{\cos \phi_S - \cos \phi_n}{\phi_n - \phi_S} \right)$$

The above expression r_n, d, and K gives the parameters for the projection.

8.7.3 Azimuthal Equidistant with two Standard Parallels

Since there won't be enough room to draw equidistant meridian for two standard parallels, azimuthal equidistant projection with two standard parallel is not possible.

8.8 Equal Area Projections with Straight Meridians

The scale along each meridian which was 1.000 in all the five previous projections is made to be compensatory for maintaining equal area. The cylindrical, conic and azimuthl cases are discussed below.

8.8.1 Cylindrical Equal Area Projection

Resembles the cylindrical equidistant in several ways. Equator is the only standard parallel and the meridians are perpendicular to it. The parallels have the same length as in the cylindrical equidistant and the east west scale factor again varies with Sec ϕ.

Along the 60th parallel scale factor is 2.000. The compensatory scale factor along the meridian must be 0.500 for maintaining equal area. The reduced scale factors in north - south direction such as 0.5000 at $\phi = 60^0$, cause compression of the perfect squares found on the equidistant. Refer to figure 8.12 (a) & 8.12 (b). The figure shows that the surface area of a sphere, or globe, happens to be equal to that of a surrounding cylinder of the same height (2R). Therefore the projection may be formed by projecting the planes of various parallels of the globe until they intersect a surrounding cylinder. Figure 8.12 (a) shows cross sectional view of a globe. By calculation the distance of equator to any parallel is R Sin ϕ.

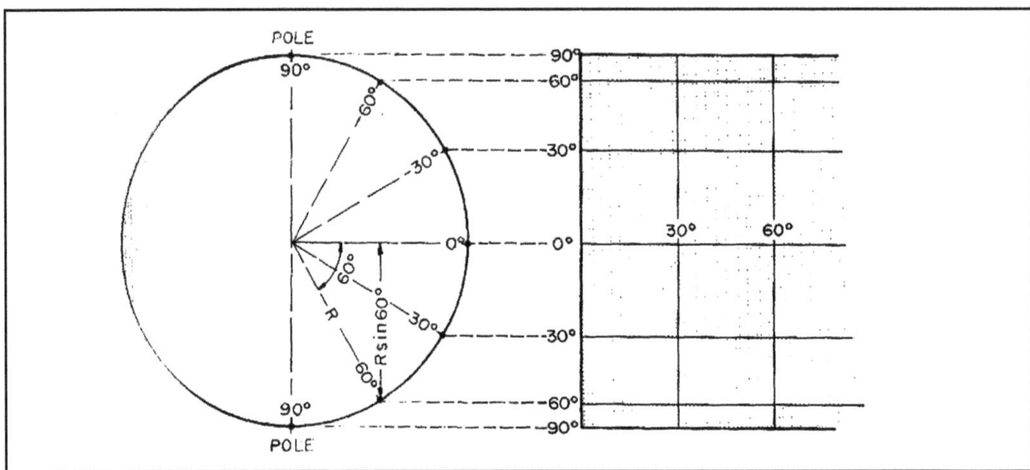

Fig. 8.12(a) Cylindrical equal area projection with its generating globe (cross section of globe).

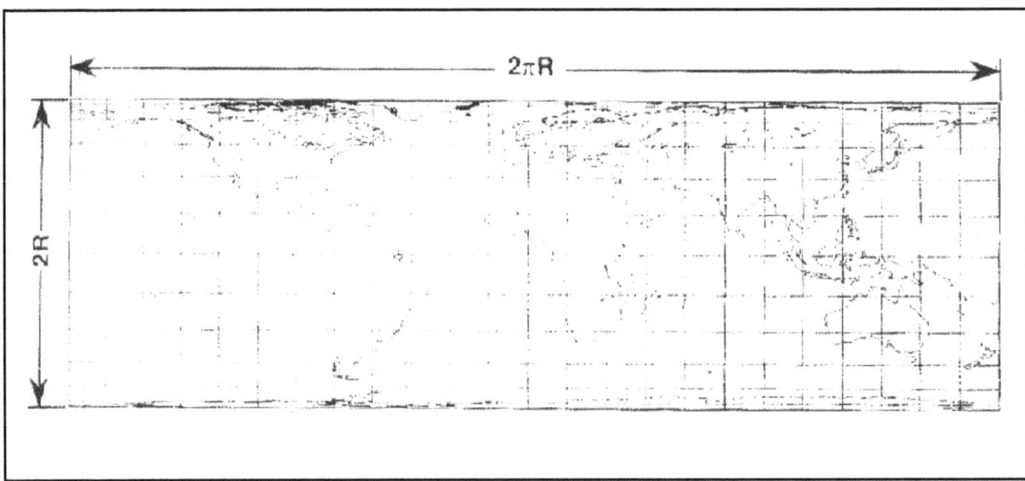

Fig. 8.12(b) World map on a cylindrical equal area projection.

8.8.2 Conical Equal Area Projection (Albers Projection)

Conical equidistant projection with two standard parallels shows less area between the standard parallels than would an equal area map. The scale factor along all meridians was 1.000 and that along parallels was less than 1.000. The product of these values also is less than 1.000, indicating a reduction in the area. The Albers projection, in order to be equal area, allows the scale factor along the meridians to be greater than 1.000 to compensate for the smaller east - west scale factors. The two standard parallels are thus further apart in the equal area map than in the equidistant map.

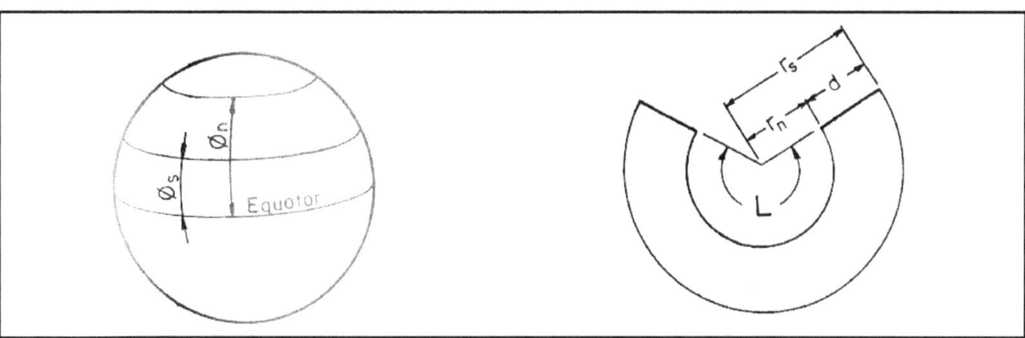

Fig. 8.13 Zone on globe and circular strip on map (equal area).

To calculate the increased distance 'd' between the standard parallels, the correct area of the zone (on the globe) is set equal to that of the circular strip to be drawn on the map, with 'd' as an unknown in the Fig. 8.13

$$d = r_s - r_n$$

Surface area between parallels $\phi_n - \phi_s$ on the globe.

$$2\pi R\,(R \text{ Sin } \phi_n - R \text{ Sin } \phi_s) = 2\pi R^2\,(\text{Sin } \phi_n - \text{Sin } \phi_s)$$

The fraction of full circle represented by L is the constant of the cone K.

If L is in degrees,

$$K = L/360^0$$

The area of the strip will be the difference of two sectors

$$K\pi r_s^{\,2} - K\pi r_n^{\,2} = K\pi\,(r_s^2 - r_n^2)$$

These two areas, one on the sphere and one on the map, are to be equal

$$K\pi(r_s^{\,2} - r_n^{\,2}) = 2\pi R^2\,(\text{ Sin } \phi_n - \text{Sin } \phi_s)$$

The lengths of standard parallels must be

$$2\pi R \text{ Cos } \phi_n \text{ and } 2\pi r \text{ Cos } \phi_S$$

While on the map the same distance must be equal to

$$K\,(2\pi\, r_n) \text{ and } K\,(2\pi\, r_S)$$

Thus from equating these distances

$$r_n = \frac{R \text{ Cos } \phi_n}{K},\; r_s = \frac{R \text{ Cos } \phi_S}{K}$$

Substituting these expressions into equal area equation

$$K\pi\left(\frac{R^2 \text{ Cos}^2\, \phi_S}{K^2} - \frac{R^2 \text{ Cos}^2\, \phi_n}{K^2}\right) 2\pi\, R^2\,(\text{Sin } \phi_n - \text{Sin } \phi_S)$$

which simplifies to

$$\frac{\text{Cos}^2 \phi_S - \text{Cos}^2 \phi_n}{K} = 2(\text{Sin } \phi_n - \text{Sin } \phi_S)$$

$$\therefore \qquad K = \frac{\text{Cos}^2 \phi_S - \text{Cos}^2 \phi_n}{2\,(\text{Sin}^2 \phi_n - \text{Sin}\,\phi_S)}$$

Which reduces to

$$K = \frac{1}{2}\,(\text{Sin } \phi_n - \text{Sin } \phi_S)$$

With K, known it is possible to obtain the radius r_1 of any other parallel ϕ_1. The area on the map between this parallel and the northern standard parallel is

$$K\,\pi\,r_1^{\,2} - K\,\pi\,r_n^{\,2}$$

This must be equal to the corresponding area on the globe

$$\therefore \qquad K\pi\,r_1^{\,2} - K\pi\,r_n^{\,1} = 2\pi\,R^2\,(\text{Sin } \phi_n - \text{Sin } \phi_1)$$

From this

$$Kr_1^{\,2} = Kr_n^{\,2} + 2R^2\,(\text{Sin } \phi_n - \text{Sin } \phi_1)$$

$$\therefore \qquad r_1 = \left\{\frac{Kr_n^{\,2} + 2R^2\,(\text{Sin}\,\phi_n - \text{Sin}\,\phi_1)}{K}\right\}^{\frac{1}{2}}$$

8.8.3 Polar Azimuthal Equal - Area Projection

In this different radii are used to draw the parallels. For any parallel the area on the map is πr^2. The surface area between the pole and the parallel will be equal to that of the strip of similar height on the surrounding cylinder. That is

$$2\pi R\,(R - R\text{ Sin } \phi) = 2\pi R^2\,(1 - \text{Sin } \phi)$$

or

$$= 2\pi R^2\,(1 - \text{Cos } 90 - \phi)$$

The two areas must be set equal to each other and solved for the unknown radius

$$\pi R^2 = 2\pi R^2\,(1 - \text{Cos } 90 - \phi)$$

$$r = R(2[1 - \text{Cos}\,90 - \phi])^{\frac{1}{2}}$$

$$= R\left(2\left[2\,\text{Sin}^2\left(\frac{90-\phi}{2}\right)\right]\right)^{\frac{1}{2}}$$

$$r = 2\,R\text{ Sin}\left(\frac{90-\phi}{2}\right)$$

This equation is used for solution of r for each parallel to be plotted.

It is also possible to find the required radius by graphic construction. The expression for r is the same as that of the length of the chord shown in Fig. 8.14. This is the chord in the globe from pole to the parallel.

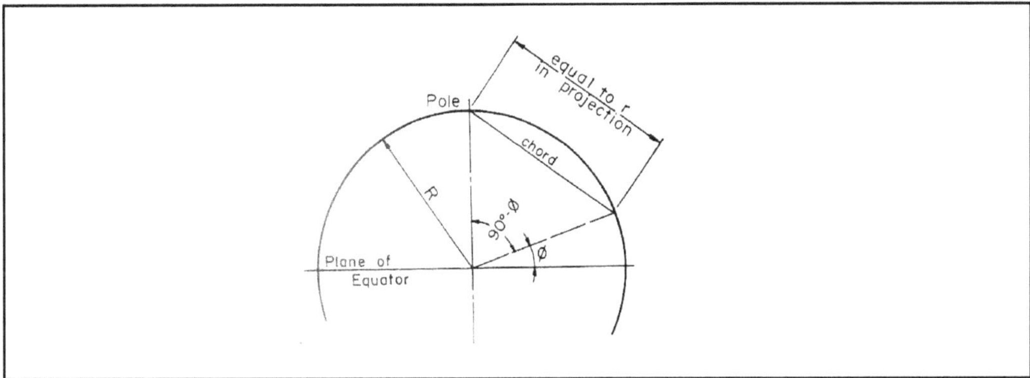

Fig. 8.14 Graphic construction for polar azimuthal equal area projection.

8.9 Projections with all Parallels Standard

All projections covered so far had scale factors of 1.000 along all meridians. They were equidistant in north - south direction. In this section four projections which are equidistant in east - west direction (all parallels are standard) will be discussed.

Cylindrical and conic projections of this type are designed mathematically. Azimuthal one in this group are true projections.

8.9.1 Sinusoidal Projection

To construct sinusoidal projection of the whole world, a standard central meridian is drawn, then straight horizontal standard parallels are constructed at equal intervals as shown in Fig. 8.15.

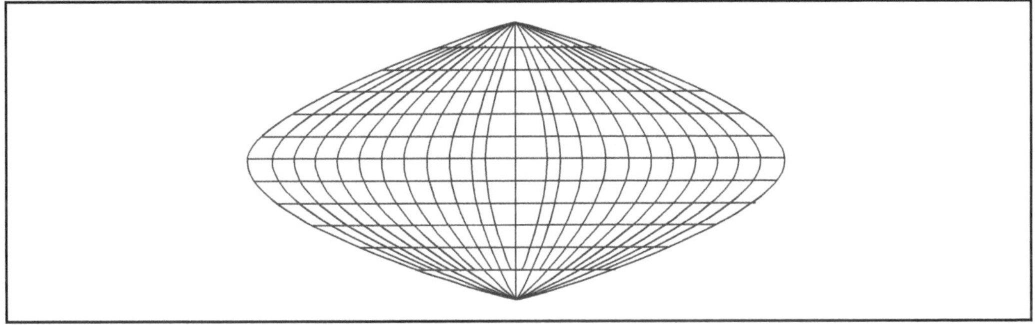

Fig. 8.15 Sinusoidal projection for the world.

The standard lengths of the parallels vary with cos ϕ. Hence the meridians are sine waves or cosine waves, hence the name *Sinusoidal*.

8.9.2 Bonne Projection

In Bonne Projection a central parallel is chosen and constructed as in conical equidistant. A central meridian is drawn vertically with standard length and remaining parallesl are added at correct spacing and with standard length. In sinusoidal case, the equator is chossen as the central parallel and is drawn with infinite radius, (making it a straight line). The apex of its cone is at infinity. In Fig. 8.16(a) and 8.16(b) shows the ralationship between sinusoidal and Bonn projections to the cylindrical and conic equidistant projections. In Fig. 8.16(c) hemisphere is shown on Bonne Projection. The resemblance to sinusoidal projection is very clear.

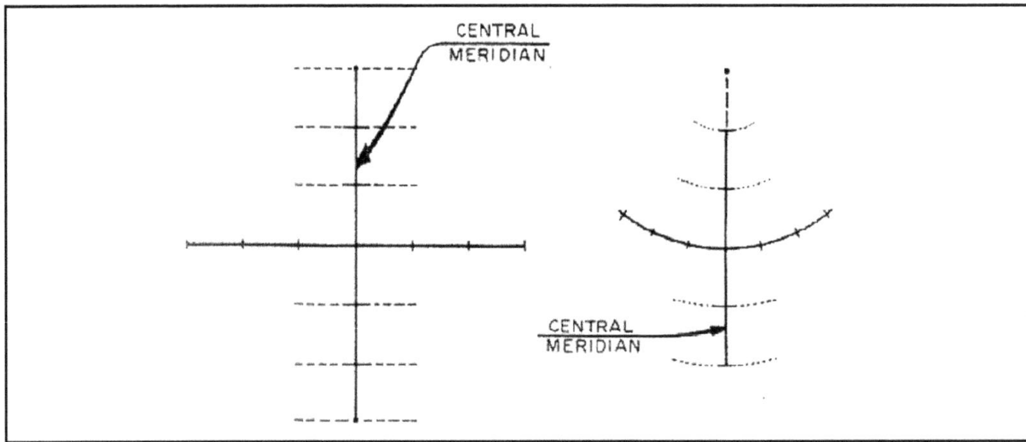

Fig. 8.16 (a) Cylindrical equidistant and sinusoidal.
(b) Conic equidistant and bonne.

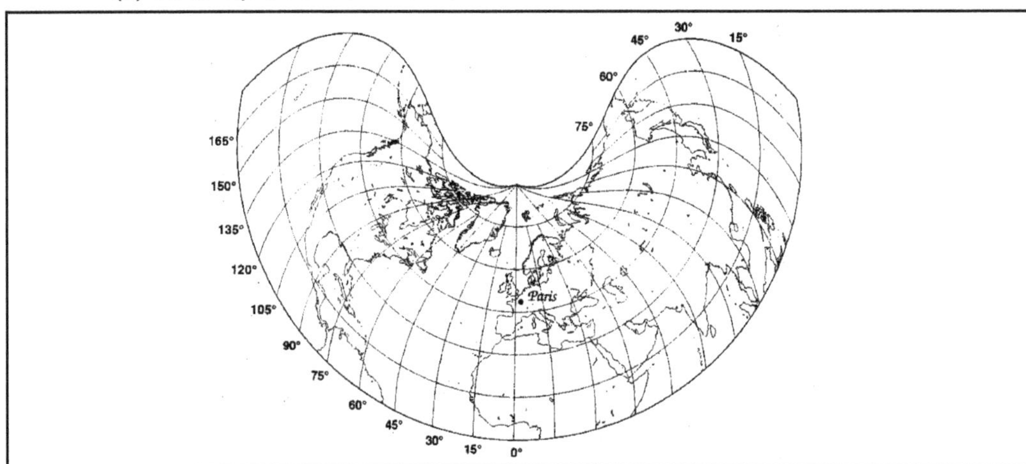

Fig. 8.16 (c) Boone projection of hemi sphere meridians are curved.

8.9.3 Polyconic Projection

In Polyconic Projection, each parallel is separately treated as if it were the standard parallel of a conical equidistant projection. Each has its own radius, R Cot ϕ called 'AT' refferred earlier.

To begin the map central meridian is drawn as a standard line. The locations of the parallels are marked at equal intervals. The centre form which to draw each circular arc is located on the extension of the cental meridian. Each centre is plotted at a distance of R Cot ϕ above the point, where the parallel is to cross the meridain. There result is shown in Fig. 8.17.

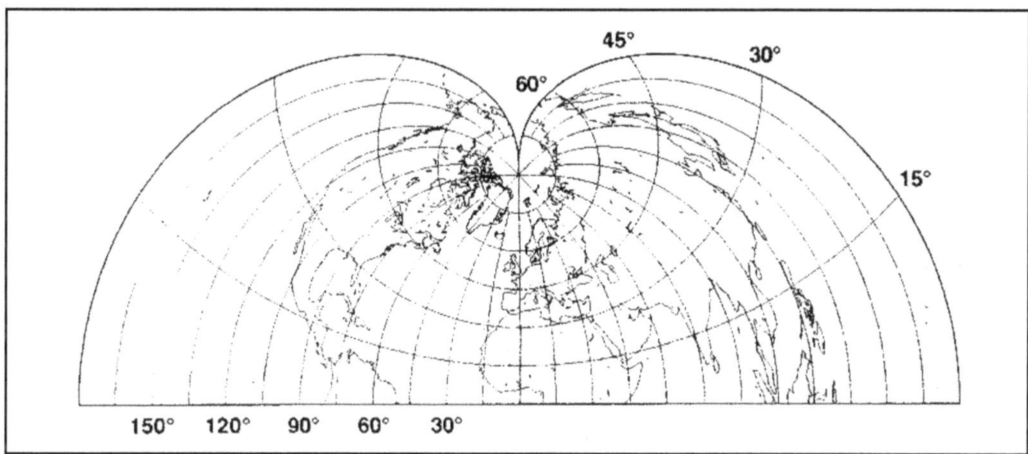

Fig. 8.17 Polconic projection of hemishphere. Meridians are curved.

Each parallel is standard. The effect is as the name implies, a projection of "many cones'. There is a separate constant of the cone for each parallel. It is equal to sin ϕ as in conical equidistant. Thus the central angle L, if an entire hemisphere is to be shown is L = 360^0 Sin - ϕ in each case.

Each parallel, being standard, is divided at equal intervals to locate the meridian crossings. The meridians are drawn with a flexible curve through the points on each parallel. As shown in the figure, the equator appears as a straight line, its cone being a cylinder and its radius being infinity (R Cot 0^0). The hemisphere appears oddly shaped in Fig. 8.17. The projection however is never used for such large region.

8.9.4 Orthographic Projection

An orthographic projection is one in which the projecting rays are perpendicular to the plane of projection. Any part of the object that is parallel to the plane of projection will appear in its proper shape and correct scale. An orthographic view is a view from infinity and this lacks perspective. Fig 8.18 (a) shows the formation of an oblique orthographic projection of a globe to a plane.

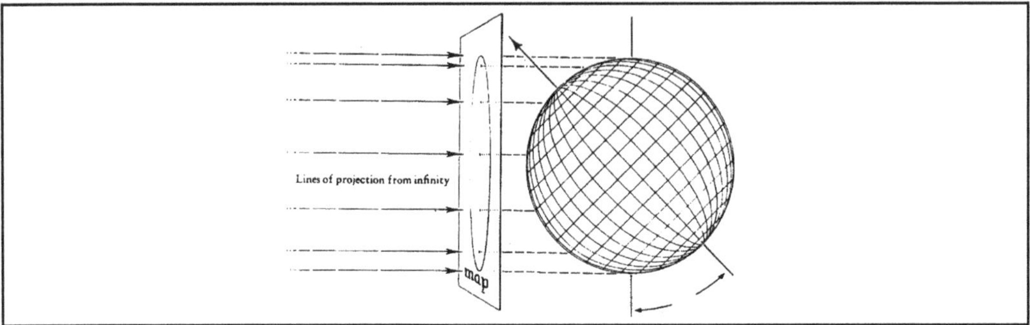

Fig. 8.18(a) Orthographic projection of globe to a plane.

A polar case of orthographic projection is formed in the same way but "looks" at the globe from directly above the north or south pole. Figure 9.4 (b) shows the northern case.

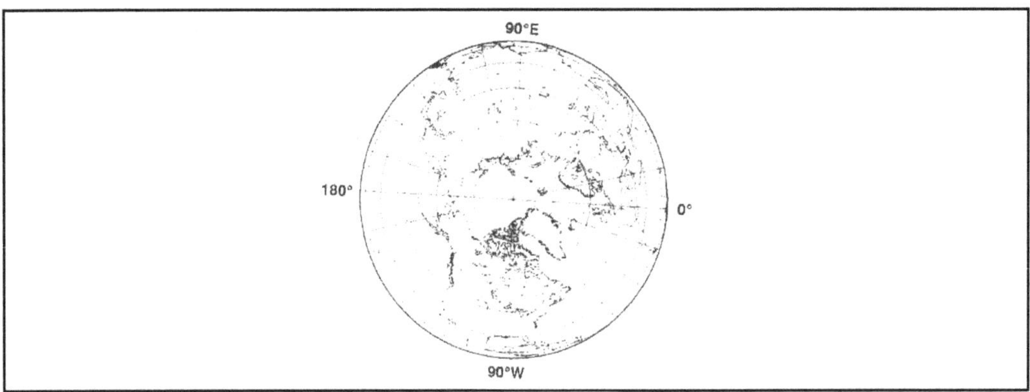

Fig. 8.18(b) Polar orthographic projection.

All parallels of latitude appear with their proper diameters; each is drawn with radius $r = R \cos \phi$ and there for a standard parallel. Orthographic Projection graphically constructed is shown in Fig. 8.18(c) - polar case.

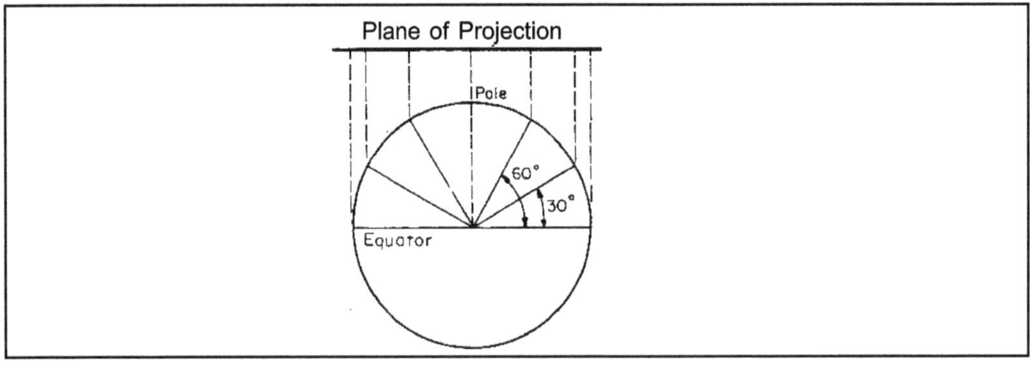

Fig. 8.18 (c) Graphic construction of the polar orthographic projection.

8.10 Conformal Projections with Straight Meridians

As earlier discussed conformal projections preserve shape, instead of size. It is already clear that no projection can be both conformal and equal area.

8.10.1 Mercator Projection

In cylindrical equidistant with its graticules of perfect squares, is not conformal. The scale factor along the meridians is 1.000, where as along the parallels it is equal to Sec ϕ, which is greater than 1.000 at all latitudes except 0^0. The Mercator projection allows the scale factor along the meridians to vary with Sec ϕ. For example at latitude of 60^0, the scale factor is 2.000 in all directions. The location of a particular parallel can be computed or taken from computed tables.

8.10.2 Lambert Conformal Conic Projection

On a conical equidistant with two standard parallels, the meridians were standard lines. The Albers projection was described as a variation in which the scale factor along each meridian varied from 1.000. So as to be compensatory with respect to the east - west scale factor. The Lambert conformal conic is a variation in which the scale factor along each meridian varies so as to be equal to east - west scale factor. This condition makes the standard parallels slightly closer together than in the other two cases. (Fig. 8.19).

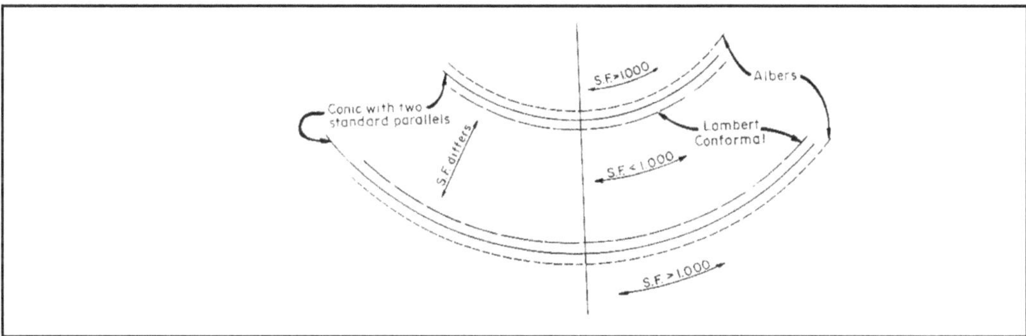

Fig. 8.19 Comparison of three conic projection.

The constant of the cone 'K' for the Lambert conformal conic is given as,

$$K = \frac{\log \text{Cos}\ \phi_S - \log \text{Cos}\ \phi_n}{\log \tan \frac{90 - \phi_S}{2} - \log \tan \frac{90 - \phi_n}{2}}$$

The radius for plotting any parallel is

$$r = \frac{C(\tan 90 - \phi)^K}{2}$$

where

$$C = \frac{R \,\text{Cos}\, \phi_S}{K\left(\tan \frac{90 - \phi_S}{2}\right)^K}$$

8.10.3 Polar Stereographic Projection

This type of projection shows meridians radiating from poles at equal angles. The spacing of parallels, instead of being equidistant, equal area or orthographic, is designed to make the projection conformal.

The Stereographic is literally projected, rather than being mathematically created. Fig. 8.20 shows that a tangent plane is used and that the rays of projection emanate from a point diametrically opposite to the point of tangency. In the polar case the rays extend from the opposite pole.

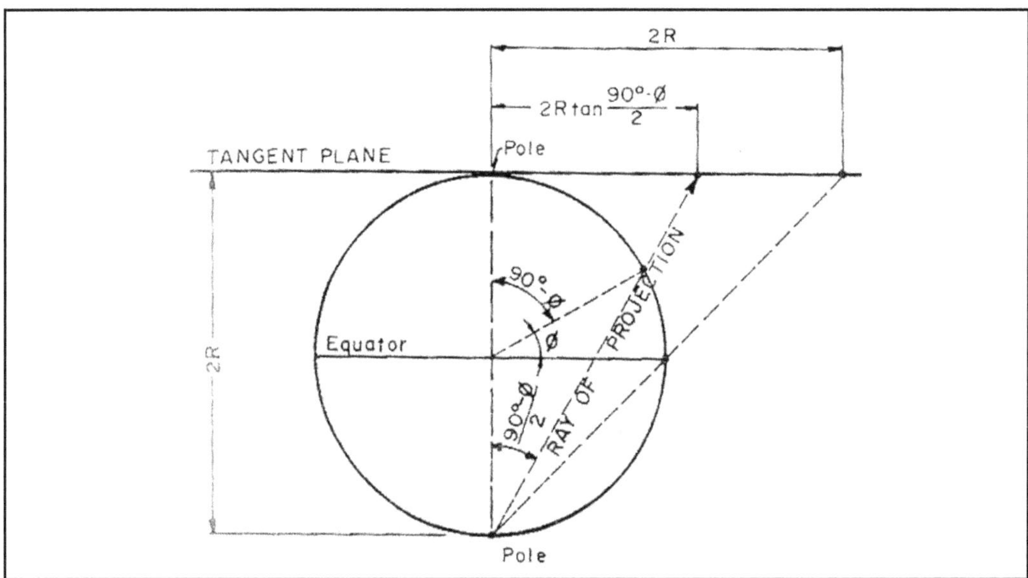

Fig. 8.20 The stereographic projection - Graphical solution.

The equator will be a circle excatly twice as large as on the globe. The radii for drawing the various parallels may be obtained graphically or mathematically calculated.

8.11 GNOMONIC Projections

Gnomonic Projections are projected from the centre of the globe or to a cylinder, cone or plane.

In Cylindrical Case, there is extreme exaggeration of areas away from the equator as shown the Fig. 8.21. Because of this the cylindrical case is rarely used. The Conic case, also is not used because other conic projections discussed in earlier sections have better advantages over a gnomonic. Gnomonic projection to a Plane is very important, which offers the advantage that all great circles appear as straight lines.

If the plane of projection is a tangent at a pole, the radius of drawing any parallel may be found graphically or mathematically as shown in Fig 8.21 (a). The general appearance of the polar gnomonic projection is similar to that of other polar azimuthal projections, but the radii are greater. Fig. 8.21 (b) and 8.21 (c) gives the comparison.

Equatorial gnomonic projection in the case of the plane is a tangent at the equator, instead of the pole. The graticules will be projected as shown.

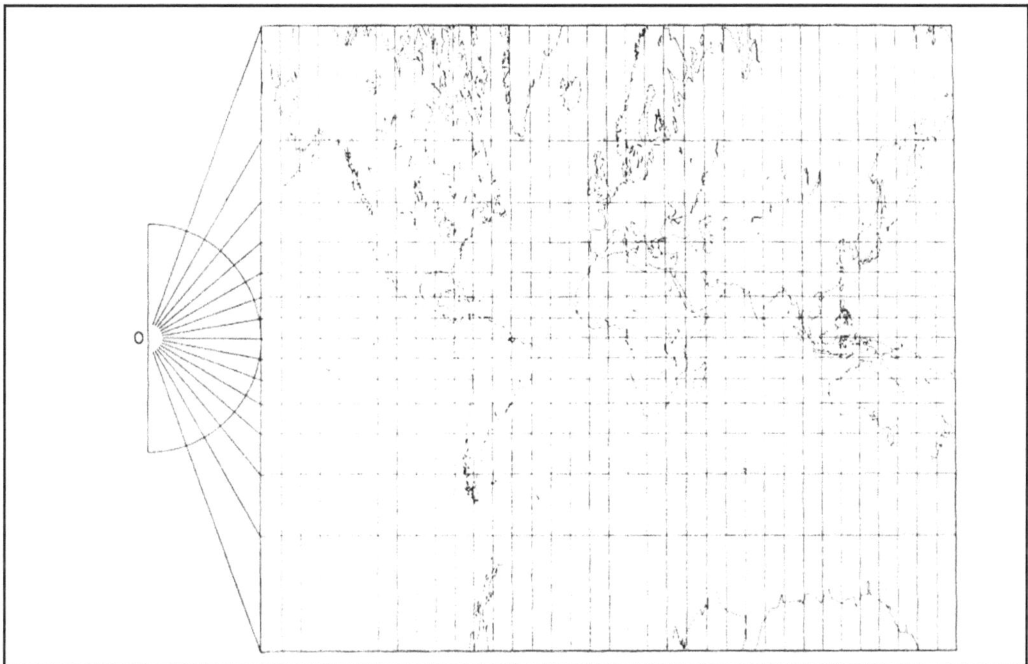

Fig. 8.21 (a) Polar gnomonic projection

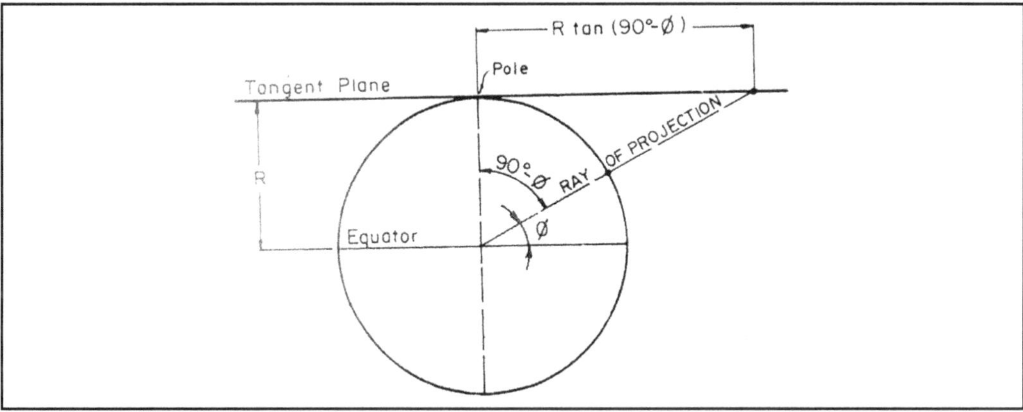

Fig. 8.21(b) Comparison between four polar azimuthal projections (generating globe).

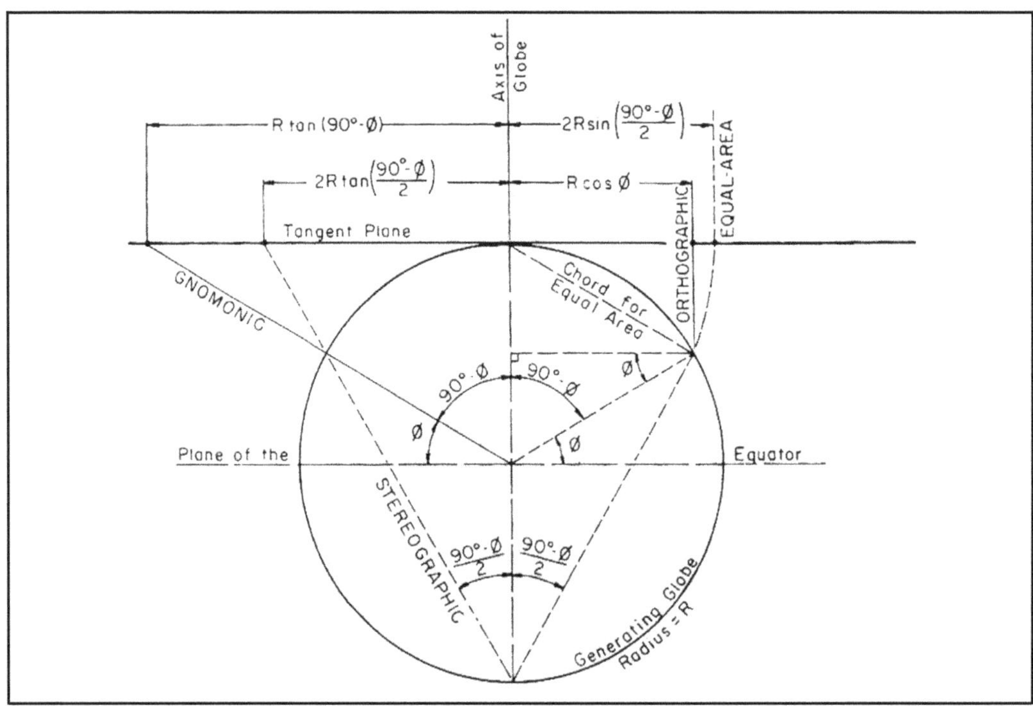

Fig. 8.21 (c) Comparison between five Polar azimuthal projection (MAP)

in the Fig. 8.22 (a). Meridians, being great circles, will all be straight lines and also the Equator. The points at which the meridians cross the Equator and those at which the parallels cross the central meridian are plotted. The figure shows (8.21(b)) the construction of circle corresponding to the 75th parallel in the Polar Gnomonic Projection.

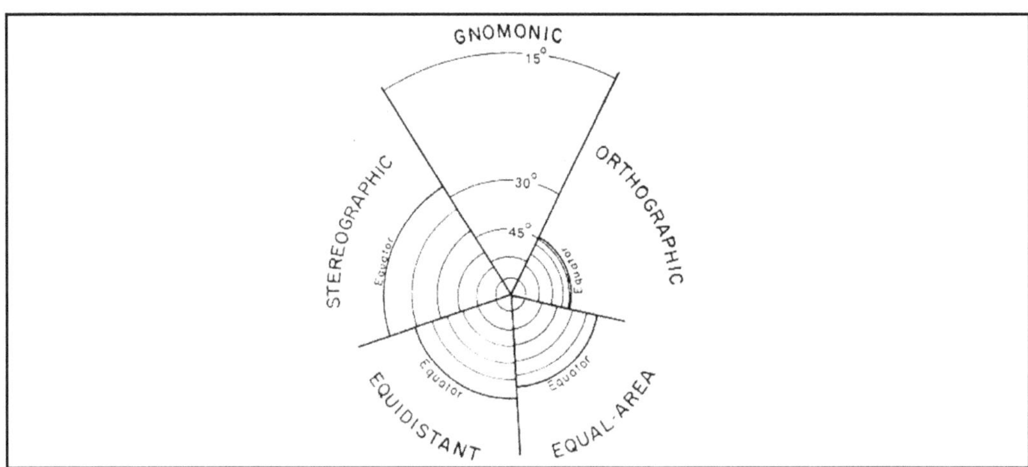

Fig. 8.22 (a) Equatorial gnomonic projection.

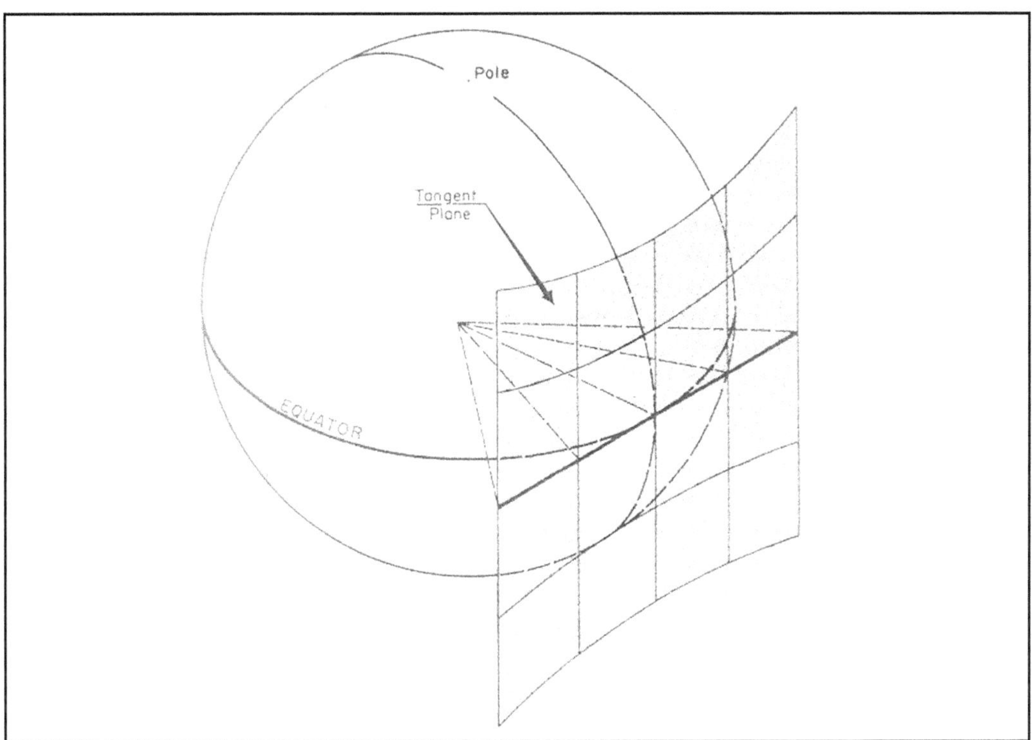

Fig. 8.22 (b) Construction of equatorial gnomic projection.

9

Geographical Information Systems

9.1 Introduction

To answer apparently simple geographical questions like : What is the population of a particular city? What are the characteristics of the soils in a particular land parcel? Are there any trends in the patterns of earthquakes in India which could help predict future quakes? How has the distribution of urban and rural population changed between the past two censuses? All these questions are pertinent and require the proper and accurate data derived from several sources, and be integrated into a consistent form. The art, science, engineering and technology required to answer these geographical questions constitute called Geographical Information System (GIS). GIS is a generic term denoting the use of computers to create and depict digital representations of the Earth's surface.

Roger Tomlinson the father of Canada GIS is credited with visualising the need for computers to perform certain simple but labour-intensive tasks associated with Canada Land Inventory.

Tomlin (1990) saw that if a map could be represented in digital form, then it would be easy to make measurements of its basic elements, specifically the areas assigned and the tedious hand-measurement of area by conducting dots on transparent overlays of known dot density. Tomlinson's cost-benefit analysis showed that computerization would be cost effective, despite the enormous costs involved and the primitive nature of the computers of the time.

It is, however, important to note that the earlier investigators also realised the importance of such a computerised analysis of the spatial data. David Bickmore (1934), the primary GIS innovator, has discovered that, GIS stems from the benefits of automating the map production process. Once information of any kind is in digital form, it is much easier to manipulate, copy, edit, and transmit. Ray Boyle invented the "free pencil " digitiser, and by 1964 Bickmore and Boyle set up the Oxford system for high quality digital cartography (Rhind 1988). At that time, major mapping agencies, including the US and other military bodies began the lengthy and often rocky process of automation. Widespread achievement of the benefits of automated cargography had to await the development of suitable mechanisms for input, display, and output of map data, but the necessary devices map digitiser, interactive graphics display device and plotter, respectively became available at reasonable cost by the early to mid 1970's and from then onwards an increasing number of organisations set out to convert all their maps into a computerised form.

The world as composed of a set of largely independent layers and spatial elements, each representing some component of the environment, and some set of environmental concerns as well. These layers might include groundwater, natural vegetation, soil and other resources. McHarg (1996) was the foremost proponent of this view, and his group at the University of Pennsylvania applied it in a long series of exemplary studies. Although the initial idea was strictly manual, the computerisation of these ideas in a layer-based raster GIS was a simple step, and many systems owe their origins to McHarg's simple model (Tomlin 1990).

9.2 Roots of GIS

GIS has its roots in the stimulus provided by the development of remote sensing, in the late 1960s and early 1970s, as a potentially cheap and effective source of earth observations. While many of the techniques for processing remote sensing data are highly specialised, more general GIS techniques become important in order to combine information derived from remote sensing with other collateral information (Star et al 1997). Today many GIS include extensive functionality for image processing, and all types of remote sensing are increasingly the primary data source, particularly for detection of landscape change, natural resource management, and, environmental monitoring and management. Similarly, GIS has many roots of evolution like map production process; one root lies in landscape architecture and environmentally sensitive planning; another root in urban and demographic data analysis; the roots of remote sensing merging with vector GIS; the roots of representational issues of space and time in GIS; the root of large scale data integration around a common data model; possibility of storing large number of layers of information on temporal changes to handle an image within a relational database environment etc, and so on.

If GIS has so many apparently independent roots, what brought them together, and why has the umbrella term 'GIS' become so widely accepted? First, there are obvious commonalities. For example, the representation of topology invented for the Dual Independent Map Encoding (DIME) system at the US Bureau of the Census, is almost identical to that incorporated in CGIS and in Australian work. The methods of raster (grid cell data) processing and storage used in remote sensing systems are almost conceptually identical to those used by the systems that have implemented McHarg's multi-layer view of the World. Second, it was easy from the viewpoint of the software engineering paradigms of the 1970s and 1980s to integrate functions around common representations. Once a raster or vector data model had been established, functions that process the data model in different ways were easy to add, so as to build large-scale integrations of image processing functions around a common raster representation.

Geographical Information is information about geography, that is, information tied to some specific set of locations on the Earth's surface including the zones of atmosphere. 'Spatial' is often used synonymously with or even in preference to, 'geographical' in this context, although in principle it might be taken to include information that is tied to frames other than the earth's surface, such as, human body (as in medical imaging) or a building. Because of this difficulty, the term 'geospatial' has become popular recently, notably in the context of the US National Spatial Data Infrastructure, the Canadian National Geospatial Infrastructure, and the UK National Geospatial Data Framework.

Today, the term GIS tends to be applied whenever geographical information in digital form is manipulated, whatever be the purpose of that manipulation. Thus using a computer to make a map is refferred to as 'GIS'. This entails using the same computer to analyse geographical information and to make future forecasts using complex models of geographical processes. The earth images collected by remote sensing satellites are geographical data, but the systems that process the images are not to be called GIS as long as they remain confined to this particular form of data in such cases. 'GIS' tends to be reserved for systems that integrate remotely sensed data with other types, or process data that have already been cleaned and transformed. Similarly, an atmospheric scientist or oceanographer will tend to associate 'GIS' with the system used more for multidisciplinary work and policy studies, and other software environments for modelling and analysis within the confines of one's own discipline.

9.3 Overview of Information System

GIS might provide the medium for studying one or more of the fundamental issues that arise in using digital information technology to examine the surface of the earth or any related systems. The Resource Information System (RIS) for agricultural management, for instance, has to be considered multidimensional with attribute dimension, spatial dimension and temporal dimension. Geographic Information System (GIS) offers capabilities of integrating multisector, multilevel and multiperiod database. GIS is a computerized database system for capture, storage, retrieval, analysis, and display of spatial data. It is a general-purpose technology for handling geographic data in digital from, and satisfying the following specific needs, among others :

(i) the ability to preprocess data from large stores into a form suitable for analysis, including operations such as reformatting, change of projection, resampling, and generalisation.

(ii) direct support for analysis and modelling, so that form of analysis, calibrations of models, forecasting, and prediction are all handled through instructions to the GIS.

(iii) post processing of results including such operations as reformatting, tabulation, report generation, and mapping.

In all these operations, the typical GIS user now expects to be able to define requirements and interact with the system through a "user-friendly" intuitive interface icons and desktop metaphors (Mark 1986). The function of an information system is to improve one's ability to make decisions. An information system is the chain of operations that takes us from planning the observation and collection of data, to storage and analysis of the data, and to the use of the derived information in some decision-making process. This brings us to an important concept that a map is kind of information system. A map is a collection of stored, analysed data, and the information derived from this collection is used in making decisions. Fig. 9.1 shows the simplified information system overview.

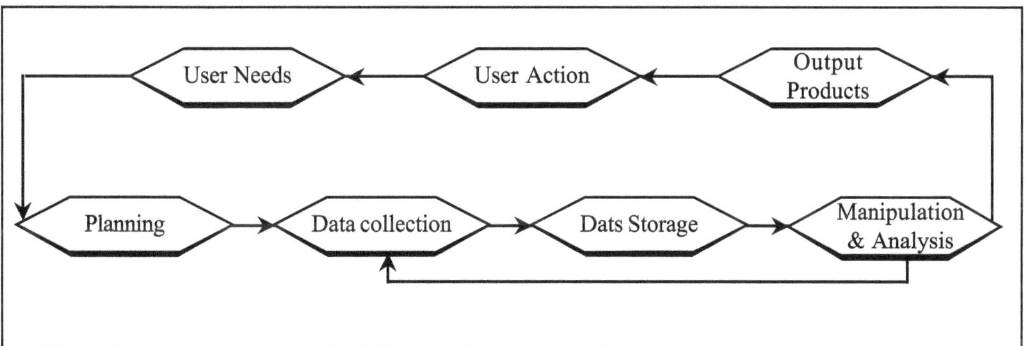

Fig. 9.1 Simplified information system overview.

9.4 The Four Ms

There are mainly four key activities that any urban planners or scientists or resource managers and others use geographic information. They observe and measure environmental parameters and develop maps which portray characteristics of the earth. They monitor changes in our surroundings in space and time. In addition, they model alternatives of actions and process operation in the environment. These, four activities are Measurement, Mapping, Monitoring and Modelling termed as key activities which can be enhanced by the using information systems technologies through GIS. Fig 9.2 explains these four Ms.

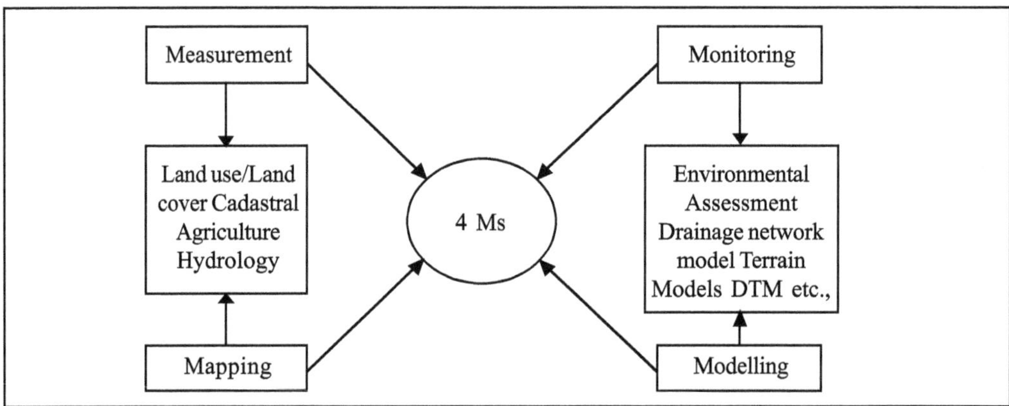

Fig. 9.2 Schematic representation of Four Ms: Measurment, Mapping, Monitoring and Modelling.

GIS technology is more different from traditional mapping and map analysis. GIS is based on a mathematical framework of primitive map analysis operations analogous to those of traditional statistics and algebra. From this perspective, GIS forms a toolbox for processing maps and fundamental concepts for spatial measurement. It provides a foundation for advanced analytic operations involving spatial analysis and measurement. Most of GISs contain analytic capabilities for reclassifying and overlaying maps. Any GIS system for the measurement of areas, distances, angles and so on, requires two components, namely, a standard measurement unit and a measurement procedure.

Another major function of GIS capability is the study of environmental surroundings and the monitoring of environmental parameters (Burrough et al, 1988). Although analytical models have been linked to GIS for spatial measurement and resource assessment, the cross fertilisation between the modules of modelling, measurement and automated mapping allows the GIS user to monitor the environment and the earth systems.

In principle, it is possible to make a clear distinction between GIS and digital cartography. Mapping technology or digital cartography deals with map features and with associated attributes of colour, symbology, name of annotation, legends, neatlines and north arrows. GIS includes the capabilities for storing, editing, and handling the relationships of attributes with their spatial entities along with the capabilities of digital cartography. A map, an ultimate product of digital cartography or GIS, is a very persuasive form of data display and a computer drawn map carries the authority of a powerful technology.

GIS applications now span a wide range, from sophisticated analysis and modelling of spatial data to simple inventory and management. They also dictate the development directions of much of the industry. However, several vendors have chosen to concentrate on the niche market for environmental applications and to emphasise support for environmental modelling. GRASS is a significant public domain GIS software developed by USA with substantial capabilities for modelling.

9.5 Contribution Disciplines

GIS is convergence of technological fields and traditional disciplines. GIS has been called an " enabling technology " because of the potential it offers for the wide variety of disciplines dealing with spatial data. Many related fields of study provide techniques which make up GIS. These related fields emphasise data collections while GIS brings them together by emphasising integration, modelling and analysis. Thus GIS often claims to be the science of spatial information. Fig. 9.3 shows the technical and conceptual development of GIS. The contributing disciplines for the evolution of a GIS (Burrough, 1998) are geography, cartography, remote sensing, surveying and photogrammetry, computer science technology, mathematics, and statistics.

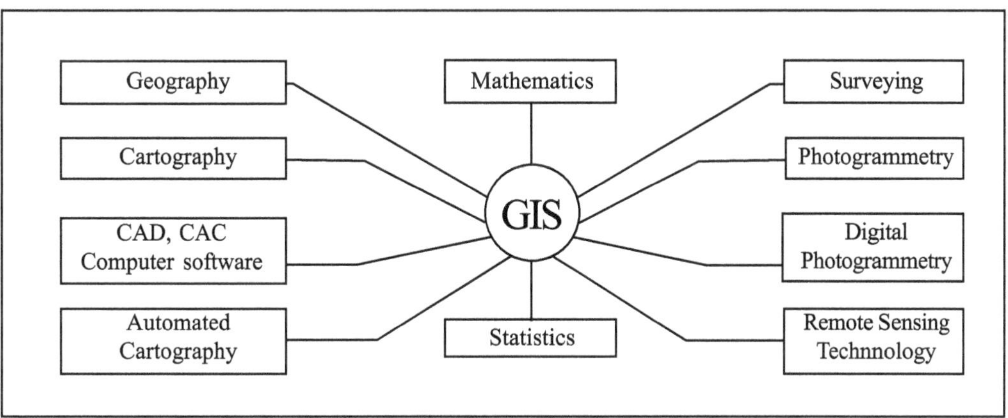

Fig. 9.3 GIS : the result of linking parallel developments in many separate spatial data processing disciplines.

Geography is broadly concerned with understanding the world and man's place in it. Geography has a long tradition in spatial analysis, and provides techniques for conducting spatial analysis and a spatial perspective on research. *Cartography* is concerned with display of spatial information. It is now the main source of input data for GIS (maps) and has a long tradition in the design of maps which is an important form of output from GIS.

Remote Sensing is becoming an important source of geographical data by providing digital images derived from space and the air. Remote sensing provides techniques for data acquisition and processing anywhere on the globe at a low cost, and consistent update potential. While integrated with GIS, remotely sensed imagery can be merged with other data in a GIS providing real-time spatial information. The first part of this book enlightens the concepts and the potential utility of remote sensing.

Surveying and Photogrammetry provide high quality data on positions of cadastral objects like land parcel and building, and topography. Aerial photogrammetry deals with the photographs taken by an aerial camera on board aircraft at different altitudes. Aerial photogrammetry is one of the most powerful data-capturing techniques for the creation of GIS spatial database. The relevant data can be extracted from the aerial photographs of various scales (Fig. 9.4), and may be used as input for GIS. Digital orthophotos provide the source of digital data. These products are scanned airphotos that have been rectified to eliminate displacement caused by variable elevation of the ground surface and the tilt of the camera. Properly registered with other digital data sets, these images can be used directly as backdrops for vector data or to provide a basemap for onscreen digitising. The user may abstract information on land use, vegetation type and other aspects of the landscape from the photograph.

Curran (1989) identifies six characteristics of aerial photographs of immense value as a data source for GIS. They are (i) wide availability, (ii) low cost, (iii) wide area views, (iv) time - freezing ability, (v) high spectral and spatial resolution, and (vi) three-dimensional perspective.

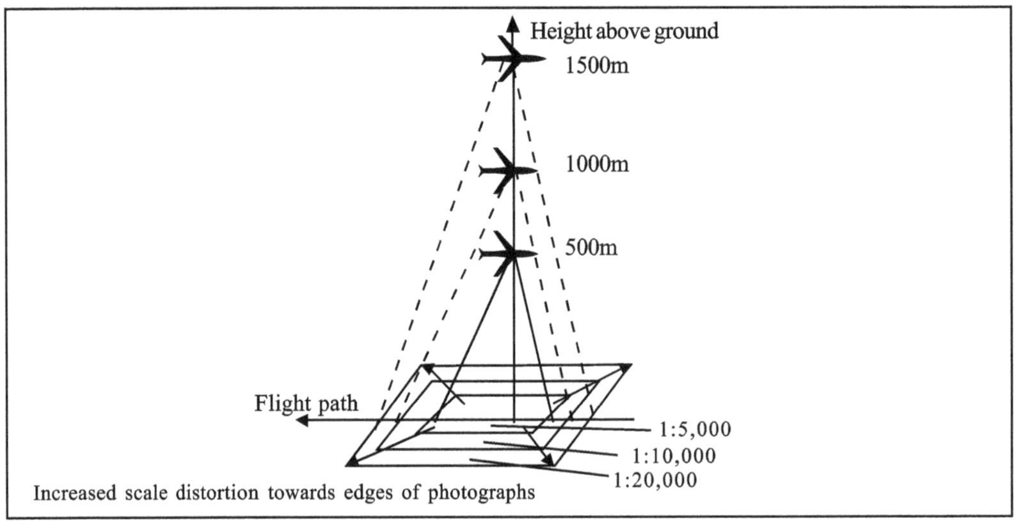

Fig. 9.4 Varying scale on Aerial photographs.

Computer Assisted Design (CAD) provides software, techniques for data input, display and visualisation, and representation, particularly in 3-dimensions. Advances in computer graphics provide hardware and software for handling and displaying graphic objects. Data Base Management System (DBMS) contributes methods for representing data in digital form and procedures for system design and update. Artificial intelligence (AI) uses the computer to make choices based on available data in a way that is seen to enhance the human intelligence and decision-making. Using AI, computer can act as an "expert" in such functions as designing maps, generalising map features and classification.

Several branches of mathematics, especially geometry and graph theory, are used in GIS system design and analysis of spatial data. Statistics is used to build models and perform spatial data analysis in GIS. Statistics is also important in understanding issues of error, quality and uncertainty in GIS data.

Availability of large quantities of spatial data in the from of digital aerial photograph, digital remote sensing imagery, advancement of computer hardware, software and software development, increasing demand of spatial information for management, and infrastructure development parameters, lead to have a system to handle all these requirements. In order to handle such data to meet these demands, to store, retrieve, handle, analyse, manipulate, and display the results, it requires computer based system. Such a system is Geographical Information System (GIS).

9.6 GIS Definitions and Terminology

GIS are decision support computer based systems for collecting, storing, presenting and analysing geographical spatial information. These systems are spatially referenced databases giving users the potentiality to control queries over space, and usually through time. GIS is much more advanced than Computer Aided Design (CAD) or any other spatial data system. The basic output of GIS or spatial data analysis system is a map. The need to analyse maps to compare and contrast patterns of earth related phenomena, is confirmed by the long standing tradition of doing so with traditional maps.

Many geographical phenomena are best described scientifically as fields. Good examples are topographic elevations, air temperatures, and soil moisture content. A 2-dimensional field may be defined as any single valued function of location in a 2-dimensional space and discrete fields, with nominal dependent variables. It appears that any geographical phenomenon can be represented either as a field or as a collection of digital objects. For example, a set of states or revence or administrative units like mandals within a country would commonly be represented in a GIS as a set of areal objects, or as a set of linear objects that form their boundaries. Fields can be digitally represented by vector approaches, but are often represented by raster data structures.

9.6.1 Geographical Entities

Spatial analysis is a technology which typically requires two types of information about spatial objects : attribute, and geographical or locational information. Given the diversity of analytical perspectives within GIS it is difficult to define spatial data analysis. The results of such analysis depend upon the spatial arrangement of events (Goodchild et al 1992). Events may be represented as geographical entities associated with the attributes. The following sections highlight these two types of information and their relationships. To the query, where the particular object is with respect to any coordinate system, the answer is the spatial or geographical or locational information. To the query, what that particular object is, the answer is the attribute data. Alternatively, it can be noted that the GIS consists of two types of data : spatial and attribute.

'Entities' are things in the real world. 'Objects' are things in the digital world. Digital objects and associated attributes and values represent geographical entities. The distinction between entity and objects makes explicit the difference between things and their representations in a formal system. The entity or the field model is more appropriate and are particularly interesting for topographic elevations. Topographic data normally are represented in GIS as fields, either through grid based digital elevation models (DEMs) or as triangular tessellations. Robinson (1958) identified four kinds of geographical quantities. They are point, line, area, and volume. There are three kinds of cartographic symbols : point, line, and area. Robinson discusses 2-dimensional data on 'Mapping quantitative point, line, and area data', and separates volume data under the title 'Mapping 3-dimensional data'.

The frequency of geographical entities with indistinct boundaries has been known for some time; yet vector GIS is tuned to represent entities with crisp boundaries, whereas raster GIS does not represent entity boundaries at all. Thus, formal methods for the representation of geographical entities with uncertain or graded boundaries is an important new area of study in GIS (Burrough and Frank 1996). Fuzzy set theory represents a possible approach to modelling entities with graded boundaries, but it has problems. Geographical entities and classification of geographical entities into categories is a well-known process both in everyday thinking and in scientific work. Various subfields of geography have developed elaborate classifications for landforms, vegetation assemblages, and settlements. In brief, we can note that locational information about the spatial objects of concern are generally described by means of their position on a map or geographical coordinate systems.

Map Features are holding the spatial information of the geographic feature entities, such as, the spatial location like latitude, longitude, x, y, z, shape of points like churches and tram stops, lines like roads and creeks, and polygons like blocks of land and parks.

9.6.2 Attributes

Attributes are the characteristics of the map features, and holding of the descriptive information about the geographic features. Attributes are the non-spatial data associated with time and area entities. They are considered characteristics of entity (Lawrini and Thompson, 1992). The GIS attributes are represented using colours, textures, and linear or graphic symbols like the gardens. The parks are shaded green, the church locations are designated using the special symbol, the bus routes are drawn with a specific line width and style, as broken lines of 12 points width, Contour lines are brown in colour, and so on. The actual value of the attribute that has been measured (sampled) and stored in the database is called attribute value. A classical example of attribute data associated with spatial entities of the environs of Hussain Sagar in Hyderabad may tell us that, a point represents a hotel, a line represents the road and area represents the boundaries of the lake. Each spatial entity may have more than one attribute associated with it, that is, a point representing the hotel may have a number of rooms, standard of accommodation and other related information.

Broadly speaking two types of attributes may be distinguished : primary attributes and secondary attributes. Socioeconomic characteristics, and physical properties of objects are some of the examples of primary attributes. Flows of information levels, districts, capitals, and mandal names are considered secondary attributes.

9.6.3 Topology

In GIS, topology is the term used to describe the geometric characteristic of objects which do not change under transformations and are independent of any co-ordinate system (Berrhardsen, 1992). The topological characteristics of an object are also independent of scale of measurement (Chrisman, 1997). Topology, as it relates to spatial data, consists of three elements, namely, adjacency, containment and connectivity (Burrough, 1986). Topology may be defined as constituting those properties of geometrical figures that are invariant under continuous deformation (McDonneland Kemp, 1995). Broadly, topology can be explained any two ways. Firstly, topologically spatial relationships with the entity which are learned by human beings at a very early age. Secondly, topology consists of metric aspects of spatial relations, such as, size, shape, distance and direction. Many spatial relations between objects are topological in nature, including adjacency, containment and overlap.

Adjacency and containment describe the geometric relationships which exist between area features. Areas can be described as being adjacent when they share a common boundary. For example, boundary of the area of municipal corporation of Hyderabad and Secunderabad is common, or may be adjacent. Containment is an extension of the adjacency that describes area features which may be wholly contained within another area feature, such as, an island within a lake. Connectivity is a geometric property used to describe the linkages between line features, like road network. The geometric relationship between spatial entities and corresponding attributes are very crucial for spatial analysis and integration in GIS.

In other cases, metric properties, such as, distance or direction, expressed either quantitatively or qualitatively, may determine the meanings of various terms, for example, both 'north of' any 'near' normally refine the 'disjoint' topological relation, and are ill-defined for non-disjoint entities. Spatial relations between disjoint entities, which neither touch nor overlap, are characterised by a system of distinctions that is essentially independent of the system used to describe and classify spatial relations for non-disjoint entities. Some of the spatial relations between disjoint objects are distance, direction, and reference frames.

Distance may be pure Euclidean distance. In natural language 'hedge' words such as 'about' are often associated with approximate numerical distance. Distance may be given in qualitative rather than metric terms, dividing distance into just three categories : 'at', 'near', and 'far'. Direction may be either qualitative or quantitative. Direction is an orientation specified relative to some reference frame. Directional relations are though of as being between points. Directions are not so straightforward between spatially-extended entities, since a large range of directions may exist, between any point in one entity and any point in the other. Reference frames are used in discourse and spatial reasoning. Geographically, in many cultures, a reference frame based on cardinal directions seems, dominant for geographical spaces, whereas viewer-centered or object-centered reference frames often dominate over bodily or tabletop ('manipulatable') spaces and entities.

9.6.4 Congnitive Models

Both entities and fields exist in cognitive models. Entities are typically conceptualised as being organised by dimensionality in points, lines, areas, volumes. Entities often have indistinct boundaries, a fact which is at odds with typical GIS representation schemes. Entities are also categorised, and since many aspects of nature from a continuum, categories may be relatively arbitrary and thus subject to disparate cultural differences. Spatial relations, on the other hand, seem to be very similar in disparate cultures and languages. Congnitive spatial relations are predominantly topological but metric factors such as distance and direction often refine the relations and characterise prototypical relations.

There is a very real sense in which all representations are cognitive. Mathematics is, after all a formalisation of how at least some people think. The cognitive view of spatial relations, however, emphasises the importance of human subjects being testing, preferably under laboratory-controlled conditions, in defining the nature of the spatial representations that are needed for geographical information systems and spatial analysis. Many geographical distributions, such as, those of soil variables are inherently complex, revealing more information at higher spatial resolution apparently without limit (Mandelbrot, 1982). Geographical data modelling is the process of discritisation that converst complex geographical reality into a finite number of database records or objects. Objects have geographical expression as points, lines, and areas, and also possess descriptive attributes, for example, collection of water samples from different wells, in which the location of the well creates point object (with the address of the location) and associated water quality parameter is its attribute value. A major difficulty arises in the case of the object models. Many geographical objects have inherently fuzzy spatial extents. One common solution to this problem is to allow objects to have multiple representations which depend on the scale. For example, a river might be a single line at scales smaller than 1:50,000, but a double line at larger scales. Six field models are in common use in GIS: (i) irregular point sampling, (ii) regular point sampling, (iii) contours, (iv) polygons, (v) cell grid modelling, (vi) triangular network models. The object models are commonly used to represent man-made facilities. For example, an underground pipe more naturally represents as a linear object than as a value in a layer. Pipes can cross each other in object model, but cause problems in a field model.

9.7 GIS Queries

As a decision support system, a GIS must provide potential for aspatial (nonspatial) and spatial queries, as well. The answer to aspatial kind of queries does not require the spatial locations of the geographical features involved. See the question : "calculate the number of churches in Hyderabad," or "Compute the percentage of grass in Hyderabad." To answer questions like this, a GIS as well as a number of statistical and spreadsheet packages, don't require the stored value of latitude and longitude, or x, y, z coordinates.

Spatial Queries carried out spatial operations and links data sets using location as the common key. Location condition trends pattern modelling. See the question: "Calculate the number of Greek Orthodox Churches in the area surrounded by the roads, SN Colony Road, Rajbhavan Road, and other roads." To answer a question like this, the GIS must know the spatial location of particular map features, in this case, the roads as line map features and churches as point map features.

9.8 GIS Architecture

According to the definition proposed by Marble and Peuquet (1983), GIS deals with space-time data, and often but not necessarily, employs computer hardware and software. GIS can be understood as the subsystem nature within the framework of a main system. According to these investigators, GIS has the following generic subsystems:

(i) A data input subsystem which is also called data capture subsystem
(ii) A data storage and retrieval subsystem
(iii) A data manipulation and analysis subsystem
(iv) A reporting subsystem.

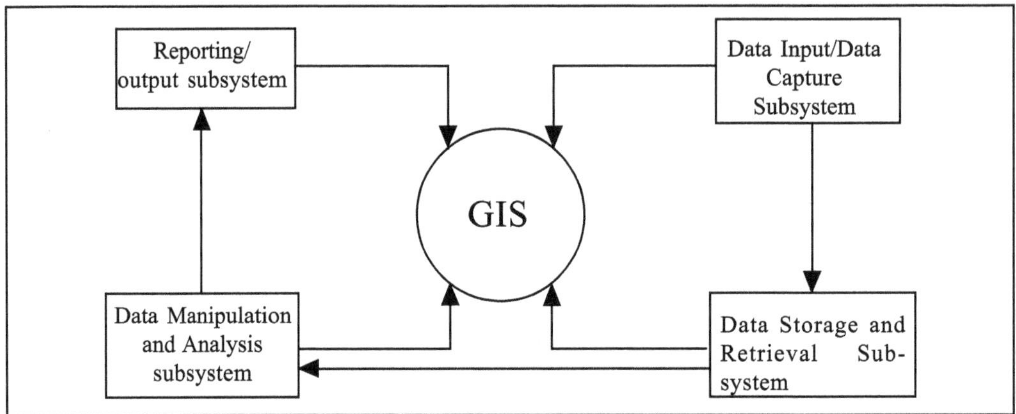

Fig. 9.5 Subsystem nature of GIS (structural prespective).

Each of the subsystems has been described in terms of functions that the respective subsystem performs. The data input/capture subsystem provides operational functions for acquiring data. The data management or data storage and retrieval subsystem stores and retrieves the data elements. The manipulation and analysis subsystem handles the transformation of data from one form to another and derivation of information from the data. The fourth subsystem output/reporting subsystem provides a way for the user to see the data in the form of diagrams, maps, and/or tables. Fig. 9.5 shows the architechture of all subsystems of GIS from a structrual perspective.

9.8.1 Components of a GIS

Geographical Information Systems have three important components, namely, computer hardware, sets of application software modules, and a proper organisational setup. These three components need to be in balance if the system is to function satisfactorily. GIS run on the whole spectrum of computer systems ranges from portable personal computers to multi-user supercomputers, and are programned in a wide variety of software packages. Systems are available that use dedicated and expensive work stations, with monitors and digitising tables built in. In all cases, there are a number of elements that are essential for effective GIS operations. These include (Burrough, 1986):

(i) the presence of a processor with sufficient power to run the software
(ii) sufficient memory for the storage of large volumes of data
(iii) a good quality, high resolution color graphics screen and
(iv) data input and output devices, like digitisers, scanners, keyboards, printers and plotters.

The general hardware components of a GIS include control processing unit which is linked to mass storage units, such as, hard disk drives and tape drives, peripherals such as digitiser or scanner, printer or plotter and Visual Display Unit (VDU). Fig. 9.6 shows the major hardware components of a GIS.

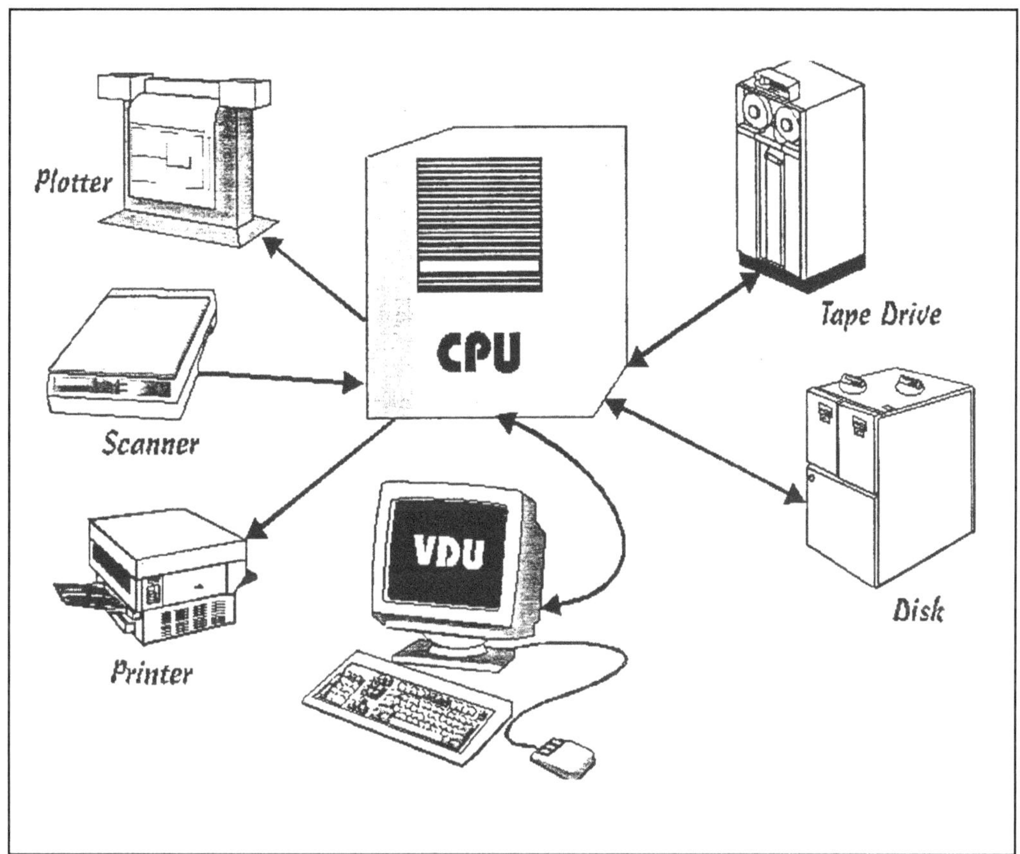

Fig. 9.6. Hardware components of GIS.

There are a number of essential software elements that must allow the user to input, store, manage, transform, analyse and output data. Therefore, the software package for a GIS consists of four basic technical modules. These basic modules are : (i) data input and verification, (ii) data storage and database Management (iii) data transformation and manipulation, and (iv) data output and presentation. The GIS software package should have the capabilities performing all the four GIS subsystems. The GIS hardware and software govern the way in which geographical information can be processed but they do not themselves guarantee that any particular GIS will be used effectively.

9.8.2 GIS Work Flow

There are five essential elements that a GIS must contain. They are data acquisition, preprocessing, data management, manipulation and analysis, and product generation. For any application of GIS, it is important to view these elements as a continuing process. Fig. 9.7 shows the work flow process of GIS in procedural perspective. Data acquisition is the process of identifying and gathering the data required for any given application. This typically involves a number of procedures. One procedure might be to gather new data by preparing large-scale maps of natural vegetation from field observations. Other procedures for data acquisition may include locating and acquiring existing data, such as, maps, aerial and ground photography, and data acquired by satellite sensing systems.

Collection, input, and correction operations concerned with receiving data into the system include manual digitising, scanning, keyboard entry of attribute information, and online retrieval from other database systems. It is at this stage that a digital map is first constructed. The digital representation can never be of a higher accuracy than the input data, although the mechanisms for its handling will frequently be capable of greater precision than that achieved during data collection. The essential preprocessing procedures include: (a) format conversion, (b) data reduction and generalisation, (c) error detection and editing, (d) Merging of points into lines, and lines into polygons, (e) Edge matching and tiling, (f) Rectification/registration, (g) Interpolation, and (h) Interpretation.

The functions of database management govern the creation of an access to the database itself. These functions provide consistent methods for data entry, update, deletion, and retrieval. Modern database management systems isolate the users from the details of data storage, such as, the particular data organisation on a mass storage medium. A modern Database Management System (DBMS) is used to create GIS database, that is, attribute database.

Storage and retrieval mechanisms include the control of physical storage of the data in memory, disk or tape, and mechanisms for its retrieval to serve the needs of the other three components. In a disaggregate GIS this data storage may be physically more from the rest of the system, and may meet the database requirements. This module includes the software structures used to organise spatial data into models of geographic reality.

The development of new derived data layers, which may form the input to further analysis, is an important function of any GIS. The list of data manipulation and analysis operations are, (i) reclassification and aggregation, (ii) Geometric Operations: as

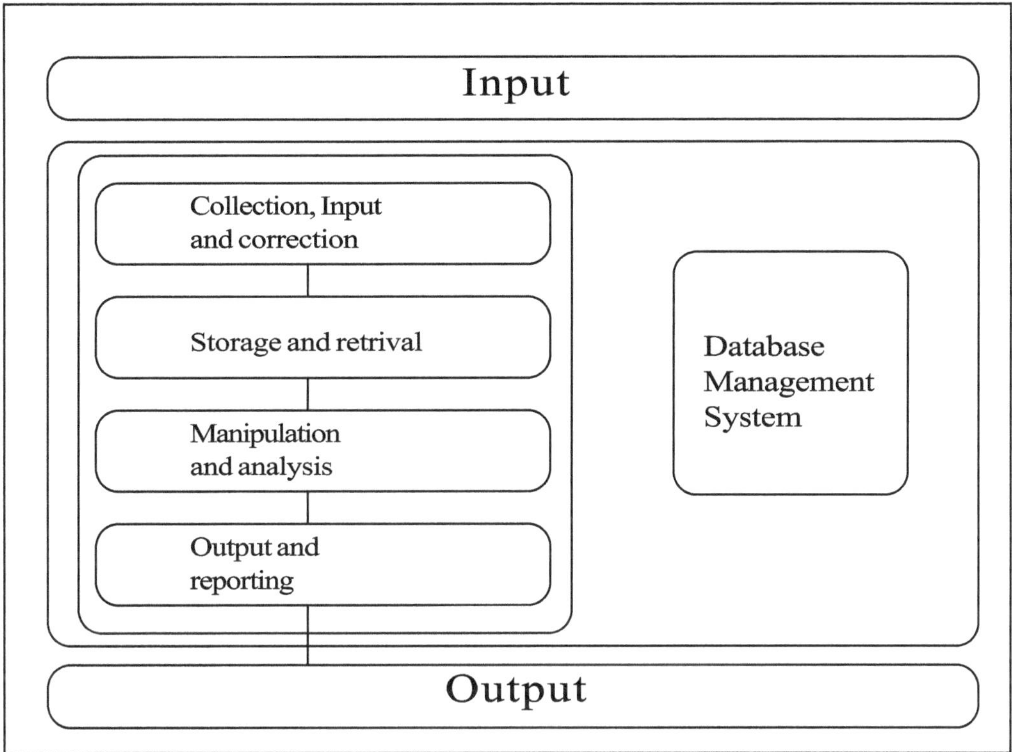

Fig. 9.7 GIS workflow process of GIS (Procedural perspective).

rotation, translation and scaling, rectification, and registration, (iii) Controlled determination, (iv) Data structure conversion, (v) Spatial operations of connectivity and neighbourhood operations, (vi) Measurement of Distance and Direction, (vii) Statistical analysis as descriptive statistics regression, correlation, and cross-tabulation, and (viii) Modelling. This operation or subsystem represents the whole spectrum of techniques available for the transformation of the digital model by mathematical means. A library of data-processing algorithms is available for the transformation of spatial data, and incorporated in new visual maps. Using these techniques it is possible to deliberately change the characteristics of the data representation in order to meet theoretical requirements. It is equally possible to mishandle or unintentionally distort the digital map at this state.

Product generation is the phase where final outputs from the GIS are created. These output products might include statistical reports, maps, and graphics of various kinds. Some of these products are softcopy images and hardcopy.

9.9 Theoretical Models of GIS

Reference should now be made to the existing models of GIS operation, which are broadly similar in nature. These may be considered in two main groups, namely, (i) the functional elements of GIS, and (ii) the fundamental operations of GIS. It will be seen that these approaches generally make little or no reference to the very important process as which have been outlined above. It is due to their failure to address these issues that they are unable to offer much help to those who would seek to apply GIS in new situations. In particular, certain socioeconomic phenomena need special treatment. Conceptual models tend only to address aspects of the operation or composition of GIS systems. They make no statement about the nature of the data representation. The components in these models are basically analogous to the main software components in any general-purpose systems. Bracken and Webster (1989) suggest an alternative classification which recognises three major components in its characterization of GIS. They are the problem-processor model database model and interface model. However, this is still an explicitly software-oriented approach to understanding GIS.

9.9.1 Functional Elements of GIS

Bracken and Webster (1987) outlined four functional elements to address the GIS technology. They are database approach, the process-oriented approach, an application oriented approach, and toolbox approach.

Database approach stresses the ability of the underlying data structures to contain complex geographical data. The process-oriented approach focuses on the sequence of system elements used by an analyst running an application. An application oriented approach defines GIS based on the kinds of information manipulated by the system and the utility of the derived information produced by the system while the toolbox approach emphasis as the software components and algorithms that should be contained in a GIS.

9.9.2 Fundamental Operations of GIS

This approach considers the functions which GIS is able to perform. The operations discussed in this section fall entirely within the manipulation-and-analysis subsystem referred to above, and are thus wholly internal to the GIS. The fundamental classes of operations performed by a GIS have been characterised as 'map algebra' (Tomlin and Berry, 1979; Berry, 1982, 1987; Tomlin, 1991) in which context primitive operations of map analysis can be seen as analogous to traditional mathematical operations. The 'classes of analytical operation' are divided into reclassification, overlay, distance/ connectivity measurement and neighbourhood characterisation of the data. These operations can be identified as follows :

(i) Reclassification operations transform the attribute information associated with a single map coverage.

(ii) Overlay operations involve the combination of two or more maps according to boolean conditions and may result in the delineation of new boundaries.

(iii) Distance and connectivity measurement include both simple measure of inter-point distance and more complex operations such as the construction of zones of increasing transport cost away from specified locations and

(iv) Neighbourhood characterisation involves the values to a location both summary and mean measures of a variable, and include smoothing and enhancement filters. Sequences of such manipulation operations have become known as 'cartographic modelling'.

9.10 Theoretical Framework for GIS

This discussion is based on an analysis of the way in which data are transformed and held as a digital model of the external world. The geographic data-processing system outlined is not intended to be a description of any specific software system, but is a model of the processes which may operate with digital geographic data. The idea of data representation used here should not be confused with work on specific data structures, either spatial, such as, vector, raster, triangulated irregular network, and quadtree, or attribute such as, hierarchical and relational which are mainly technical issues. Actual software systems may be identified which perform all or some of the principal transformations to a greater of lesser degree, but only those containing some capacity for input, manipulated and output digital spatial data in some form are considered to be 'GIS' in the present context.

The ground work for this approach has been laid by cartographers seeking to understand the relationships between the world and the map as a model. We have seen how the digital map is related to the analog map, and we must now modify and extend the existing theoretical structure to replace the paper map with a digital map sitting within a GIS. The process of analog map production may be modelled as a series of transformations between the real world, raw data, the map, and the map image (Fig. 9.8).

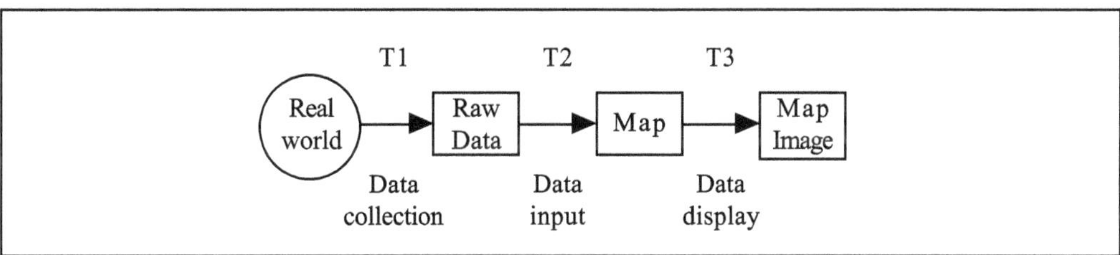

Fig. 9.8 Transformation stages in the tradional cartographic process.

The significance of these transformations is that they control the amount of information transmitted from one stage to the next. The cartographer's task is to devise the very best approximation to an 'ideal' transformation involving a minimum of information loss. A detailed explanation of the 'transformational' view of digital cartography is given in Clarke (1990).

In the context of GIS, we may add an additional transformation stage which sits entirely within the GIS (Fig. 9.9.). The sequence of transformation stages illustrated in the figure forms the basis for the discussion of GIS techniques later in this book. In the first transformation (T_1), data are selected from the real word, for example, surveying measurements or census data. These are then input for the GIS in some form (T_2) to provide the basis for its digital map representation of the real world. Within the system, a vast range of manipulation operations are available to further transform the data and store the results (T_3), and these may be communicated as tabulated or graphic images by means of a hardcopy or screen (T_4). It is worth noting that each of these transformation stages may actually involve several physical operations on the data. For example, T_1 may involve both collection and aggregation, and T_3 will almost always consist of a whole series of data-processing operations. If the 'thematic and spatial' charateristics noted by Berry (1987) are to be understood, it is necessary for us to carefully consider the way in which these four transformations may effect the digital representation of the real-world objects.

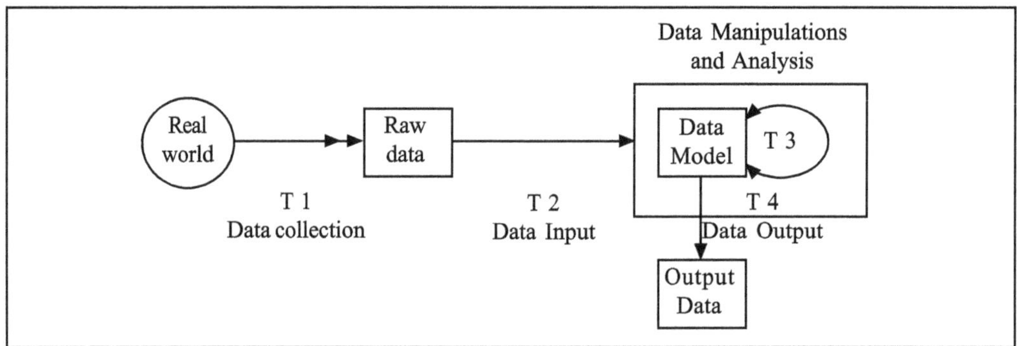

Fig. 9.9 A transformation based view of GIS operation.

9.11 GIS Categories

For more than two decades, the GIS applications were dominated by the so-called "map geometry", that is GIS data with only x, y, coordinates. If an entity or a point/line is represented by two values, for example, x-coordinate; y-coordinate, thus it is said to be 2-D GIS. Efficient GIS products are offered by vendors which, besides the two-dimensional x, y-geometry, intergrate the "third-dimension" using the "normal case" Digital Terrain Model (DTM) equation $z = f(x, y)$ for the height, or any other predefined simple equation for an attribute like soil type. This is called 2.5 D GIS.

In the 2.5-D case, for the same pair of x, y co-ordinates only one spatial location (point's height) or attributes or attribute value (soil type) exist, and this can be calculated using a predefined formula.

Users' perception is three-dimensional. Therefore, there is an existing need for 3-D data, but in this case the data handling is more difficult, and requires more complex algorithms. In the 3-D case, for the same pair of x, y co-ordinates a number of spatial locations (points) with a different z-co-ordinate could exist because the z co-ordinate is calculated independently from the x, y pair. Hence, these systems are suitable for any kind of terrain modelling. Also, in cases where the third dimension is referred to an attribute, the actual attribute value is indepedent from the x, y spatial location and it is calculated using complex mathematics. 4-D GIS or Temporal GIS is the fourth category of GIS. In this case, the fourth dimension namely, time, is introduced. For this reason, the kind of GIS applications depending on time are regraded as temporal GIS systems.

9.12 Levels/Scales of Measurement

So far, we have examined the fundamental considerations and, various components of GIS. The entities associated with them, a set of coordinates that allow to locate the positions and the corresponding descriptions of each location, can be called attribute value.

All these spatial features or entities, contain information not only about how they occupy space but also about what they are and how important they are. The additional non-spatial information that helps us to describe the objects we observe in space comprises the feature's attributes. The character of attribute data themselves can influence the utility of data sets in GIS analysis. One characteristic which is of considerable importance is the scale of measurement used to record and report the data. There is already a well-established measurement frame work for nearly all forms of data including geographic data. These are called levels of geographic data measurement (Fig. 9.10). They illustrate the levels of measurements in terms of commonly used geographic features.

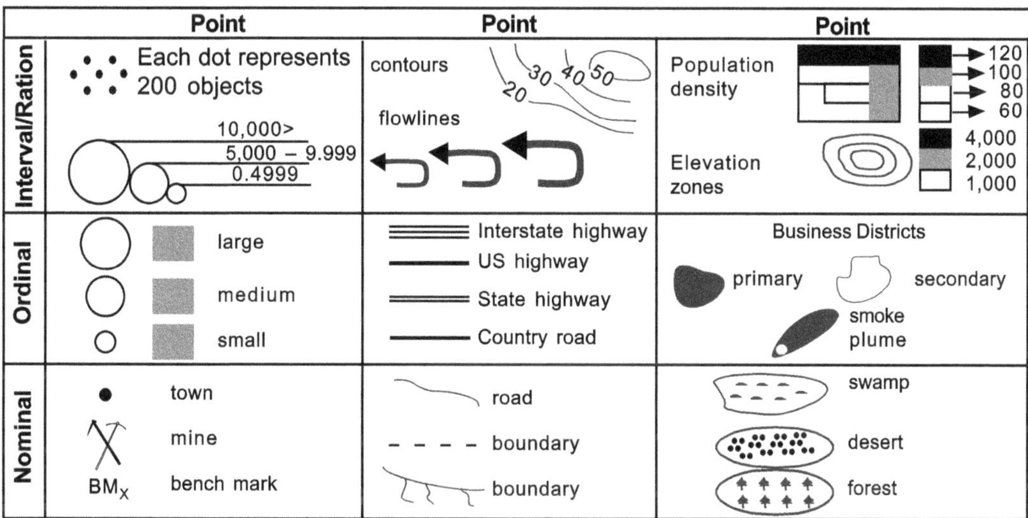

Fig. 9.10 Levels of measurments.

The four commonly refered levels of measurement are, (i) nominal scale, (ii) ordinal scale, (iii) interval scale, and (iv) ratio scale. Nominal scales are those variables which are described by name, with no specific order, for example, land use, parks, residential areas and so on. These are 'named' data. The system allow us to make statements about what to call the object, but it does not allow direct comparisons between one named object and another.

Table 9.1 Scale (level) of measurement

Data	Unit of measurement	Scale/level
Hotel name	Text	nominal
Status of Hotel	Three Star	ordinal
Average Tariff	in Rupees	Interval
Size of the Hotel	m^2	Ratio

Ordinal variables are those variables which are lists in discrete classes but with an inherent order, for example, classes of streams, like I order and II order. In the interval scale of measurement., the numbers are assigned to the items measured. Data measured can be compared as in the case of ordinal scale of measurement and example have a natural sequence of temperature measured in degrees. The ratio variables have the same characteristic as interval variables, but they have natural zero or a starting point. Rainfall per month is an example of ratio variable. Table 9.1 shows an example of level (scale) of measurement.

9.13 GIS Software Producers and their Main Products

A Geographic Information System (GIS) is a computer system for capturing, storing, querying, analyzing, and displaying geographic data. Like any other information technology, GIS can be divided into the following four components :

- **Computer System :** The computer system includes the computer and the operating system to run GIS. Typically the choices are PCs that use the Windows operating system (e.g., Windows 2000, Windows NT) or workstations that use the UNIX operating system. Additional equipment may include monitors for display, digitizers and scanners for spatial data input, and printers and plotters for hard copy data display.

- **GIS Software :** The GIS software includes the program and the user interface for driving the hardware. Common user interfaces in GIS are menus, graphical icons, and commands.
- **Brainware :** Equally important as the computer hardware and software, the brainware refers to the purpose and objectives, and provides the reason and justification, for using GIS.
- **Infrastructure :** The infrastructure refers to the necessary physical, organizational, administrative, and cultural environments for GIS operations. The infrastructure includes requisite skills, data standards, data clearinghouses, and general organizational patters.

GIS is not new. Since the late 1960s computers have been used to store and process geographic data. Early examples of GIS - related work include the following :

- Computer mapping at the University of Edinburgh (Coppock 1988), the Harvard Laboratory for Computer Graphics (Chrisman 1988), and the Experimental Cartography Unit (Rhind 1988).
- Canada Land Inventory and the subsequent development of the Canada Geographic Information System (Tomlison 1984).
- Publication of Ian McHarg's Design with Nature (McHarg 1969) and the map overlay method.
- Introduction of an urban street network with topology in the U.S. Bureau of the Census' DIME (Dual Independent Map Encoding) system (Broome and Meixler 1990).

For many years, though, GIS has been considered to be too difficult, expensive, and proprietary. The advent of the graphical user interface (GUI), powerful and affordable hardware and software, and public digital data has broadened the range of GIS applications and brought GIS to mainstream use. The following is a list of GIS software producers and their main products.

- Environmental Systems Research Institute (ESRI), Inc. (http:www.esri.com/) : **Arc Info, Arc View**
- Autodesk Inc. (http://ww3.autodesk.com/) : **Auto CAD Map**
- Baylor University, Texas (http://www.baylor.edu/~grass/) : **GRASS**
- Clark Labs (http://www.clarklabs.org/) : **IDRISI**
- International Institute for Aerospace Survey and Earth Sciences, the Netherlands (http://www.itc.nl/ilwis/): **ILWIS**

- Map Info Corporation (http://www.mapinfo.com/) : **MapInfo**
- Think Space Inc. (http://www.thinkspace.com/) : **MF works**
- Intergraph Corporation (http://www.intergraph.com/) : **MGE, Geo Media**
- Bentley Systems, Inc. (http://www.bentely.com/) : **Microstation**
- PCI Geomatics (http://www.pcigeomatics.com/) : **PAMAP**
- TYDAC Inc. (http://www.tydac.ch/) : **SPANS**
- Caliper Corporation (http://www.caliper.com/) : **Trans CAD, Maptitude**
- Northwood Technologies Limited (http://www.northwoodtec.com/) : **Vertical Mapper**

According to a published survey (Crockett 1997), ESRI Inc. and Intergraph Corporation have dominated the market for GIS software. Two main products from ESRI Inc. are Arc View and ArcInfro 8. A desktop GIS package, Arc View uses extensions and stand - alone utility programs to extend its capabilities. Arc Info 8 consists of Arc Info Work station, Arc Catalog, Arc Map, and ArcToolbox. Arc Info Workstation, which is basically the same as ARC/INFO version 7.x and referred to as ARC/INFO hereafter, operates on UNIX, Windows 2000, or Windows NT, whereas the other three are limited to Windows 2000 or Windows NT. Intergraph Corp. also has two main products: MGE and GeoMedia. MGE consists of a series of products and operates on Windows 2000, Windows NT, or UNIX. GeoMedia is a desktop GIS package and is compatible with standard Windows development tools.

9.14 Strategic Partnerships in the GIS Industry

Small GIS companies often deploy strategic partnerships to strengthen their roles on the GIS market (Box 1.1). GRASS is unique among GIS software packages because it is free. Originally developed by the U. S. Army Construction Engineering Research Laboratories, GRASS is currently maintained in both the United States (Baylor University) and Germany (University of Hannover). TransCAD is a GIS package specially designed for use by transportation professionals. Microsoft and Oracle, two giant companies in information technology, have also entered the GIS industry, Microsoft Map Point targets business analysis who need to analyze and map geographic data (http://www.microsoft.com/). Oracle Spatial provides the features to store, access, and manage spatial data in Oracle8i, a relational database management system (http://www.oracle.com/).

Along with the proliferation of GIS activities numerous GIS textbooks have been published (see Appendix A), and several journals and trade magazines are now devoted to GIS and GIS applications.

9.15 GIS Applications

Over the years, new applications have been found by intergrating GIS with other technologies. Among the examples are precision farming, interactive mapping on the Internet, and location - based services. Precision farming refers to site - specific farming such as herbicide or fertilizer application. Technologies for precision farming include GPS to mark the location of weeds or plots, GIS to store and process data and to prepare maps, and advanced computer-controlled sprayers.

Interactive mapping lets Internet users select map layers for display and make their own maps. Websites set up for interactive mapping usually provide the functionalities of pan, zoom, and simple query. The U.S. Geological Survey (USGS) maintains a website for interactive mapping using the National Atlas of the United States (http://www.nationalatas.gov/).

The U.S. Census Bureau also has a website called Tiger Mapping Service, where Internet users can map public geographic data of anywhere in the United States (http://tiger.census.gov/).Typically, GIS provides the database and the query and mapping functionalities for interactive mapping on the Internet.

A new application that has been promoted by GIS companies, especially Map Info and Intergraph, is called location-based services (LBS). This application allows users to deliver and receive information that is related to a particular location through wireless telecommunication. For example, an LBS user may use a web-enabled mobile phone to be located and to access, through mobile location services, information that is relevant to the user's location such as nearby ATMs and Chinese restaurants.

Since the beginning, GIS has been important in natural resource management, such as land use planning, timber management, wildlife habitat analysis, riparia zone monitoring, and natural hazard assessment. In more recent years GIS has been used in emergency planning, market analysis, facilities management, transportation planning, and military applications. Integration of GIS with other technologies such as Global Positioning System (GPS) and the Internet has introduced new applications (Box 1.2). In summary, GIS has become one type of information system, specializing in information that has a geographic or spatial component.

The ability of GIS to handle and process geographically referenced data distinguishes GIS from other information systems. *Geographically referenced data* describe both the location and characteristics of spatial features on the Earth's surface. Roads and land use types are spatial features and are precipitation and elevation. In describing a road, for example, we need to refer to its location (i.e., where it is) and its characteristics (road classification, traffic volume, etc.). GIS therefore involves two geographic data components: *spatial data* relate to the geometry of spatial features and *attribute data* give the information about the spatial features.

9.16 GIS Data and Models

9.16.1 Spatial Data

Mapmakers represent spatial features on the Earth's surface as map features on a plane surface. The locations of map features are based on a Cartesian coordinate system with intersecting perpendicular lines along the x-axis and y-axis, whereas the locations of spatial features are based on the geographic grid expressed in longitude and latitude values. The transformation from the spherical geographic grid to a plane coordinate system is called map projection. Hundreds of map projections have been developed for mapmaking. Every map projection preserves certain spatial properties while sacrificing other properties.

A basic principle in GIS is that map layers to be used together must be based on the same coordinate system. Otherwise, map features from different layers will not register with one another spatially.

Spatial features may be discrete or continuous. Discrete features are those that do not exist between observations, form separate entities, and are individually distinguishable (Dent 199). Wells, road, and land use types are examples of discrete features. Continous features exist spatially between observations (Dent 1999). Precipitation and elevation are examples of continuous features.

GIS uses two basic data models to represent spatial features: vector and raster the vector data model uses points and their x-, y- coordinates to construct spatial features of points.

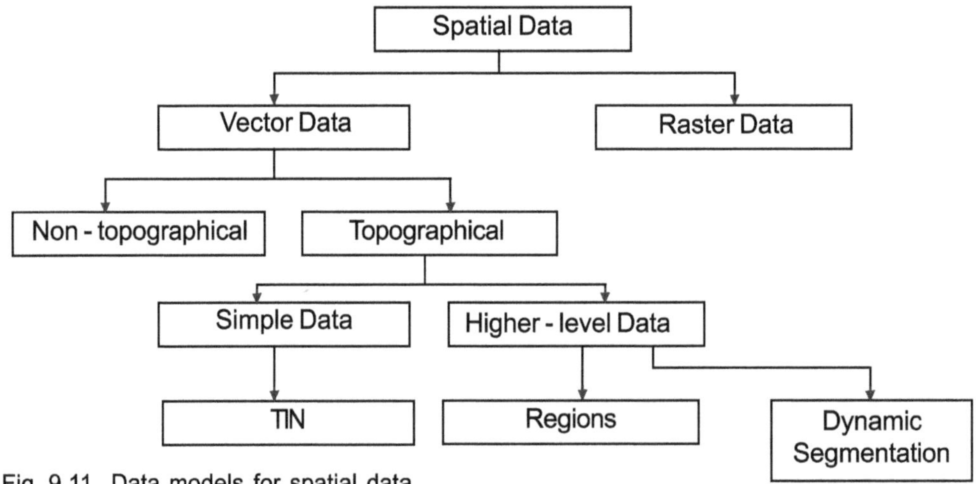

Fig. 9.11 Data models for spatial data.

Lines and areas : Vector-based features are treated as discrete objects over the space. The raster data model uses a grid to represent the spatial variation of a feature. Each cell in the grid has a value that corresponds to the characteristic of the spatial feature at that location. Raster data are well suited to the representation of continuous spatial features like precipitation and elevation.

The data model determines how the data are structured, stored, processed, and analyzed in a GIS. Many GIS functions are either vector-based or raster-based. So are different modules of a GIS package. Arc View provides its raster-based functionalities through the Spatial Analyst extension, and ARC/INFO offers its raster-based commands in the GRID module. Likewise, MGE include the Grid Analyst to work with raster data. Some GIS packages started as either raster-based or vector-based and later acquired capabilities of working with the other data model through strategic partnerships.

The data model is not really an issue to the GIS user nowadays. Raster and vector data can be displayed simultaneously. Raster data can be converted to vector data, and vice versa. A GIS has in fact become a useful tool for integrating raster and vector data.

Vector data are categorized as topographical and non-topological (Fig. 9.1). *Topology*, as used in GIS, expresses explicitly the spatial relationships between features, such as two lines meeting perfectly at a point and a directed line having an explicit left and right side. Topology is useful for detecting digitizing errors in digital maps and is necessary for some map overlay operations and network analysis. Non-topological data display faster and, more importantly, they can be used directly in different GIS software packages. GIS users should know whether their data are topological or not. ESRI, for example, uses the term *coverage* for topology-based digital maps and *shape file* for non-topological vector data.

Topological vector data can be further divided into simple and higher-level data. Simple vector data consist of points, lines, and polygons. Higher level data are built upon simple points, lines, and polygons. Higher level data are built upon simple points, lines, and polygons. The triangulated irregular network (TIN) approximates the terrain with a set of non overlapping triangles. Each triangle in a TIN consists of points and edges (lines) that connect these points to form triangles. A region is a collection of polygons, which may or may not be connected, and regions may overlap with one another or form a nested set. Dynamic segmentation is a data model that is built upon lines of a network and allows the use of real-world coordinates with linear measures such as mileposts.

Some GIS packages have recently introduced the object-oriented data model. The object-oriented data model uses objects to organize spatial data. An object has a set of built-in properties and can perform operations upon requests. To represent spatial features, objects may take the form of points, multi points (a collections of points), polylines (a set of line segments), or polygons (including disjoint and nested polygons). As discussed in the next section, the object-oriented data model is different from the above data models because the geometries of objects are stored as an attribute called geometry.

9.16.2 Attribute Data

Attribute data describe the characteristics of spatial features. The amount of attribute data to be attached to a spatial feature can vary significantly depending on the feature type and the application. The Map Unit Interpretations Record (MUIR) attribute database, for example, contains properties, interpretations, and performance data in a series of tables for each soil map unit surveyed by the Natural Resources Conservation Service (NRCS) in the United States (http://www.statlab.iastate.edu/soils/muir/).

Many commercial GIS packages store attribute data separate from spatial data in a split data system, often called the geo relational model. Spatial data are stored in graphic files and managed by a file management system. Attribute data, on the other hand, are stored in a relational database. A relational database is a collection of tables, also called relations, which can be connected to each other by attributes whose values can uniquely identify a record in a table. Spatial data and attribute data in a GIS are typically linked through the feature IDs.

GIS users can use a relational database for data search, data retrieval, data editing, and creation of tabular reports. A relational database has two distinctive advantages in GIS applications. First, each table in the database can be prepared, maintained, and edited separately from other tables. Second, the tables can remain separate until a query or an analysis requires attribute data from different tables be linked together. Because the need for linking tables is often temporary, a relational database is efficient for both data management and data processing.

The relational database model is the standard model in GIS and management information system. With GIS increasingly becoming part of an organization's much larger information system, attribute data needed for a GIS project are likely to come from an enterprise-wide database. If spatial data distinguish GIS as a special type of information system, attribute data tend to connect GIS with other information systems.

Instead of storing spatial data and attribute data in a split data system, the object-oriented data model stores both data in a single database. A field called geometry is used to store spatial data. The object-oriented data model eliminates the complexity of coordinating and synchronizing two data file systems and actually brings GIS closer to other information systems.

9.17 GIS Operations

This section provides an overview of GIS operations. Although GIS activities no longer follow a set sequence, to explain what GIS users do, GIS activities can be grouped into spatial data input, attribute data management, data display, data exploration, data analysis, and GIS modeling (Fig. 9.12).

Spatial Data Input	1. Data entry: use existing data, create new data
	2. Data editing
	3. Projection and re-projection
	4. Geometric transformation
Attribute Data Management	1. Data entry and verification
	2. Database management
Data Display	1. Attribute data query
	2. Spatial data query
	3. Geographic visualization
Data Analysis	1. Vector data analysis: buffering, overlay, distance measures, map manipulation
	2. Raster data analysis: local, neighborhood, zonal, global
	3. Terrain mapping and analysis
	4. Spatial interpolation: global, local
	5. Regions-based analysis
	6. Network analysis
GIS Modeling	1. Binary models
	2. Index models
	3. Regression models
	4. Process models

Fig. 9.12 A classification of GIS activities.

9.17.1 Spatial Data Input

The most expensive part of a GIS project is database construction. Two basic options for database construction are (1) use existing data and (2) create new data. Digital data clearing houses have become commonplace on the Internet in recent years. The strategy for a GIS user nowadays is to look at what exists in the public domain before deciding either to create new data or to buy data from private companies.

New GIS data can be created from satellite images, GIS data, or paper maps. A variety of maps such as land use, land cover, and hydrography can be extracted from processing satellite images. Using satellites in space as reference points, a GPS receiver can determine its precise position on the Earth's surface, which can then be used to determine the location and shape of spatial features.

There are two methods for converting paper maps to digital maps : digitizing by using a digitizing table or a computer monitor, also called manual digitizing, and scanning. Scanning is preferred over manual digitizing in most cases, because scanning uses the machine and computer algorithm to do most of the work, thus avoiding human errors caused by fatigue or carelessness.

No matter how carefully it is prepared, a newly digitized coverage always has some errors. Digitizing errors can be removed through data editing, a part of database construction. One common type of digitizing error relates to the location accuracy of spatial data, such as missing lines or polygons or distorted lines. Another common type consists of topological errors, such as dangling arcs and unclosed polygons, which are caused by the failure of digitized features to follow the topological relationships among points, lines, and areas.

A newly digitized map has the same measurement unit (e.g., inches) as the source map used in digitizing or scanning. Therefore it must be converted to real-world coordinates by using a set of control points with known real-world coordinates and a process called geometric transformation.

9.17.2 Attribute Data Management

To complete database construction for a GIS project, attribute data must be entered, verified, and managed. Similar to spatial data, attribute data entry and verification also involve digitization and editing. Attribute data are usually managed in a relational database in GIS, whether attribute data are separate from spatial data or not. Because a relational database consists of separate tables, some of which may even reside at remote locations, the tables must be designed to facilitate data input, search, retrieval, manipulation, and output.

Two basic elements in the design of a relational database are the key and the type of data relationship. A key is a common field between two tables, which can establish a connection between corresponding records in the tables. The type of data relationship dictates how the tables are actually joined or linked. Expressed in the number of records, the type of relationship may be one-to-one, one-to-many, or many-to-one.

9.17.3 Data Display

Data display through maps, tables, and charts is a common part of a GIS project. As a visual tool, maps are most effective in communicating spatial data, whether the emphasis is one the location or the distribution pattern of spatial data. Either individually or grouped together, maps are important for visualization and query. Maps are also plotted to show results of GIS analysis.

A map consists of such elements as the title, subtitle, body, legend, north arrow, scale, acknowledgment, and border. These elements work together to bring spatial information to the map user. The first step in data display is to assemble map elements. Windows-based GIS packages have simplified the process of putting together a map. The GIS user can choose from program menus conventional symbols for qualitative and quantitative data, and can point and click the graphic icons to construct a map.

The second step in data display is map design. A well-designed map can help the mapmaker communicate spatial information to the map reader. A poorly designed map can confuse the map reader and even distort the information intended by the mapmaker. A creative process, map design cannot be easily performed by computer codes.

9.17.4 Data Exploration

Data exploration is data-centered query and analysis. Data exploration can be a GIS operation by itself or a precursor to formal data analysis. Data query allows the user to explore the general trends in the data, to take a close look at data subsets, and to focus on possible relationships between data sets. The purpose of data exploration is to better understand the data and to help formulate research questions and hypotheses.

An important component of effective data exploration consists of interactive and dynamically linked visual tools, including maps, graphs, and tables. The GIS user can perform data query using a map or table. Following a query, the GIS user can view both the spatial and attribute components of the data subset, thus facilitating information processing and synthesis.

Geographic visualization (Mac Eachern 1995) is functionally similar to data exploration except that it is map-oriented. Tools for geographic visualization include data classification, data aggregation, and map comparison.

9.17.5 Data Analysis

Data analysis in GIS is closely related to the data model. The basic data models are vector and raster each having its own set of analytical functions. Common vector functions include buffering, map overlay, distance measurement, and map manipulation. Raster data analysis can be conducted at the level of individual cells, or groups of cells, or cells within an entire grid. Based on the level of computation, raster data operations are commonly grouped into local, neighbourhood, zonal, and global operations. Although some GIS concepts, such as map overlay and buffering, are the same for vector and raster data, the operational procedures differ.

The difference between vector and raster data is not an issue for terrain mapping and analysis because both types of data can be used. Raster data are ideal for the representation of surfaces, and the raster data structure is well-suited to intense computation often required for terrain analysis. But vector data, such as elevation points, contour lines, and streams and roads that represent changes of the terrain, are important inputs for terrain mapping and analysis. Moreover, the TIN is a basic data model in terrain analysis.

This terrain has been the object for mapping and analysis for hundreds of years. Mapping techniques such as contouring, profiling, hill shading, hypsometric tinting, and perspective views have been developed to portray the land surface. Measures of the land surface, such as slope, aspect (the directional measure of slope), and surface curvature, have also been developed. These land surface measures are important in studies of forest inventory, soil erosion, wildlife habitat suitability, site analysis, and many other fields. Terrain analysis also includes viewshed analysis (to determine areas of the land surface that are visible from an observation point) and watershed analysis (to derive flow direction, watershed boundaries, and stream networks).

The terrain with its undulating, continuous form is a familiar spatial feature to GIS users. Other surfaces used in GIS applications may not be physically present but can be visualized in the same way as the land surface. Cartographers called these types of surfaces statistical surface have been concentrated on developing spatial interpolation methods.

Spatial interpolation describes a process of using control points with known values to estimate values at other points. For example, from known temperature readings at nearby weather stations, interpolation can estimate the long-term average July temperature at locations that have no weather stations. GIS applications typically apply spatial interpolation to a grid (raster data) and make estimates for all cells in the gird. Spatial interpolation is therefore a means of converting point data to surface data. Like terrain analysis, spatial interpolation integrates vector and raster data.

Spatial interpolation methods can be grouped into global and local. A global method uses every control point available in estimating as unknown value. Trend surface analysis and regression modeling are two well-known global methods. A local method uses a sample of control points for estimation. GIS applications often use such local methods as Thiessen polygons, density estimation, inverse distance weighted, thin-plate splines, and kriging.

Higher-level vector data offer additional analytical functions that are not possible with simple pints, lines, and polygons. Because regions may overlap and occupy the same area, they can be built as separate theme layers, each with its own attributes, in an integrated polygon coverage. An integrated coverage is therefore a collection of theme maps, which can be queried by different selection criteria. In this way, the regions data model allows map overlay (up to 32 coverages in ARC/INFO) to be combined with attribute data query in a single operation. With each region layer representing a spatial scale, the regions data model can also provide the frame for analyzing the effect of spatial scale on ecological systems and studies based on census data at the block group, census tract, and county levels.

A network is a connected line coverage with the appropriate attributes for the flow of objects such as traffic. Shortest path analysis is based on a network, with the objective of finding the path with the minimum cumulative cost in either time or distance between points on the network. Other network-related analyses include allocation, which studies the spatial distribution of resources, and location-allocation, which studies the spatial distribution of supply and demand.

Dynamic segmentation combines a network and a linear measurement system such as the mile post system to form higher-level objects. Because transportation planners and engineers use linear measures for events such as accident locations, speed limits, traffic volumes, and pavement conditions, dynamic segmentation allows the linking of those events to a road network for analysis.

9.17.6 GIS Modeling

A model is a simplified representation of a phenomenon or a system. GIS modeling refers to the use of GIS in building analytical models with spatial data. A very useful GIS operation for modeling is map overlay, which combines spatial and attribute data of different variables into a composite map. Because each map feature on the composite map represents a select set of data characteristics by location, the composite map can be further processed to extract new information for modeling purposes.

Analytical models differ in the degree of complexity. This book presents four types of GIS models : binary, index, regression, and process models.

A binary model turns a composite map into a binary map, separating map features that satisfy a set of selection criteria from map features that do not. An index model produces a ranked map with index values calculated using the attributes of a composite map. A regression model shows the statistical relationship between a dependent variable and independent variables from a composite map. A process model integrates existing knowledge about the environmental processes in the real world into a set of relationships and equations for quantifying the processes.

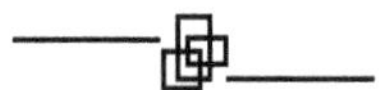

10
Forest Resources Management

10.1 Introduction

We live in a landscape and fulfil our most of the needs from its diversity. The increasing human population satisfies their demand from various kinds of economic activities carried out on the landscape, which ultimately, along with various geomorphic and biological processes, give the birth of spatial and structural changes. Renewable natural resources, especially forests, are conserved for three major sets of purposes: the maintenance of ecological processes and life support systems; the sustained use of the resources for consumptive and social benefits (AKCA., A., 1971). and the conservation of biodiversity units own right for ethical, moral, aesthetic and evolutionary reasons. Only if all three of these can be met and integrated into the socio-political process will true sustainable development achieved, to meet them requires quantification of the current status and of changes on the resources.

Forests, 'all lands bearing a vegetation association dominated by trees of any size, exploited or not, capable of producing wood or other products, of exerting an influence on climate or on the water regime or providing shelter for livestock and wildlife", are the most vulnerable renewable resource of the 'biodiversity' as the plants basically support the survival of each and every member of the biome. The forest resource is large in volume and diverse in character, on the other hand, the excess population pressure intensifies the demand of forest and forest products side by side. But unfortunately, this natural resource has been overlooked by the human kind regarding the aspect of degrading trend of forest resources all over the world.

The forest resources and forest lands need to be managed in such a way that they can meet the needs of not only this generation but for future too. Stringent strategies are to be evolved to conserve the existing forests afforest the state owned forest land as well as degraded private lands. To develop such strategies quick, repetitive and accurate information about forest cover is required at the local, regional, state and national levels. The temporal and spatial distribution of forest cover can be derived using remotely sensed data and the strategies can be developed using GIS technology by integrated studies.

The key factors involved in developing forest resources management strategies are - large economic development programs involving resettlement, agriculture and infrastructure, pressure for increased food production conversion of forest land to agriculture, increase in the demand and decrease in forest area, and future demand for forest products.

Vegetation is an important dynamic characteristic and very prominent feature of earth's surface and forms a major element of the landscape and is either aquatic or terrestrial. The natural vegetation on land is generally grouped as forest, grassland, scrub etc. Forest provides two benefits: direct and indirect. The direct benefits are harvesting timber, fuel and other products whereas the indirect benefits include conservation of soils, water and air, controlling and regulating the global cycling of oxygen, nitrogen and carbon.

10.2 Geomatics in Forestry

Mapping and monitoring of forest cover is one of the foremost requirements for its planning, management and conservation. Remote Sensing, GIS and GPS play vital role in mapping of existing forest resources information at a particular period or time. The multi-spectral data obtained from remote sensing satellites like Landsat, IRS, SPOT, IKONOS have been used to study forest features either by visual interpretation or by processing digital data using digital computers. The spectacular development in space technology, particularly, the repetitive satellite remote sensing across various spatial and temporal scale, offers the most economic means of assessing, planning, managing and monitoring the forest resources. It provides excellent capability to monitor composition of forest ecosystem and impact of management and degradation processes. Sometimes the Remote Sensing provides unique data that cannot be obtained by any other source, but much more often, the collected data facilitates fieldwork and enables tasks to be completed at lower cost and much more quickly. It is unlikely, nowadays, forest inventory would be undertaken without remotely sensed data.

Remote sensing is a useful tool for assessment of environmental conditions, either in an existing forest or prior to planting a new one in accordance with the climatological

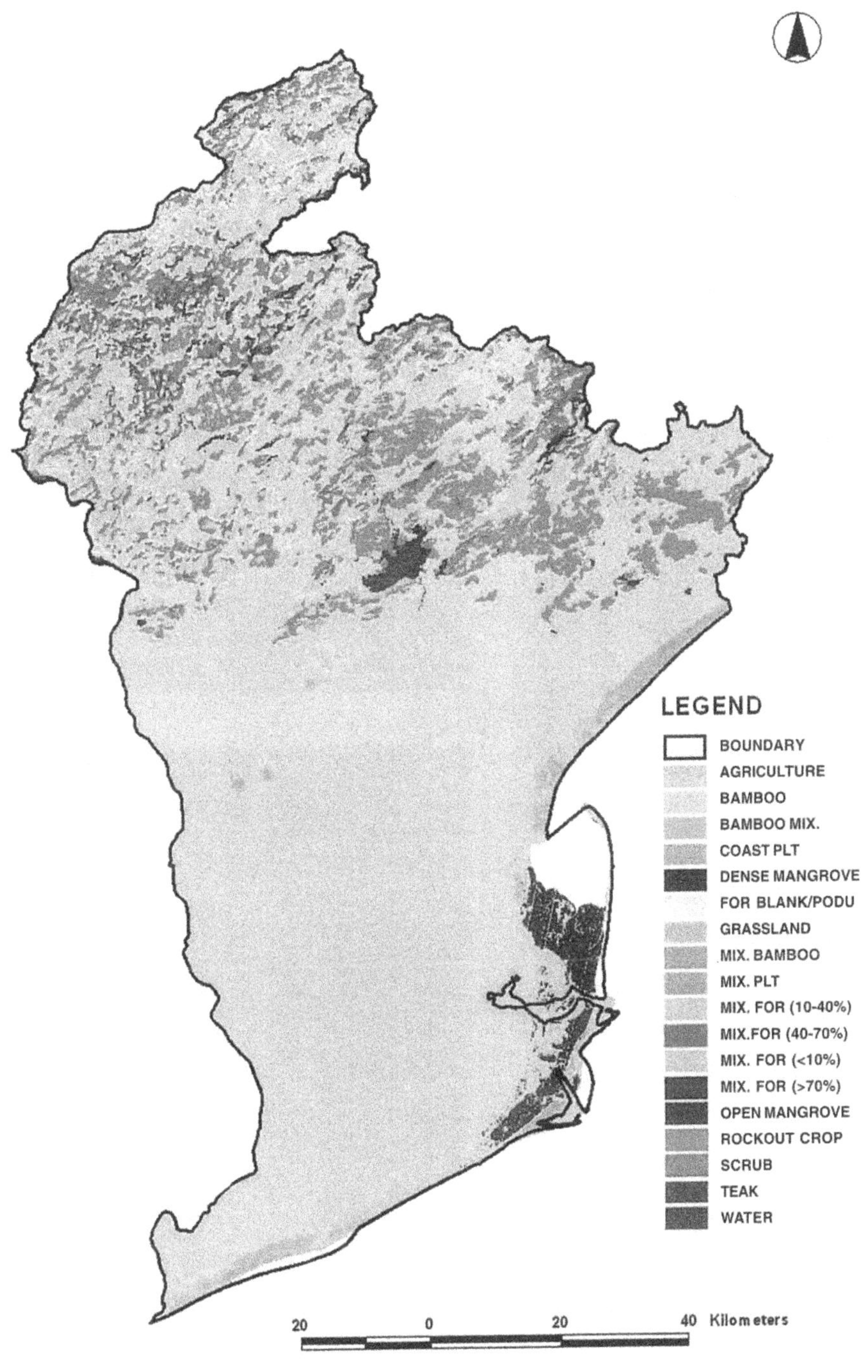

Plate 10.1 Stock Map of East Godavari District of Andhra Pradesh State, India.

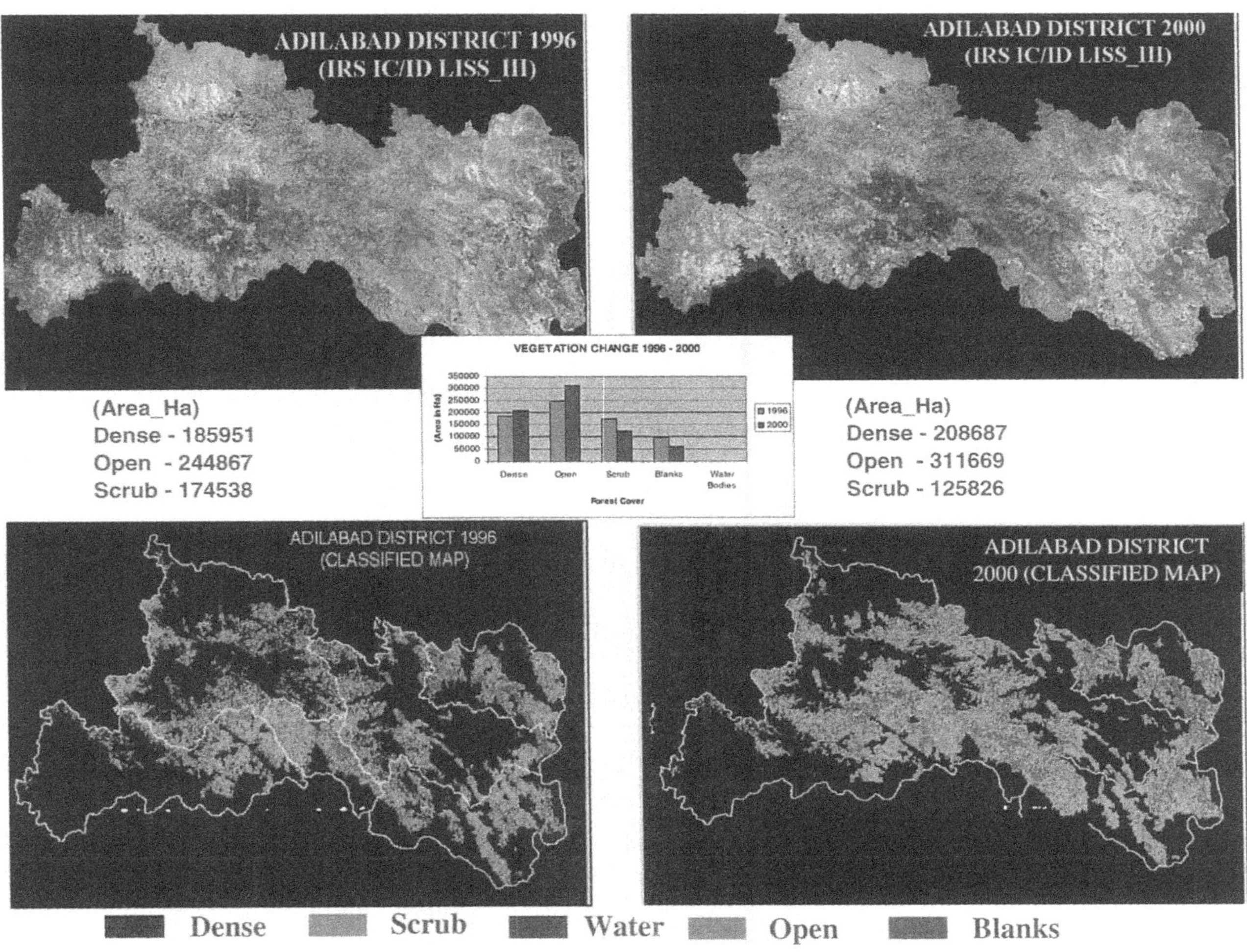

Plate 10.2 Forest Density and Vegetation change map of Adilabad District of Andhra Pradesh State

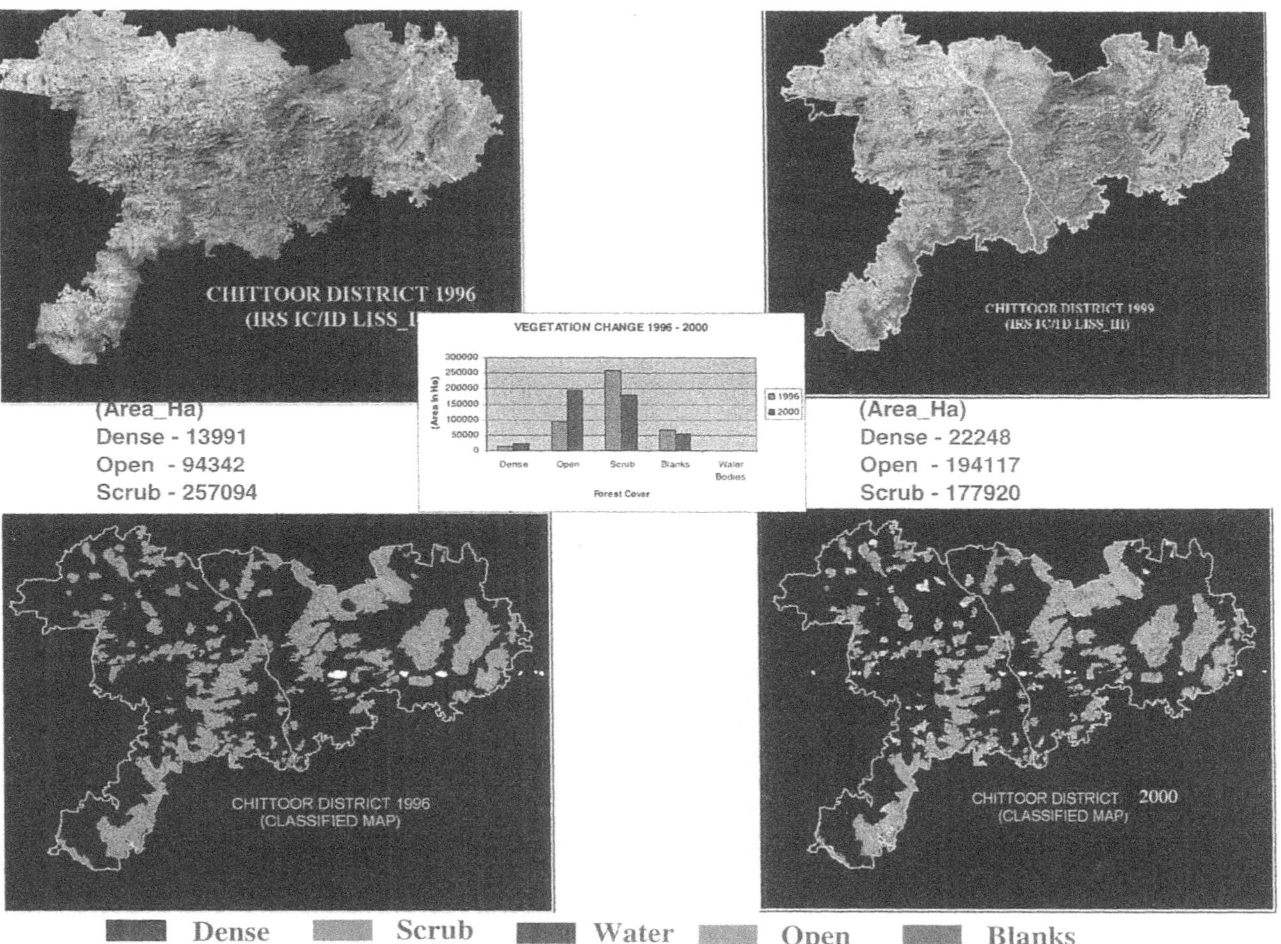

Plate 10.3 Forest Density and Vegetative change map of Chittor District of Andhra Pradesh

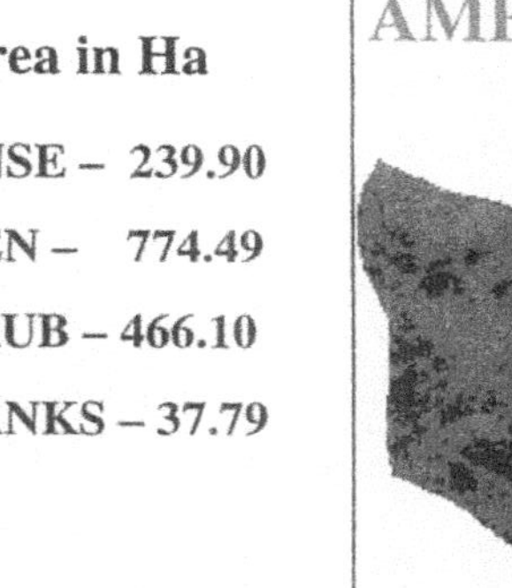

Area in Ha

DENSE – 239.90
OPEN – 774.49
SCRUB – 466.10
BLANKS – 37.79

Area in Ha

DENSE – 367.20
OPEN – 884.62
SCRUB – 249.24
BLANKS – 17.22

Area in Ha

DENSE	341.22
OPEN	900.05
SCRUB	194.74
BLANKS	71.36

AMEENGUDA 2000

DATE OF PASS: 6th Nov'2k

Area in Ha

DENSE	380.79
OPEN	842.63
SCRUB	235.29
BLANKS	53.97

Plate 10.4 Vegetation change in single Beat

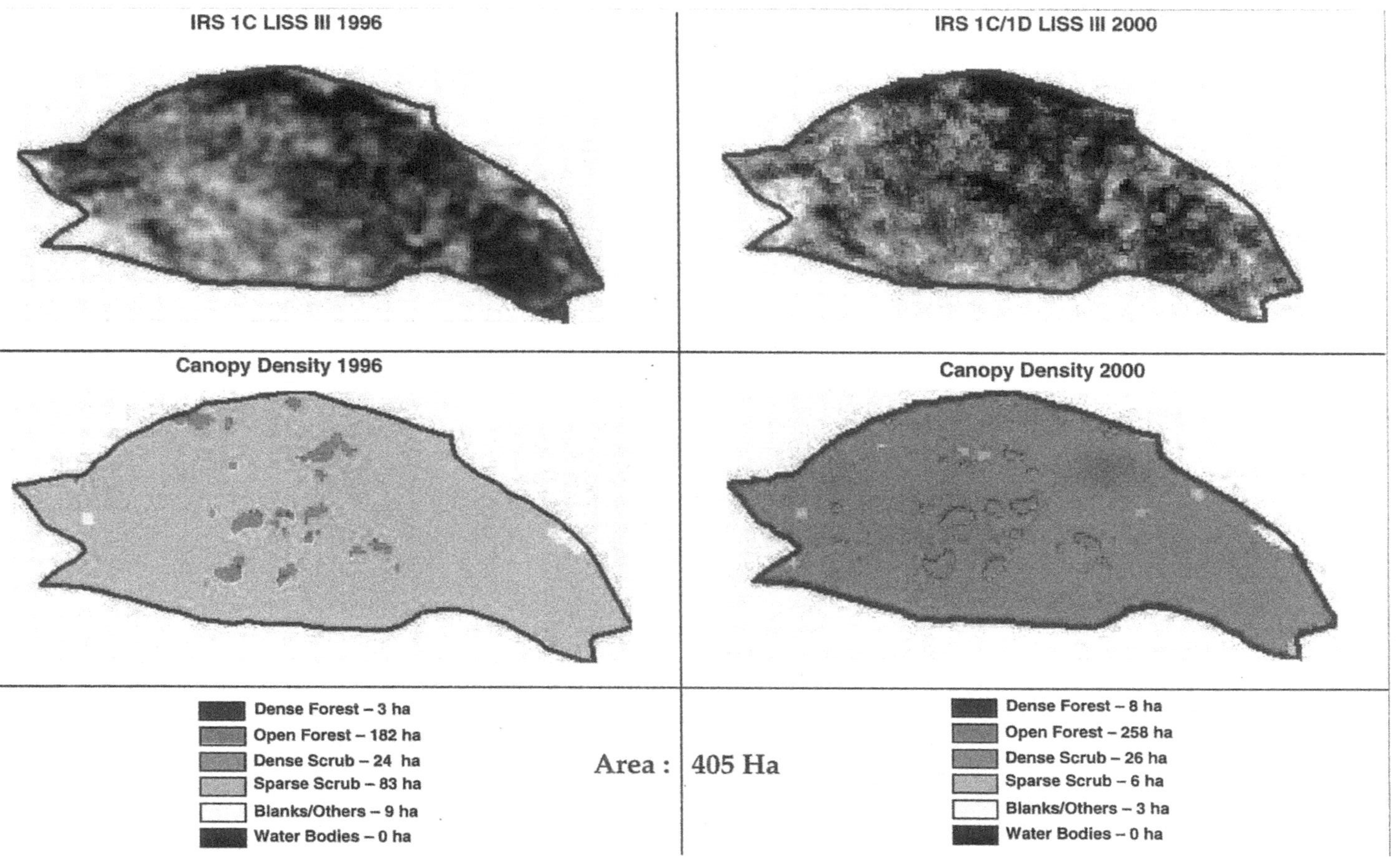

Plate 10.5 Vegetation Status Kotha thanda VSS

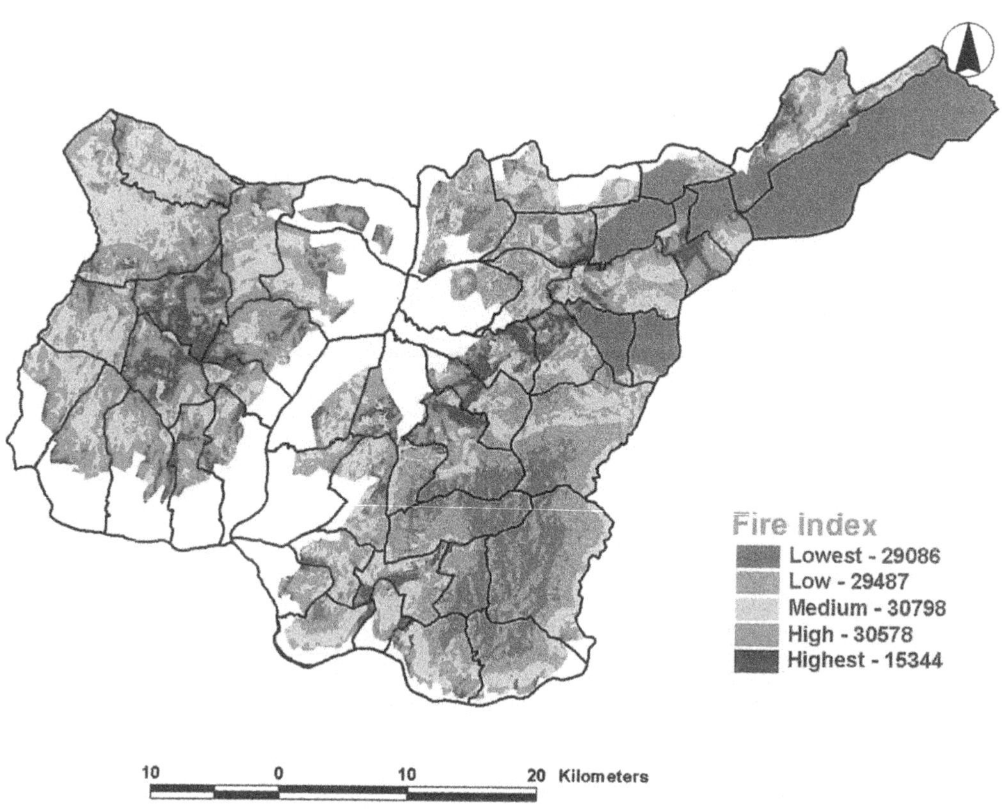

Plate 10.6 Fire Risk Zonation Map of Bhadrachalam South Division

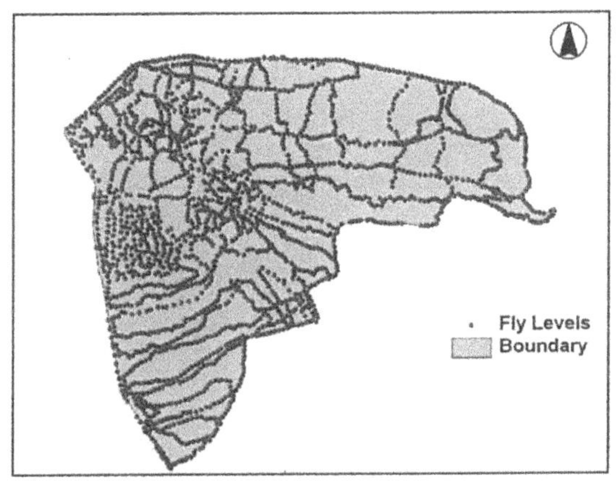

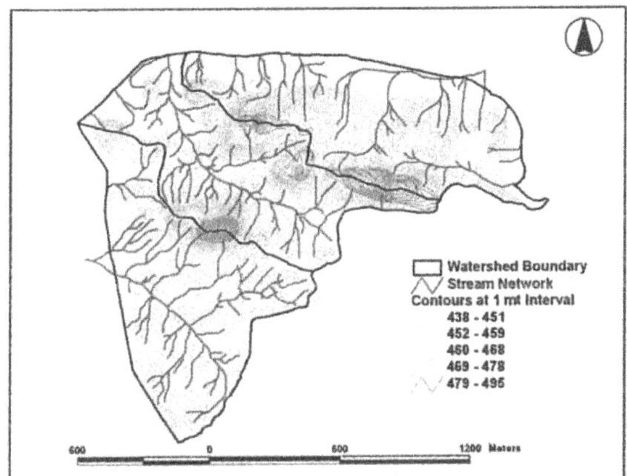

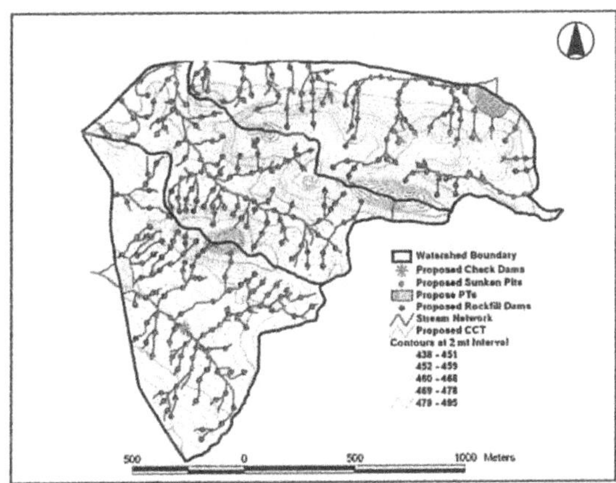

Plate 10.7 Spatial data for forest resource management
(a) DGPS data (b) Contour of stream network and
(c) Watershed management plan

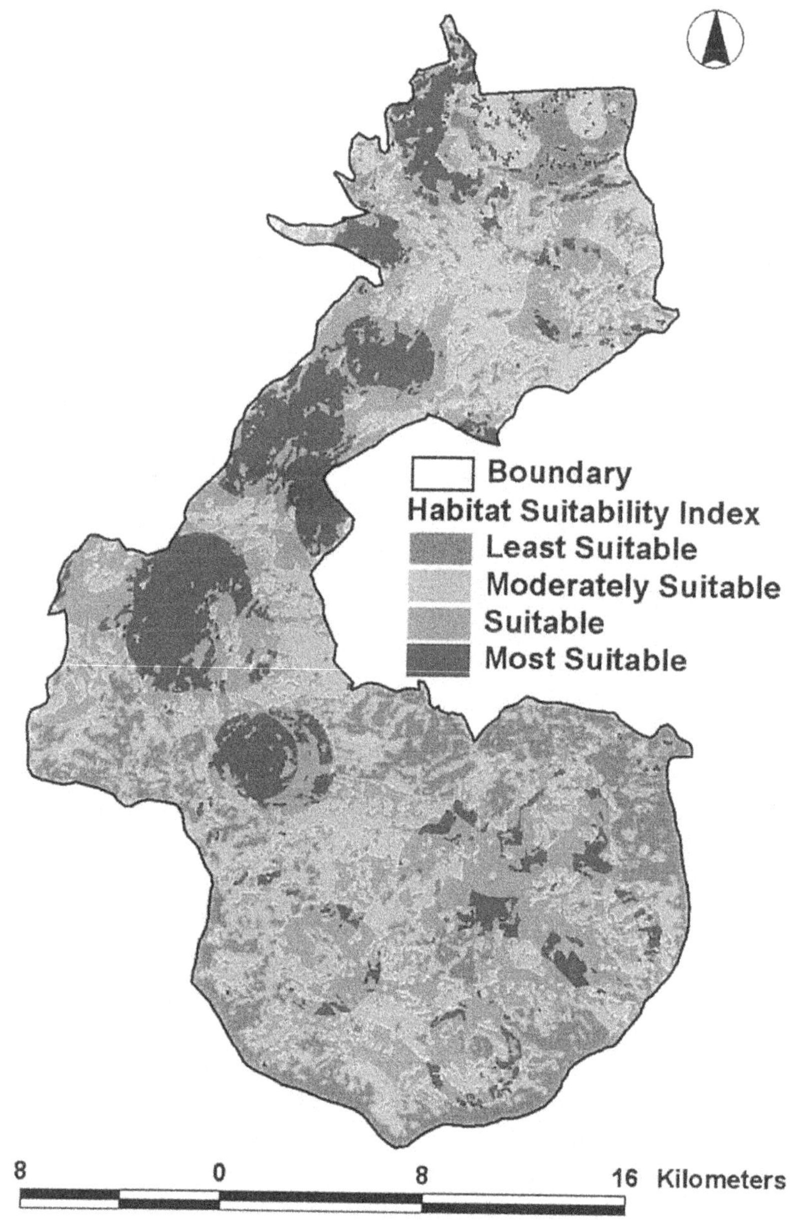

Plate 10.8 Habitat Map of Mannaur Range

conditions. Data for climate analysis can be obtained through microwave or radar sensing, which is ideal for gathering information about the atmosphere. The microwave data provides an uninterrupted flow of accurate data for interpretation, and such data can be used to track rainfall and winds, to measure water and ozone content in the air to determine the compatibility of the climate with forest growth, and to predict potential problems in forest growth. It also gives the regeneration status using SAR imagery (C band-VV Polarization). Remote sensing contributes to fire-fighting efforts, as well. Radar and thermal sensing allow for constant observation of fires, unaffected by clouds, smoke, or other conditions that hinder aerial observation. After a fire, damage can be quickly and in expensively assessed by using the data obtained by thermal sensors like AVHRR, HCMM. This accurate spatial information of the burn scar can be used to know the amount of biomass destroyed, the amount of smoke produced and the air pollution, by intensity of forest managers. The Geomatics can be used to predict the fire risk areas, which will be useful in taking effective management decisions (Agarwal, N. K. and Sharma, M. K. 1983).

During every stage of forest management, foresters can use remote sensing data to estimate future urban spread and population growth. Then, forest management can be planned taking into account the future needs of settlements. Urban planning data can also be applied to the management of urban forestry, to create inventories of trees in parks and on streets.

IRS-IC/ID LISS-III multispectral scanner with 23.5m resolution enables discrimination of different forest types, grasslands, plantations etc. The discrimination of forests into deciduous and evergreen depends on the use of optimal season satellite data based on phenology and different band combinations. The dry period (Feb-May) satellite data is used for delineation of evergreen & deciduous forests. The use of middle IR data facilitates delineating vegetation types such as evergreen, semi-evergreen, moist deciduous & dry deciduous. The LISS-III data of March help in delineating the deciduous forest being in leaf fall condition with other vegetation areas. Pan data with 5.8m resolution is useful in delineating the vegetation into various canopy density classes and also provides micro level details for stock mapping. Hence the fused Pan and LISS III data will be useful in revision and updation of stock maps. Wide Imaging Field Sensor (WiFS) payload with repeativity of 5 days permits rapid forest cover monitoring and mapping. It provides forest cover details at global/regional level, used for annual forest cover mapping which saves adequate time and money. WiFS data is used for forest fire monitoring and to account damages and suggest control measures for large scale forest fires because the WiFS data can be acquired for every 5 days. NOAA Satellite AVHRR sensor data is extensively used for large area vegetation monitoring, the spectral bands used for this purpose are visible band (0.58-0.68um) and NIR (0.73-1.I, um). Various mathematical combinations of multi-spectral data have been found to be sensitive indicators of the presence and condition of green vegetation. These mathematical quantities are called as vegetation indices.

GIS technology has been widely accepted by public as well as private forestry agencies at local, regional and global level. In large part, this has been a result of the benefit of using GIS technology over current forest maps. It is used to assess the existing forest resource and develop harvest schedules and treatment programs to project future timber supplies and for other operational planning activities. Forest inventory data can be collected using remote sensing techniques. The conventional forest inventory is in general, done progressively with a small portion of the forest being inventoried each year. To update a forest cover map could take 20 years or more with expensive manual drafting. With GIS, the forest cover maps can be updated on a constant basis and it provides the forest managers more current data than what was previously available. The inventory can be more effectively designed by stratifying the satellite imagery into homogeneous units and the sampling intensity and number of sampling points in each unit can be estimated based on their probability of existence. With GIS technology, the average age of the information in the forest database could be reduced from 20 years to only a few months. The time factor alone has led to a wide acceptance and large demand for GIS applications in forestry.

In itself, the use of GIS to update the forest inventory maps is not much more than automated cartography but it is the analytical power of GIS that sets it apart from cartography. The GIS can store and analyze the forest information in ways that could not be previously done. It can be used to calculate the harvestable timber in the forest or model the spread of a forest fire.

GIS technology provides the means to collect and use geographic data to assist in decision making for natural resources management. A digital map is generally of much greater value than the same map printed on paper as the digital version can be combined with other sources of data for analyzing information. GIS makes it possible to synthesize large amounts of different data, and to manage and retrieve the data in a useful manner. With GIS, different layers of information can be combined and analyzed to better study and understand the complex relationships between the Earth's ecosystems. So, the integration of all these new technologies will give integrated information and help in effective planning with accurate and valid information. The integrated information can assist in evolution of appropriate action plans for initiating sustainable forest management.

10.3 Forest Cover Mapping and Change Detection

Deforestation, desertification, pollution and natural resources depletion are now serious problems in the world, which threaten the survival of half of the living things in the world. These environmental change processes, are to be understood to conserve renewable and non-renewable resources for future generations. The forested landscape is not only that part of the earth's surface dedicated to the production of timber, but also extends to areas of woody vegetation sustaining other resources and providing protective benefits. Based on the landscape that has been threatened or influenced by man's activities for thousands of years and supports the current development functions, the forests may be categorized as industrial forestry, environmental forestry, rural development forestry and urban forestry. The impact of the increasing human population pressure on the forested landscape and deforestation accelerates the indirect and direct threats of pollution. This causes the major changes in the patterns of land use, soil erosion, declining soil fertility, decreasing productivity, climate change and in some regions desertification.

Forest change detection and monitoring has been confined mainly based on comparing two or more temporal sets of data, comparing old aerial photographs and present satellite imagery and comparing satellite imagery with existing old maps. Monitoring from satellite sensing system includes the detection and evaluation of sudden forest damage and continuous forest inventory. The temporal satellite sensor data is used to monitor changes which are sudden fire damage, wind blow, gradual over a short period like bark or beetle damage and gradual over a long period of time like forest decline. These change detection studies can be done by using visual interpretation and digital analysis techniques with the help of satellite data acquired on two different dates for the same ground area taken on an identical scale. The digital analysis of the satellite data has the advantage of being fast and able to handle large quantity of data whereas the visual interpretation is based on the tonal variation of different features observed on imagery. By allocating a single color to each multidate black and white transparency which is called as overlay analysis, the imagery can be compared which has been taken on several different dates. The major limitation in using satellite imagery for change detection of forest cover is the spatial and temporal resolution of the sensors. The high spatial resolution obtains low temporal resolution like thematic mapper sensor of Landsat, which cannot detect rapid change, and low spatial resolution obtains high temporal resolution like AVHRR of NOAA satellite and WiFS of IRS ID which cannot detect detailed spatial change. To understand causes of changes and deforestation process, the relationships between deforestation and landscape attribution has to be analyzed using GIS by generating a deforestation risk map.

GIS can be associated with satellite based forest cover monitoring since it is not only relatively easy to store and recall earlier satellite data sets for comparison with future datasets and to mask out those areas that are not relevant, but also the spatial changes can be associated quickly with other categories of field data such as climatic factors, soils, transportation routes environmental parameters which in turn help in the generation of a final change detection map of any project area. A flowchart showing the step-by-step process of preparing a final forest change detection map is given in Fig. 10.1.

Observations of temporal changes through the forest change detection map can be helpful for forest development activities like forest damage assessment, afforestation working plans, identification of areas of new plantations, separation of deciduous and evergreen forest cover and forest canopy studies (Heller, R.C. and Ulliman, J. J. 1983).

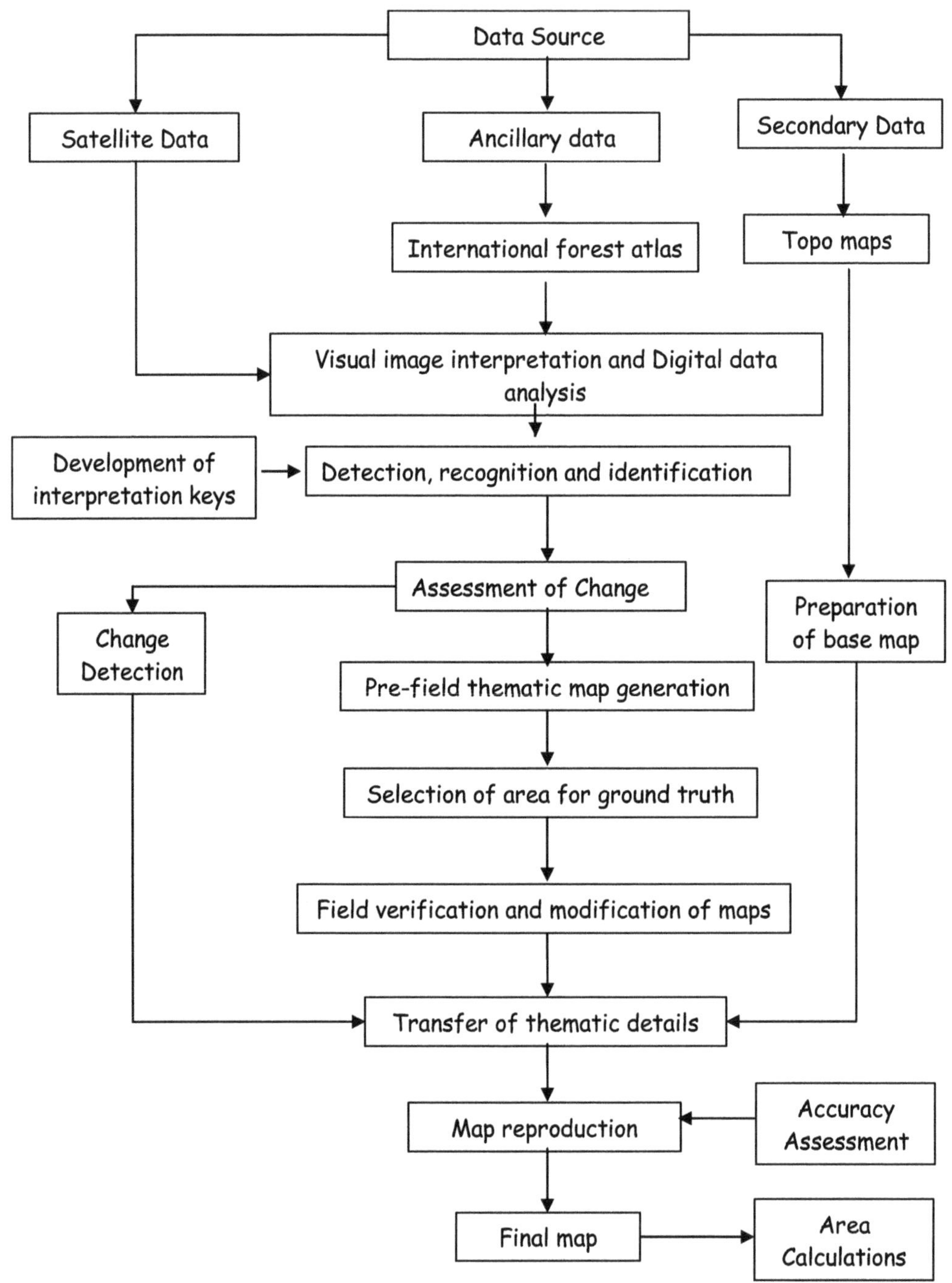

Fig. 10.1 Flow-chart showing the methodology for change detection.

10.4 Forest Inventory (Stock Mapping)

Forest stock maps are defined as the maps which contain detailed spatial information on extent of recorded forest lands including the administrative jurisdiction at various levels, infrastructure and communication facilities, water resources and the status of forest vegetation. The status of forest vegetation include categorization of forest into density classes, species distribution, assessment of growing stock, growth data for various species and age distribution. The forest stock maps depict forest type, density, encroachments, cultivation patches, human habitats, regeneration status and provide an idea about available resources. The purpose of forest inventory is to assess and monitor to the extent, state and sustainable development of nations forests in a timely and accurate manner by considering all the relevant parameters.

The disadvantages of conventional method of enumeration of the forest inventory are uncertainty in working plan staff, and lack of generation of real time information highly time consuming unrealistic. By the time enumeration of the forest stock of one division is completed and the working plan is prepared for the next 10 years or so, there may be a drastic change on the ground reality and the forest working plan may have become obsolete. The other disadvantage of using conventional method for forest inventory is that the forest stock maps are not geo-referenced so that, not possible to analyze the data on GIS platform even after converting the data into digital mode under computer-based GIS.

These disadvantages can overcome with the use of Geomatics technology. Remote sensing with its multi-spectral, multi-temporal and synoptic view has the potential to provide both spatial and temporal information on forests. This is possible because of the characteristic differences in structure, phoenological state, internal biochemical and biophysical properties of different vegetation types which result in reflection properties unique to each vegetation type. The spatial and spectral resolution of satellite data is of great advantage in precise spatial mapping of forest type distribution.

Recent studies conducted by Forest Department of Government of Andhra Pradesh and National Remote Sensing Agency (NRSA), Government of India have shown that the interpretation of high-resolution satellite data like IRS-ID PAN and LISS III can be used to prepare forest stock mapings and the possibility of achieving more informative stock maps than conventional ones. Forest density is one of the most important components in revising stock maps. With the advent of high-resolution satellite data it is now possible to obtain better information on stand density as shown in Plate 10.1. The stock maps and the forest density maps derived form satellite remote sensing and GIS provide the data for global changes, information on episoci events like conservation strategies, land use planning, availability of resources, master plan for afforestation programs, environmental impact assessment, planning for developmental activities, presenting the management options, reforestation of degraded forests and management of protected areas for biodiversity conservation. A classical example of forest density map of Adilabad District for the year 2000 is shown in Plate 10.2 and

10.3. Plate 10.4 and plate 10.5 show the vegetation change in single and vegetation status of Kotha thanda which can be used to protect the forest environment.

10.5 Dynamics of Forest Ecosystem and Forest Canopy

Forests cover approximately 30% of the earth's land surface and represent important sites for evaporation, transpiration, photosynthesis and the cycling of carbon and nitrogen. These processes control the exchange of energy and matter in vegetation and they are directly linked to net primary production. Forests represent a major source for terrestrial carbon and change in forest area, particularly the effects of forest clearance in the tropics, may be partly responsible for the increase in global atmospheric Change in the magnitude of the flows of energy or matter in forest ecosystems may have widespread implications for the earth. Clearly, there is a need for information on these forest ecosystem processes, and models have been developed to describe the dynamics of such systems. The application of these models, however, requires the input of quantitative information on a range of vegetation canopy variables. The multi-spectral reflectance properties of forest canopies are known to be affected by a wide range of biophysical, biochemical and structural variables. The most important factors affecting canopy reflectance are :

- Leaf area index
- Leaf optical properties reflectance, transmittance and absorption
- Leaf location in three dimensional space: vertical and horizontal distribution of leaves and degree of clumping
- Leaf orientation in 3 dimensional space: leaf inclination angle, distribution and leaf azimuth angle distribution
- Reflectance of the under storey vegetation, soil or other background
- Geometry of illumination and view

10.5.1 Leaf Reflectance

The spectral reflectance of individual leaves or needles is controlled by a number of factors, the most important of which are absorption by photosynthetic pigments and water. The diffuse spectral reflectance caused by internal scattering and specular reflectance caused by leaf surface. The shape of the reflectance invisible region (0.4 p - 0.7 μm) is based on the composition and concentration of the photosynthetic pigments, chlorophyll and the cartotenoids Leaf reflectance in near infrared region (0.7 μm - 1.3 μm) is mainly due to the internal structure of the leaf, the number of air spaces and their arrangements. The reflectance in the near infrared and middle infrared (1.3 - 3.0μm) region are the effects of absorption by leaf water content, and biochemical constituents of leaves. It can be observed that the spectral signatures or spectral signatures or spectral reflectance curves of dry and deciduous tree leaves are similar in nature and hence the wavelength region of (1.0 μm - 2.5μm) electro-magnetic spectrum is more appropriate for estimating canopy biochemsitry of forest cover. The reflectance of leaf varies with the variation of stress factors like air pollution, the presence of heavy metals, the drought conditions, Environmental and physiological factors, the concentration of pigments, water, and chemicals in leaves and maturity of the leaf.

Variation in leaf or needle spectral reflectance will also occur from species to species as the spectral reflectance of broad leave species is different to those of conifer needles.

10.5.2 Leaf Area Index (LAI)

LAI is a quantitative measure of the surface area available for the interpretation of photosynthetically active radiation (PAR) and for transpiration and is the single most useful characteristic of vegetation for quantitative analysis of interest in global ecology. Vegetation canopy reflectance depends on the number and size of the foliage elements and their arrangement on individual plants. The relationship between canopy LAI and reflectance are well known for many agricultural crops and grassland areas but they are less well known for forest canopies. In the former, there is generally a negative relationship between red reflectance and LAI up to an asymptote which, for agricultural crops is normally reached at an LAI of 1 to 2. Interaction of near infrared with successive leaf layers results in a positive relationship between with LAI reaching an asymptote at an leaf area index of 3 to 4. The relationship between middle infrared reflectance and LAI is similar to that for the visible wavelengths. In unstressed canopy water content is normally correlated with LAI and the form of the relationship between LAI and middle infrared reflectance is likely to be negative and asymptotic. Negative relationship between LAI and red and infrared reflectance, and positive relationship with NIR were found early in the season when there was full canopy over.

10.5.3 Effect of Canopy cover

Forest canopy cover plays a key role in determining the spectral response of forest canopies because it controls the amount of under storey vegetation, which is visible to the sensor. In agricultural crops there is normally a simple functional relationship between LAI and canopy cover but in forest canopies this covariance cannot be assumed as forest canopy cover and forest canopy LAI may be independent variables and changes in forest LAI may be confounded by change in canopy cover.

10.5.4 Effect of Leaf Location and Orientation

The directional reflectance properties of the canopies are similar between those of agricultural crops, natural grass land and forest canopy. In the more sparse canopies the clumping of vegetation gave rise to increased forward scatter and decreased back scatter. The directional reflectance of forest canopies is highly dependent on the structure of the canopy elements.

10.5.5 Effect of Illumination and viewing geometry

The target illumination and viewing conditions are also important. These are the solar zenith angle, the solar azimuth angle, the sensor zenith view angle and sensor azimuth view angle.

10.5.6 Forest Stand Measurements

The term forest stand is often loosely used as a synonym for forest crop and precisely it may be defined as descriptive of an aggregation of trees sufficiently uniform in composition to be distinguishable from adjacent crops with different species composition, age classes and size classes.

10.5.7 Stand Measurements using Aerial Photographs

There are three types of Stand measurements can be possible using aerial photographs, namely, (i) Stand height (ii) Crown closure/Crown cover and (iii) Stocking.

Stand Height: the stand heights of a small sample of the trees are measured using the photographs, which is sufficient to provide a reasonable estimate of the mean height of the forest stand. Determination of stand height of forest cover can also be done using stereo-photogrammetry, which deals with the 3-dimensional measurements from a stereo-pair. But under many circumstances, this overcomes the problem that the physiognomy of the forest does not permit the stereoscopic measurement of the height of all trees recorded on stereo model permits the determination of stand height stereo-photography of the individual dominant and co-dominant trees using image parallax method discussed in the previous chapter. Thus the mean stand height' determined using the parallax formula is that of the co-dominants and dominants and not the mean stand height of all trees, as could be measured on the ground. The application of image parallax method may not be possible to measure stand height in rain forest, since the ground is seldom seen and the crowns coalesce and are entwined with lianas.

Crown closure / Crown cover

This is defined as the proportion of unit area of the ground covered by the vertical projection onto it of the overall tree crowns. Based on parallax methods in combination with stand height, it is used to stratify the forest into timber volume classes which is measured on aerial photographs using either stereo gram or dot grids or crown cover percentage cards.

Stocking

Stocking is defined as the number of trees per unit area. The stocking and crown closure are the two direct measurements of stand density obtainable from aerial photographs. Counting of trees from stereo pairs of photographs depend on the skill of the interpreter scale of photograph, physiognomy of forest and the age of the stand.

10.5.8 Stand Measurements using Satellite Remote Sensing

The role of Satellite remote sensing will remain mainly contemporary and its attractiveness will be the comparative low cost per unit land area of data acquisition, and the lower cost per unit land area of image analysis. Satellite remote sensing can be used particularly in large area surveys and for monitoring frequently at national, continental and global level changes in forest cover, density and landuse.

Visual image analysis of satellite imagery can be applied to forest stands. The advantage of visual analysis is that it can be intimately associated with field work and local knowledge. These studies will be qualitative in approach as compared with aerial photographic interpretation. The smallest classifiable areas are likely to be a group of nine pixels or probably 25 pixels.

Most forest activities, involving the computer processing of remotely sensed digital image data have been carried out in many countries like China, India, Thailand, Latin America etc. This digital data analysis is used to provide informative image classification to improve geometric accuracy, to enhance the imagery for field use, to update maps and to provide layered information as input to GIS. That means, the output of visual as digital image analysis is the input data to GIS for subsequent spatial mapping. Various steps involved in the process are :

- selection and review of input data
- decision on classification level and type of desired classes
- radiometric and geometric corrections of digital imagery
- image enhancement and unsupervised classification scene-wise
- probability proportionate stratified random sampling to derive the number of sample points to be ground verified in each class
- ground truth data collection and error correction
- supervised classification
- post classification filtering of classes and seamless mosaicking
- ground checking and error correction
- classification accuracy analysis
- planimetry of classes & cartographic completion and production of hard copy outputs.

10.6 Forest Damage Assessment

Forest damage assessment can be performed using low altitude remote sensing data obtained from a light winged aircraft or a helicopter or a light aircraft using small format cameras. The damaged areas in Forest lands can be mapped using photo interpretation techniques qualitatively and metric photogrammetric techniques can be used to make the quantitative measurements with the help of these low altitude remote sensing data products. Generally, the count dead trees will be done on 1:4,000 scale and the spatial information about the subtle differences in stress on 1:1000 scale using these photographic products, more specifically, the colour infrared photographs. So far satellite remote sensing has been modest in mapping damaged areas as the coverage of each scene is very large but very much useful for the aerial reconnaissance survey. The usage of GIS techniques is in primitive stage and it attracts the positive momentum in using these technologies because of the availability of very high resolution satellite data like IKONOS and PAN and LISS III of IRS.

The fundamental principle involved in using the satellite data for assessing the forest damage is occurrence of major differences in the spectral signatures between the healthy and unhealthy vegetation covering extensive areas at various wavelength regions of the spectrum. It is possible not only to discriminate the damaged forest from healthy forest cover, but also to recognize the host forest type. But the forest species identification using satellite sensor data needs research to establish the methodology based on a number of environmental considerations (Townshend, J.R. and Justice, C., 1981). Since damage is associated with individual tree species and differences in stand density, the classification of the forests using satellite data needs to be better than level II and that the major constraint to the wider use of satellite data in disease monitoring and insect infestation monitoring is the failure to be able to identify the host tree species at an acceptable level of accuracy.

10.6.1 Disease and Insect Damage

The assessment of the forest damage due to disease and insects is performed using colour infrared film with a yellow filter. In general, in the ensuing years until 1950s, panchromatic black and white aerial photography is little used and confined to severe physiognomic damage including fire and wind storms. Heavily defoliated and dying trees appear conspicuously in dark tone on black and white IR aerial photographs, but severe defoliation could be interpreted on black and white panchromatic photographs. Variation in the spectral reflectance in the visible and near infrared between healthy and diseased foliage in color aerial photographs is now being extensively used to study the classes of damage caused by insects and disease. In general, color IR film has the advantage of providing a record of spectral reflectance in both the visible and near IR. Although on processed imagery, healthy foliage often appears red/reddish and dead or dying foliage appears blue/bluish-green, there is no exact hue, value and chroma indicator of disease.

10.6.2 Pollution Damage

Increasing consideration is being paid to using color aerial photography and high resolution satellite data in assessing water, air and soil pollution, which leads to foliage damage. This can result in the death of the trees along roadsides in urban areas and in wide-scale forest damage in the countryside. Infrared film and the satellite sensor data in infrared region of electromagnetic spectrum is used to identify the forest damage due to air oxidant. Air oxidant injured trees can be easily identifiable on colour photographs and false colour composites of satellite data. This is possible because of the absence of lush foliage on injured trees and the high spectral resolution data at an appropriate wavelength permit the identification and discrimination of positively affected trees. According to the earlier investigations, it is revealed that healthy conifers usually retain their needles for upto five years, appearing green/ green yellowish on normal color prints, whereas smog-affected foliage may have only a current growth of short needles at low density, appearing yellow/yellow-reddish on the prints. One of the most extensive and severe occurrences of pollution damage of forest stands is caused by sulfur dioxide. This damage can be readily seen and monitored through satellite imageries like IRS IC/ID LISS and Land Sat TM. Industrial pollution in urban areas causes the decline of forest health and a gradual process of tree deterioration in condition and viqour and the interaction of a number of hidden parameters like climate changes, changes eventually lead to the death of the trees.

10.7 Parameters of Forest Inventory

Forest inventories are the main source of data for the national forest inventory. Management agencies are continually updating and upgrading their forest inventories. The age of their inventories is always changing. As a result, the system of forest inventory is a snapshot of different aged information, collected to a number of different standards.

The current approach to national inventory is cost effective in that it is based on existing data. The process of developing forest inventory system is a well established and accepted by the contributing agencies internationally. The approach provides detailed information about forests that are consistent with forest management information. The inventory also contains location-specific information on the characteristics and quantity of the forest resource, providing mapping and spatial analysis capabilities. The forest inventory system should accomplish the information on the nature and rate of changes to resources, current status of the forest, vegetation attributes, data related to complete and consistent area coverage, stratification of sampling points, species diversity, wood volumes, attributes related to land use, ownership, protection status, human influence conversion of forest lands, origin of exotic trees and spatial maps of all these parameters.

10.8 Development of Working Plan

Management of forest Resources in the country primarily require growing stock maps which is generally used in the states for working plan preparation. Working Plan is the document of the forest department, written for each forest division, which incorporates the spatial, quantitative and qualitative information of the forests under the control of the division. The Working Plan is an extremely vital document since it presents the various strategies for the systematic management of forest, to be followed for the next ten years or so. Working Plan prescriptions are prepared on the basis of forest stock map, which contain information on extent, density, forest blanks, topography, soil etc. Generally for preparation of working plan maps, information is required at forest block/compartment levels. Using remote sensing techniques it is possible to identify density classes at the compartment level by overlaying the forest compartment boundaries. The study of spatial distribution of forest types would help greatly in forest management for accounting the changes that have resulted in the changing quality and composition of forests. This can be best achieved through use of multi date satellite data since conventional ground surveys are time consuming and strenuous. The use of satellite data during the optimal green season and deciduous season would help in understanding the distribution of various types on a timely and cost effective manner. The two important components of working plan are (i) description of forest boundary (Forest Map) and (ii) description of forest vegetation (Forest inventory). Survey of the forest boundary to check about their correctness and encroachment etc. is done mainly by the staff of Working Plan Divisions. The conventional plain table survey method adopted in this case is not only time consuming but also perennial in nature, since detection of any change of boundary or encroachment can be confirmed only through another detailed field survey like DGPS.

GIS has tremendous potential in storing the above voluminous spatial and non-spatial data. The socio-economic data can be combined with this data on village basis for better understanding of the human dependencies. Using GIS analysis the forest areas can be categorized into various regions defining the ecological and social role as Production forest, Protection forest, Natural reserves, Wildlife sanctuaries, Village forestry, Tree lands etc. that needs different approaches in the forest management for maximizing the benefits from forest and maintaining ecological balance.

The WP suggests the working / management of the forest as per their structure and resource availability and divides the forests into various working circles by handling complex data sets through GIS techniques by taking into consideration of locality factors and Socio-Economic conditions of an area as Production WC, Protection WC, Afforestation and Reforestation WC, Bamboo WC, Village wood WC, NTFP WC, Grass WC, Rehabilitation of Degraded Forest WC -JFM WC, Agro Forestry WC etc.

10.9 Forest Management Information System (FMIS)

The FMIS is envisaged as an integrated system which will be used to support the planning, implementation and monitoring of multi-objective forest management activities. FMIS has the ability to maintain current forest inventories and generate maps of spatially oriented data such as attributes of entities depicted on a map such as population of a village, whose location can be fixed on a map. FMIS requires the integration of information from various sources, information technology, and in particular the integration of database management system, remote sensing, image processing and GIS provides a broad range of tools for forest area mapping, monitoring and management to achieve optimization in their conservation. The elements of FMIS are shown in Flow chart (Fig. 10.2). The FMIS can be used for strategic, tactical and operational planning and implementation, and operational control in and across administrative units and levels of the organisational hierarchy. Functional areas addressed by FMIS are

(i) Administration
(ii) Research and extension
(iii) Protection
(iv) Land management
(v) Forestry project
(vi) Production
(vii) Plantation
(viii) Wild life
(ix) Development and planning
(x) Working plan
(xi) Accounts
(xii) GIS

The purpose of creation of FMIS is to :

- Monitor the status and evaluate forest resources
- Coordinate in case of forest fire
- Make effective decision and for everyday planning
- Collect information from various sources and map with integrated information
- Communicate results and findings to a broad audience in effective way

The critical issues that need to be addressed while implementing FMIS are :

- Linking the planning at all levels (strategic, tactical and operational), and providing facilities to reflect the changes in the actual situations and their implications for the plan.
- Providing a framework for the integration of GIS, remote sensing and GPS to resolve the problem of non-availability of the latest map or spatial information.
- Helping the decision makers to effectively achieve the social, environmental and economic objectives of the forest department's plan in a multi-objective decision making environment.
- Planning for and ensuring the availability of trained people during the entire life cycle of FMIS.

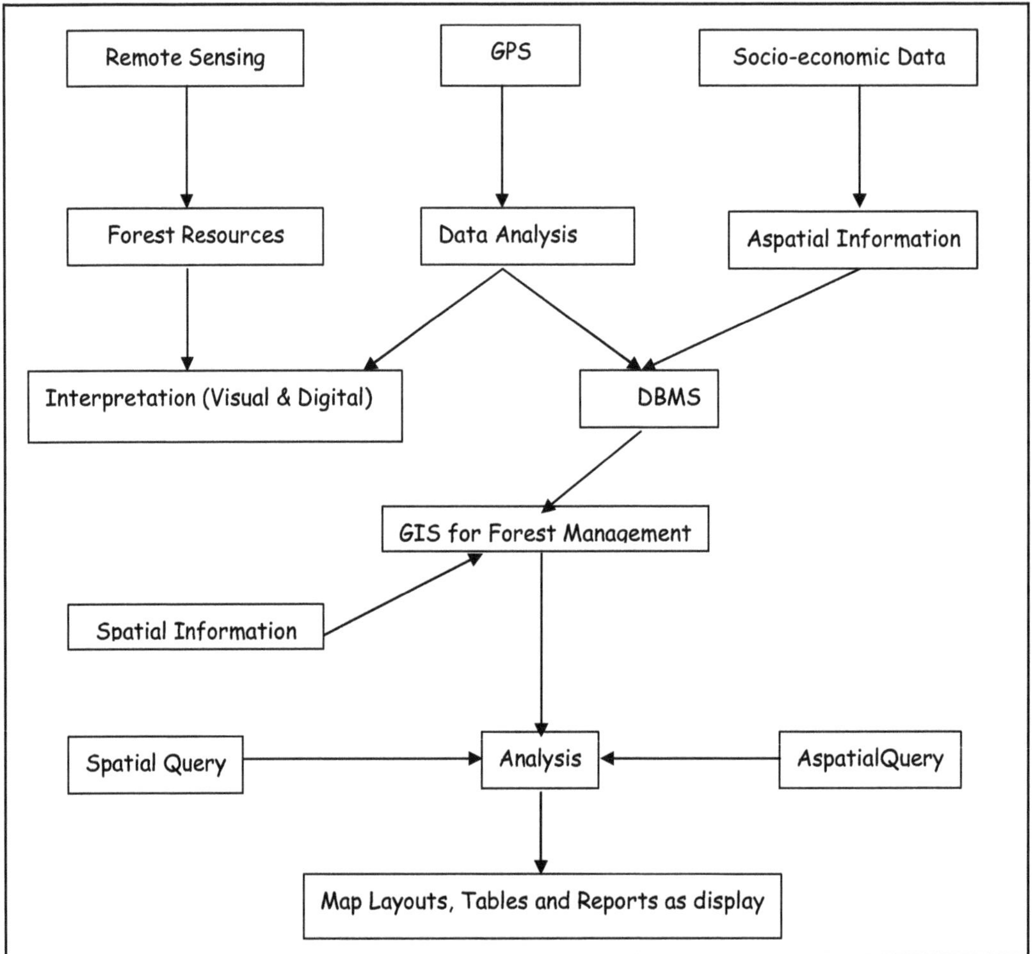

Fig. 10.2 Elements of FMIS.

10.10 Forest Fire Forecasting and Risk Area Mapping

Forest fires have an important influence on the vegetation cover, animals, plants, soil, stream flow, air quality, microclimate, and even general climate. The loss of timber is obvious and so is the damage to life and property. The loss of recreation value of the forest and the destruction of wildlife habitat are also consequences of forest fires.

The causes of the forest fires can be classified in three main categories (i) Natural, (ii) Intentional/deliberate due to man and (iii)Unintentional/accidental due to man.

Natural fires occur mainly due to lightning or sometimes due to rolling stones and rubbing of dry bamboos with each other in the strong wind. Nothing can be done to avert such fires. Intentional/deliberate forest fires are caused by the villagers, who set fire to drive away the game destroying their crops and to catch the wild animals the enmity with forest staff, neighboring rural population, villagers for collecting minor forest produce (MFP) like honey, mahua, flowers etc. that fall on the ground, and sparks from coal-fired railway locomotives. Unintentional/accidental fires occur in a forest due to careless throwing of match- sticks and burning ends of cigarettes/bidis, carrying of naked fire by people passing through forests, spread of fire from labor camps and recreation or picnic sites. These are the factors governed by parameters like vicinity to settlements and distances from roads. For any area to be prone to fire depends on many factors such as vegetation type/density, soil type, humidity of the area, vicinity to settlements, distances from roads and the host of others. While it is not possible to account for the natural occurrences or the deliberate attempts, the areas prone to accidental fires can be mapped.

10.10.1 Forest Fire Model

Geomatics can be used to detect fire spots, assessment of forest fires & burnt areas, fire hazard rating system. The layers used for such fire hazard zonation are the Forest cover map, Forest density map, Contour map, Road network map which give the required indices. The inflammability index or fuel type index is generated from the forest type and density map; elevation, slope, aspect indices from the contour map; accessibility index from the settlements and the road & communication network map. The layers and other ancillary information like wind, temperature, humidity etc., can be integrated in GIS environment to generate forest fire index map. The forest department, government of Andhra Pradesh, with the aid of Ministry of Environment and Forests (MoEF), Government of India GOI has completed the fire risk modeling for 15 districts in-house and remaining districts are to be completed. A sample map showing the zones of five risk can be seen in Plate 10.6. The objectives of the department are

- forest fire risk modeling
- mapping forest fire burnt areas
- designing of preventive measures for alleviating the forest fires
- detection of fire spots if equipment for recording NOAA AVHRR data
- rational utilization of funds.

10.10.2 Fire Behaviour Prediction Model

Researchers and scientists have long been trying to predict the behavior of a forest fire. Computer modeling has been the effort of many scientists using high resolution remote-sensing satellite imagery, powerful software, and GIS. In order to model a forest fire, the techniques for obtaining, analyzing and displaying spatial information in a timely and cost-effective manner are needed. As forest fires are spatial, GIS is used as a tool for modeling. A fire simulation program called FIRE! has been developed using ARC/INFO. "The model puts the power of comprehensive fire behavior prediction into the hands of qualified ground resource managers where it can be most effectively applied." (Weinstein, Green, Campbell, and Finney). Through the integration of fire behavior models with GIS models, new insights in the fire danger situation can be gained. One example is the damage potential that arises from fires starting at a certain point to sensitive objects and areas like buildings, railway lines, fire-sensitive ecosystems. Proximity is a concept which is used in a great many GIS-related models. The need for accurate and dependable tools for forest fire management has led to the demand for GIS in fighting forest fires.

In the past, the fire behavior prediction tool used was BEHAVE. BEHAVE is a non-spatial tool that inputs fire fuel types, topography data, weather data, and initial fuel moisture data. While BEHAVE is useful for predicting fire characteristics in a given area, it cannot give spatial output. It is unable to predict spread rates, flame lengths, fire line intensities, and heat calculations when these parameters change and they do constantly during a fire. Whenever there is a change, the entire BEHAVE model must be recalculated. Obviously, this is a very inefficient way to model the behavior of a forest fire. However, with the use of GIS technology, the prediction of fire behavior has not only become possible, but incredibly efficient and effective.

The GIS-based forest fire model puts the power of sound, accurate and efficient information into the hands of forest fire management personnel charged with coordinating the containment and extinguishing of wildfires. The model can now become one of the most effective tools for managing personnel, financial and equipment resources for battling one of our most destructive and dangerous forces of nature." (Weinstein, Green, Campbell and Finney).

10.10.3 Fire Group

To improve local decision support to integrated fire management through improved information and knowledge, we think that the documentation of events, in combination with local information and insight, must be customized to local needs to support improved knowledge, and hence, efficient decision making. Customized information improves local understanding of fire issues, allowing better definition of strategies, better management of resources, and better assessment of their implementation.

The Fire Group has expertise in fire monitoring, forestry, socio-economics, remote sensing and GIS, and decision support. It is also part of the Natural Resources Institute (NRI), which is an internationally recognized centre of expertise in research and consultancy in the environment and natural resources sector.

Most of NRI's work supports development assistance programs and contributes to environmentally sound management of physical and biological resources ('natural capital') for use by future generations. We can offer experience in low-cost technologies for the Local Application of Remote Sensing Techniques (LARST) and state-of-the-art expertise in the extraction of locally relevant fire management information.

10.11 Bio-Diversity Characterization

We live in a landscape and fulfil our most of the needs from its diversity. The increasing human population satisfies their demand from various kinds of economic activities carried out on the landscape, which ultimately along with various geomorphic and biological processes give the birth of spatial and structural changes.

Renewable natural resources, especially forests, are conserved for three major sets of purposes: the maintenance of ecological processes and life support systems; the sustained use of the resources for consumptive and social benefits; and the conservation of biodiversity units own right for ethical, moral, aesthetic and evolutionarily reasons. Only if all three of these can be met and integrated into the socio-political process, true sustainable development will be achieved, to meet them requires quantification of the current status and of changes on the resources. To ensure political and public support for conservation and wise use of such resources may require economic value.

Sustainable development of natural resources has become a key issue for survival of planet earth. In recent times, conservation of biodiversity has been put to the highest priority through Global diversity convention. There has been severe threat to the ecosystem by human interference and unscientific exploitation of natural resources. Therefore, there is an urgent need to conserve the gene pool in situ before it lost forever. The most appropriate method to that would be, to assess the ecological susceptibility by way of understanding the ecosystem/landscape complicities and their uniqueness. The landscape elements will be useful to generate scientific basis for further decision-making.

10.11.1 Scientific rationale

The landscape change spatially and temporally due to the tectonic, climate, biotic interaction with the environment. Of these the influences brought by the human beings through indiscriminate utilization of the natural resources and action, without due regards to the environmental impacts and sustainable development, can be prevented and controlled by scientific and technical know how, if proper data base is built. The landscape characterization has been proved to be effective for evaluation of landscape ecologically, and for the sustainable utilization of natural resources, to maintain the essential ecological processes and the life support system, to conserve the biological diversity and to prioritize the conservation areas and creation of network of protected areas.

- Study of structure, function and change in landscape elements.
- Study of patch characterization for habitat suitability analysis.
- Prioritization of natural and biological conservation areas by identifying "hotspot" regions and creation network of protected areas for present and future generations, to follow the WCS, and NCS.

- Study of fragmentation and disturbance indices for the attention of policy makers and the managers of the ecosystems.
- Study of landscape patterns for better management practices.
- Acquisition and preparation of long term sets for perception of global climatic change, deforestation, forest dieback, desertification, acid precipitation etc., for management intervention.
- To meet the objectives of sustainable developments.

10.11.2 Landscape ecology

Landscape can be defined as heterogeneous land surface composed of a cluster of interacting ecosystems that is repeated in similar from throughout. Therefore its study focuses on -

Structure : The relationships among the distinctive ecosystems or elements present i.e. distribution of energy, materials, and species in relation to their sizes, shapes, numbers, kind and configuration of the ecosystem.

Function : The interactions among spatial elements

Change : The alternation in the structure

The above three characteristics are closely related to each other, by studying one of them it is also possible to obtain information and insight concerning the others. Landscapes differ functionally according to their structure. Also changes in the structure of the landscape are related to changes in this functioning, and vice versa. Finally, the overall structure of the landscape influences the biodiversity, the species flow and the energy flow.

Landscape ecology focuses on the relationship between the various spatial units. Each landscape is formed by several landscape elements, which appear as patches and vary markedly in size, type, heterogeneity and boundary characteristics. Each of these characteristics has its own significance in the ecosystem and is important in evaluating landscape structure.

In addition, landscape ecology involves the application of these principles in the formulation and solving of real-world problems. Landscape ecology considers the development and dynamics of spatial heterogeneity and its affects on ecological processes, and the management of spatial heterogeneity.

There have been recent advances in computer hardware, remote sensing, GIS information theory, hierarchy theory, fractal geometry, and mode developing. Geographic information systems (GIS) are computer-based systems for the manipulation and analysis of spatially distributed data (Plate 10.7).

10.11.3 Remote Sensing and Landscape Ecology

Remote sensing plays an important role in analysis of environmental, cultural and natural resource management characteristics and because of its multi-spectral capabilities, it presents a unique perspective for observation and management of biophysical characteristics. Remotely sensed data can be collected at multiple scales and at multiple times, thereby offering the opportunity for analysis of various phenomenons synoptically from local to global scales throughout the time. These attributes make remote sensing data appealing to landscape ecology. Moreover, remote sensing data can be used as an important source of data for the development and refinement of models and can be used to validate models that predict landscape change.

Remote sensing promises to bridge the gap between the ecological research and the better planning and management of landscape and the distinct advantages are :

- Observe interference is prevented because of the large distance between the sensor and the object.
- Regional or global measurements can be done repeatedly, and
- There are a wide variety of spectral ranges and sensors available to provide remotely sensed data.

The remote sensing techniques are effectively useful on landscape ecology viz., inventory and mapping of resources, qualification of environmental characteristics, describing the flows of matter and the energy in the ecosystem, and evaluating change and alternative solution for ecosystems management etc. It is also useful in identifying community characteristics from a synoptic perspective; the ability to detect and observer temporal changes in habitat conditions.

10.11.4 Landscape Ecology and GIS

The GIS as a tool can be used to :

- Analyze temporal change,
- Determine spatial coincidence of physical and biological features,
- Determine spatial characteristics such as proximity, contiguity, patch size and shape
- Analyze the direction and magnitude of fluxes of energy; organisms or materials produce graphic outputs
- Interfaces with simulation models to generate new spatial data.

GIS capabilities with respect to landscape analysis are :

- Provide a database structure for efficient storing and managing ecosystems data over large regions.
- Enable aggregation and desegregation of data between regional, landscape plot scales,
- Assist in location of study plots and/or ecologically sensitive areas,
- Support spatial analysis of ecological distributions.
- Improve remote sensing information extraction capabilities, and
- Provide input data/parameters for ecosystem modeling

Geomatics can be used to evaluate landscape parameters from the analyzed raster data that is generated from the remote sensing data and also to generate many metrics useful in determining landscape pattern and structure like **Basic measures - Area statistics** -Patch statistics, Perimeter statistics, Information theoretic measures, Contagion, Diversity, Dominance, Evenness; **Probabilistic measures** – Electivity, Adjacency; **Shape measures -** Average perimeter to area ratios, Fractal dimension; **Textural measures -** Angular second moment, Inverse difference moment, Lacunarity; **Structural measures** – Connectivity etc., which will be useful in evaluating the biodiversity value, the disturbance index, the ecosystem uniqueness and finally the biological richness of the area.

10.12 Wildlife Habitat Mapping

Habitat analysis is considered most important for planning and management of protected areas. Wildlife Habitat includes wide variety of factors viz., vegetation, cover characterization including human influence on all these, soil, and topography and water availability. Since each species required a particular habitat to meet the requirement of space, food, cover and other needs of survival, the database with respect to above parameter can be used to identify the species needs for particular wildlife for Habitat suitability analysis.

Remote sensing has great potential in wildlife management because of its role in providing information about three most important parameters viz., cover, and food value and water resources and the fourth physiography indirectly required for Habitat suitability analysis. Besides other analysis like home range, tourist route planning and carrying capacity analysis can also be made. The change in the habitat that is mainly due to change in cover characteristics, degradation due to reduction in canopy density, fragmentation of forested areas, degradation due to fire, over grazing and cultivation can also be analyzed. A hapitat map of Mannanur Range of Andhra Pradesh State in seen in Plate 10.8.

The Geomatics will be helpful in :

- Monitoring of Physical setting of the protected area and land cover changes in the vicinity.
- Wildlife habitat suitability mapping for carrying capacity analysis and creation of corridors.
- Monitoring of the corridor of the protected area, important population migration and genetic exchange
- Monitoring of Positive and negative impacts on the habitat conditions.
- Management and monitoring of various changes in flora and fauna.

11
Watershed Management

11.1 Introduction

Soil, water, and vegetation are the most vital natural resources for the survival of man and his animals. To obtain the maximum and optimum production of vegetation, all the three resources have to be managed efficiently. For their efficient management, one has to look for suitable unit(s) of management so that these three resources are handled and managed effectively, collectively and simultaneously. Soils can be managed on the basis of a soil series or type or any other convenient unit of land; vegetation can be managed on forest type/forest sub-type or some similar classification for trees/grasses etc., water can be managed if a watershed is taken as a unit. Since soil and vegetation can also be conveniently and efficiently managed in this unit, the watershed is considered the ideal unit for managing these three vital resources of soil, water and vegetation.

Watershed conditions affect and influence water quality, quantity and regime. These three water parameters are vital for human and animal consumption, irrigation, industry, power generation, recreation, transportation and in many other ways. By manipulating different watershed characteristics the quality, quantity and regime can be regulated according to requirement.

Watersheds should be properly managed through scientific planning to prevent loss of top soil and water that results in low productivity, to prevent land from deterioration, to mitigate flood hazards in the downstream reaches, to prevent rill/ gully formations even at low rainfall events etc., which can be achieved by adopting suitable soil and water conservation measures to intercept and reduce runoff, and induce larger and extensive adoptions of rain water, to trap eroded material and thus reduce sediment transport into streams and in turn into reservoirs, to restore/reclaim degraded lands into productive lands, to increase total productivity and provide a stabilized environment.

Many countries all over globe are facing acute water problem due to population explosion, urbanization, conversion of forests to agriculture, indiscriminate use of surface and ground water. Water on earth goes through a hydrologic cycle. Rain water is the major source for agricultural production for major parts of the world and especially the tropical countries. Major part of the total cropped area in these countries is under rain fed/unirrigated. Hence effective rain water management is vital to increase and stabilize their yields. Realizing the fact the Andhra Pradesh State Government, India is implementing massive Neeru-Meeru (Water and We) (Anji Reddy., M. 2002) program for harvesting the rain water (Fig. 11.1).

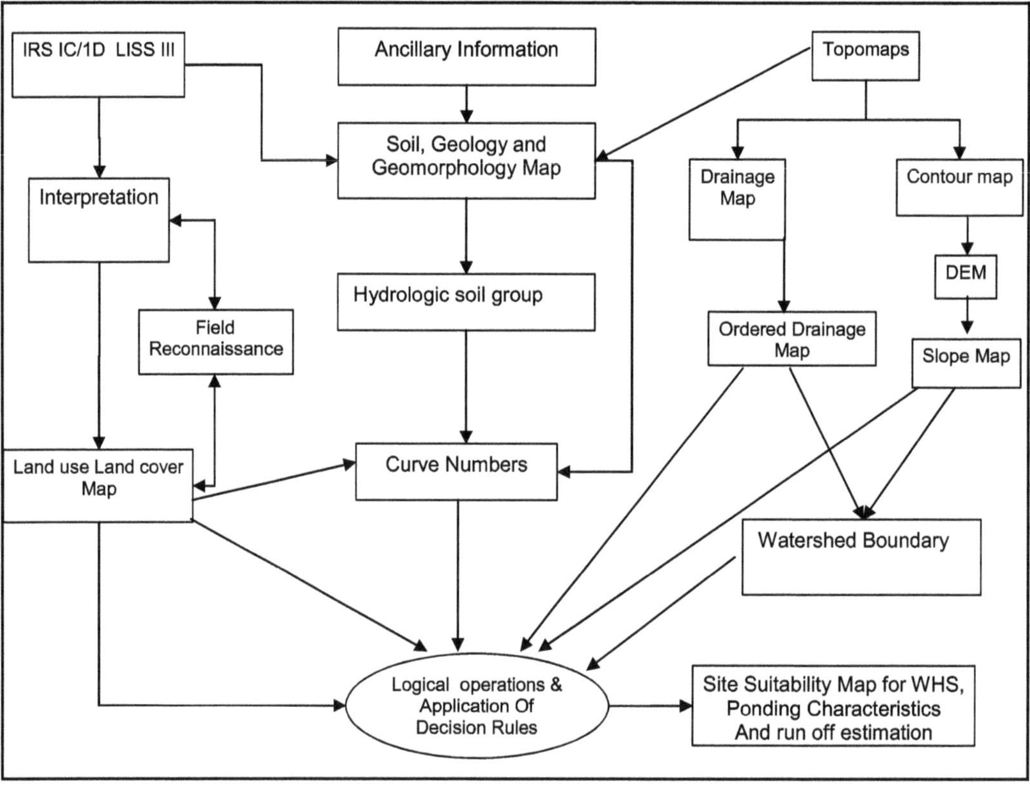

Fig. 11.1 Schematic representation of preparation of site suitability map of water harvesting structures.

The traditional rain water management practices have not yielded satisfactory results. GIS, RS, GPS and various analytical tools associated with them, are proven to be capable, providing sources of data and means of manipulating data appropriate for large scale management of natural resources i.e., forests, water, soil etc., and appropriate planning on sustainable basis. The DGPS technique in conjunction with the GIS and RS can be used for site suitability analysis of water harvesting structures.

Water is the elixir of life. The first civilisations, Mesopotamia and Harappa (Indus valley Civilisation) flourished on the banks of rivers. Thus the importance of the water was known to the first civilisations of our world. Even though groundwater was discovered long before Christ, no technology could replace the necessity of ground water. Thus the exploration for water is still continuing even in the twenty-first century. Science and technology are growing at such a rapid pace that we can communicate with a person who is on the other side of the earth or in space, but the scarcity of even drinking water is still persisting.

The growing pressure on land for food, fibre and fodder in addition to industrial expansion and consequent need for infrastructure facilities due to ever-increasing population, has given rise to conflicting demands on finite land and water resources. India is a land which has been blessed with a number of perennial rivers like the Ganga, and the Bramhaputra in the north; the Godavari and the Krishna in the south, and so on. The water from these rivers can be put to many uses, such as, drinking, agriculture, domestic and industrial uses, and navigation and power generation.

Rainfall is the source of water in our country. About 175 Mha of land in India, consisting about 53% of her total geographical area, suffers from deleterious effect of soil erosion and other forms of land degradation. A country like India that supports 16% of world population on 2% of the global land area, the problem is serious enough. Keeping in view of the ever increasing population and need for food security for future generation, it is realised that the water and land resources need to be developed, used and managed in an integrated and comprehensive manner.

It has already been realised that the soil and water conservation measures carried out on a watershed basis play a prominent role in the strategy of comprehensive land and water management. A watershed is an area from which runoff resulting from precipitation flows past a single point into a large stream, river, lake, or ocean. Thus watershed is the surface area drained by a part of the totality of one or several given water courses and can be taken as a basic erosional landscape element where land and water resources interact in a perceptible manner. Watershed management is the process of formulation and carrying out a course of action involving modification of the natural system of watershed to achieve specified objectives. It implies the proper use of land and water of watershed for optimum production with minimum hazard to natural resources environment. Remote Sensing and GIS techniques are being effectively used in recent times as tools for watershed management.

Watershed management is an integrated approach to the management of water, land, and vegetative resources for multiple objectives. Watershed being a natural hydrological entity responds effectively to various engineering, biological, and cultural treatments designed to maximum production. The quantitative analysis of drainage system is an important aspect for characterisation of watershed. Earlier studies indicate a relationship between cumulative stream length, stream order, bifurcation ratio, drainage density, stream frequency, and different topo elements, such as, texture ratio, form factor, circulatory ratio, and elongation ratio, and the prioritisation depending upon the Sediment Yield Index (SYI).

India is well endowed with rainfall and is one of the wettest countries in the world with annual precipitation estimated at around 400 millions hectare-meters; however, a vast area of the country is drought prone. The arid and semi-arid regions of India covering 53.9% of the geographical area are drought areas. Precipitation is confined to three or four months a year, and most of the rainfall is lost as runoff and evapo-transpiration (Seshagiri Rao, K. V., 2000).

In developing countries like India, the major thrust of any Government is the sustainable development of rural areas. This is possible only with proper planning of watershed, which is essential for the conservation of land and water resources management. Characterisation and analysis of watershed are prerequisite for watershed management. Watershed characterisation and its productivity can be achieved by using watershed concept. Watershed characterisation requires generation and gathering of information on a number of parameters both static and dynamic in nature comprising geology, geomorphology, hydrology, soils, landuse/landcover, soil erosion, and climate. Sustainable development aims at maintaining an equilibrium between human needs and economic development within the parameters of environmental conservation through efficient use of natural resources to ensure trade-offs between desired production-consumption levels within the natural ecosystem.

Watershed characteristics

- Physiography: size, shape, land slope, drainage density and pattern
- Edaphic conditions: soil and geology, land use, vegetative cover, rainfall
- Socio-economic factors: Demographic profile, social stratification, farm structure, attitudes and behavior, land tenure systems
- Organizational structure: people's participation
- Peak runoff rate

11.1.1 Concept of Sustainability

The concept of sustainable development drew worldwide attention after the publication of Bruntland report. Sustainable development is defined as a development that meets the needs of the present without compromising the ability of the future generation to meet their own needs. Turner (1992) emphasised that sustainable development should be based on economic development, and depends upon interactions among three systems : the biological system, the economic system, and the social system. The goal of biological system is maintenance of genetic diversity and biological productivity. For the economic system, it is the satisfaction of basic needs (reduction of poverty), equity enhancement, and an increase in useful goods and services. The aim of the social system is to ensure cultural diversity, institutional sustainability, social justice, and people's participation.

In order to make use of the existing technologies for the sustainable development of any part of the developing countries, the management of local institutions should adopt a three tier system. This system consists of (a) group level (b) community level and (c) locality level (Rajora, 2000). Group level consist of persons in a small residential area based on occupation, age, and gender. The user groups of Andhra Pradesh at micro watershed level are an instance in this regard. The community level consists of persons from self-contained socio-economic residential unit as watershed development committee of Andhra Pradesh. The third is the locality level with a multi-disciplinary team drawn from local institutions, district level institution and project implementing agencies.

11.2 Philosophy and Concept of Watershed

The significance of watershed strategy is traced back to two independent historic processess. The attempt to restore Alpine in the last quarter of nineteenth century and the conservation movement of the 'U. S' have pushed the watershed to the surface. However, it is FAO, which has been playing the catalytic role for the last five decades at the international level in campaigning for the watershed approaches. The land restoration and torrent control techniques developed in Europe and the vegetation manipulation, and soil and water conservation techniques of U.S along with conceptual and theoretical inputs have formed the basis for the FAO's watershed strategy.

Watershed is a spatial unit with complex relations among its entities and is defined by stream pattern where the rainwater is collected through surface and sub-surface runoff and culminated at the common point down the stream. The hydrological regime of watershed has inalienable relationship with flora and fauna, soil, climate, and population, specific to local eco-system. Human interface with nature on spatial and temporal scale in terms of the resource extraction/exploitation is a right strategy to understand the forces and relations of development. The indiscriminate exploitation of the resources during the phase of modernisation clearly demonstrated the historical negation of resources.

The multi-dimensional nature of the watershed impinges upon the strategies to be multiple in nature and will be invariably tackled by interdisciplinary approaches. It is increasingly recognised that watershed management not only tackles soil erosion, soil percolation, soil moisture, surface and sub-surface runoff, conservation of water and soil, deforestation, contour trenching checked on development but also poverty, unemployment, housing, illiteracy, soil inequalities, economics and social justice, cells for people's mobilisation towards environmental awareness, and right to access for unpolluted resources.

The concept behind planning and management of any watershed involve in developmental activities using Natural Resources available in accordance with socio-economic conditions of that area. The two general natural resources for the development of watershed are : land and its related resources and both surface and subsurface water resources and the details of cultivation and other infrastructure facilities. Some of the developmental activities in any watershed in general comprises forestry along ridge lines and steep slopes; pasture development in the forest areas along ridges; soil conservation needed in most of the area spreading from forest boundary up to common areas; dry farming practices in the rainfed cultivated areas; water conservation measures including construction of percolation tanks in the rainfed cultivated areas; construction of irrigation tanks across minor streams; groundwater exploration works where groundwater use is available; sericulture under well ayacut; and making people aware of the importance of watershed management.

If all these developmental activities are implemented on watershed basis instead of administrative boundaries, in a planned manner they have a greater impact on the economic benefits since most of the activities support each other.

11.2.1 Classification of Watersheds

Classification of watersheds becomes important for the development of countries at village level for its sustainability on a systematic, scientific, and rational approach to watersheds. In order to apply scientific and technological results to planning and developmental activities, it is essential to have the hierarchical system of delineating bigger hydrologic units into watersheds of different sizes. The size of the watershed may vary from a few square meters to thousands of square meters. On the otherhand for the creation of an action plan for land and water resource development or for the construction of a percolation tank, a few hectares of watershed are sufficient.

Therefore for all practical applications on land and water resources development, all the hydrological units have to be classified as having a workable size of the watershed. The size of the watershed is also affected by afforestation, grass land development, cultivation, and so on. Many physiographic features like valley, undulating hillocks, and rugged hilly terrain also influence the size of the watershed. Based on the developmental activities, physiographic parameters, geomorphometric characteristics, aims and objectives of a particular developmental programme, and planning and implementation of an action plan, Bali (1980) has identified the hydrologic unit taking their dimensions into consideration. (Table 11.1)

Table 11.1 Terminology against the size of the watershed

Terminology	Size
Sub-watershed	100-500 sq. kms.
Milli-watershed	10-100 sq. kms.
Micro-watershed	1-10 sq. kms.
Mini-watershed	<1 sq. kms.

The Central Water and Power Commission (CWPC), delineated India into 6 distinct Water Resource Regions (WRR). They are :

Region (i) Rivers flowing into the Arabian Sea, excluding Indus system.

Region (ii) The Indus basin in India.

Region (iii) Rivers flowing into the Bay of Bengal, other than the Ganga and the Brahmaputra systems.

Region (iv) The Ganga system.

Region (v) The Brahmaputra system.

Region (vi) Rajasthan.

The number of hydrologic units at different stages of delineation of our Country are shown below :

Water Resources Regions - 6

Basins - 35

Catchments - 112+

Sub catchments - 500+

Watersheds - 3237+

As shown in Table 11.2 watersheds in India are delineated on the basis of size Table 11.3 shows the terminology used for the watershed, symbolism for the boundary configuration and respective code.

Table 11.2 Terminology, size and digital code adopted for watershed delineation

Delineation order	Suggested Terminology	Mean size (+50%) sq. km.	Example of Digital code
(i)	Water resources Region	5,00,000	4000000
(ii)	River Basin	50,000	4060000
(iii)	River Sub-Basin	5,000	4060400
(iv)	Watershed	500	4060480
(v)	Subwatershed	50	4060486
(vi)	Mini-watershed	5	Open
(vii)	Micro-watershed	Below. 2.5	Open

Table 11.3 Hydrologic Units, Boundaries and Code Numbers

Terminology	Symbology	Code
Region	═══════	②
Basin	▬▬▬▬▬	Ⓓ
Catchment	—✕——✕—	[4]
Sub-catchment	▬ ▬ ▬ ▬ ▬	[D]
Watershed	————	8

Water resources regions are assigned Arabic numbers 1, 2, 3,....

Basins are assigned letters as A, B, C,..

Catchments are assigned Arabic numbers 1, 2, 3,....

Subcatchments are assigned letters as A, B, C,...

Watersheds are assigned Arabic numbers 1, 2, 3,....

Thus watersheds will have the codes like 1A1A1, 2B2A3, 3A5C4,...

11.3 Technology Vectors and Social Dynamics

Deforestation, poor management of top soil, water resources, and total neglect of environment have intensified poverty in this area. The extinction of biodiversity caused by anthropogenic impact on the environment is destroying the rich natural heritage built up through decades of evolutionary process. The long-term vision of a sustainable and equitable world requires harmonisation of cultural diversity with human solidarity to achieve a social revolution with global planetary consciousness.

Science and technology which alone can usher in a quantitative transformation of society and ensure its very survival, is helpless unless the society as a whole is able to imbibe the scientific temper and make science and technology an integral part of its cultural fabric.

Long-term solutions for solving the root causes are suggested to meet basic human requirements within the constraints imposed by limited land resources and sustainability of environmental integrity (Venkateswarlu., P. 2000).

The Ideal Integrated Watershed Management Model (IIWMM) is presented in this book by examining a case study. This study is concentrated to meet the micro level priorities and needs by incorporating the ridge-to-valley concept keeping in mind the dictum of thinking globally and acting locally.

11.4 Role of Remote Sensing and GIS

The concept of watershed as a planning unit for development of land and water resources, has gained importance since 1974, when the Ministry of Agriculture, Government of India, initiated various developmental programmes like Drought Prone Area Programme (DPAP) and Desert Development Programme (DDP). Therefore it is necessary to delineate watershed boundaries at various levels of hierarchy to identify developmental potentials under various activities in each watershed. The plans for watershed development should bear in mind the fragile economic system of a region and ensure user's participation in developmental activities.

Remote Sensing through air borne/space borne sensors have been gainfully used for watershed characterisation, assessment of watershed (priority) evaluation problems and potentials, management requirements, and periodic monitoring. Satellite image interpretation and analysis of digital raster/vector data has brought out substantial improvement in watershed survey and mapping techniques during the last two decades. Remote Sensing invites integrated satellite inputs of various land information units for micro level watershed management. GIS technique can be used to integrate all the themes obtained from remote sensing data interpretation for integrated multi-discipline watershed approach for its proper management for sustainable development of rural areas.

Space technology in the form of remote sensing can play useful role in watershed management. Modern technologies employed in watershed management are remote sensing and GIS. The advent of Remote Sensing and GIS techniques has opened new possibilities for the study of various components of watershed management including land resources, water resources and evaluation of greenery status. The spatial data obtained by remote sensing satellite along with the attribute data about the watershed can be integrated on GIS platform for the overall management of any given watershed. On this basis a plan on micro level watershed management for sustainable rural development of drought prone areas using Remote Sensing and GIS techniques, is presented in this chapter for understanding the concept of watershed approach by examining a model watershed.

The Geomatics can be used :

- To create Digital Elevation Model / Digital Terrain Model and to estimate slope percentage (DGPS will be employed to collect the fly levels for specific micro watershed areas to generate contours at 1 mt interval) and to generate the stream network (upto streamlets) and to delineate the watershed boundaries
- To derive the various watershed parameters from the satellite remote sensing and other ancillary data.
- To integrate the watershed parameters in GIS environment by framing various decision rules to identify suitable sites for water harvesting to adopt appropriate soil and water conservation measures and their prioritization as per feasibility and local importance and to prepare a comprehensive watershed management plan.
- To characterize the geometry of flow and ponding at a selected site, to estimate the timing and magnitude of surface water flow and accumulation, to establish the actual content and extent of surface water ponding at the selected site at two or more specific times, to compare the results of simulated estimates of water distribution with the measured data on actual ponding.
- To assess the impact of the watershed management practices.

Remote Sensing, GIS and GPS technology will provide much information and can be used to estimate the runoff or used as an aid for estimation of equation coefficients and model parameters and also in selecting suitable sites for water harvesting structures and prioritization according to criteria and budget availability. Satellite data can be interpreted to derive a number of thematic information in land use, soil, vegetation, surface water, snow cover, stream network, land form, erosion intensity etc., which combined with conventionally measured climatic parameters (temperature, precipitation, evaporation etc.,) and topographic parameters (height contour, slope) should provide the necessary inputs to the models, both empirical and conceptional, are in practice reliable and timely estimation of runoff (Fig. 11.2).

The DGPS system will be employed for sensitive micro watersheds to collect the fly levels and the information will be utilized for generating Digital terrain model which in turn helpful in generating the contours at 1 mt interval, slope and stream network. Basing on various criteria rules framed with the help of contours, slope, soil, drainage pattern etc., and other collateral information, the site suitability analysis for WHS will be carried out. The runoff will be estimated using other parameters governing the watershed development.

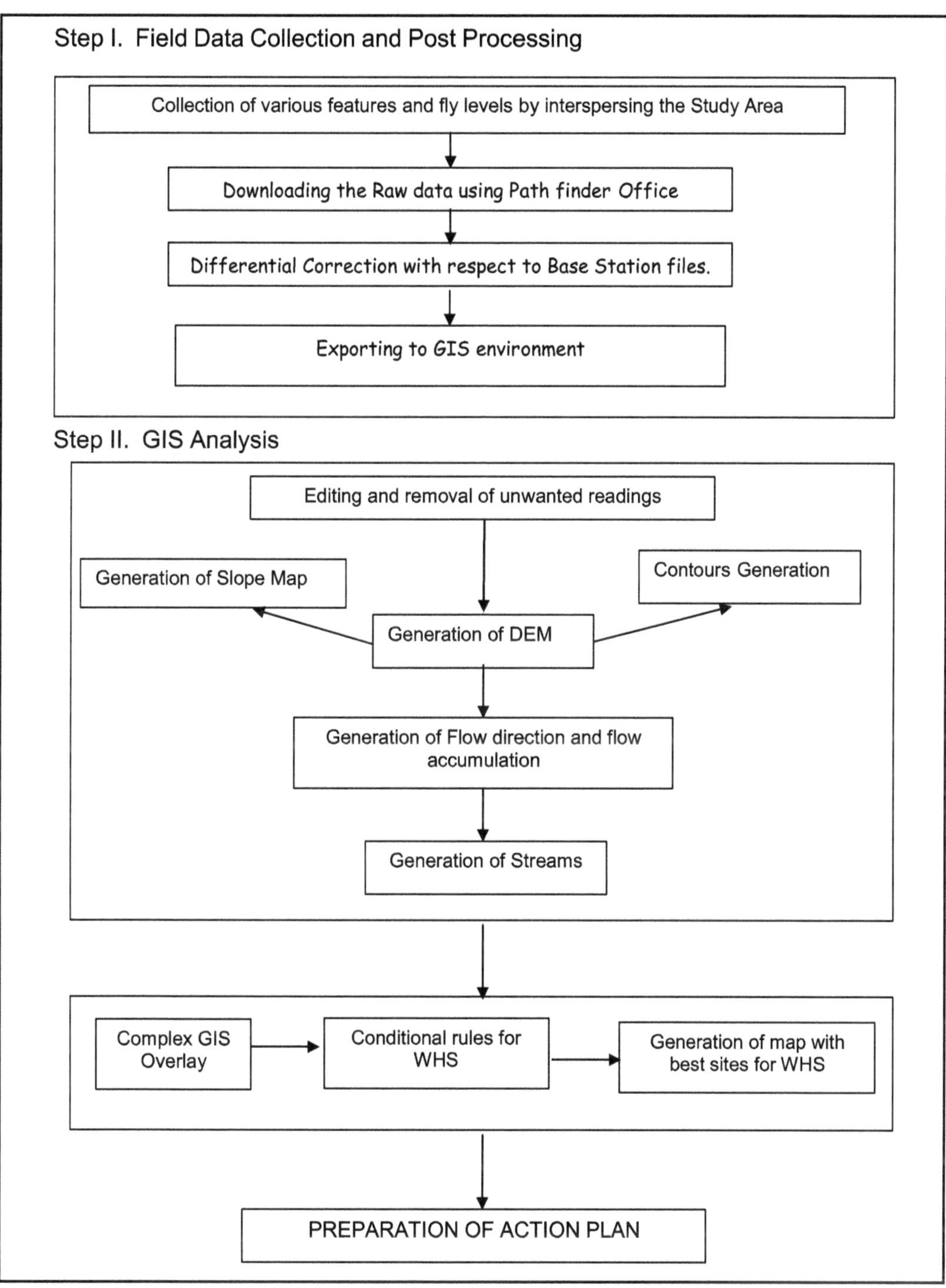

Fig. 11.2 DGPS techniques for preparation of action plan for water harvesting structures.

11.5 GIS Database for Watershed Management

The database required for the management of watershed is furnished below.

(a) Natural Resource data bases :

Soils, geology, forest, land resources
Rainfall
Surface water sources (canals, tanks, rivulets, lakes, and rivers)
Ground water (open wells, tube wells, depth to water table - & water quality) and
Minerals / other natural resources

(b) Population :

Occupation
Male, Female
Age-wise distribution
Employment status
Per capita income
Crops grown, kharif / rabi / summer

(c) Infrastructure :

Education :

Primary, Secondary, Higher and Technical Institutions

Health :

Primary health centres
Local fund hospitals
General hospitals

(d) Communication :

Roads
Post & Telegraphs
Railways
Waterways

(e) Marketing :

Banking facilities
Markets and
Administrative units

11.5.1 Suggested Developmental Plans

Development plans include :

Scale of operation
Extent of peoples participation
Expenditure involved
HRD benefits anticipated
Self-employment
Employment generation
Value of benefits
Built-in checks for ensuring sustainability
Maintaining & following mechanism

11.5.2 Technical Inputs

Aerial photographs
Remote sensing data analysis
Database
GIS representation of basic data
DEM/DTM analysis

Hydrologic analysis: soil erosion/loss, water conservation measures, and associated structural measures - sizes, costs involved and benefits likely to occur.

The problems encountered in the watershed management study are : (i) drinking water problem, (ii) problems related to dry land agriculture, (iii) degradation of forest land due to commercial activity, human interference and so one (iv) soil leaching due to erosion of soils, (v) problem of salinity, and (vi) problem of fluoride and so on.

11.6 General Objectives of Watershed Management Program

Earlier investigators have revealed that soil and land degradation in the catchment areas pose ecological problems like sedimentation of reservoirs causing reduction in the irrigation potential of the command areas, low hydro-electric power output, floods, and drought. Realising the seriousness of the problem of sedimentation following soil erosion, a theme of micro-level watershed management is initiated to prevent soil erosion, and develop watershed. Keeping this in view, the following objectives have to be designed for the development of any area on the basis of watershed for its sustainability:

(i) To evaluate the natural resources of the watershed using Remote Sensing satellite data for the extraction of feasibility condition for its development.

(ii) To derive required attributes of each sub-water drainage basin with the help of morphometric analysis.

(iii) To estimate sediment yield for the watershed and prioritisation of sub watersheds within the micro-level catchment area. The requirements of prioritising the sub-watershed are mainly based on the sediment yield, drainage characteristics, intensity of eroda bility, and topography of the area.

(iv) To integrate the information obtained from the analysis of the satellite data, results from various characteristics and the quantity of sediment of sub-watershed within the catchment area and socio economic factors into GIS for providing indicators of developmental activities and to prepare a master action plan for its sustainable development.

11.6.1 Operational Objectives

(i) To study the existing natural resources, both land and water, for their assessment, management, proper utilisation, and feasibility condition for their development. The natural resources mapping package consists of hydrogeomorphology, geology, groundwater potential zoning, soils, land use/ land cover, drainage network, slope, and surface water bodies.

(ii) To promote the economic development of the project area which is directly or indirectly dependent on the watershed through (a) optimum utilisation of the watershed's natural resources to mitigate the adverse effects of drought and prevent further ecological degradation. (b) employment generation and development of the human and other economic resources at village level in order to promote savings and other income generation activities.

(iii) To create a digital database information system for the watersheds under study consisting of spatial and non-spatial information of natural resources and socioeconomic data.

(iv) To introduce a physical drainage network model based on basin geo-morphometry for the development of action plans for land and water resources and application of this model for prioritising to sub-watersheds in the study area for sustainable development.

(v) To formulate an ecofriendly and economically viable physical model for integrated watershed management, which includes the generation of action plans for land and water resources development. This model is an ideal integrated watershed management model. The formulation of this model is based on (a) existing natural resources database, (b) drainage basin geomorphometric parameters, and (c) socioeconomic status of the study area.

11.7 Research Approach

The approach to research based on the use of remote sensing and GIS begins with acquiring the satellite imagery and toposheet of the required study area. The following steps are adopted for the proper management of watershed using remote sensing and GIS techniques. The conceptual frame work of integrated mission for sustainable development is presented in Fig.11.3.

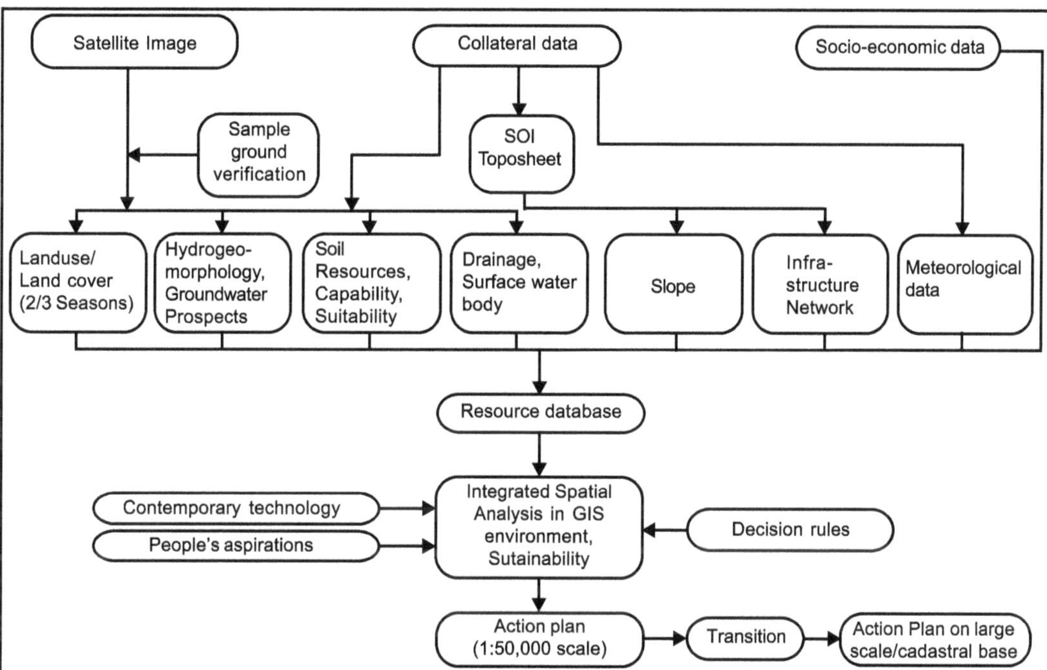

Fig. 11.3 Flow chart showing the steps involved in integrated mission for sustainable development (IMSD).

Step (i) : Land use/Land cover map is prepared to know the various drainage characteristics like drainage density, stream frequency, and basin slope for each micro-water shed.

Step (ii) : Land use/Land cover map is prepared to know the various uses of the land, vegetation, water bodies, rock/soil, artificial cover and others which are due to land transformation. This map is also used to find the areas of the crop land, forest, and watershed for each sub-watershed for calculating Sediment Yield Index and vegetative cover factor.

Step (iii) : Hydrogeomorphology map is used for finding out ground prospects to suggest water harvesting structures, such as, percolation tanks.

Step (iv) : Slope map is used to know the terrain properties like steep slope, and plains.

Step (v) : A GIS digital system like ARC/INFO is used for input, manipulation, and creation of error-free digital database for all the natural resources.

Step (vi) : Precipitation data of the watershed is used in finding Sediment Yield Index.

Step (vii) : Depending on the combination of the above-mentioned resource themes, action plan for land and water resources and treatment plan for catchment area are generated for the development of water resources.

11.8 Model Watershed

This section of the book highlights the fundamental procedure for the study that deals with a comprehensive evaluation and integrated analysis of Shivannagudem watershed. The watershed is situated in the south-west corner of Nalgonda district of Andhra Pradesh state at a distance of 25 km. from Chautuppal. The watershed under study forms a part of Musi catchment which inturn forms a part of Krishna river basin. It lies between North latitudes 16° 58' 20" and 17^0 I' 40" and East longitudes 78^0 and 78° 57' 48". Chautuppal and Devarakonda are the nearest towns to the watershed (Fig 11.4). It is connected by a convenient network of both metalled and unmetalled roads to the mandal head quarters. This watershed is covered by parts of Indurti, Shivannagudem, Kottala, Namapuram, Saran and Lankalapalli villages. Agriculture is the principal avocations of the people living in the area. The total extent of the watershed is 3589.00 hectares.

The area has been categorized into 10 land forms/geomorphic units. Pediplain with moderate weathering and pediplain with shallow weathering are the most prominent units in the model watershed area. Groundwater occurs in these geomorphic units under both unconfined and semi-confined conditions from the weathered and fractured/ jointed rocks. Rainfall is the principal source of recharge in the study area. The concentrations of flouride vary from 3.5 to 3.8 mg/l, and so the water in this area is not potable.

11.8.1 Objectives of Model Study

(i) To study the existing cropping patterns on the available water resources of the study area using Remote sensing data.

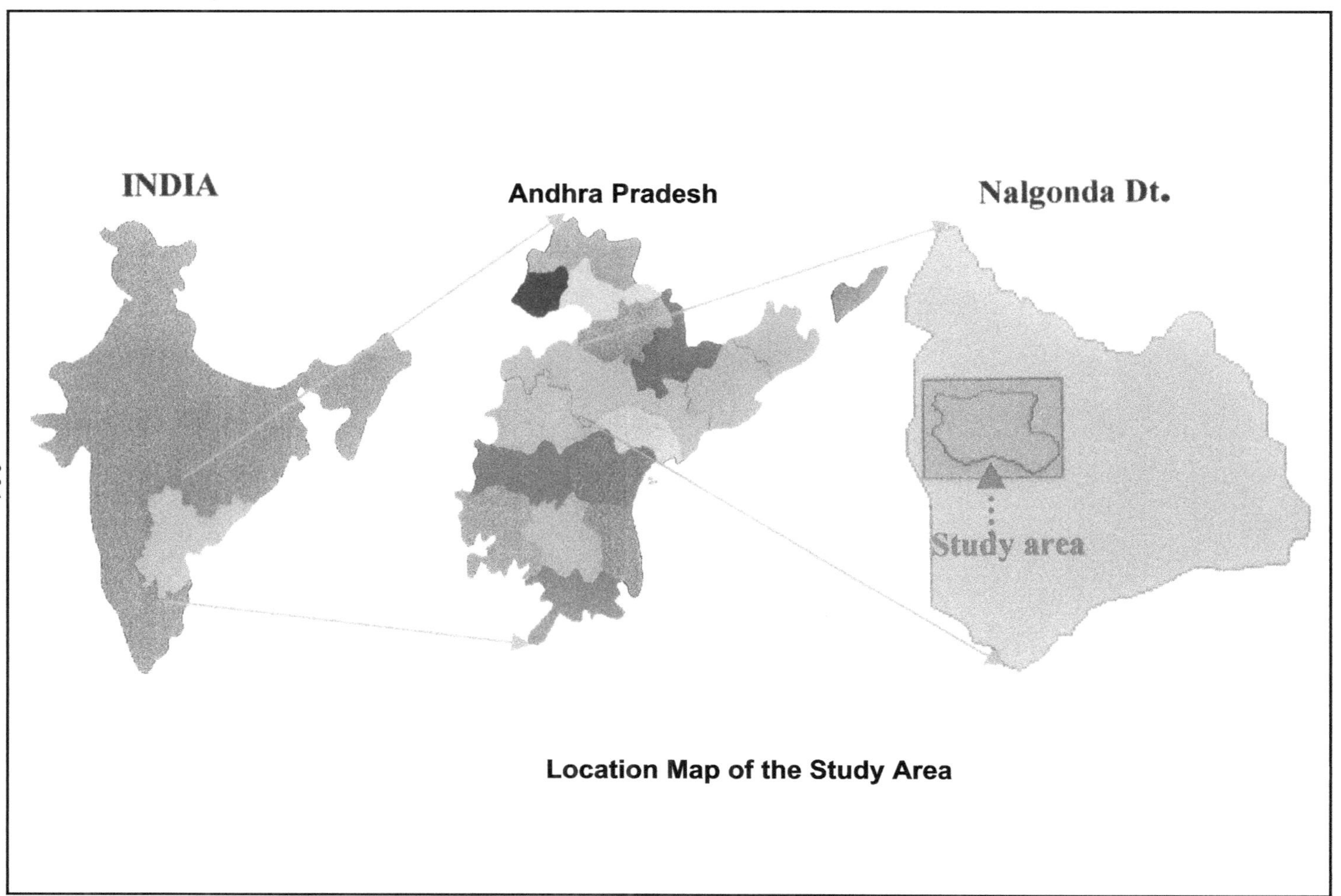

Fig. 11.4 Location map of model watershed.

(ii) To map all the thematic information by interpreting the satellite imageries and SOI toposheets and production of maps through ARC INFO GIS software.

The Thematic information includes:

(a) Land Use/land Cover

(b) Slope analysis

(c) Soils and their classification

(d) Hydrogeomorphology

(e) Drainage

(f) Groundwater prospects

(iii) To select a suitable cropping pattern for the study area, an action plan for land resources is prepared by the integration of the above themes.

11.8.2 Method used in Model Study

Watershed is made up of its natural resources, both physiographical and hydrological, and human resources. A comprehensive development of watershed requires the analysis of the present status or conditions of its natural resources, such as, soils, topography, climate, drainage, land use, water resources, and the present trend of land degradation. So, the basic prerequisite is the availability of timely, accurate, and detailed information on the watershed resources. With the advent of modern information systems, such as, satellite remote sensing, and owing to its repetitive coverage in time and space, accuracy, and cost effective nature, it has become possible to apply watershed models efficiently. The required database combines remote sensing data and other ancillary data in the form of toposheets, reports, maps, and groundtruth data. The important watershed resources considered in the present study include climate, slope, drainage and surface water bodies, Hydrogeomorphology, land use/land cover, and soils.

Realising the importance of comprehensive and integrated approach to the problem, watershed resources were evaluated from different sources using new ideas, techniques, and tools. The thematic mapping of various watershed resources is done on 1 :25,000 scale. A flow chart showing the steps involved in the development of action plan for watershed management is given in Fig. 11.5.

The watershed area of Shivannagudem is delineated and a base map, slope map, and drainage map are prepared from the SOI toposheets of 1:25,00 scale. The land use and land cover, soil and groundwater prospect maps are prepared from 2,3, and 4. False Colour Composite imageries on 1:25,000 scale. The maps are then scanned using Logitech Scanner, and digitised using Autocad 14. Then the drawing output is given, using Arcview GIS software.

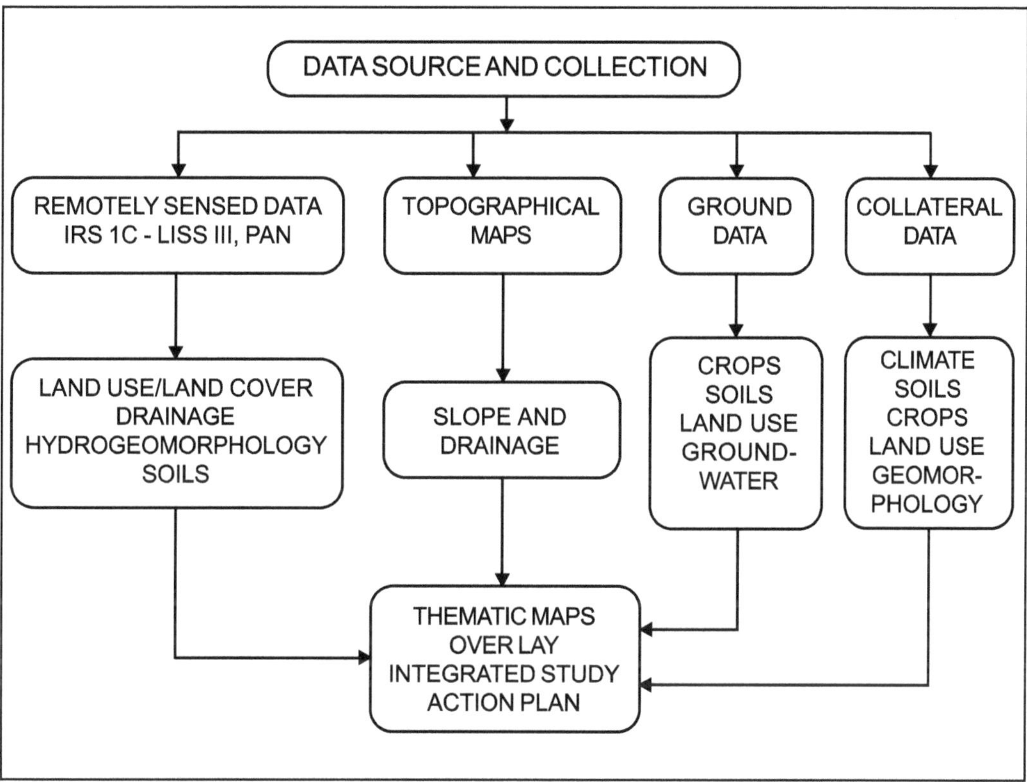

Fig. 11.5 Flowchart showing the methodology adopted in the model study.

11. 9 Land Use and Land Cover

A comprehensive information on land use and land cover is the basic prerequisite for land resource evaluation, assessment, utilisation, and management. Today, with increasing population pressure on land and the resulting changes in the land use pattern and processes, a considerable degree of land transformation and environmental deterioration is being witnessed. Therefore, it is important to understand the cause and effect of the change through a scientific study.

Preliminary interpretation of satellite data has been carried out on 1:25,000 scale using visual interpretation techniques for delineating various land use/land cover classes in the study area. The satellite data related to kharif and rabi seasons are used for mapping the land use/land cover classes. The doubtful areas identified during the interpretation are physically verified on the ground during the field work, and necessary modifications and corrections of the misinterpreted land use/land cover boundaries are made. The final thematic details are transferred on to the base map prepared on 1:25,000 scale. The various land use/land cover classes (Plate 11.1) delineated in the watershed include (i) Built up land, (ii) Agricultural land, (iii) Wasteland, (iv) Water bodies, and (v) Others like quarrying. The area-wise estimates of land use/ land cover classes of the study area are given in Table 11.4.

Table 11.4 Area wise estimates of land use/land cover of Shivannagudem watershed

S. No.	Class	Lu/Lc	Area	% area
1.	Built-up land	Town/village	60.60	1.69
2.	Agricultural land	Cropland		
		Kharif irrigated	58.40	1.63
		Kharif Unirrigated	2188.00	60.97
		Double Crop	1000.00	27.86
3.	Wastelands	Land with Scrub	2.80	0.08
		Land without Scrub	133.40	3.72
		Barren rocky/stony waste/sheet rock area	23.40	0.65
4.	Water Bodies	River/Tank	110.20	3.07
5.	Others	Quarrying	1.50	0.04
		Total	3589.00	100.00

11.10 Slope Analysis

Slope is one of the most important terrain characteristics and plays a vital role in geomorphologic and runoff processes, soil erosion and land use planning, and watershed management. It is very important to have an understanding of the spatial distribution of slopes for the development, planning, and management of both land and water resources.

The slope map of Shivannagudem watershed has been prepared on 1: 25,000 scale using Survey of India Topographical maps. The contour lines drawn on toposheets have been used in the preparation of the slope map. The nature of the slope of a terrain is explained by the horizontal spacing of the contours. The closely spaced contours correspond to steeper slopes and the spare contours reveal gentle slopes. The slope categories are expressed in percentages. The slope map (Plate 11.2) has been prepared by following the guidelines of All India Soil and Land Use Survey on slope categories. The details of slope classes that correspond to the categories and percentages are presented in Table 11.5.

Table 11.5 Details of Slope classes corresponding to the categories and percentages

S.No	Slope Class	Slope Category	Slope %
1.	1	Nearly level	0-1
2.	2	Very gently sloping	1-3
3.	3	Gently sloping	3-5
4.	4	Moderately sloping	5-10
5.	5	Strongly sloping	10-15
6.	6	Moderately steep to steep sloping	15-35
7.	7	Very steep sloping	> 35

The study area has been categorized into six slope classes as 1, 2, 3, 4, 5, and 6. Slope class 1 is the most predominantly occurring class in the study area. This Slope class which corresponds to nearly level ground is generally associated with pediplain of granite gneiss with shallow weathering and pediplain of granite gneiss with moderate weathering.

Slope class 2 which represents nearly very gently sloping land corresponds to pediments of granite gneiss and dolerite dykes. This is generally associated with barren rocky/stony waste/sheet rock area land with scrub and land without scrub. This slope class is observed in the study area in the northern and north-eastern parts of the study area.

Slope class 3 which represents gently sloping land is observed in the north-western part of the study area. Slope classes 4, 5, and 6 generally correspond to hilly terrain of residual hills and inselberg of granite gneisses. These are generally observed in the study area southern side of Namapuram, north-west and south-west parts of the watershed. In general, the slope of the watershed is towards the east. The nearly level ground corresponding to the pediplain of granite gneiss with shallow and moderate weathering is the most predominant unit in the watershed.

11.11 Soil Mapping

Geological forms and geomorphological variations, influenced by the varied physiographical setups, have their own say in soil formation. In fact, the soil survey conducted envisaged assessment of the transformation of basic rock to solum and the influences of climate, microbial actions and the conditioning by the gradient over a period of time, all collectively accelerating the pedogenic process. The area under investigation is located in a semi-arid environment and is under the influence of hyperthermic soil temperature regime.

The study area comprises moderate to high hills ranging from 700-800 mm. Dykes have 400-500 m elevations and valleys are located in 300 m-MSL range in the existing topographical system. Thus the soil catena starts from about 700 m and extends up to 300 m with increases in textures and depths gradually from hill to valley. The runoff passing through streamlets and rivulets has been harnessed by small to medium tanks of various sizes and capacities under which irrigation is practised. Except for very small areas, no hydromorphic soils are encountered. Rest of the area located between hills and valleys is an undulating terrain with the elevational variations ranging between 300-500 m.

The soil legend has been initially compiled during the visual interpretation of satellite data duly combining with the information derived from Survey of India topographical maps. This constituted the initial legend, and the validity of boundaries has been checked in the field to ascertain the associated soil composition. The methodology followed in case of soil mapping is briefly discussed in the following sections.

11.11.1 Soil Mapping Procedure

As the source is remote sensing data and the bases are on 1:25,000 scale, detailed reconnaissance type of soil survey technique is deployed for mapping soils in the survey area. Under this system of survey specific areas are covered by detailed survey and information extrapolated to other areas where reconnaissance type of survey is carried out.

Discriminating features, as manifested by varying tones, textures, shape, size, composition, location aspect, and other relevant components available in the imagery have been identified and segregated. Differences are connected to topographic features available in the toposheets and integrated with local knowledge of the terrain. Thus, an initial legend was formulated for all the themes like soils, landuse/landcover, and Hydrogeomorphology. This legend formed the basis for further operations in all the surveys. Depending upon the variations arrived at during preliminary interpretation of satellite imagery suitable sample strips and sample study areas have been selected in the watershed.

Semi-detailed soil survey of low intensity is carried out in selected areas and extrapolation done to the rest of the area. Profile pits were excavated in almost all the mapping units of all sample strips, up to murrum or 100 cm. Profile descriptions are made following the FAO guidelines. Stipulations laid down in the Soil Survey manual of All India Soil and Land Use Survey have been closely followed throughout the period of field work. Apart from the profile study, a number of shallow pit observations are made to confirm the percentage of soils at the series association level.

11.11.2 Soils of the Area

Eight mapping units have been identified in the survey area. In total, 11 soil series have been recognised and mapped. Description of each mapping unit along with the series and taxonomic classification is provided in the following sections (Plate 11.3).

The details of mapping units identified in various physiographic positions along with the soilscapes are compiled and presented in the form of a Final Soil Legend in Table 11.6.

11.11.3 Suitability for Sustainable Agriculture

Due to inherent characteristics of soils, surface features and conditions related to water holding and nutrient supply, differences in adaptability to various crops exist under rainfed conditions. In the total area 2188.00 ha. is the area under rainfed agriculture. The rest of the area is either barren, hilly or under irrigated agriculture. The rainfall is received in two spells, most of it occurring in rabi seasons. Failure of rains is frequent. Soils are leached and moderately deep in the rainfed zone. Slopes are moderate and surface textures are coarse in most cases. Keeping in view all these factors, an attempt is made in this section to asses the suitability of each mapping unit to some principal crops that are in vogue in the survey under the classification for suitability to rainfed agriculture. In all, five classes have been envisaged.

(i) Soils with very few limitations for growing crops under rainfed conditions indicated as Very good (VG)

(ii) Soils with few limitations for growing crops under rainfed conditions indicated as Good (G)

(iii) Soils with moderate limitations for growing crops under rainfed conditions indicated as Fair (F)

(iv) Soils with severe limitations for growing crops under rainfed conditions indicated as Poor (P)

(v) Soils with severe limitations for growing crops under rainfed conditions indicated as Unsuitable (US).

The irrigability classification aims at assessing the inherent properties of the soils that express their degree of limitations for development under irrigation agriculture. If water is available with varied inputs and costs, any soil can be brought under irrigation.

Table 11.6 Soil taxonomic classification in Shivannagudem watershed

MU	PHYSIOGRAPHY	ASSOCIATION OF SOIL SERIES	TAXONOMIC CLASS
1	Hills	Rocky	
		Cherlagudem	Loamy skeletal, mi, hyp, Typic Ustorthents
2	Dykes	Rocky	
		Indurti	Loamy skeletal, mi, hyp, Lithic Ustorthents
3	Upper pediplains	Paluvayi	Coarse loamy, mi hyp, Typic Ustorthents
		Kottala	Fine loamy, mi, hyp, Typic Ustochrepts
4	Middle pediplains	Kottala	Fine loamy, mi, hyp, Typic Ustorthents
		Parlapalli	Fine loamy, mi, hyp, Typic Ustochrepts
5	Lower pediplains	Gudalakunta	Fine, loamy, mi, hyp, Typic Ustochrepts
6	Buried pediplains	Vinjamur	Fine, montomorillonitic, hyp, Leptic Haplusterts
		Battugudem	Fine, mont, hyp, sodic Haplusterts
7	Narrow valleys	Godkondala	Fine loamy, mi, hyp Typic Ustorthents
8	Broad valleys	Shivannagudem	Fine, mixed, hyp, Aeric Halaquepts
		Kudasapalli	Fine, mixed, hyp, Typic Ustochrepts

But the core idea of classification is to indicate the best soils suitable for irrigation with limited management and sustained use of the soils under irrigation. The following soil properties are generally considered in classifying the soils under varied degrees of suitability for sustained use under irrigation. A generalised picture of these various parameters are to be considered before deciding any soil on its suitability for irrigation. Since the consideration is for sustained use under irrigation, considering the present status of the soils or the present condition is not simply sufficient.

The long-term behavior of the soils under irrigation should also be considered by evaluating the following soil properties :

Depth of soil
Texture at surface
Water holding capacity
If salts present, the amount and nature
pH of the soil
Cation exchange capacity
Depth of groundwater

Major soils are moderately eroded except the areas under irrigation, where lands are levelled and erosion is moderate in rainfed areas and severe in Gully. Taming with loose boulder structures, contour trenching, gaps filing with afforestation, and grassland development as under cover to plantations, are some of the measures to be taken up in the model study area.

11.12 Hydrogeomorphological Mapping

The delineation of various geomorphic units and the knowledge about their groundwater prospects play an important role in the optimum utilisation of land resources. Besides, the information on land forms and geological information like lithology and structures also help in identifying the groundwater potential zones. The Hydrogeomorphological map has been prepared on 1: 25,000 scale by sensing techniques (Plate 11.4).

Visual interpretation techniques have been adopted in the delineation of various geomorphic features. Besides, the linear features like dykes and structures like lineaments/fractures are also delineated. The interpreted geomorphic units/land forms and structures have been transferred onto the base map which was already prepared using Survey of India topographical map. A detailed programme of field work has been planned to verify the doubtful areas and arrive at the information regarding the groundwater prospects of various land forms/geomorphic units.

Based on the groundtruth information, corrections and modifications of the misinterpreted units were made and the final thematic details were transferred onto the base maps. The final drawings were prepared and each Hydrogeomorphic unit/ structure was assigned a particular symbol. The various geomorphic units occurring in the study area were described in the legend and presented in Table. 11.7.

Table 11.7 Legend of Hydrogeomorphology of Shivannagudem watershed

Mapping symbol Prospects	Geomorphic unit	Litho Stratig-raphy	Structure	Description	Ground Water
PPM	Pediplain With Moderate Weathering	Archean gneissic	Fractures/ Lineaments	Flat and smooth surface of pediplain with 8-20 m. Thick weathered material and covered with red soil	Moderate to good. Very good yields along fract-ure/lineament.
PPMA	Pediplain With Moderate Weathering covered shall-ow weathering	- do -	-do-	Flat and smooth surface of Pediplain with 8-20 m. Thick Weathered Material and Covered with Alkaline Alkaline soils	Moderate to good. Quality-wise not suitable for drink-ingand agricultural Soils purposes.
PPSA	Pediplain With Shallow Weathering Covered With Alkaline Soils	- do -	-do-	Flat and smooth surface of pediplain with 0-8 m. Thick weathered material and covered with black soil	Poor to moderate. Moderate yields are expected along fractures/lineam-ents. Quality-wise not suitable for drinking and agri-tural purposes.
BPPS	Buried Pediplain With Shallow Weathering	- do -	-do-	Flat and smooth surface of pediplain with 0-8 m. Thick weathered material, covered with black soil	Poor to moderate. Moderate yields are expected along fractures/ lineaments
P	Rocky Pediment	- do -	-do-	Gently sloping smooth surface of erosional bed rock between hill and plain with never of detritus.	Negligible to poor. Poor yields are expected along the fractures.
I	Inselberg	- do -	-do-	Isolated hill occupying small area	Poor
RH	RESIDUAL Hill	- do -	-do-	Isolated hill occupying considerable area	Poor
	Dyke/Dyke Ridge	DOLER-ITE	-----	Massive, linear feature cutting across various geomorphic units.	Poor
----	Fracture	----	-----	Linear vertical/subvertical features generally associ-ated with drainage pattern	Moderate to good

11.13 Groundwater Prospects Map

In recent years, the increasing use of satellite remote sensing has made it rather easy to define the spatial distribution of different groundwater prospect classes on the basis of geomorphology and other associated features. Visual interpretation of image is in hydrogeomorphological mapping, and each landform was studied and mapped (Plate 11.5). Depending upon the geology, structure, and tectonic phenomenon that have caused development of secondary porosity have been coupled with elevation and drainage information from the SOI toposheets and potential zones for groundwater development are delineated as shown in Table 11.8.

Table 11.8 Legend of groundwater prospects

Mapping symbol	Geomorphic Unit	Groundwater Prospects
PPM	Pediplain With Moderate Weathering	Very Good
PPMA	Pediplain With Moderate Weathering Covered With Shallow Weathering	Moderate - Poor
PPS	Pediplain With Shallow Weathering	Moderate - Good
PPSA	Pediplain With Shallow Weathering Covered With Alkaline Soils	Moderate - Good
BPPS	Burried Pediplain With Shallow Weathering	Moderate - Poor
P	Rocky Pediment	Poor
I	Inselberg	Nil
RH	Residual Hill	Nil

The ground water potential zones were deduced after the integration of hydrogeomorphological and lineament maps through GIS. Various hydromorphic units are grouped into potential zones, moderately potential zones and poor zones. Generally there will be more recharge in alluvial plains, flood plain, buried pediplain and Infield valleys. Other parameters are also taken into consideration like lineaments, particularly joints, fractures and their intersection to enhance the potential of these units. Intersecting lineaments are also the potential zones for ground water exploration and development in the study area.

11.14 Drainage Mapping

Drainage pattern is a plan which the individual stream courses collectively form. It refers to both the spatial relations of individual streams and the overall pattern made by the individual drainage lines. In the present study, the drainage map is prepared using satellite imagery and SOI toposheets on 1:25000 scale. The drainage map is scanned using Logitech deskscanner and digitised by AutoCAD 14. Then it is exported to Arc/Info 3.5.1 in which the editing is done. The output is given using the Arcview

3.0 software. This process is the same for the preparation of all the thematic maps. This drainage map is used as an action plan for water resources and is shown in (Plate 11.6).

The drainage pattern in the Shivannagudem watershed is found to be of a dendritic pattern. As it is a hard-rock granitic terrain, drainage follows the weak zones in which the erosion is, much, like joints and fractures The dendritic pattern is the most common type of pattern formed by a main stream with tributaries branching and rebranching freely in all possible directions. It generally develops upon homogeneous rocks of uniform resistance, such as, horizontally bedded sedimentary rocks and, massive igneous terrain offering no variation in the resistance to the flow of water. The resulting stream runs in all directions with no definite preference to any particular direction.

11.15 Action Plan Generation

The base map, slope map, drainage and watershed map have been prepared from SOI toposheets. The land use/ land cover, hydrogeomorphology and soil maps are prepared by the visual interpretation of IRS-IC satellite image. These maps are digitised using AutoCAD after scanning the above cartographic maps. The digitised maps are then converted into a form compatible to Arc/INFO GIS software. These digital data are transformed through editing and analysing these layers.

A total of six tanks were constructed across the drainage in the watershed. 8 soil mapping units and 12 soil series were identified in the watershed. The land use/land cover analysis indicates that 90% of the watershed area is covered by crop land of which 61% of the area corresponds to Kharif unirrigated land. The wastelands in the form of land with scrub, land without scrub and barren rocky/stony waste/sheet rock area amounts to about 4% of the total watershed area. Based on the technical guidelines given by the Department of Agriculture, Govt. of India, various water resource development and management techniques, soil and water conservation measures, optimal land use patterns were suggested for the overall sustainable economic development of the watershed.

Action plan for Water Resources is developed from the integrated study in which three check dams are recommended based on the slope, soil, land use and land cover, and hydrogeomorphological criteria, to obstruct the surface water flows and thereby increase its influence over the command area and the groundwater levels. A percolation tank is also suggested on a major stream in the watershed with the stream overflow and area criteria under consideration. In the remaining part of the project area, the slope is gentle and the major streams have more area of influence. Action plan for Land Resources is made from the integration of the prepared thematic maps and suitable crops have been recommended. (Plate 11.7) Through the implementation of the recommendations made, the study area can be well developed.

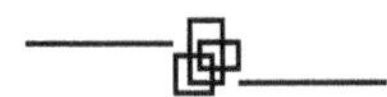

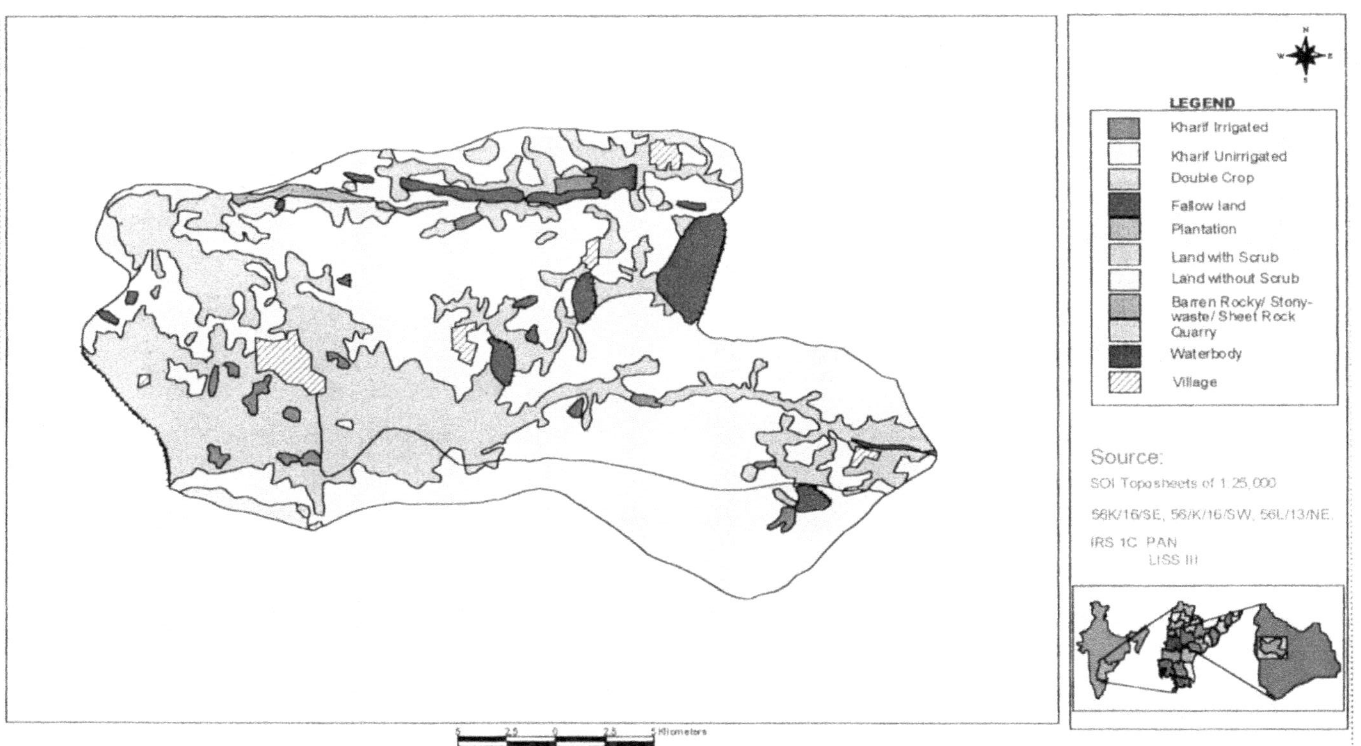

Plate 11.1 Landuse and Landcover map of Shivannagudem watershed, Nalgonda Dt., (A.P) INDIA.

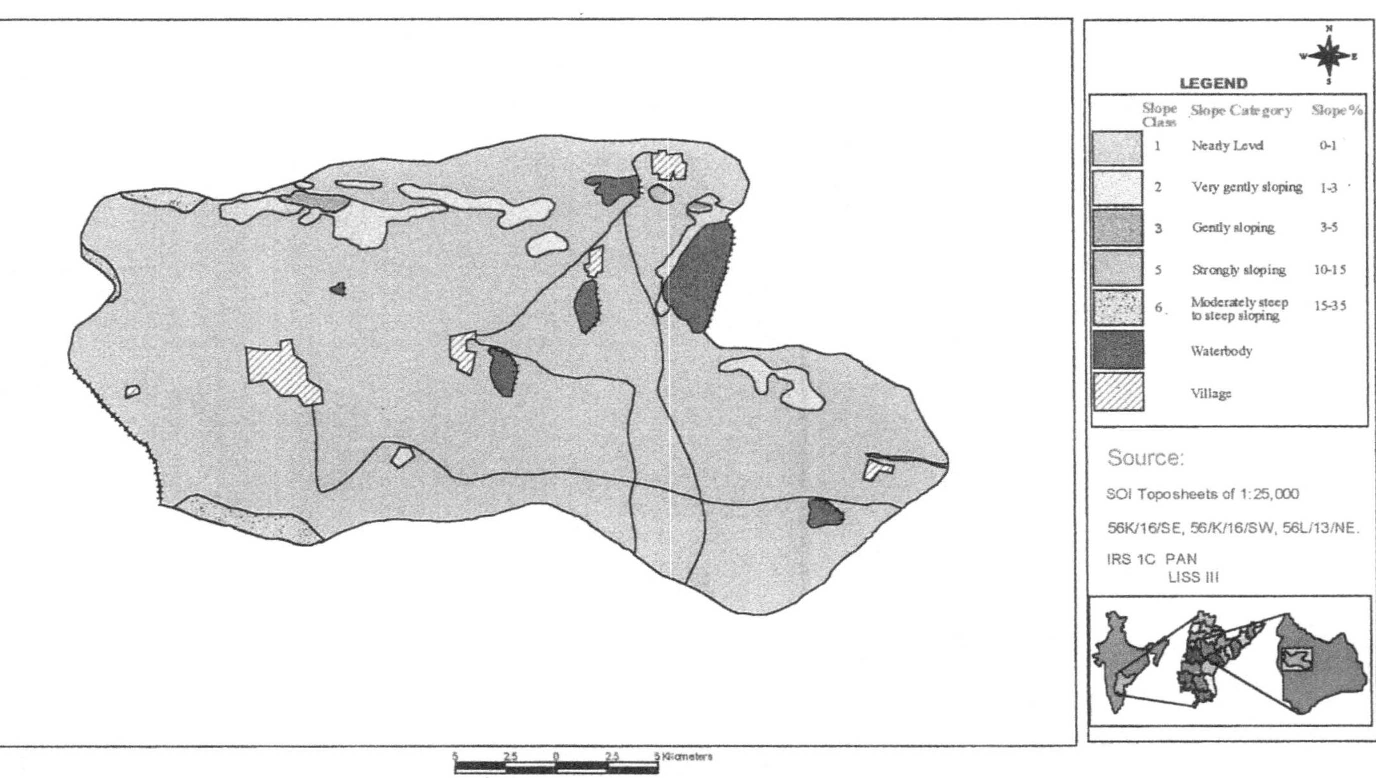

Plate 11.2 Slope map of Shivannagudem watershed, Nalgonda Dt., (A.P) INDIA

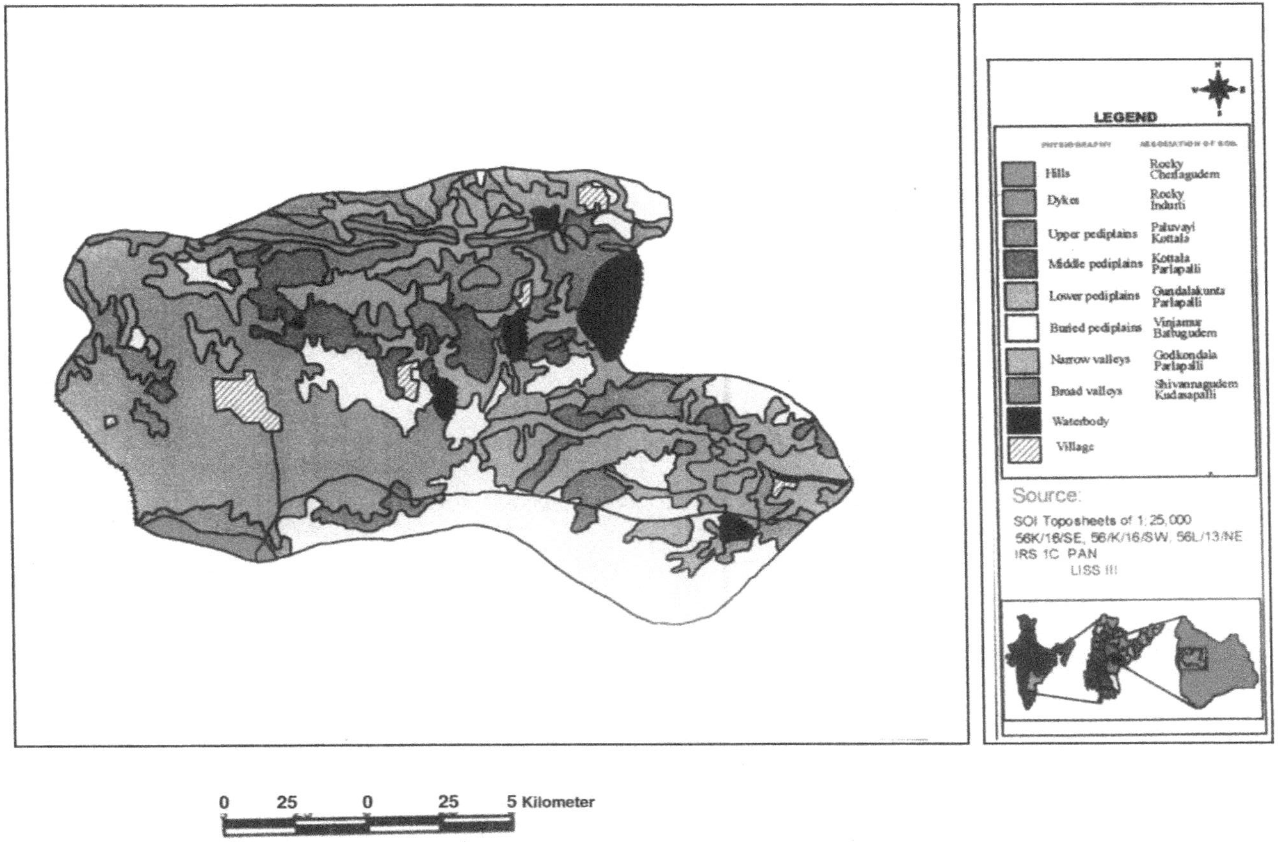

LEGEND

PHYSIOGRAPHY — ASSOCIATION OF SOIL

Hills — Rocky Cherlagudem
Dykes — Rocky Indurti
Upper pediplains — Paluvayi Kottala
Middle pediplains — Kottala Parlapalli
Lower pediplains — Gundalakunta Parlapalli
Buried pediplains — Vinjamur Battugudem
Narrow valleys — Godkondala Parlapalli
Broad valleys — Shivannagudem Kudasapalli
Waterbody
Village

Source:
SOI Toposheets of 1:25,000
56K/16/SE, 56/K/16/SW, 56L/13/NE
IRS 1C PAN
LISS III

Plate 11.3 Landuse and Landcover map of Shivannagudem waterhsed, Nalgonda Dt., (A.P) INDIA.

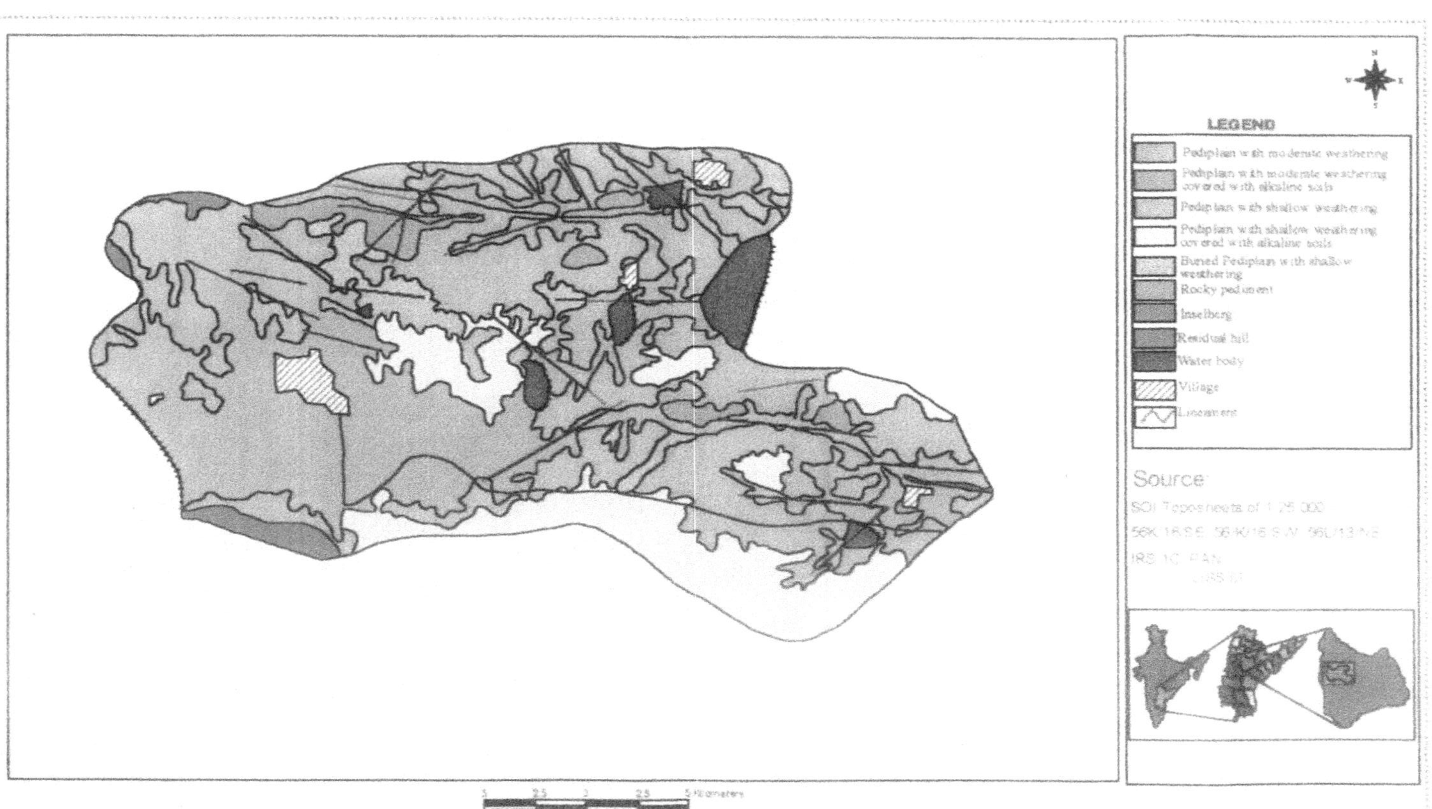

Plate 11.4 Hydrogeomorphological map of Shivannagudem waterhsed, Nalgonda Dt., (A.P) INDIA

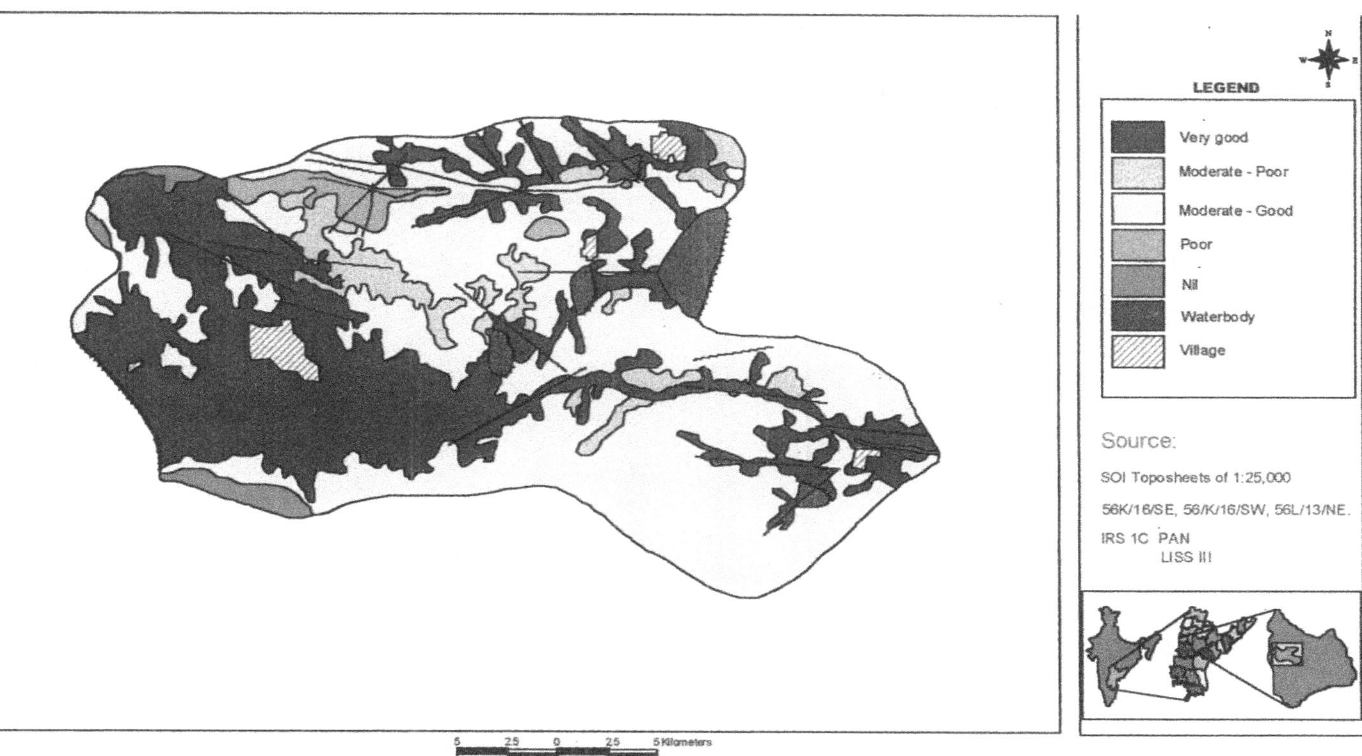

Plate 11.5 Groundwater prospectus map of Shivannagudem waterhsed, Nalgonda Dt., (A.P) INDIA

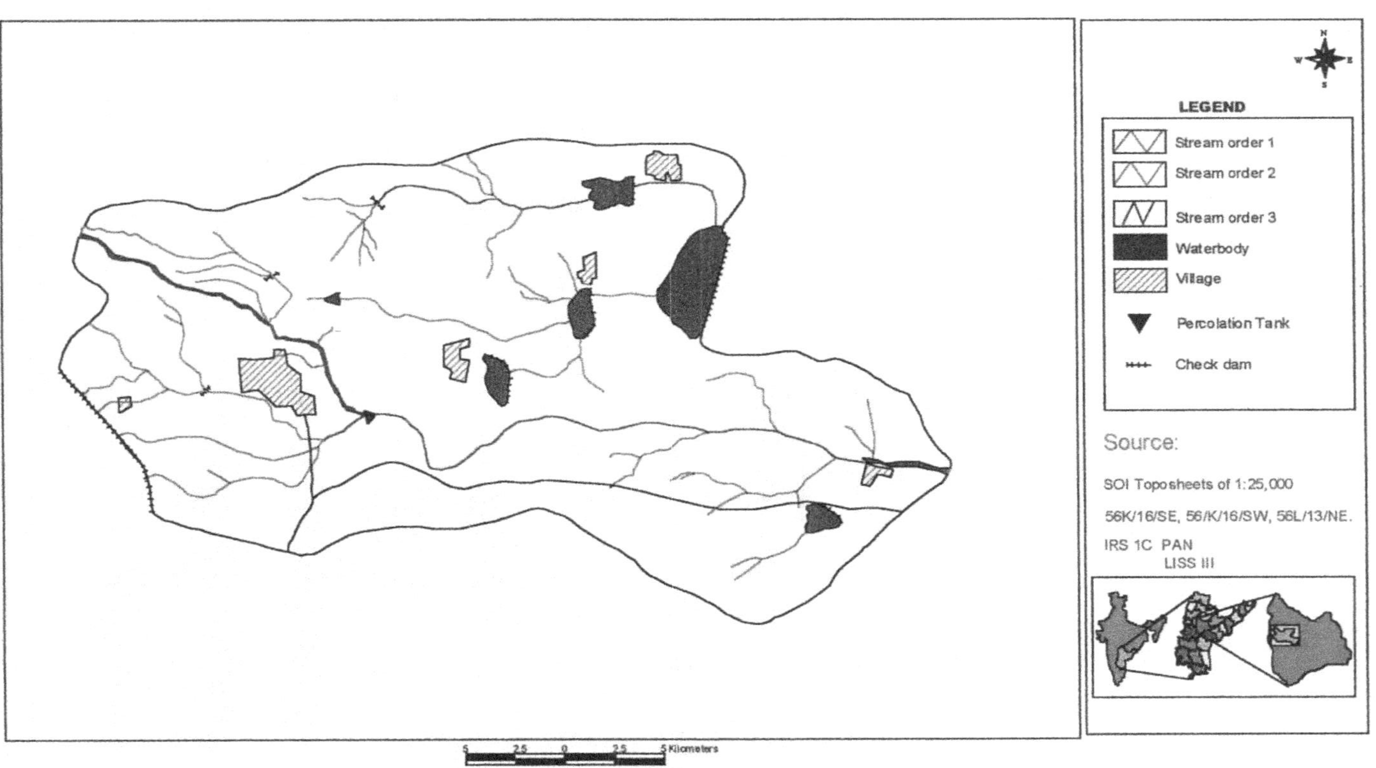

Plate 11.6 Action plan for Water Resources of Shivannagudem watershed, Nalgonda Dt., (A.P) INDIA

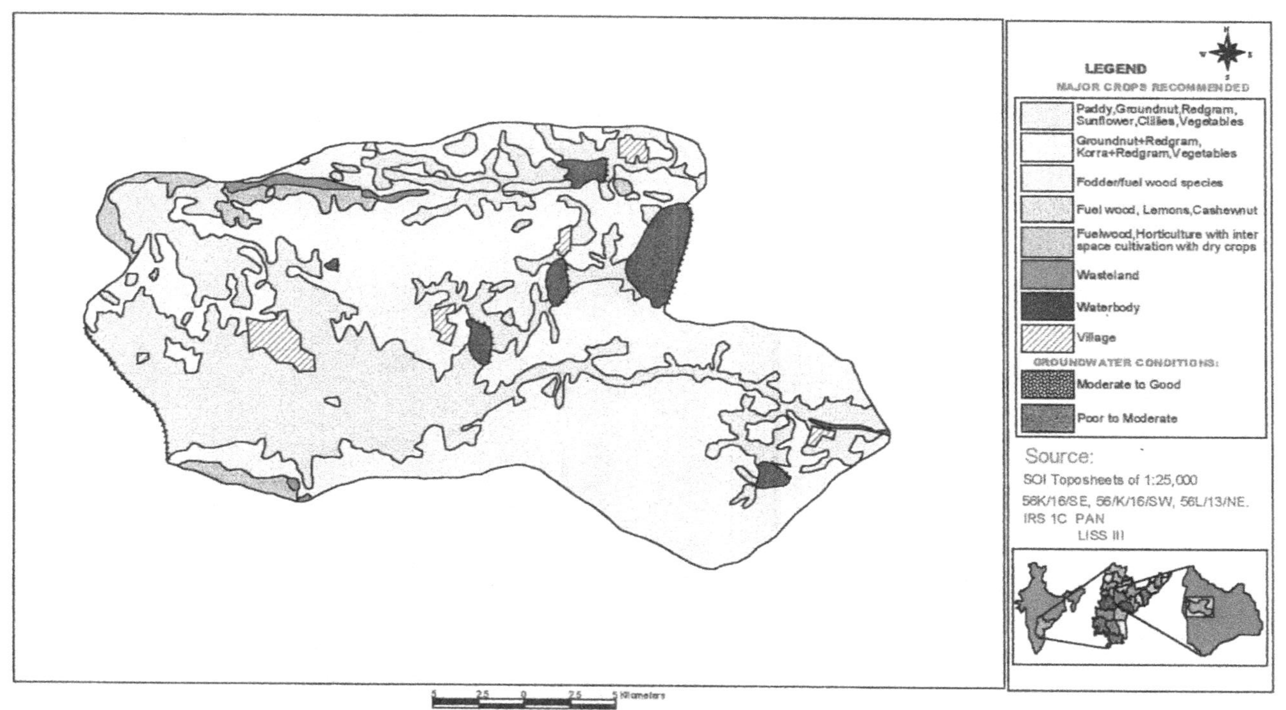

Plate 11.7 Actionplan for Land Resources : Major crops recommended for Shivannagudem watershed, Nalgonda Dt., (A.P) INDIA

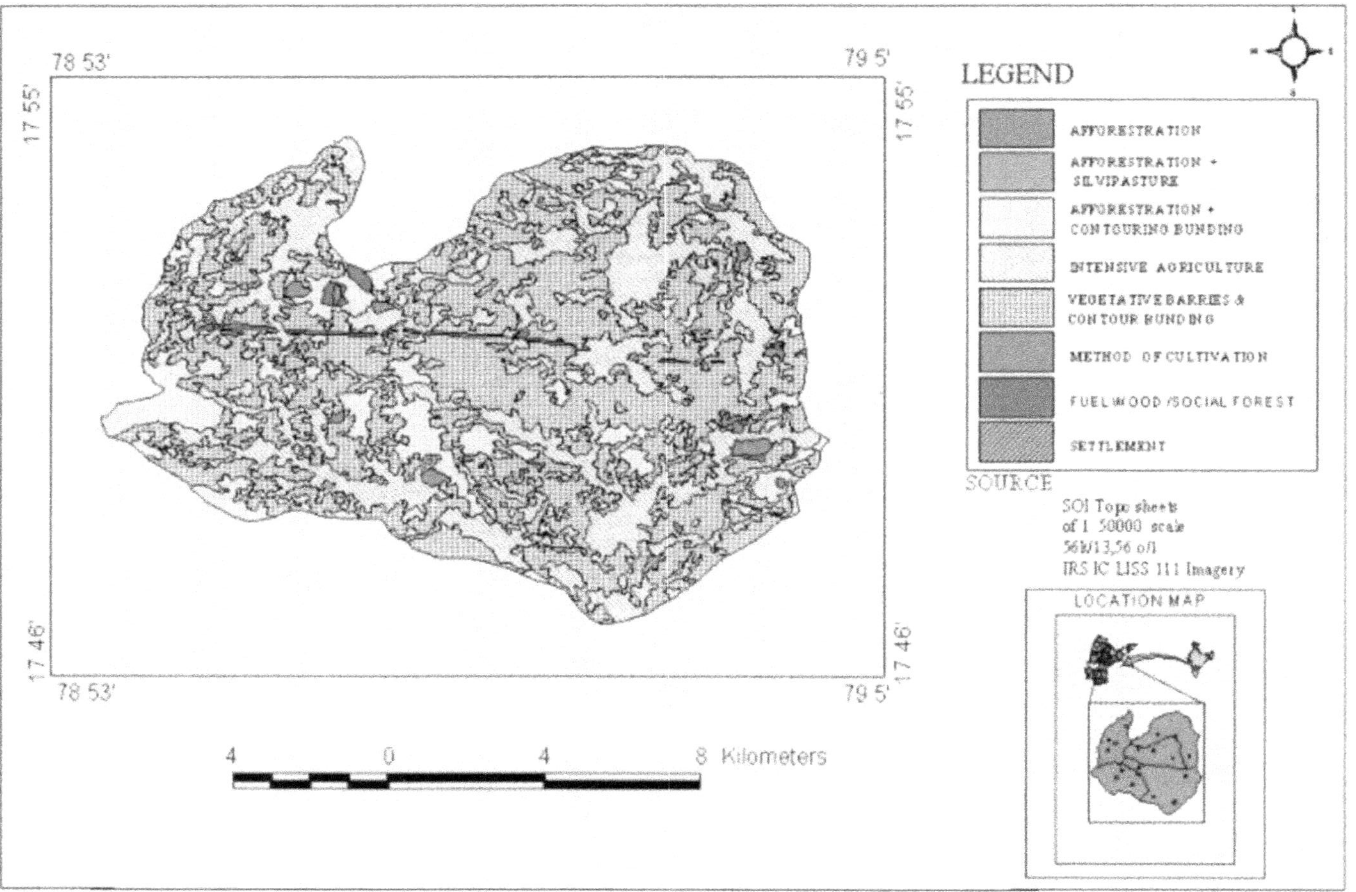

Plate 11.8 Action plan for land resources development map

12
Salt Water Intrusion Modelling

12.1 Introduction

Saltwater intrusion refers to the movement of salty water into freshwater estuaries. However, saline ground water can occur naturally in coastal areas, and has similar implications upon ground water use. Saltwater intrusion has been a major ground water resource problem in coastal environments for decades. Andhra Pradesh being predominately an agricultural state and the increase in demand for irrigation at a dependable level necessitates the practice of conjunctive use of surface and ground water in its deltas. Pennar canal system in Nellore district is no exemption. If the saltwater travel into the well fields, underground water supply becomes useless. Moreover the aquifer becomes contaminated with salt that may take years to remove even the adequate fresh ground water available to flush out the saline water. Nellore and other areas, which depend upon groundwater consumption, have added to the problem as the saltwater intrusion moves further inland in response to pumping. This critical problem requires proper management to protect our groundwater resources. At present there is an unplanned use of ground water through construction of wells, private shallow wells and deep tube wells. Therefore this practice may not be useful in the deltas. To improve the present practices, a realistic assessment of the surface and ground water resources be used for proper environmental planning over the study area.

Monitoring coastal aquifers is important in studying the problem of seawater intrusion through a model study. Such model can be helpful in finding out the features/sources of saltwater/freshwater interface and the pattern of its movement and the chemical characteristics in the seawater and the fresh water are actually miscible fluids and therefore the zone contact between them takes the form of a transition zone caused by hydrodynamic conditions.

12.2 Saline Water Intrusion (SWI)

Water is one of the most important resources of the country. But today we are wasting more water than we are actually using whether it is in our irrigation, urban system or anywhere else. Agriculture, which is a major part of the Indian economy and the backbone of our self-reliance, is dependent primarily on water. Water that was considered to be in plenty has now come to be realized as limited resources, which is further accentuated by the failure monsoons and recurrence of droughts. No water resource management or development, whether it be for the purpose of water supply for the population, agriculture, industry, or energy production, is possible without an assessment of quantity and quality the water available. Information on the existing condition and trends of the country's water resources are required as a basis for economic and social development and for maintenance of environmental quality through a proper perception of physical processes controlling the hydrological cycle in time and space.

To improve the management of water resources, greater knowledge about their quantity and quality is required. There is a need for regular and systematic collection of hydrological data, together with a system for processing the quantitative information for the various types of the water bodies. Moreover keeping the adequate inventory of the water availability is one of the desirable pre-requisition for the quantification of water user rights, for the formulation of the water quality standards. At present there is an unplanned use of ground water through construction of wells, private shallow wells and deep tube wells. Therefore this practice may not be useful in increasing the command area and it may invite saltwater intrusion problems in the deltas to improve the present practices. A realistic assessment of the surface and ground water resources and also proper planning for their use would be required.

Lateral or depth wise increase in the salinity of ground water cause due to man's interference with the ground water regime. Intrusion of salt water into heavily exploited aquifers is a serious problem faced in coastal areas and other places in the interior whereever fresh water aquifers adjoin saline waters of connate are meteoric origin. Seawater intrusion, in coastal aquifers, occurs through aquifers out cropping along the continental shelf. While in coastal areas the intruding water is generally of seawater composition in the interiors the quality may vary widely from brackish to concentrated brines, the lateral quality being typical basins in arid regions. The interface between the salt and fresh water moves towards the fresh water zone in hydraulic gradient and is established from the saline water zone to the fresh water zone, as a result of ground water development or by a rise in the head of saline water relative to that of fresh water during high tide. A wide aeration may occur in the disposition of fresh and saline ground water bodies, depending upon hydrogeologic and hydrologic conditions. In the context of intrusion of saline water the following situations to be considered :

- Saline water underlies fresh water, in a homogenous and isotropic medium under water table conditions.
- Saline water underlies fresh water, the two zones being separated by an impervious or semi-previous layer
- Saline water over lays fresh water, the two zones being separated by a semi-previous or impervious layer.
- Fresh water laterally grades into saline water, occurring either underwater table, confined or semi-confined conditions.
- Fresh water aquifers alternate with saline water zones.

12.2.1 Salinity Influx in Estuaries

Most of estuarine rivers are contaminated, during high tides by the influx of sea water up to distance which, depending upon the river bed slope, width, depth and shape, quantity of discharge, tidal level and fluctuations, wind velocity and direction, may range from less than a kilo meter to tens of kilo meters. The tidal fluctuations is accompanied by a cyclic diurnal change in water quality due to the mixing, in continually varying proportions of original river water with sea water in the lower reaches of estuary, the water quality resembles that of sea water and, at the farthest point inland ,that of primary river water. The damming effect of the sea water causes a fresh water tide in the upper reaches, where, although the river stage responds to tidal effects, the quality remains unaffected. Within the tidal reach, the change in the water quality depends on the composition of the original river water and sea water (which is nearly constant), and the relative quantities involved in mixing. A zone of mixing and diffusion migrates up and down the river, the maximum contamination occurring during high tide and the minimum, during low tide. During the high river stage in the rainy season, the tidal effect on quality is minimal because of large discharge of fresh water. In the lean months when the river stage is at its lowest, the quality is most adversely effected by tides due to the pre dominance of sea water. In perennial streams, the low flows of rivers are sustain largely by base flow. In placid streams, salt water, being denser, most beneath the fresh water, the movements often being in opposite directions. Depending upon the degree of turbulence, the water may be thoroughly mixed or stratified.

Within the estuarine reach of the stream, saline water may encroach into the underlying fresh water aquifers in a variety of ways, depending on the relative positions of the water table and river stage and degree of interconnection between the stream and aquifer. Periodical monitoring of the quality of water in the tidal reach of streams will yield useful information on the usability of various beneficial purposes, and also on the proportion of the quantities of water of different sources making up the mixture.

12.2.2 Occurrence of Saline Water Intrusion

Wherever overdraft condition occur in coastal aquifers connecting with ocean sea water intrusion can result by lowering the water table in unconfined aquifers or the piezometric surface in unconfined aquifers, the natural gradient sloping downward towards the ocean is reduced or reversed. Because two fields of different densities are involved, a boundary surface or interface is formed whenever the fields are in contact. A hydrodynamic balance of the fresh water and salt waters governs the shape and movement of the interface

Sea water intrusion can develop and it would be accentuated wherever a direct artificial access exits between sea water and groundwater. One of the earliest reports of intrusion was published by Bran Waite (1855), who described the increasing salinity of waters pumped from wells in London and Liverpool, England. Intrusion of sea water into coastal aquifers is a natural consequence of the density contrast between fresh and saline water. The denser sea water often forming a deep wedge that can be extend for many kilometers inland. Although this can cause problem of saline upconing when the fresh groundwater reserves are heavily exploited by individual coastal wells.

12.2.3 Composition in Saline Water Intrusion

The average of chloride content of rainwater in 1969 was 3.3 ppm. The total dissolved solids and specific conductance of groundwater vary linearly with chloride content. The relationship was observed by Vishor and Mink (1964) and Carr (1969). The ratio of chloride content to total dissolved solids in sea water remains nearly constant, when the sea water is mixed with freshwater. Most of the chloride is derived from the simple mixing of fresh and sea water. The chloride content was therefore used as an index of chemical quality in determining the extent of mixing of fresh and intruded sea water.

12.2.4 Effects of Saline Water Intrusion

The derivational effects of saline intrusion are not always as serious as one might expect, and in some cases the saline inflow may prove beneficial in enhancing the recovery of fresh ground water reserve. More serious problems occur when aquifers is over developed on a regional scale. This results in general lowering of the fresh water potentiometer surface throughout the area and a progressive and extensive invasion of the aquifer by the sea water. At first appearance, this scenario seems highly undesirable as water quality is being compromised and fresh groundwater reserves are being depleted.

This developed fresh groundwater usually has a chloride concentration in the range of 10 to 100 mg/l, although in places groundwater contains 250 to as mush as 500 mg/l utilized by necessity. However, salty groundwater is defined as that contains more than 250 mg/l chloride concentration.

Salt-water encroachment into existing freshwater aquifers is usually the results of some individual activity, both vertical and lateral encroachment can be caused by

- Reduction of freshwater head due to heavy pumping
- Drilling top deep and intercepting salt water zones
- Imposing greater head on salt water zones from injection of water brine, representing of oil producing zones, and hydraulic fracturing of rock units at oil and gas operations.

In areas of agriculture activities, where large amounts of freshwater are extracted, a reduction in the head of freshwater aquifers can cause upward migration of the saltwater interface. As the result, the sea water tends to migrate into the fresh water zones of lower head. Examples of fresh groundwater development near salty ground water induced both natural occurrence of salt water and unnatural encroachment of saltwater caused by stresses placed on the subsurface environment by the man. In costal regions, surface water sources can become saline due to the tidal influence of the sea .As the high tide moves into the coastal areas, sea water moves into streams and drainage canals and travels inland. This upstream migration of seawater alters the quality of water in affected streams and drainage canals significantly. This phenomenon is also observed during times of drought. Apparently the artificial raising of the pizeometric level of the natural brine triggered by salt water contamination because freshwater aquifers are hydraulically connected to the salt water zones. It is important to note that most of the unnatural salt water aquifers has been detected only because of the deterioration of the water pumped by the wells.

Industrial and community expansion will increase groundwater use in most areas. Development of new well fields over large areas and reduction of pump age in the existing well fields near salt water zones will help to reduce encroachment of salt water and conserve freshwater aquifers in critical areas. In order the over development may be detected before deplatation or salt water encroachment become critical, pump age inventories and chemical analysis should be made periodically and correlated with records of the groundwater storage.

The freshwater flows field is affected by the rise of the salt water interface. The solution of the model produces a profile of the interface in the case of the stable up coning or it unstable. The model shows that the interface breaks through the bottom of the river In the case of the unstable upconing salt water is being fed directly to the river system from the salt water reservoir. In most irrigated valleys, lands adjacent to the river or a turbidity, where the first to be irrigated, later water elevations an excessive water was applied on higher elevation lands, drainage water from these lands caused high water tables and salt problems on the lower lands.

Before production wells are planned, it is important to know the depth of freshwater in each prospective test site. Ideally the property should not be purchased until an adequate water supply is assumed. Because salty groundwater also occurs naturally at relatively shallow depths, it has been difficult to determine unnatural conditions except after characterizing the extreme variations in chloride content with the short periods. The available historical data for specific aquifer areas indicate no large-scale natural variations in salt content.

12.3 Applications of Geoinformatics

Geographic Information System (GIS), a computer based system to capture, display manipulate and edit geographically referred information. GIS is expected to capture the information into a digital form. With the possibility of using vector and raster information for GIS there has been impetus to use the remote sensing data at least as one of the inputs. Remote Sensing is the collection of data by sensors not in direct contact with the phenomenon of interest. Analysis of remotely sensed images from many different sensors is being applied in numerous fields of study system that depends on electromagnetic radiation requires 4 elements source, the sensor, interactions with the atmosphere and interactions with the earth's surface. Remote sensing is the science and art of obtaining information about an object, area or phenomenon through the analysis of data acquired by a device that is not in contact with the object, area or phenomenon under investigation. Remote sensing provides database for GIS manipulation, both of which combined into powerful management tool for a wide spectrum of use.

At the outset, one must be able to understand the format, which is compatible for the GIS conventionally. The principal difference between remote sensing and GIS technology was the difference between images and digitized maps. Images are based on measured electromagnetic wave reflections (emissions) from real world surfaces and are most often stored in raster format (with pixels). GIS technology has become more widely used management tool chiefly because its ability to manipulate digitized information from known maps for the purpose of creating higher forms of information.

The future of both technologies is in their integration. Aerial photographs, images that overlay on maps, are the bridge between two technologies.

12.4 Case Study : Pennar Basin

The study area proposed for the investigation extends from the Nellore Ayacut on Pennar up to Bay of Bengal about 50 Km down stream of Nellore. All the overflow of wastewater from Nellore and its surrounding are presently joining the River. The extent to be covered all area is included in Surrey of India toposheet 57N/4, 57N/15, 66B/2 and 66B/3 on 1:50000 scales between longitude and latitude 80^0 0', 14^0 45' NW 80^0 15' 14^0 25' SE and the study area is under laid by alluvium of recent age. A small patch of laterites of sub-recent age occurs under unconfined and semi-confined conditions. The ground water is abstracted through filter points mostly, also through the bore wells and dug wells,. The proposed study area is experiencing man-made pollution with over exploitation and the development of aquaculture.

Pennar basin, the River has lowest average flow because of only moderately low annual rainfall (620mm at Kadiri, 750mm at Cuddapha and 1060mm at Nellore). The meagre water wealth of the basin has been utilized only for limited irrigated use. The pennar and its tributaries do not have enough flow to support direct abstraction for larger towns, so infiltration wells are used for most of the riverside town.

12.5 GIS Database

The data required for the identification and detection of the problematic areas and to correlate land use with environmental degradation in the study area are basically environmental quality parameters and its associated thematic data which include land use/land cover, slope, drainage, geomorphology, and maps showing the ground water quality. The different data products required for the studies include the 57N/4, 57N/15, 66B/2 and 66B/3 toposheet which are obtained from Survey of India and fused data of IRS – 1C, and IRS – 1D LISS-III satellite imagery obtained from National Remote Sensing Agency (NRSA), India.

The data input capture subsystem converts data from their existing form into one that can be used by the GIS. Georeferenced data are commonly provided as paper maps, tables of attributes, electronic files of maps and associated attribute data and satellite imagery. The data to be entered in a GIS are of two types namely spatial data and attribute data. Spatial data represent the geographic location of features i.e., various thematic maps, and attribute data provides descriptive information such as the water quality data. During data input, spatial and attribute data are entered and correctly linked. The attribute data is commonly entered by keyboard that is manually entering the data at a computer terminal whereas the spatial data in the form of maps is entered through manual digitization using AutoCAD software. The raster data in the form of imagery is converted to vector data by the process of scanning using desk A0 flatbed scanner. The output data is mainly created by integration of Remote Sensing and GIS techniques. IRS ID LISS III data is used for Land use/Land cover classification. The software used for creation of database for this study is Arc info and Arc view. The geographic database has two basic components, namely, spatial database and attribute database.

Spatial data relevant for this study consists of thematic maps of natural system which includes

- Drainage network Map
- Land use/Land cover Map.
- Slope map
- Geomorphology

The attribute data base generated mainly water quality parameters like, ph, alkalinity, chlorides, nitrates, sulphates, total dissolved solids, hardness, sodium, potassium and salinity.

12.6 Spatial Database

The thematic maps which are drawn manually are scanned and digitized to produce a vectorised coverage which is then processed for errors such as dangles constituting the overshoots and undershoots and the labels for polygons using Arc/info. Finally the coverage is being processed for topological modeling.

12.6.1 Drainage Network Map

Drainage network map is to know the drainage pattern and for calculating various characteristics like drainage density, basin slope etc. According to Watershed Atlas of India, the Central Board for the Prevention and Control of Water Pollution (CBPCWP) at New Delhi, in collaboration with State Control Boards, has setup the monitoring network for all major river basins in India. They are classified the inland waters into five categories (A to E) on the basis of designated best use. The quality requirement becomes progressively lower from A to E. Pennar basin, the river has lowest average flow because of only moderately low annual rainfall (620mm at Kadiri, 750mm at Cuddap and 1060mm at Nellore). The merge water wealth of the basin has been utilized only for limited irrigated use. The Pennar and its tributatries do not have enough flow to support direct abstraction for larger towns, so infiltration wells are used for most of the riverside town. The Pennar Delta is having four major, fourteen medium and forty-one minor drains. The main drain basins are Maldevi, the Alluri swamp, Nakkalavagu and Paideruvagu. The Paideru traversing for a distance of 43.2 Km and joins the sea (Bay of Bengal) at Ponnapudi village of Vidavaluru mandal. The rest three drains, River Pennar, Chippaleru and Uppateru, which finally empties into sea. Drainage pattern and texture seen on space images are indicators of landforms and bedrock type and suggest soil characteristics and drainage condition. (Plate 12.1) shows the drainage map of the case study area derived from toposheet and digitals prepared using ARC/INFO GIS software.

12.6.2 Land Use / Land Cover Map

Comprehensive information on land use / land cover is the basic prerequisite for land resources evaluation, assessment, utilization and management. Today, with increasing population pressure on land and the resulting changes in the land use pattern and processes, a considerable degree of land transformation and environmental deterioration is being witnessed. Therefore it is important to understand the cause and effect of the changes through scientific studies. An analysis of mechanisms of land use changes plays an important role in forecasting environmental changes. As the population increases, infrastructure develop, and economy blooms in the study area and its vicinity. These cause changes in the coastal environment and directly effect on the benefits of the farmers. Hence, we consider the factors of land use changes from the viewpoint of human impact.

The main objectives of land use map are :

- To utilize as a basic database, which provides the information for allocating new land use practices.
- To incorporate demographic, economic and environmental impact, which have occurred in an area.
- To indicate intensive development already taken place and where there is open land suitable for future expansion, but it will also make it possible to determine special areas, such as prime agricultural lands.
- To serve as a basis for monitoring land use change and extraction of temporal changes.
- To serve as a base in the integrated overall planning of agricultural and industrial development of the region.

Land Use / Land Cover Classification

The USGS devised a land use and land cover classification system for use with remote sensing data in the mid-1970. The basic concepts and structure of this system are still valid today. The USGS classification system was devised according to the following criteria : (Anji Reddy. M. 2000).

(i) The minimum level of interpretation accuracy using remotely sensed data should be at least 85%

(ii) The accuracy of interpretation for the several categories should be about equal

(iii) Repeatable results should be obtainable from one interpreter to another

(iv) The classification system should be applicable over extensive areas

(v) The categorization should permit land use to be inferred from the land cover types

(vi) The classification system should be suitable for use with remote sensing data obtained at different times of the year

(vii) Categories should be divisible into more detailed subcategories that can be obtained from large scale imagery or ground survey

(viii) Aggregation of categories must be possible

(ix) Comparison with future land use and land cover should be possible

(x) Multiple uses of land should be recognized

Levels I and II, with classifications specified by the USGS are principally of interest to users who desire information on a nation-wide, inter-state, or state wide-basis. Levels III and IV can be utilized to provide information at a resolution appropriate for regional or local planning and management activities (Table 12.1). The analysis of satellite images revealed that the major land use changes whcih are mapped are as follows :

- Changes from paddy to shrimp farming,
- Changes from mangrove forest to shrimp farming,
- Changes from paddy to abandoned paddy.

Land use categories are distinguished on the color composite of IRS LISS III Sensor data, based on differences in color and texture. These categories are recognized as: standing crop (paddy) kharief unirrigated, aqua culture, built up land, brackish water, salt ponds, land with scrub, land without scrub and others. Paddy is the major agricultural crop in the area. Both irrigated and non-irrigated paddy fields cover about 49 % of the total study area, second major culture is aqua. Active shrimp farming areas, along the coast lines, along the Pennar river, are seen clearly from both combinations of IRS LISS III. Standing crop can be recognized as bright red colour with fine texture along river Penna and Major canals, shown in dark blue to bright blue colour with regular patches along both sides of the river, mainly southern parts of basin which shows aqua culture. Dark blue colour shows the brackish water at North - East corner of the Image, and nearby brackish water white colour patches observes as salt ponds, light red to dark brown colour tells as Kharif unirrigated land and some green patches are recognized as land with scrub (Plate 12.2).

12.6.3 Slope Map

Slope aspect and altitude are important parameters from land utilization point of view. Among the three, slope is very vital one for land irrigability and land capability assessment. The density of contours on map is used for preparing slope map which gives various graphs or categories of slopes. Slope map is derived from SOI toposheet using the concept given below :

Slope % is calculated by the formula:

$$\text{Slope \%} = (h / L) * 100$$

where h = Height,

L = Length

Based on this classification, the present study area is categorised as nearly level and is low lying area, which is shown in (Plate 12.3).

Table 12.1 USGS Land Use / Land Cover Classification System

S.No.	Level I	Level II
1.	Urban or built-up land	11 Residential 12 Commercial and service 13 Industrial 14 Transportation, communications and utilities 15 Industrial and commercial complexes 16 Mixed urban or built-up land 17 Other urban or built-up land
2.	Agricultural land	21 Cropland and pasture 22 Orchards, groves, vineyards, nurseries and Ornamental horticultural areas 23 Other agricultural land
3.	Rangeland	31 Herbaceous rangeland 32 Shrub and brush rangeland 33 Mixed rangeland
4.	Forest land	41 Deciduous forest land 42 evergreen forest land 43 Mixed forest land
5.	Water	51 Streams and canals 52 Lakes 53 Reservoirs 54 Bays and estuaries
6.	Wetland	61 forested wetland 62 Nonforested wetland
7.	Barren land	71 Dry salt flats 72 beaches 73 Sandy areas other than beaches 74 Bare exposed rock 75 Strip mines, quarries and gravel pits 76 Transitional areas 77 Mixed barren land
8.	Tundra	81 Scrub and bush tundra 82 herbaceous tundra 83 Bare ground tundra 84 Wet tundra 85 Mixed tundra
9.	Perennial snow or ice	91 Perennial snowfields 92 Glaciers

Table 12.2 Slope Classification and Percentage

Slope Classification	Slope %	Upper and Lower limit of Contour Space
Nearly Level	0 – 1 %	More than 4 cm
Very gently sloping	1 – 3 %	1.33 cm – 4 cm
Gently Sloping	3 – 5 %	0.8 cm – 1.33 cm
Moderately Sloping	5 – 10 %	0.4 cm – 0.8 cm
Steeply Sloping	10 – 15 %	0.26 – 0.4 cm
Moderate Steeply sloping	15 – 35 %	0.11 – 0.26 cm
Very Steep Sloping	> 35 %	Less than 0.11 cm

12.6.4 Geomorphology map

The earth has a high complex and variable surface whose topographic relief and material composition reflect the bedrock and unconsolidated materials that under lie each part of the surface as well as the agents of change that have acted on them. Each type of rock, fracture or other effect of internal movement and each erosonal and deposition feature bear the imprint of process that produced them.

Persons who describe and explain earth materials and structures must understand geomorphologic principles and able to recognize the surface expressions of the various materials and structures through the process of visual image interpretation for geological and soil mapping. These materials and structure can be identified and evaluated using satellite imagery.

Geomorphologic mapping will always require a considerable amount of field exploration, but mapping process can be greatly facilitated by the use of Visual image interpretation. The geomorphology in the case study area clearly shows the coastal environment, the main categories are beach, coastal plain, channel islands observed in river mouth and brackish water areas, salt flats in brackish water and in sides of river has alluvium plain deep and then alluvium plain is moderate (Plate 12.4).

12.7 Attribute Data Base Creation

Attributes are the characteristics of the map features and hold the descriptive information about the geographic features. Attributes are the non-spatial data associated with time and area entities. The groundwater quality data is considered as attribute data base for this study.

Table 12.3 Ground Water Quality Parameters And Instruments/ Methodology Adopted for the Estimation of These Parameters

S.NO	PARAMETER	INSTUMENTS/ METHODOLOGY
1.	pH	Electrometric Method
2.	Electrical Conductivity	Conductivity Meter
3.	Chlorides	Argentometric Method
4.	Sulphates	Spectrophmeteric method at 420 nm
5.	Alkalinity	Titrimetric Method
6.	Nitrates	UV -Spectrophotometer at 220nm
7.	Hardness	Titrimetric Method
8.	Total Dissolved Solids	Analytical Method
9.	Sodium	Flame Photometry Method
10.	Potassium	Flame Photometry Method

12.7.1 Ground Water Quality Database

Fieldwork is conducted and 95 ground water samples are collected at predetermined locations in the study area. The water samples were analyzed for various physico-chemical parameters like pH, chlorides, sulfates, nitrates, sodium etc. Table 12.4 shows the results of laboratory analysis of these 95 samples.

12.8 Generation Spatial Distribution Maps of Water Quality

GIS offers many options for creating customized maps and reports which are useful for decision-making. Summarizing and presenting understandable results of GIS analysis, both graphic and tabular are the key factors for planning and management and will help to determine effect of our analysis in decision-making. By creating the topology between the spatial and attribute data, the maps showing the spatial distribution of water quality parameters are prepared. The maps are shown in (Plate 12.5). These maps are then critically analysed, discussed and the results are presented in the following paragraphs.

ph is the negative logarithm of hydrogen ion concentration. ph is the measure of acidity or alkalinity present in water samples. Acidic waters have a ph of 0 to 7,whereas alkaline waters have a ph of 7 to 14. The ph value present in the water samples can be determined by electrometric method. ph value should be made to keep the ph as close as possible however the permissible value according to the Bureau of Indian Standards may range between 6.5 - 8.5.

Table 12.4 Attribute Database

SAMP_LOC	CONDUCTION	ALKALIN mg/l	SODIUM mg/l	POTASSIU mg/l	TOTAL_HA mg/l	CHLORID (mg/l)	TDS (mg/l)
1. Isakapalli	2392	550	298	101	590	425	1542.000
2. Ragadachilaka	649	280	80	25	290	35	421.850
3. Khammapalem	1115	290	195	24	292	213	724.750
4. Allur Road near Rail cross r	1667	623	214	29	292	213	1083.550
5. Allur N H	1354	330	142	32	270	254	880.100
6. Allur NH	622	160	72	45	200	110	404.300
7. Ulvapalli	497	208	71	32	208	28	323.050
8. Peddapalem	737	203	122	33	280	127	479.050
9. Singarupalli	1265	310	195	49	280	254	822.250
10. Peddapalem	321	202	89	22	80	37	208.650
11. Peddapalem Road	293	105	42	21	80	30	190.450
12. Allur Town	1007	402	48	33	300	120	654.550
13. Inupur	3490	432	593	38	520	764	2268.500
14. Purini	5450	390	929	39	1120	1428	3418.120
15. Gogulapalli Road	7380	980	1227	44	876	1602	4797.000
16. Gogulapalli towards Allur	1459	490	198	15	440	164	948.350
17. Sunnapubatti	680	208	68	25	280	73	442.000
18. Gugulapalli (East) near Nand	1172	210	182	41	292	255	761.800
19. Gogulupalli	5500	310	789	35	1200	1680	3575.000
20. Near Ramatheertham Road	1747	450	340	37	300	268	1135.550
21. Mopuru	2420	340	308	49	980	596	1573.000
22. Racherlapadu	362	122	42	8	120	37	235.300
23. Dandigunta	3170	600	596	42	294	752	2060.500
24. Towards Chakacherla Road	1077	401	152	36	294	165	700.050
25. Dandigunta	1180	221	171	28	180	253	767.000
26. Gadidladine	1688	270	240	7	180	418	1097.200
27. Gadidladinne	1676	240	260	10	197	402	1102.000
28. Chekacherla Cheruvu (opposit	893	340	176	14	160	135	580.450
29. Chandrasekharapuram	251	60	47	8	80	37	163.150
30. Vavalla Road	1925	526	369	43	120	465	1251.250
31. Padmanabuni Satram	1857	520	198	34	590	289	1207.050
32. Guvvagunta	1083	320	191	10	280	198	703.950

Table 12.4 *Contd...*

SAMP_LOC	CONDUC-TION	ALKALIN mg/l	SODIUM mg/l	POTASSIU mg/l	TOTAL_HA mg/l	CHLORID (mg/l)	TDS (mg/l)
33. Pallipalem	1420	486	187	18	428	168	923.000
34. Ramco Seed Manufacturing Comp	1160	205	175	28	276	227	754.000
35. Kodavaluru	2480	358	316	37	960	554	1612.000
36. Gundammapalem	1349	330	139	28	272	246	876.850
37. Bodduvaripalem near N H	1137	406	196	41	220	168	739.050
38. Bodduvaripalem	1801	230	218	104	500	508	1170.650
39. Ramudupalem	2669	608	702	36	670	412	1734.850
40. Kudithipalem Limits toward R	5960	410	998	142	978	1380	3874.000
41. Kudithipalem Limits	765	215	122	31	256	118	497.250
42. Kudipadu	3880	418	592	49	683	1008	2522.000
43. Muduvarthipalem	885	321	165	24	152	148	575.250
44. Pallipalem Gangapatnam Cross	1466	497	170	17	418	160	952.900
45. Near Muduvarthipalem	1384	350	141	21	288	247	848.900
46. Pallipalem	1402	304	165	28	280	234	899.600
47. Nagaraju Thopu	1128	237	165	28	280	209	733.200
48. Nagaraju Thopu Limits	612	150	89	15	189	102	397.800
49. Pattapalem	5008	585	870	72	1140	1432	3251.120
50. Ganga Patnam	1448	406	172	18	427	168	941.200
51. Narsapur	1555	409	192	32	580	253	1010.750
52. Nagaraju Thopu (Pallipadu)	807	270	143	11	150	97	524.550
53. Near Marripadu Road	1234	230	160	30	268	240	802.100
54. Mothalu	492	206	65	11	185	38	319.800
55. Pallipadu	670	215	86	10	131	88	435.500
56. Devispet	506	209	96	9	140	55	328.900
57. High Depth Kothakaluva limit	753	210	118	21	246	105	489.450
58. Near Podurupadu	752	221	110	12	235	96	488.800
59. Podugupadu Irrigation	884	330	149	24	149	138	574.600
60. Indukurpeta	942	340	200	42	289	149	612.300
61. Kotturu	950	350	188	49	260	170	617.500
62. Penna River Bank	1012	390	156	33	305	125	657.800
63. Indukuru Limits	809	275	142	32	165	105	525.850
64. Nellore towards Narayanaredd	556	220	82	36	120	65	361.400
65. Prasanthinagar	520	167	78	38	106	72	338.000
66. Navalakura Thota	992	381	144	34	300	108	644.800

Table 12.4 *Contd...*

SAMP_LOC	CONDUC-TION	ALKALIN mg/l	SODIUM mg/l	POTASSIU mg/l	TOTAL_HA mg/l	CHLORID (mg/l)	TDS (mg/l)
67. 4th Mile near Kothakaluva	1213	220	165	39	284	251	788.450
68. Kothakaluvu	496	201	68	13	179	40	322.400
69. Near Narkur Road	1583	403	199	32	590	241	1028.950
70. Allipuram after locks	558	208	79	38	124	61	362.700
71. Near Allipuram Road	723	203	121	18	229	92	469.950
72. Narkur Cross Roads	1314	320	164	38	299	245	854.100
73. Allipuram towards Nellore	475	201	65	11	185	32	308.750
74. Nagaraju Thopu (Pallipadu)	765	220	115	25	261	96	497.250
75. Ramco Seed Manufacturing Comp	1352	330	145	32	292	228	878.800
76. Vavalla Road	1186	224	149	16	174	263	770.900
77. Kudipadu	1485	490	199	39	540	169	965.250
78. Mopuru	583	229	83	18	190	89	378.950
79. Singarupalli	1204	274	159	38	274	238	782.600
80. Indupur	2811	421	536	27	596	694	1827.150
81. Allur Town	1403	360	159	18	424	148	911.950
82. Isakapalli	1899	528	198	41	590	267	1234.350
83. Near Muduvarthipalem	958	348	192	37	282	169	622.700
84. Indukuru Limits	748	203	128	36	251	106	486.200
85. Ulvapalli	1035	373	164	9	254	144	672.750
86. Racherlapadu	461	201	69	17	183	45	299.650
87. Ramco Seed Manufacturing Comp	1592	427	198	47	599	207	1034.800
88. Ramco Seed Manufacturing Comp	1151	244	189	36	294	209	748.150
89. Penna River Bank	817	264	168	32	165	102	531.050
90. Isakapalli	2345	510	228	104	594	420	1524.250
91. Near Ramatheertham Road	2814	412	556	28	590	692	1829.100
92. Indukurpeta	840	280	138	24	144	113	546.000
93. Devispet	794	215	145	29	264	118	516.100
94. Devispet	876	308	194	28	184	129	569.400
95. 60.Peddapalem Road	2643	603	758	34	710	407	1717.950

1. Acidic Range: 0-7
    - Cause: Presence of mineral acids like Sulphuric Acid, Iron, Cadmium (Industrial Effluents).
    - Effects: Tuberculation & Erosion.
2. Alkaline Range:7-14
    - Cause: Presence of Bicarbonate, Calcium, Magnesium, Potassium, Sodium etc.,
    - Effects: Indigestion, Sedimentation deposition

The maximum pH values are found to be 8.5 in brackish water.

Alkalinity is a measure of water to neutralize acids. It is the measure of the quantity of ions that will react to neutralize ions.

- Sources

  It is mainly caused due to the salts of carbonates, hydroxides, ammonia etc. High alkaline waters are usually unpalatable. Alkalinity reacts with cat ions of water and forms precipitate which damages the pipes and other valves. Alkalinity values provide guidance in applying proper doses of chemicals in water and wastewater treatment processes, particularly in coagulation, softening and operational control of anaerobic digestion.

- Effects

  When the alkalinity exceeds the hardness, the presence of basic salts generally sodium and potassium has been indicated. If the alkalinity has been less than hardness then calcium and magnesium would be present in association with sulphates, chlorides and nitrates.

The permissible value of alkalinity as recommended by Bureau of Indian Standards is 200 mg/l as caco3. The alkalinity concentrations of the water samples in the present study area are ranged from 60-608mg/l.

The total dissolved solids present in water are regarded as Total Dissolved Solids.

- Constituents of TDS

  Mineral dissolved gases and Organic constituents

- Source

  Seawater Intrusion, Irrigated drainage water domestic and sewerage are the main source for this quality. Dissolved minerals, gases and organic matter.

- Effects

  Presence of TDS depletes dissolved oxygen, distress to live stock and displeasing color, odour and taste.

  The total permissible value of solids in water Bureau of Indian Standards is

500 ppm. TDS value of the area ranged from 190 ppm to 4792 ppm. The high TDS concentrations are observed in brackish water and salt pond areas.

Chlorides are generally present in the water as sodium chloride. Nacl is the main chloride concentration in the water. A chloride present in the water samples has been determined by argenometric method.

- Sources

  Discharge from agricultural drainage water, seawater intrusion, sewage disposal and domestic sewage.

- Effects

  Imparts salty taste to the water. Corrodes concrete. There is no evidence that chloride constitute is human hazard.

The permissible limit for chlorides in drinking water as per Bureau of Indian Standards is250mg/l. The chloride concentration in groundwater of the study area varied between 30mg/l-270.mg/lt. Chloride concentrations of the study area exceed the permissible limit in the several areas.

Total hardness in water is the characteristics, which prevents the formation of sufficient foam when such hard water is mixed with soap. It is caused by the presence of certain salts of calcium and magnesium dissolved salts in water which form scum when reaction with soap. Hardness of the water samples has been determined by titrimetry method (Table 12.5)

Table 12.5 Total Hardness Classification

Degree of Hardness	Mg/l as caco3
Soft	0 - 50
Moderately hard	50 - 100
Hard	150 - 300
Very hard	> 300

- Effects
  - Not suitable for production of ice soft drink
  - Increases cardio-vascular diseases.
  - Makes food tasteless.

The desirable hardness limit as per Bureau of Indian Standards is 300mg/l as CaCo3.The hardness values ranged from 80-1140 mg/l. Areas such as Nagarau Topu, showed a concentration >1140mg/l. Hence these waters can be considered as very hard water.

Sodium was estimated by flame photometry. The principle behind flame photometer deals with the excitation of electrons from ground state to the higher energy state and come back to its original state with emission of light.

- Source

  Agricultural drainage water, sewage, industrial effluents, sea water intrusions in coastal areas, and the use of sodium compounds for corrosion control and water softening process.

- Effects

  In general sodium salts are not toxic substances, because of their much efficiency mature kidneys will excrete sodium chloride causes vomiting.

The permissible limit for sodium as per Bureau of Indian Standards is 200 ppm. Sodium values obtained showed no variation for the present study area and they are considered within the permissible limits. The value present in the study area varies from 42 – 1227 mg/l.

Potassium was estimated by flame photometry. The principle behind flame photometer deals with the excitation of electrons from ground state to the higher energy state and come back to its original state with emission of light.

- Source

  The major natural source of potassium in groundwater is Feldsparn, some Micas and clay minerals.

- Effects

  Potassium is corrosive to metals.

The permissible limit for potassium as per Bureau of Indian Standards is 200 ppm. Potassium values obtained showed no variation for the present study area and they are considered within the permissible limits. The value present in the study area varies from 8- 142 mg/l.

12.9 SWI Model Development

The location of the sea water/freshwater interface is important to groundwater Management in coastal regions. The exact locations of the freshwater/saline water interface and the transitional zone are difficult to determine, either theoretically or by field testing. There are many approaches to delineating the depth to the extent of a saline water body. First it is necessary to define what is meant by saline water. Terms relating to the degree of salinity were suggested by the USGS (Table 12.6).

Table 12.6 Saline Water Classification

Terms describing degree of salinity as used by USGS	
Description	**TDS (mg/l)**
Fresh	< 1000
Slightly saline	1000 - 3000
Moderately saline	3000 - 10000
Very saline	10000 - 35000
Brine	> 35000

12.9.1 Determination of Extent of SWI

The case study shows that the lower Pennar basin has severely effected by the saline water intrusion. Delineation of saline water areas based on the TDS (mg/l). This study area was fallen in 3 categories i.e., Fresh, Slightly saline, Moderately Saline, (Plate 12.5) shows the villages that are effected by saline water intrusion. Moderately saline observed nearby brackish water and poor drainage areas ranging from 3000above to 6000. Slightly saline waters are observed in aqua culture areas, and fresh water zones are observed in reaming parts of study area. The Villages effected by the Saline Water Intrusion are as follows :

- **Moderately Saline :** Indupur, Purani, Gogulapalli, Pattipalem, Yerugunta, Iskapalli, Peddapalem Pallepalem and Near by Villages
- **Slightly Saline :** Alluru, Ramathertham, Mopuru, Dandigunta, Gadiladinne, Padmanabhuni sdatram, Boduvaripalem, Ramudu palem , Kudipadu, Mudivartipalem, Kodavuluru, Ganga patnmam, Narasapur, Narkur road, and nearby Villages
- **Fresh water :** Reaming villages are fallen in this zone

From the present study it is clear that salinity is the main problem in the study area. Sea water from the Bay of Bengal is the main contributor to salinity, and this is caused by the reduction in the Pennar water flow, which lowers the river and canals water level and ultimately paves the way for saline water intrusion in to the fresh water zone of the study area. In fact, no effect has yet been made to determine ground water removal as a cause of sea water intrusion in the fresh water aquifer because ground water development began only during the last decade and removal is still only small scale. The simple mechanism that accounts for the sea water intrusion into the groundwater of the area is that when the surface water level drops, the nearby groundwater level also drops and thereby sea water intrudes the aquifer through the channels by lateral or horizontal replenishment. That is why shallow aquifer water is more saline than that from deeper aquifer. The problem gets severe during the dry period when the Pennar and its tributaries chocked because of the drastic fall of fresh water supply in the River and its canals. This situation is increasingly worsening and in the future it will lead to a terrible environmental hazard unless a suitable ambitious approach is made. The project has developed, tested and finalized saline water intrusion zones identification methodologies suitable for three different environments and based on satellite remote sensing, GIS and field investigations.

From the results of this study, it is recommonded that :

- A detailed salinity – discharge velocity study should be carried out for the different time periods and for the different rates of fresh water outflow during selected tidal cycles in order to represent the salinity –river discharge relationship.
- Geomatics can be used to determine the saline – fresh water interface and the degree of both surface and groundwater contamination by saline water intrusion. More ground and surface water monitoring stations should be set up on a grid basis. Moreover, geophysical surveys should be performed for comparison with hydro chemical studies.
- the construction of surface water reservoirs in upstream areas to maintain the annual flow system in an equilibrium condition.
- Emphasizing more salinity-tolerant crops so as to get maximum output should modify the cropping pattern in an area.
- Unavoidable drainage waters should be intercepted, isolated, and reused to irrigate a succession of crops of increasing salt tolerance possible including eucalyptus and halophyte species, so as to reduce drainage water volumes further and to conserve water and minimize pollution, while producing useful biomass.
- Co-ordination among geoscientists, engineers, environmentalists and planners/ policy makers is important for the whole spectrum to be a multidisciplinary approach.

There is usually no single way to achieve salinity control in ground and surface waters. Many different approaches and practices can be combined into satisfactory control systems, the appropriate combination depends upon economic, climatic, social, as well as edaphic and hydro geologic situations. Thus, no procedures are given for selecting "the" appropriate set of control practices for different situations. Rather, some important goals, principles and strategies of salinity management, at both on-form and project levels, that should be considered in the selection and implementation of control practices.

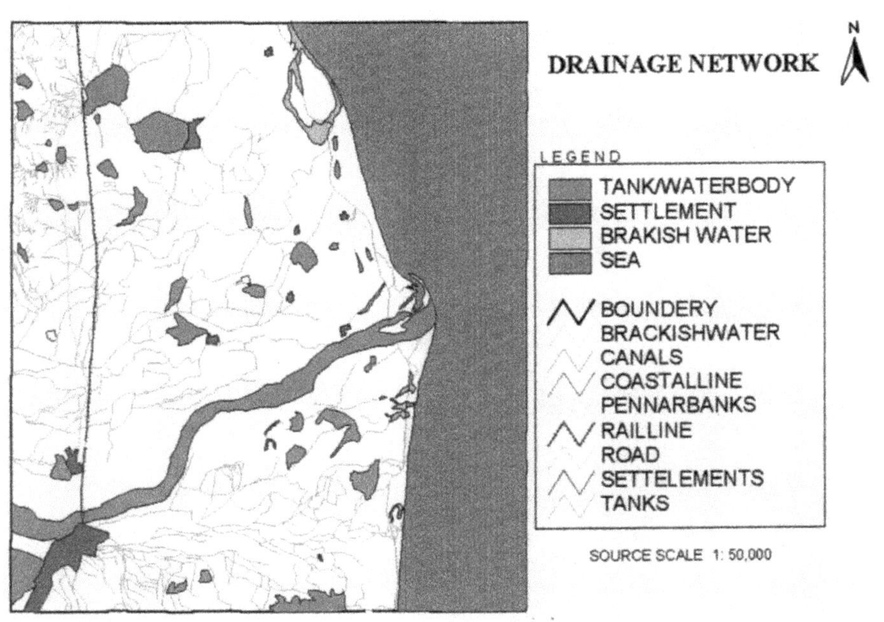

Plate 12.1 Drainage Network map of case study

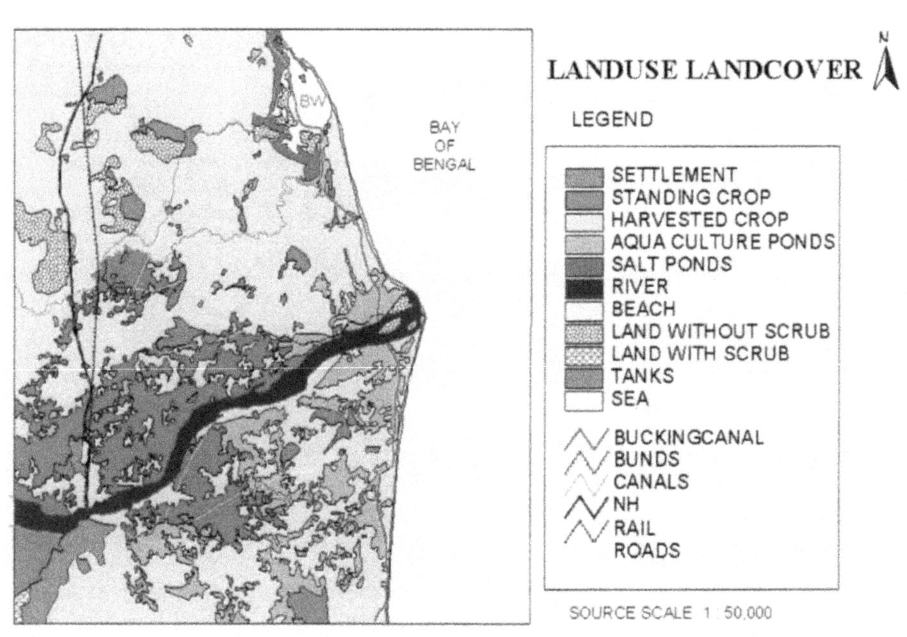

Plate 12.2 Land use / Land cover map

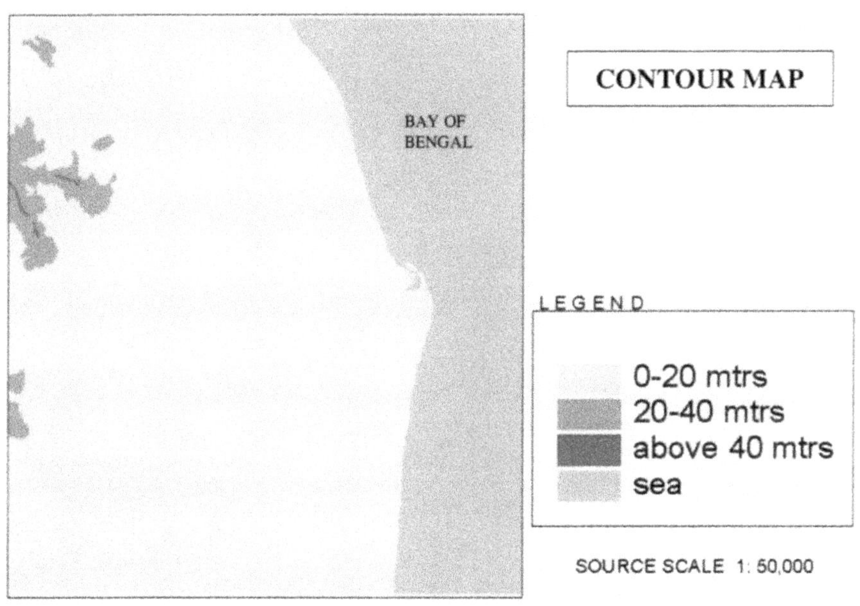

Plate 12.3 Contous Map

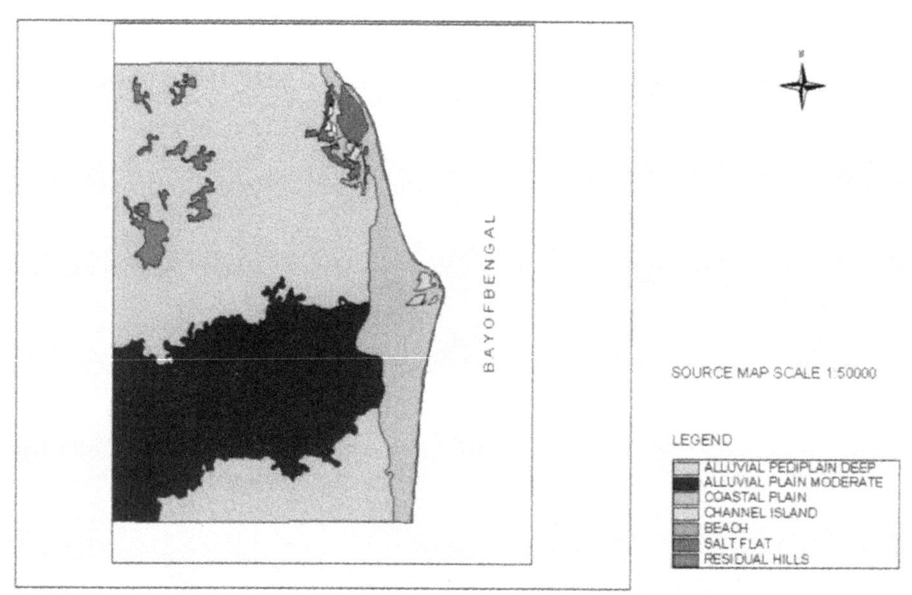

Plate 12.4 HYDROGEOMORPHOLOGY

13 Water Quality Mapping and Modelling

13.1 Introduction

Fresh water resources get polluted due to increasing urbanisation, industrialisation, and modern agricultural practices. Wastewater contains objectionable and hazardous constituents, which exert direct toxicity or bring about physio-chemical changes in the aquatic environments. The groundwater quality around any area is steadily deteriorating due to seepage of pollutants. The extent of this effect on groundwater depends upon the level of the water table. The higher the level of the table the more will be the impact. The distinction between water quality criteria and water quality standards is subtle but important. Water quality criteria can be defined as statements concerning the limiting levels of concentration or intensity of key quality parameters established by the intended use of the water. On the other hand, water quality standards can be defined as regulations limiting the levels of concentration of key quality parameters established by regulatory authority for the purpose of protecting a water resource for one or more uses. In this chapter, two case studies are presented to appreciate the technology of geoinformatics, more specifically, remote sensing and GIS in mapping and modelling the water quality data for its monitoring and management. The two case studies are Hyderabad city and Visakhapatnam city. Arc/Info and SPANS GIS softwares are used to map the water quality of Hyderabad and Visakhapatnam industrial areas respectively.

The purpose of providing chapter is, how the technology of Geoinformatics can be used to find out the correlation between impact of landuse and groundwater level and corresponding water quality of the area. A database is created on GIS platform in order to execute the pollution related project to fulfil the specified objectives. Such database can be used on GIS platform by any decision maker who is involved in implementing the operational project to restore the environmental quality of the area.

13.2 Role of Remote Sensing and GIS

An analysis of the nature and rate of environmental change over recent decades is essential for a proper understanding of why present environment problems have arisen. It is necessary to allow formulation of accurate productive models of environmental change. GIS and remote sensing with their unique capabilities play an important role for this purpose. Remote sensing is one of the excellent tools for the natural purpose. Remote sensing as it possesses unique possibilities of vividly displaying the natural resources that are occuring on the earth's surface owing to its synoptivity. In addition, its multispectral nature provides appropriate contrast provides information on the change that are taking place over the natural resources and their environment. While integrated with GIS remotely sensed imagery can be merged with other data in a GIS providing real time spatial information. Geographical Information System is a way of looking at data from our environment within a spatial context. GIS involves mapping data and interpreting the relationships among that data and making inferences. GIS data such as Arc/Info and Arc view. GIS data forms to demographics. Source data comes from public database maps, aerial photos, satellite scans, etc.

Photographs of the earth taken from aircraft or satellite at various electromagnetic wavelength ranges can provide useful information regarding ground water conditions. The satellite sensed photographs give spectacular information on the size, shape, pattern, aerial extent and spatial distribution of the water bodies. Aerial photographs can also reveal fracture patterns in rocks, which can be related to porosity, permeability and ultimately well yield. On the basis of such data the surface water budgeting can be done regularly. Maps classifying an area into good, fair and poor ground water yields can be prepared by integrating the data in GIS.

The present chapter mainly concentrates on mapping of ground water quality in Hyderabad area using remote sensing and GIS. Groundwater is an important, yet, often neglected global source of water supply. Enhanced utilization of groundwater for irrigation, industries and domestic purposes is a natural consequence where perennial surface water sources are absent. The quality along with quantity of ground water used for these purposes is more important. Due to the rapid pace of change in land use and uncontrolled unbanization and industrialization the groundwater resources in Hyderabad are adversely affected. In order to minimize such impacts, an integrated study including an overview on source of pollution supported by appropriate monitoring procedures is needed. One such obvious tool, GIS is used to present accurate and existing data on the land use / land cover and to study its impacts on the groundwater quality.

IRS-IC PAN and IRS-ID LISS-III satellite imageries obtained from NRSA, Hyderabad are georeferenced using the ground control points with SOI toposheets as a reference and further merged to obtain a fused, high resolution (5.8m of PAN) and colored (R, G, B bands of LISS - III) output in EASI/PACE Image processing software which is further interpreted for the generation of thematic maps. Two types of database (spatial and attribute) are created using GIS. Thematic maps like base map, road network map, drainage network and land use / land cover maps are prepared from the satellite imagery and SOI toposheets on 1 : 25,000 scale using AutoCAD and Arc/ Info GIS software to obtain a baseline data. Fieldwork involving collection of groundwater samples based on the land use and drainage network maps and further analysis of the samples for various physico-chemical parameters adopting standard protocols is attribute database generated are integrated for the generation of spatial Chlorides, Sulphates, Nitrates, TDS, Total hardness and Fluorides. The water quality data (attribute) is linked to the sampling location (spatial) in Arc/Info and maps showing spatial distribution are prepared to easily identify the variation or change in concentrations of the above parameters in the groundwater at various locations of the study area using curve fitting technique of Arc/View GIS software. These maps are then overlaid on the satellite imagery to correlate the groundwater quality with the land use changes. Water quality index is also calculated to assess the suitability of water for drinking purpose.

13.3 Case Study of Hyderabad City

An attempt is made to assess the magnitude of pollution in the groundwater with the corresponding levels in the water table using GIS with remote sensing data as input. Two types of databases : spatial and attribute, are created by using GIS software package, Arc/info. The spatial data consists of all the necessary thematic maps, which are converted into digital format compatible to Arc/info. The attribute data consists of groundwater quality of all the areas under study. Broadly, Remote Sensing and GIS can be used (i) to identify and detect the problematic areas of water quality, (ii) to study the impact on the landuse, and (iii) to study the correlation between the groundwater level and the corresponding water quality index by creating buffer zones using GIS. Remote sensing and GIS technology are used to study the pollution of groundwater of a part of Hyderabad city. The area considered for this study is Katedhan which is the major industrial area of Ranga Reddy District, Andhra Pradesh state in India. The population of Katedhan shows a steady to rapid growth from 1915 onwards and as per 1981 and 1991 census stands at 8.5 lakhs. The two main reasons for this study are rapid growth in population, and uncontrolled industrialization and unchecked in migration. The unabated spectacular sprawl and fluid spread of the study area is due to the absence of any natural barrier. The study area is the non-municipal zone of Hyderabad located towards its southern side. The study area covers an area of 15.061 Sq. km. and is situated at 17^{0}22'30 c" N latitude and 78^{0}27' 30" E longitude and is at an altitude of 570 meters above mean sea level. The location map is presented in Fig. 13.1. The Katedhan has become totally industrialised for 20 years. Just before 20 years, the area was completely under residential use. The study area is now densely populated with very less vacant space left. The landuse/landcover map is shown in (Plate 13.1) and is prepared from IRS ID LISS III data using visual interpretation, scanning and digitisation processes and GIS techniques. Table 13.1 shows the area statistics of landuse/landcover.

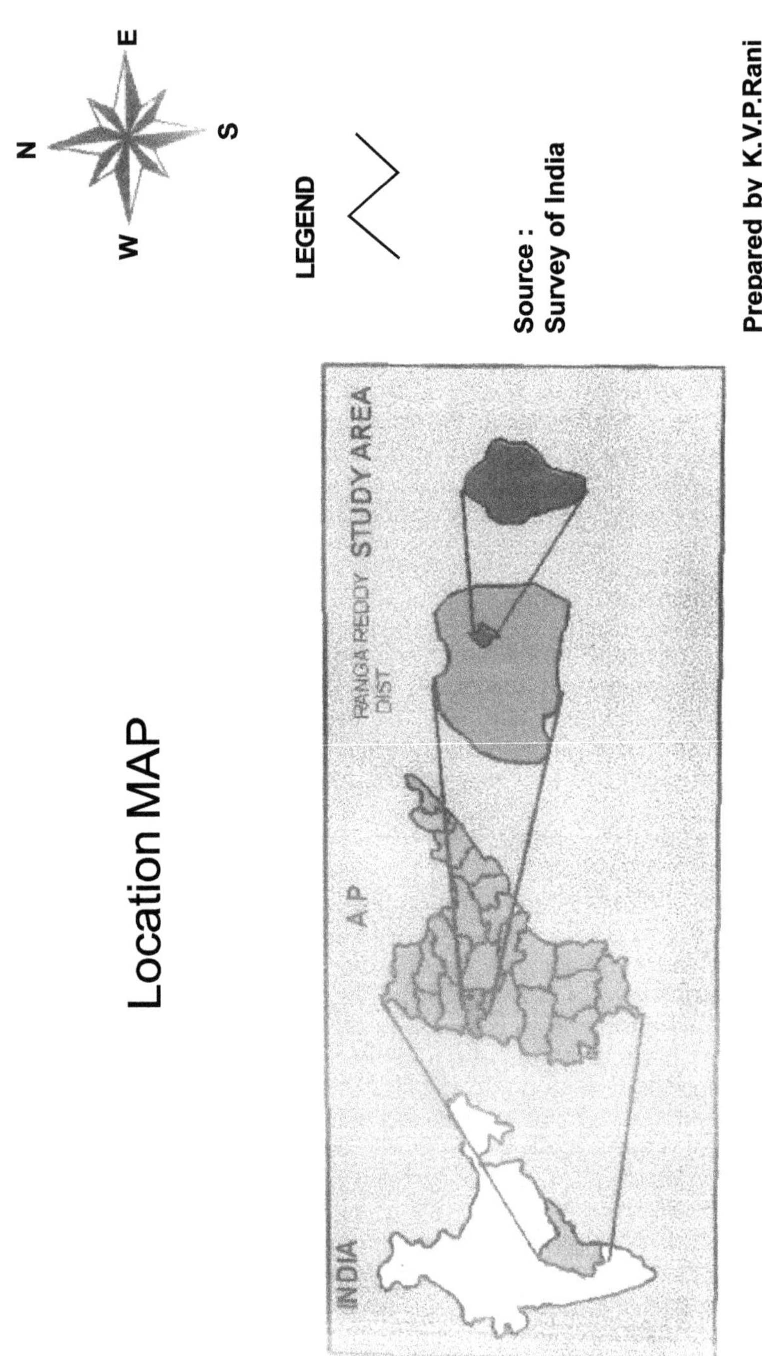

Fig. 13.1 Location map of Katedhan industrial area.

Table 13.1 Existing Land Use of Study Area

Landuse	Area (Sq. Km)
Residential area	1.612
Double crop area	1.964
Plantations	0.132
Industrial area	3.271
Water tanks	1.459
Land with scrub	3.962
Land without scrub	1.082
Barren waste/stony	0.784
Waste/sheet rock area	
Quarries	0.006
Total	**15.061**

13.4 GIS Data Analysis

The major difference between geographic information systems and a system for computer-assisted cartography is the provision of capabilities for transforming the original spatial data in order to answer particular queries. The GIS provides a much larger range of analysis capabilities to operate on the topology or spatial aspects of the geographical data on combined non-spatial and spatial attributes. The GIS analysis functions, are classified into four groups : (i) maintenance and analysis of spatial data, (ii) maintenance and analysis of attribute data, (iii) Integrated analysis, and (iv) output formatting.

To examine the groundwater pollution of the Katedhan study area, water samples are collected at 16 sampling stations from different wells at different depths. These samples are analysed and the results obtained from the analysis of these 16 samples are extrapolated for assessing the problems of pollution. The GIS data analysis is performed on these 16 samples by creating buffer zones. Buffer Zones are created with respect to these samples by considering groundwater quality index for each of the years 1996, 1997, and 2000. Table 13.2 shows the standards of water quality index for three different categories. Table 13.3 shows the water quality variation in the study area with corresponding to water quality index (WQI) of various places. The water quality variation of the places in the study area which are derived from the analysis on the Arc/Info GIS platform is discussed in the follwing paragraphs.

Katedhan : The Water Quality Index (WQI) of Katedhan industrial area of the place varied from (–4.024 to 22.572), with the corresponding water levels being varied between 5.04 and 7.5 meters. The water quality Index had shown a maximum

Table 13.2 Standards of Water Quality Index

Category	Water Quality Index	Description
I	<150.00	Tolerable
II	150-200	Moderately Polluted
III	200-250	Excessively Polluted

Table 13.3 The water level and water quality variation from 1996-2000

Name	Year	Water Quality Index (WQI)	Water Level (meters)
Katedhan	1996	159.56	5
	1997	215.68	7
	2000	22.572	6.65
Sivarampally	1996	217.78	6.1
	1997	163.93	5.89
	2000	176.25	5.04

during the year 1997. Since the place is bounded by disposal site, the change in water quality from year to year solely depends upon the water table level of the area. It can be seen from the table during the period that as the groundwater level decreases the groundwater contamination increases.

Sivarampally is basically residential with a part of small scale industrial area. The water quality Index of the area varied between (176.29-11.818) with a high corresponding water quality index during the year 1996 and it stood about 217.78 with corresponding groundwater level being 6.1 meters. During the year 1996, the value is very high probably because the percentage change in the groundwater table is also high during this period and the level increased from 5.04 to 6.11 meters.

Buffers are geographic locations and are created by the area surrounding each buffer for which the centre of the point/circular buffer is the sampling location. The radius of the buffer is the value of WQI. The operation of creating these buffers performed by the buffer command of Arc/Info generates one or more polygons surrounding the existing geographic features, and these buffer zones are used to determine the spatial proximity and the effect of pollution on the surrounding areas.

Each buffer zone depicts the condition of the groundwater quality of the respective samples and the information obtained from this is used to determine the condition of the surrounding sample, which fall in the buffer zone. In this experiment, the radius of each buffer is considered with respect to water quality Index as mentioned above.

The sample having the highest water quality index is given the highest buffer radius. All the samples are overlaid with landuse/landcover features for cursery examination of the pollution levels and the corresponding landuse/landcover through the correlation study. The buffer zones created on the basis of the water quality data are shown in Fig. 13.2. From the above discussion it is clear that the study area contains a mixed type of groundwater qualities. The water quality index maps are shown in Fig 13.3, 13.4, 13.5. The water level maps are shown in Fig. 13.6.

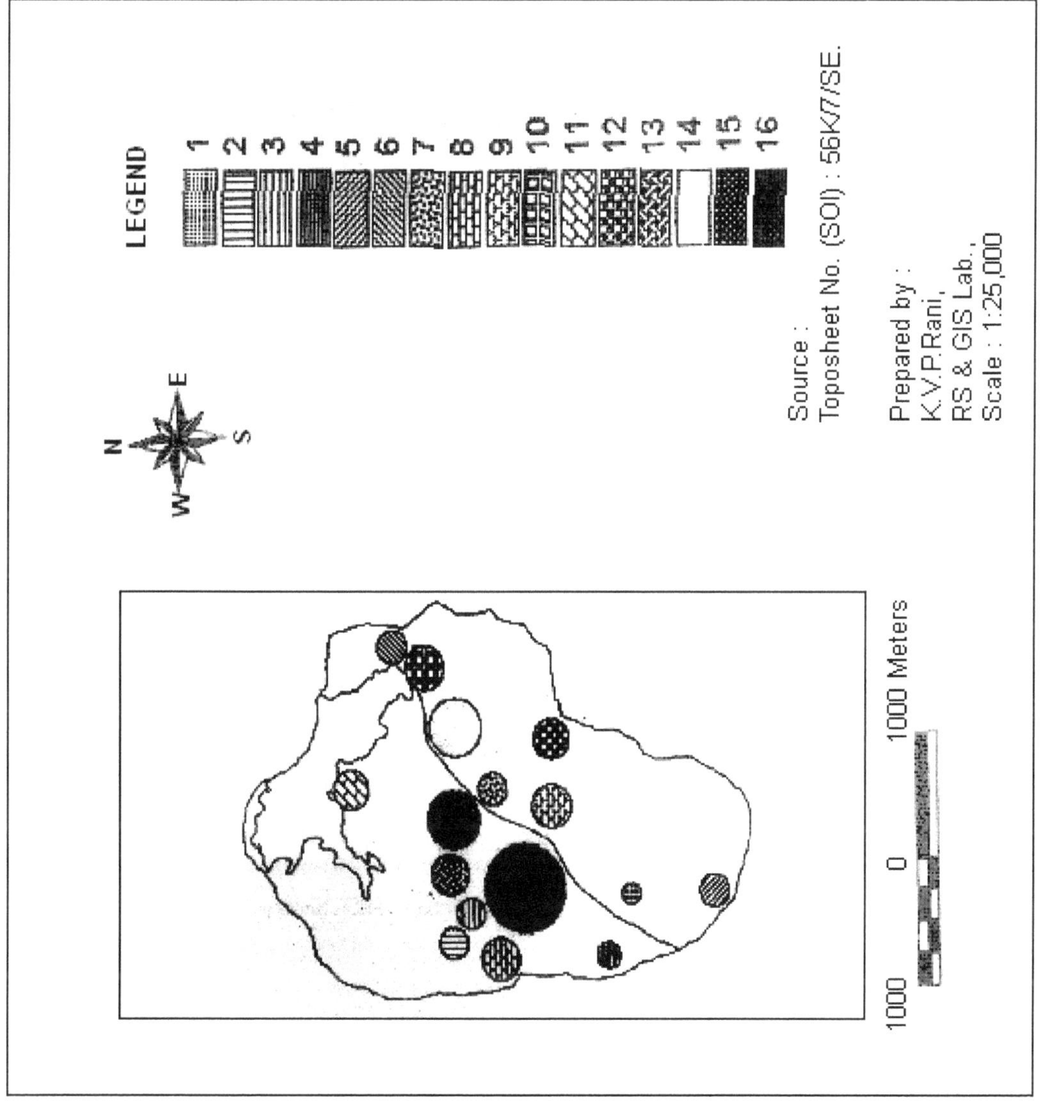

Fig. 13.2 Buffer Zones Created based on the water quality data.

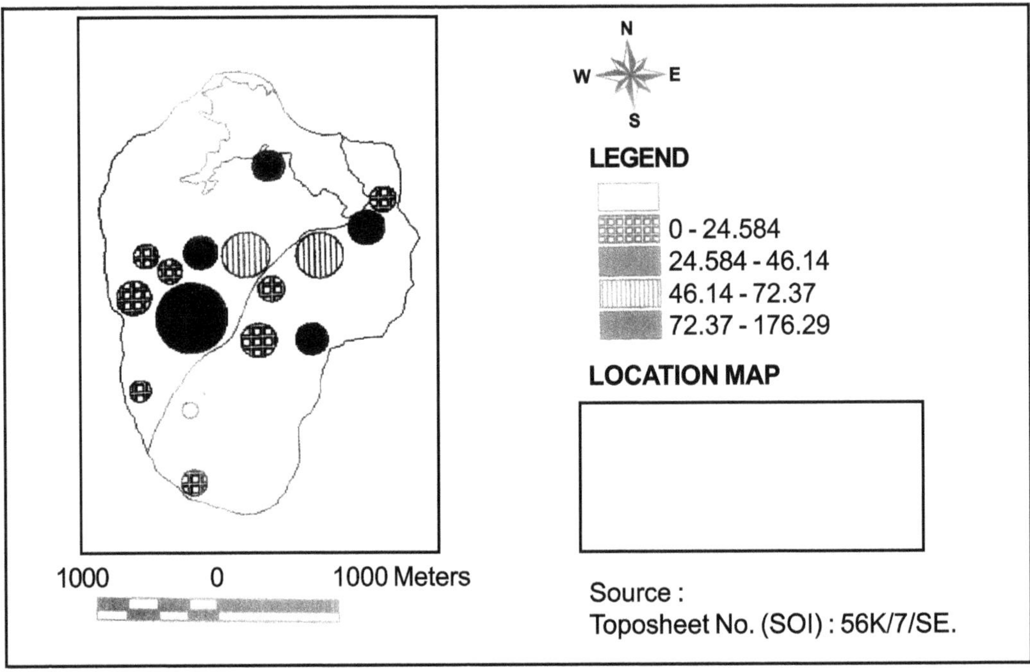

Fig. 13.3 Buffer zones created based on the water quality index for the year 2000.

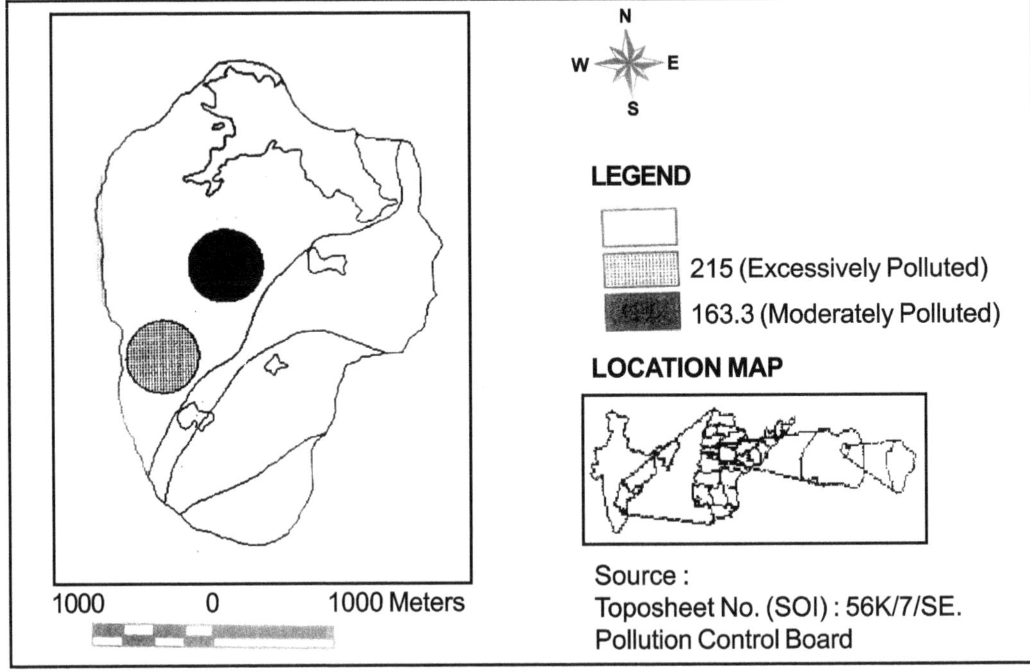

Fig. 13.4 Buffer zones created based on the water quality index for the year 1997.

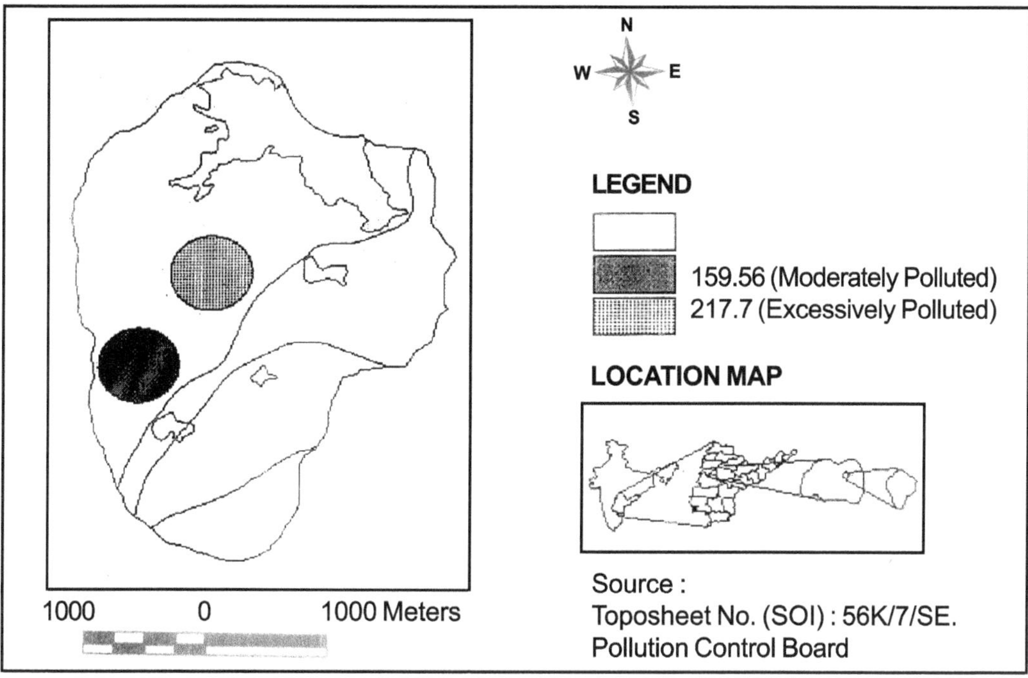

Fig. 13.5 Map showing Water Quality Index of 1996.

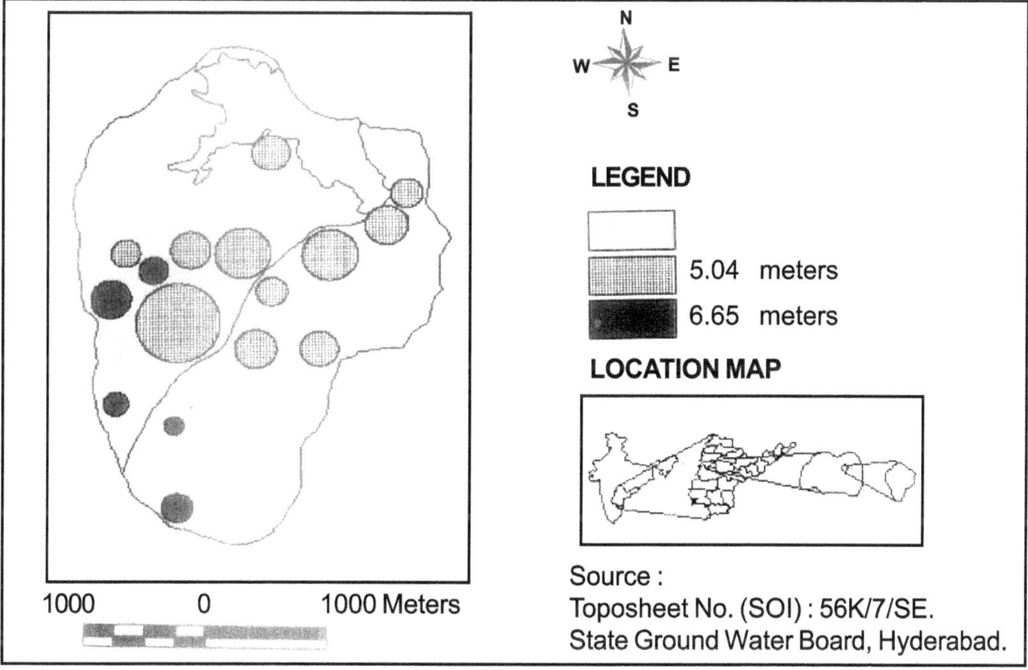

Fig. 13.6 Map showing Ground water levels in 2000.

The water quality index of Katedhan has increased from 159.56 to 215.68 during the period 1996 to 1997; this may be due to the decrease in the water level from 5 to 7 meters. The water quality index has decreased from 215.68 to 22.572 during the period 1997-2000. Therefore the seepage of the pollutants may be very less. The water quality index of Sivarampally decreased from 217.78 to 163.93, during the period 1996-1997. This may be due to the decrease in the water level from 6.1 to 5.98. The water quality Index has increased from 163-93 to176.25 for the period 1997 to 2000, this may be due to the collection of samples in the month of June when the water level increases. From the above results it can be seen that the water quality index is an important cumulative indicator of pollution and this can be evaluated, assessed, and mintored through the GIS technique.

13.4.1 Evaluation of Water Quality

Total dissolved solids are one of the important water quality parameters, which indicate the amount of the dissolved solids in the water samples. The total dissolved solids, as already mentioned, can be categorised into two kinds as fresh water and brackish water. When the total dissolved solids are greater than 1000, the water is known as brackish water. This area falls under the Category II of the brackish water. Therefore the area of Sivarampally falls under the brackish water category. An excessive fluoride level increases the occurrence and severity of dental fluorosis, while decreasing the incidence of decaying and missing of filled teeth. Table 13.4 shows the variation of ***fluoride*** for the period 1996 to 2000,

Table 13.4 The variation of fluorides during 1996-2000

Name	Year	Fluoride (ppm)	Category
I Katedhan	1996	1.5	Very good
	1997	2.0	Tolerable
	2000	0.82	Good
I Sivarampally	1996	0.75	Good
	1997	1.25	Very good
	2000	1.91	Tolerable

The fluoride content has increased gradually from 0.75 to 1.91, from 1996 to 2000 in the Sivarampally area. The fluorides content has increased from 1.5 to 2.0 from 1996 to 2000 and there is a decrease from 1997 to 2000, from 2.0 to 0.82 in the Katedhan industrial area.

13.5 Correlation Between Water Quality and Groundwater Level

Water quality of an area is related to the level of the groundwater table. The higher the depth of the water table below the ground level, the less is the corresponding change to environmental degradation. It is observed from the water table level and water quality index of the study area that the lower the water quality index, the safer are the corresponding areas. The areas, which are safe during one season, may change to other category when the conditions of the water table depth are changed. Taking this into consideration the ratio map of water quality index to water table of the study areas that are presently needed and are produced, can act as an instrument for decision making process. Table 13.5 shows the variation of water qualiy.

Table 13.5 The variation of Water Quality Index during 1996-2000

Name	Year	Water Quality Index	Category
I Katedhan	1996	159.56	Moderately Polluted
	1997	215.68	Excessively Polluted
	2000	12.68	Tolerable
Sivarampally	1996	217.78	Excessively Polluted
	1997	163.96	Moderately Polluted
	2000	176.29	Moderately Polluted

13.6 Correlation Between Water Quality Index and Land Use

Water quality of an area is related to the Landuse of that area. The more the landuse, the more will be the area under environmental degradation. It is observed from the landuse/landcover map (Plate 13.1) and Water Quality Index Map that the water quality will deteriorate according to the landuse. The wastelands have less water quality Index compared to the industrial area. The industrial area has less water quality Index compared to waste disposal site. Therefore the landuse plays the major role on the groundwater quality. In the present study area, the water quality varies from a tolerable to moderately polluted. The area where it is moderately polluted may due to industrial location. The Industrial area is situated on gently sloping and the other area is nearly sloping, therefore the effluents reach to the nearly sloping area from gently sloping area, where the more water quality Index is reported.

Based on the calculated values of quality index, the impact of the areas under environmental stress is identified. The areas are shown under environmental stress with respect to water quality. Groundwater invariably contains dissolved solids since the rain water percolating through the soils strata dissolve salts like chlorides, fluorides, nitrates. The deeper the water percolates, the higher will be the dissolved solids in groundwater. Technically it is easier to remove suspended solids than dissolved solids either physically or chemically. Removal of dissolved solids has been tried with coagulation, ion exchange, reverse osmosis and electrodyalisis.

The analysis of results of this study is revealed that the analysis of

- Decrease surface water usage and wastewater generation through the promotion of water conservation, and wastewater treatment, and reuse.
- Use of constructed wetlands to control non-point sources.
- Use of techniques to decrease soil erosion by planting rapidly growing plants and various types of grasses.
- Rotation of landuse practices in the area to facilitate natural recovery without the continuing stress related to any one landuse practices.
- If groundwater usage decreases then the impact will naturally lessen, so disposal sites should not be located near potential groundwater resource areas.

Industries should have their own effluent treatment plants to treat the effluents according to the effluent disposal standards and then release them into the drain. Industries can even lay the pipeline to drain the treated effluent into Musi river located North of the study area. The North side of the study area is nearly sloping compared to south side, which is gently sloping. The Government may construct the Common Effluent Treatment Plant (CETP) in Katedhan industrial area. The construction of common effluent treatment plant will be economic since there are many more large scale Industries than small scale Industries in this region.

13.7 Groundwater Quality Studies using SPANS

The spreading of water borne diseases with drinking water of a poor quality is a massive problem in today's world, particularly in developing countries like India. The densely populated areas with limited access to water and sanitation are increasing due to industrialisation and urbanisation. Many problems in developing countries occur because of the disposal of huge quantities of urban wastewater and industrial effluents on the ground surface. Elements, such as, lead and cadmium in groundwater have recently received special attention as indicators of pollution. Generally these elements occur in a very low concentration and are also below the detectable levels in the groundwater, because of their occurrence and the nature of geochemical behavior. Consuming water with lead and cadmium can cause serious health hazards. The latest innovative technologies like Remote Sensing and Geographical Information Systems are used for the evaluation of the impact of lead and cadmium as sources of pollution. (Madhusudan Reddy, 2000)

An attempt is made to study the impact of landuse/landcover of Visakhapatnam area on groundwater quality using Remote Sensing data and its integration with Geographical Information Systems (GIS). The groundwater quality parameters selected are lead and cadmium. The land use/land cover map is prepared based on the interpretation of Satellite (IRS-1C) imagery and Survey of India (SOI) toposheet. Thirty groundwater-sampling locations were selected on the basis of the existing landuse/landcover patterns keeping the possible sources of groundwater pollution. A map showing the locations of all thirty water samples is prepared using SPAN-GIS software. Based on the analysis of groundwater quality data and its correlation with land use/land cover patterns, the source of pollution and the areal extent of its contamination are mapped using the concept of buffering technique using SPANS. The correlation study of sources of pollution and contaminated areas with the chemistry of industrial effluents revealed that the zinc smelter causes pollution of groundwater with cadmium. The cursory examination of the highly polluted points and the corresponding land use/land cover revealed that the Hindustan polymers, zinc smelter, heavy traffic zones, industrial development corporation areas are causing the groundwater pollution with lead. It is also observed that some of the industries located in this area are discharging their waste directly onto the ground without proper treatment. The concentrations of lead and cadmium in the groundwater influenced by the industries are higher than the maximum permissible limits for drinking purpose. Prominent health hazards observed in the study area are nervousness, and heart, skin, gastrointestinal, and respiratory disorders and miscarriage. Keeping these health hazards in view it is suggested that there is the need for regular monitoring of chemical quality of groundwater and status of health for implementing the necessary remedial measures to improve environment.

13.7.1 Case Study : Visakhapatnam City

Visakhapatnam (lat. 17^0 41' 30" and 17^0 45' N and long. 83^0 21' E) located on the East Coast of India (Fig. 13.7) and was almost quiet until two decades ago, has been growing by leaps and bounds into a sprawling urban complex humming with industrial activity. Though urbanisation and industrialisation in the Visakhapatnam area have been progressing at a very fast rate, the primary needs of proper legislation programs, such as, physical planning, landuse, water resource management, waste disposal, and basic public health, infrastructure have not been proportionately increased. It leads to deterioration of groundwater quality in some areas. Earlier study on the concentration of trace elements such as lead and cadmium did not detect these elements.

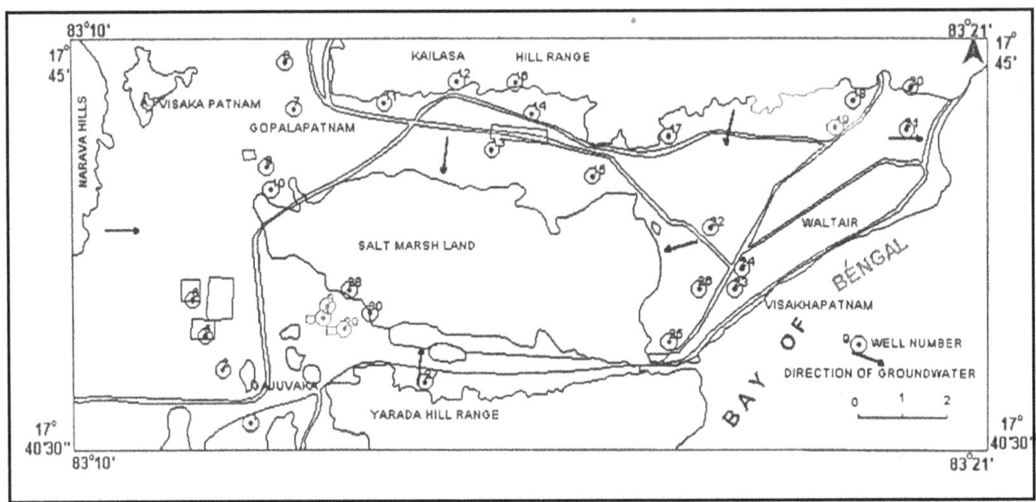

Fig. 13.7 Map showing TDS variation in 2000.

From the groundwater point of view, the rocks in the area are referred to as "crystallines". They are weathered and fractured. Generally, thickness of soil, and weathered and fractured zones ranges from 1 to 8 m, 1 to 18 m and 14 to 60 m with an average of 2 m, 11 m and 35 m in depth, respectively, followed by unfractured rock zone. Groundwater exists both in the clayey weathered zone at shallow depth and in the fractured zone at deeper levels. Dugwells extend to a depth of 2m to 14m below ground level (bg*l*) with an average of 9 m. Water levels range between 1 m and 12 m bg*l* and average 7m. In case of borewells, the depth ranges from 20 to 60 bg*l* and the potentiometric surface is between 3 m and 25 m bg*l* with an average of 24 m and 9m, respectively. The direction of groundwater flow is generally towards the marshland, located centrally from surrounding high lands. In the north costal part of the area, however, the flow is towards the east and south-east. Topography exerts a dominant control over the movement of groundwater.

13.7.2 Methodology

The entire study is carried out in the following steps.

(i) Interpretation of satellite data : IRS-1C LISS III data are obtained from NRSA, India, and Survey of India toposheet No.: 65 0/2 and 0/6 interpreted visually for mapping existing landuse/landcover patterns. A map of landuse/landcover is prepared and scanned using Abakos AO colour scanner. The scanned data is then digitised with the help of AutoCAD - 2000 software and edited to control the quality of spatial patterns is created.

(ii) Thirty groundwater samples are collected from dugwells in different parts of the study area (Fig. 13. 9). They are analysed for lead and cadmium detection. The limits of lead and cadmium are 0.% X 10.4 and 0.3 X 10.5 mg/L respectively. The sampling locations are mapped using AutoCAD 2000 software and SPANS.

(iii) Spatial maps for groundwater areas affected with lead and cadmium are prepared using buffer zoning concept of SPANS. (Fig. 13.8(a) & (b)).

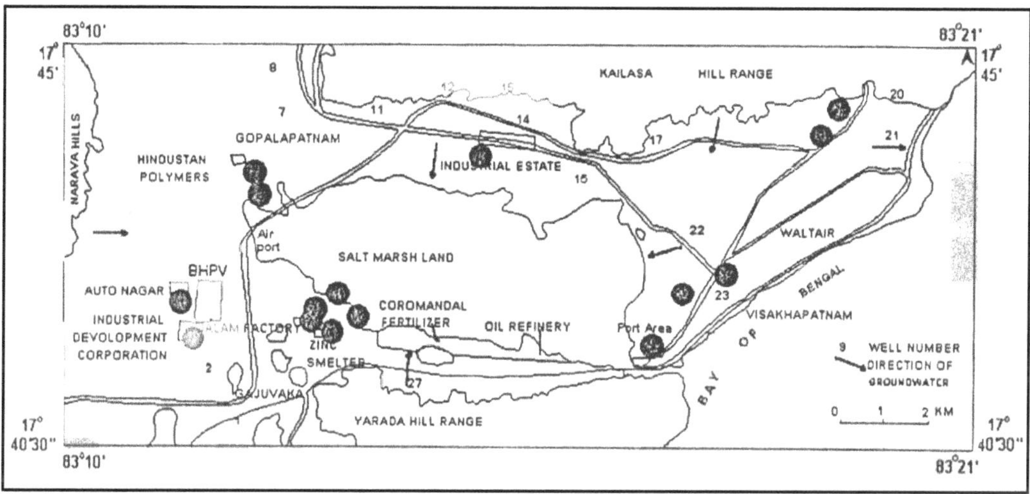

Fig. 13.8(a) Ground Water Affected Areas with Lead

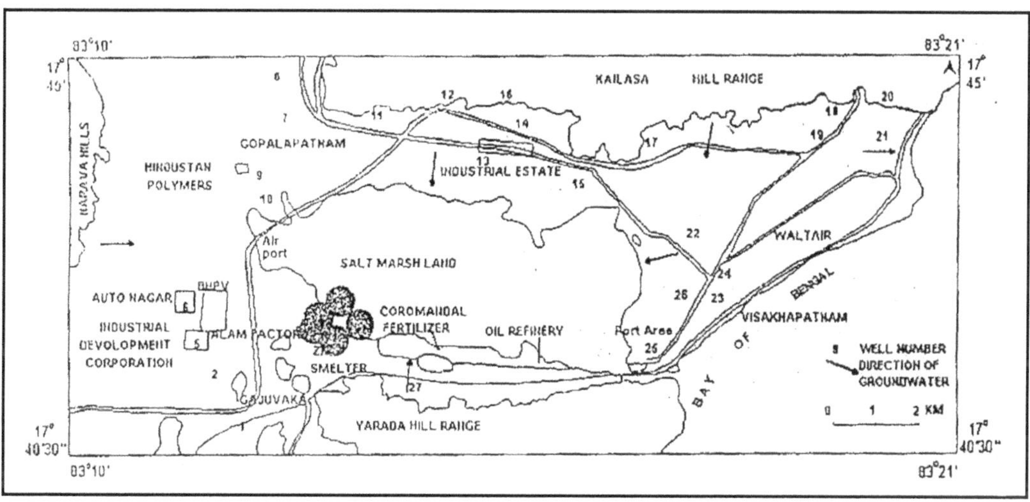

Fig. 13.8(b) Ground Water Affected Areas with Cadmium

(iv) The polluted sampling locations are transferred to landuse/landcover map and prepared an overlaid map (Fig. 13.9) to identify the sources of pollution by using SPAN-GIS techniques.

(v) Petrochemistry of the study is also determined in order to correlate with groundwater chemistry. Published data of industrial effluent chemistry are used to confirm the sources of pollution.

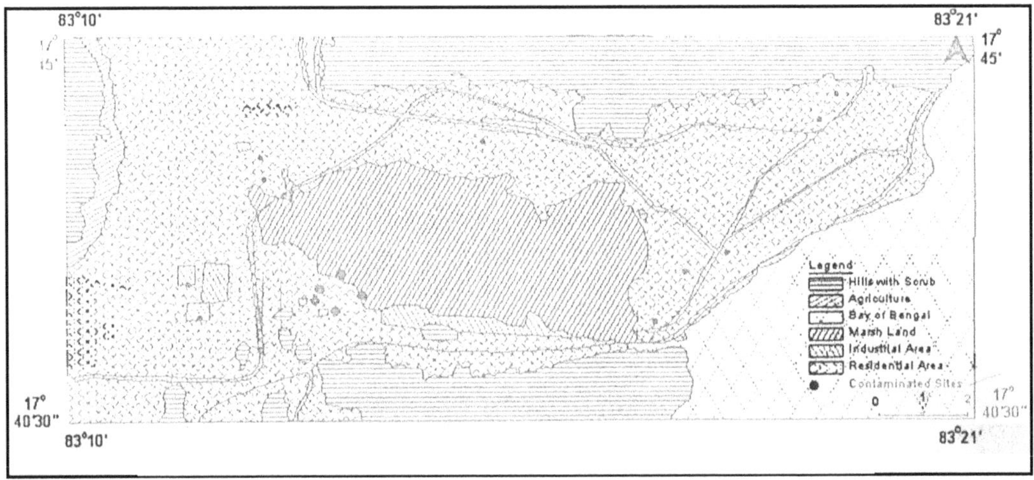

Fig. 13.9 Land use/Land cover with contaminated sites

13.7.3 GIS Data Analysis and Output

Table 13.6 shows the concentrations of lead and cadmium in the groundwater. It is clear from this table that lead and cadmium in the groundwater of the area, except in some locations, occur in very low concentrations and are also below the detectable levels in the groundwater because of (a) their rare occurrence and (b) different geochemical behavior (solubility and mobility). The correlation of groundwater chemistry with petrochemistry has no significance because the rocks of the study area are showing lead and cadmium concentrations below the detectable limits. The results of this study indicate considerable deterioration in the groundwater quality of well waters in the study area due to the influence of Hindustan Polymers in Gopalapatnam and the Zinc Smelter-Alum factory marshland. Thus the enrichment of chemical constituents follows the chemistry of industrial effluents. As the Coromandel Fertilizers and Oil Refinery are discharging their effluents directly onto salt marshland, their impact on groundwater quality is nil. This is due to their location adjacent to salt marshland.

Table 13.6 Lead and Cadmium in groudwaters of Visakhapatnam area

Well Nos. As shown in Fig. 2	Lead mg./L	Cadmium mg./L	Polluted by
1.	N.D	N.D	-
2.	N.D	N.D	-
3.	0.050	0.070	zinc Smelter
4.	0.055	0.150	do
5.	0.082	N.D	Industrial Development Corporation
6.	0.071	N.D	Auto Industries
7.	N.D	N.D	-
8.	N.D	N.D	-
9.	0.050	N.D	POLYMERS
10.	0.050	N.D	POLYMERS
11.	N.D	N.D	-
12.	N.D	N.D	-
13.	0.050	N.D	Industrial Estate
14.	N.D	N.D	-
15.	N.D	N.D	-
16.	N.D	N.D	-
17.	N.D	N.D	-
18.	0.020	N.D	Heavy Traffic
19.	0.019	N.D	Heavy Traffic
20.	N.D	N.D	-
21.	N.D	N.D	-
22.	N.D	N.D	-
23.	N.D	N.D	-
24.	0.060	N.D	Heavy Traffic
25.	0.053	N.D	Heavy Traffic
26.	0.090	N.D	Heavy Traffic
27.	N.D	N.D	-
28.	0.190	0.720	Zinc Smelter
29.	0.180	0.490	Zinc Smelter
30.	0.180	0.300	Zinc Smelter

Groundwater near heavy traffic locations in the study area show low concentration of lead ranging from 0.019 to 0.090 mg./l (Table 13.6). It may be attributable to the combustion of gasoline on highways. At industrial locations, groundwater shows high concentrations of lead and cadmium range between 0.050-0.190 and 0.070 mg/l respectively. There are many industries, such as, Zinc Smelter, and those of fertilisers, paints, metal plantings, plastics and storage batteries. Some of them discharge their waste on to the ground without any proper treatment. Hence high concentrations of lead and cadmium could be the result of polluted waters.

According to the World Health Organization (WHO, 1917) and the analysis of results drawn from laboratory techniques, it is observed that, at the sample locations No. 3, 4, 28, 29, and 30 the cadmium levels are above the standards of groundwater levels that is, 0.01 mg/l. To understand the impact of these high levels of pollution, a spatial map the aerial extent of contaminated areas with cadmium is prepared using buffering concept. The maps have clearly depicted the contamination of groundwater quality with respect to lead and cadmium. Therefore, a detailed study is carried out to identify various sources of this contamination using the concept, overlay, of SPAN GIS Software. The polluted sampling locations of lead and cadmium are transferred to the landuse/ landcover patterns and then examined.

The concentrations of lead and cadmium in the groundwater influenced by the industries are higher than the maximum permissible limits recommended for drinking purpose, while the lead is within the safe limit in some locations influenced by the highways. Prominent health hazards observed during this study in the area are nervousness, heart, skin, gastrointestinal and respiratory disorders and miscarriage, the percentage of affected persons increasing in the industrial locations. As per the information of the long-standing residents of these locations, the health hazards have started only after the establishment of these industries. Therefore, this study suggests the need for regular monitoring of the chemical quality of groundwater and status of health, to take necessary remedial measures for clean environment in this area. The specific conclusions of this GIS based study are :

(i) The groundwaters of the study area are influenced by landuse/landcover patterns especially at industrial locations and high traffic areas.

(ii) Lead and cadmium levels are higher than permissible limits for drinking purpose at Mindi and Gopalapatnam because of zinc smelter and Hindustan Polymers.

(iii) Oil Refinery and Coromandal Fertilisers are not showing their impacts on groundwater quality because they are discharging effluents directly on to salt marshland.

(iv) The health hazards are also observed from the polluted areas like nervousness, heart, skin, gastrointestinal and respiratory disorders, and hence regular monitoring of groundwater and status of health of the residents of the study area is necessary to implement remedial measures.

13.8 Evaluaton of Impact of Landuse/Land Cover Changes on Ground water Quality

Anthropogenic activities associated with changing urban sprawl, natural resources and land use changes may invariably result in the depletion of water bodies, deterioration of ground water quality, contamination of drinking water and various health hazards. The present case study deals with the evaluation of the impact of land use changes on the ground water quality of zone-VI under Municipal Corporation of Hyderabad (MCH) by integrating Remote Sensing, GIS and field studies. The total area of MCH is 179 Km^{2}s and divided into 11 zones wherein the area covered by the municipal zone-VI (present study area) is 28.5 Km^{2}s. The study area is situated between 78^{0}22' and 78^{0}28' East Longitudinal and 17^{0}21' and 170^{0}24' North Latitude and is at altitude of 580m above Mean Sea Level (MSL).

IRS-IC PAN and IRS-ID LISS-III satellite imageries obtained from NRSA, Hyderabad are georeferenced using the ground control points with SOI toposheets as a reference and further merged to obtain a fused, resoution (5.8m of PAN) and colored, (R,G,B bands of LISS-III) output is EASI/PACE Image Processing Software. The study area is then delineated and subsetted from the fused data based on the latitude and longitude values and a final hard copy output prepared which is further interpreted for the generation of thematic maps. Two types of databases (spatial and attribute) are created in the present study.

13.9.1 Spatial Database Creation :

Thematic maps like base map, drainage networkm and land use/land cover map (Plate13.2) are prepared from the satellite imagery and SOI toposheets on 1:25,000 scale using AutoCAD and Arc/Info GIS software to obtain a baseline data. A base map is generated consisting of various features like major road network, railway tracks, settlements, water bodies, canals and their drainage pattersn, vegetation etc., which are delineated from the toposheet. Land use/Land cover map of the study area was prepared using visual interpretaiton technique from the fused satellite imagery (IRS-IC PAN + IRS-ID LISS-III) and SOI toposheets along with ground truth analysis.

13.9.2 Attribute Database Creation

Fieldwork was conducted and ground water samples were collected from 36 predetermined locations based on the land use and drainage network maps in the study area. The water samples were analyzed for various physico-chemical parameters adopting standard protocols. The water quality data thus obtained forms the attribute database for the present study.

13.9.3 Integration of Spatial and Attribute Database

The spatial and the attribute database generated are integrated for the generation of spatial distribution maps of selected water quality parameters like pH, Alkalinity, Chlorides, Sulphates, Nitrates, TDS, Total hardness and Fluorides. The water quality data (attribute) is linked to the sampling location (spatial) in Arc Info and maps (Plate 13.3) showing spatial distribution are prepared to easily identify the variation in concentrations of the above parameters in the groundwater at various locations of the study area using curve fitting technique of Arc/View GIS software.

Plate 13.4 shows the map showing the spatial distribution of groundwater quality index of 11 municipal zones of Hyderabad.

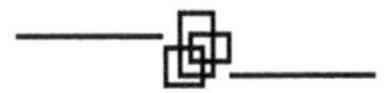

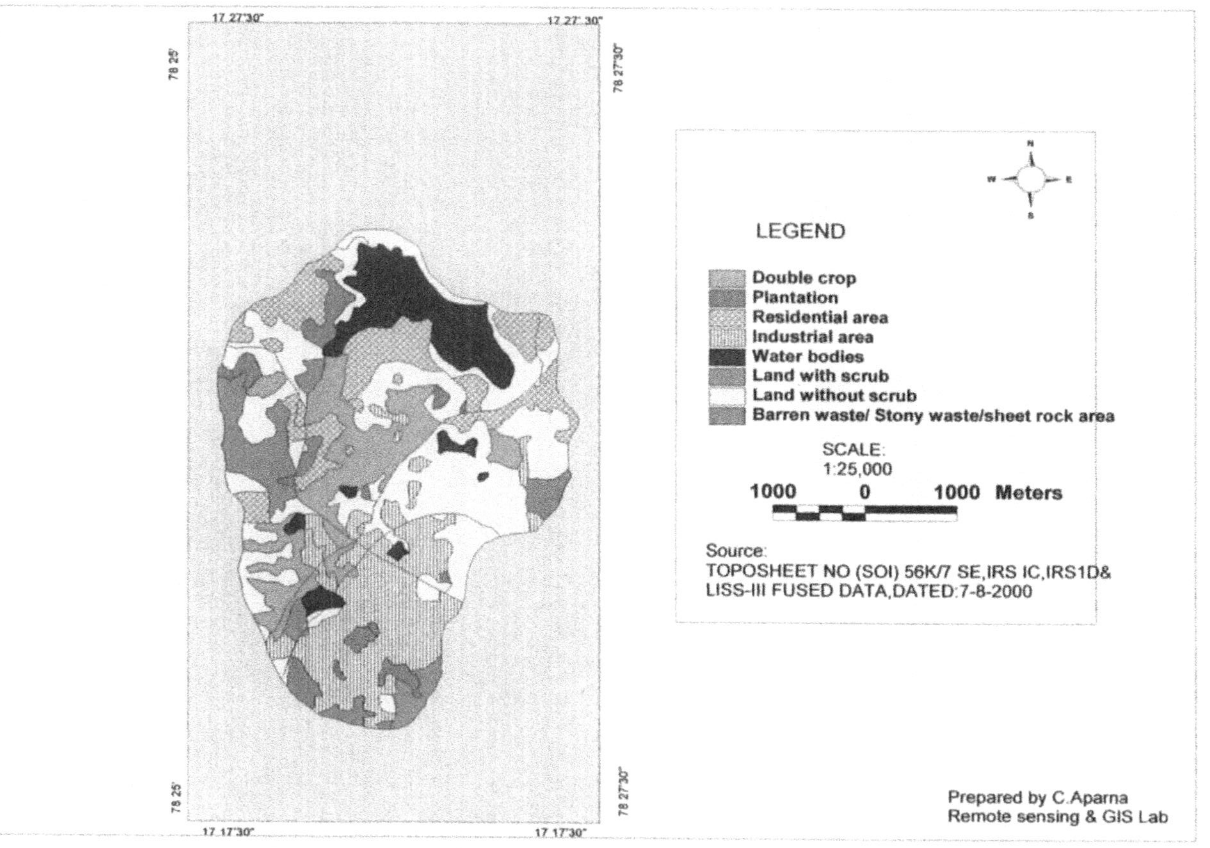

Plate 13.1 Map showing Land use / Land cover of Kattedan Industrial Area

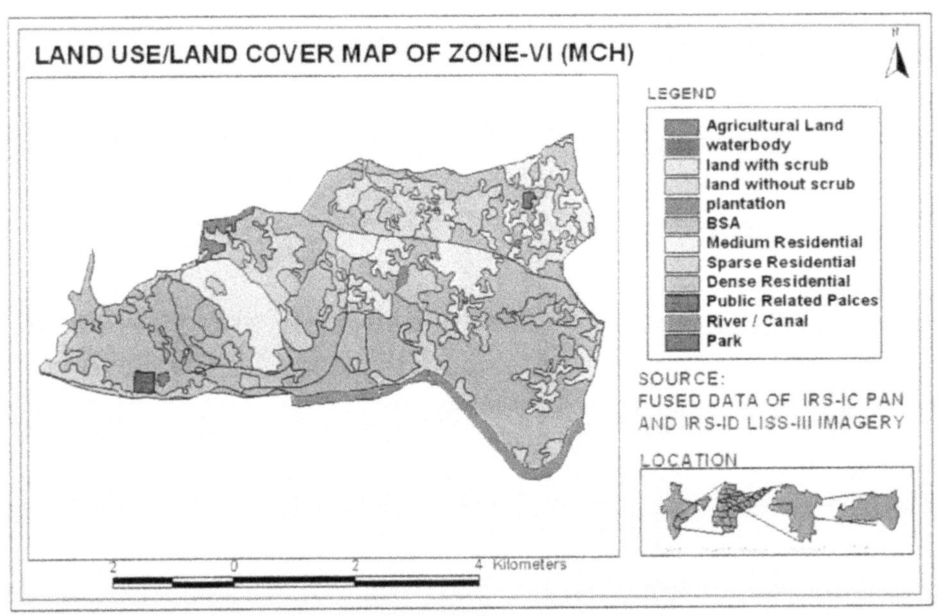

DISTRIBUTION OF LAND USE/LAND COVER IN ZONE-VI OF MCH

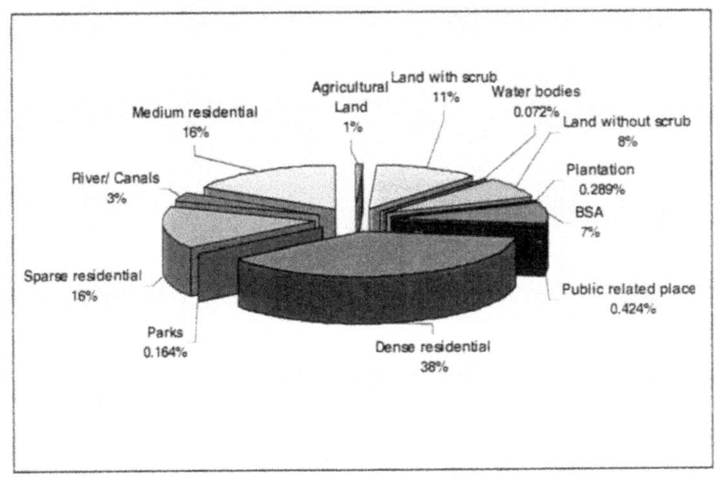

Plate 13.2

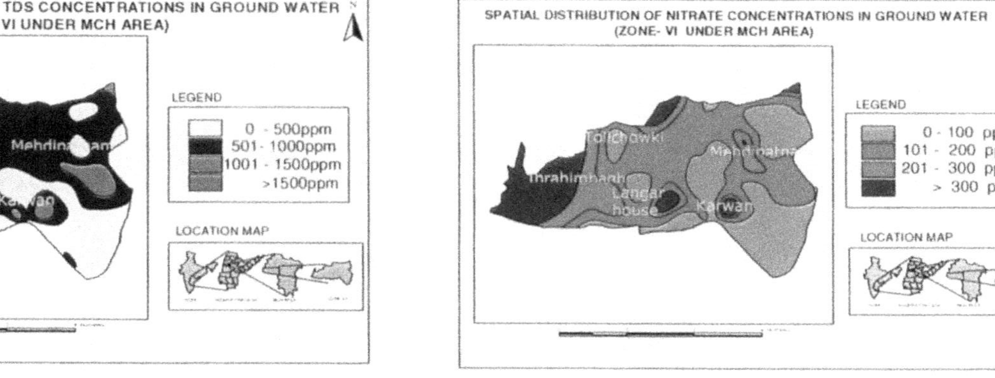
SPATIAL DISTRIBUTION OF TDS CONCENTRATIONS IN GROUND WATER
(ZONE- VI UNDER MCH AREA)
LEGEND
0 - 500ppm
501 - 1000ppm
1001 - 1500ppm
>1500ppm
LOCATION MAP
Langar house
Karwan
SPATIAL DISTRIBUTION OF NITRATE CONCENTRATIONS IN GROUND WATER
(ZONE- VI UNDER MCH AREA)
LEGEND
0 - 100 ppm
101 - 200 ppm
201 - 300 ppm
> 300 ppm
LOCATION MAP
Tolichowki
Langar house
Karwan

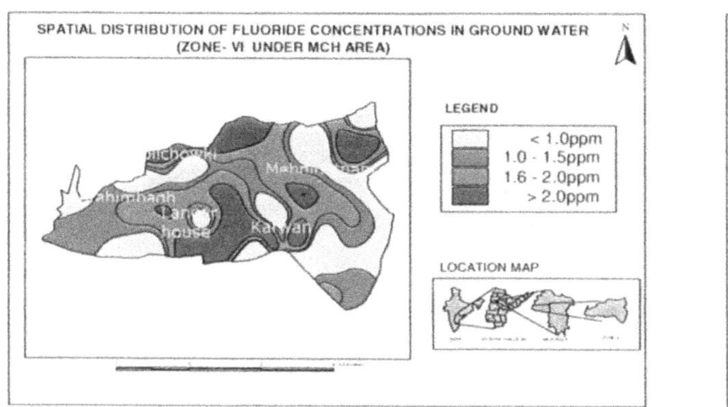
SPATIAL DISTRIBUTION OF FLUORIDE CONCENTRATIONS IN GROUND WATER
(ZONE- VI UNDER MCH AREA)
LEGEND
< 1.0ppm
1.0 - 1.5ppm
1.6 - 2.0ppm
> 2.0ppm
LOCATION MAP

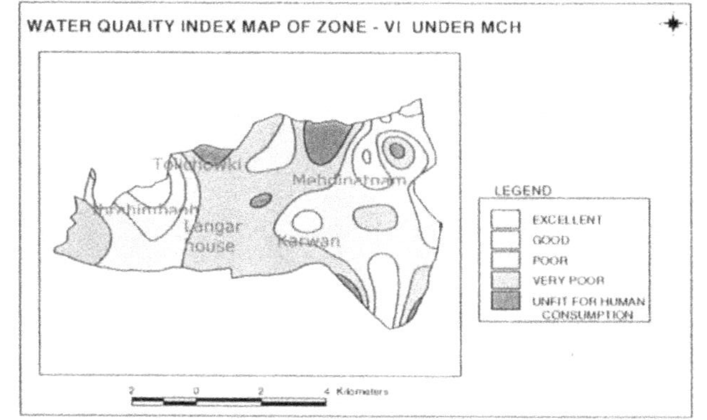
WATER QUALITY INDEX MAP OF ZONE - VI UNDER MCH
LEGEND
EXCELLENT
GOOD
POOR
VERY POOR
UNFIT FOR HUMAN CONSUMPTION
Tolichowki
Mehdinatnam
Langar house
Karwan
2 0 2 4 Kilometers

Plate 13.3

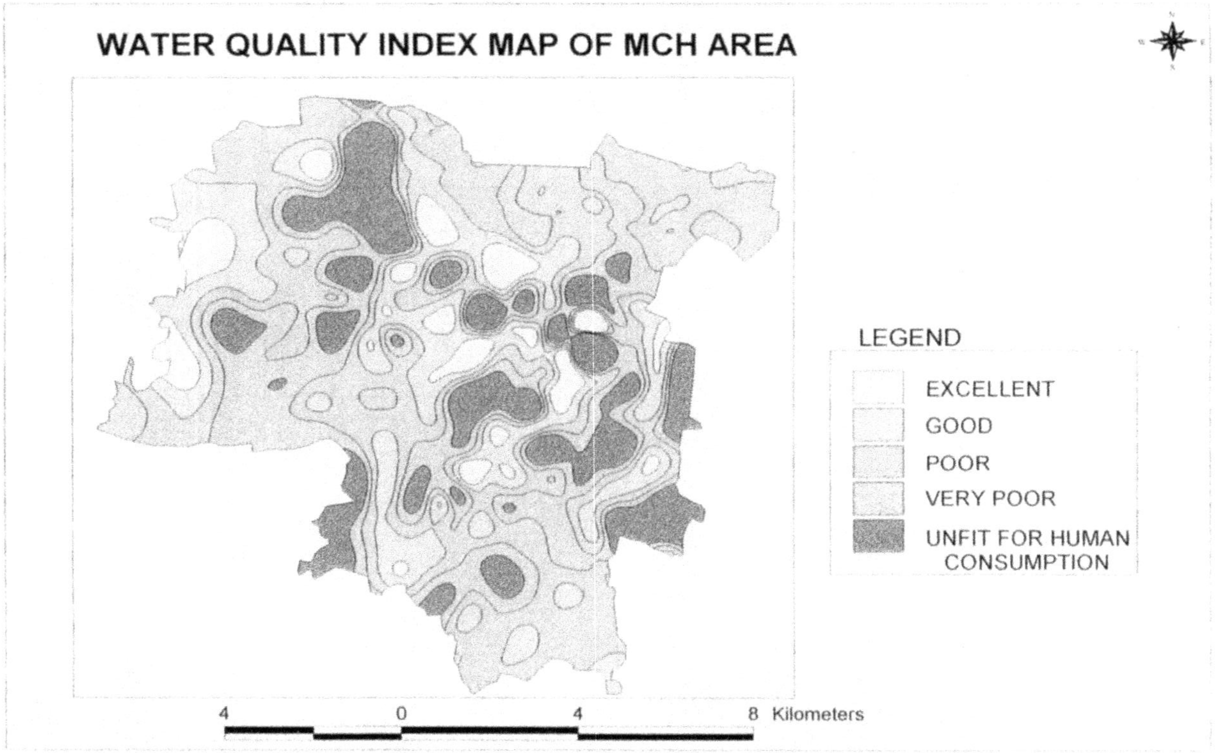

WATER QUALITY INDEX MAP OF MCH AREA

LEGEND

EXCELLENT
GOOD
POOR
VERY POOR
UNFIT FOR HUMAN CONSUMPTION

4 0 4 8 Kilometers

Plate 13.4

14
Solid Waste Management

14.1 Introduction

"Any unwanted or discarded material from residential, commercial, industrial, mining and agricultural activities that cause environmental problems may be termed as solid waste". The term 'solid waste management' is used to describe the collection, storage, transport, treatment, recovery and disposal of solid waste materials, keeping in view the economics, public health and the environment. Most of the municipal solid waste generated in India are disposed in open dumps. Generally, the low-lying areas and out skirts of the towns and cities are used for the purpose. The open dumps are, however, a menace to the ambient environment. They become source of objectionable smoke and odours and serve as breeding grounds for flies and mosquitoes. Identification of sites for solid waste disposal, thus become as a management issue and as such the selection should be based on a number of considerations. The selection of disposal site involves a multi disciplinary approach and broad spectrum of considerations.

There has been a significant increase in Municipal Solid Waste (MSW) generation in media in the last few decades. This is largely because of rapid population growth and economic development in the country.

14.2 Classification of Solid Waste

The solid waste is categorised in different types based on the sources are:

(a) Municipal solid waste
(b) Hazardous and Industrial Waste
(c) Medical Waste
(d) Putrescible Waste

14.2.1 Municipal Solid Waste

Wastes generated from household sources - may include hazardous or putrescible waste. The municipal solid waste mainly consists of food wastes, rubbish , ashes and residues, demolition and construction waste, special wastes, treatment plant wastes. The municipal solid waste mainly consists of food wastes, rubbish, ashes and residues, demolition and construction wastes, special wastes, treatment plant wastes. Type of Solid Waste Components included in it are :

Food Waste	The animal, fruit or vegetable residues resulting from handling preparation, cooking, eating. Because of food wastes are putrescible, they decomposed rapidly in warm weather.
Rubbish	Combustible and Non-combustible solid wastes such as paper, plastics, textiles, rubber, leather, non-combustible consists of items such as glass, crockery, constructional waste.
Ashes and Residues	Materials remain after burning of wood, coal, coke and other combustible wastes. Residues from power plants.
	This includes dirt, stones, bricks, plasters and plumbing, heating and electric parts.
Demolition and Constructional wastes	Wastes from sweepings, roadside litter, debris, dead animals and abandoned vehicles.
Treatment plant wastes	The solid and semi solid waste from water, waste water and industrial waste treatment facilities.

14.2.2 Hazardous Wastes

Any waste containing significant quantities of a substance which may present a danger to the life or health of living organisms when released into the environment is hazardous waste. These wastes may include both medical and radioactive wastes.

Industrial and hospital waste is considered hazardous as they may contain toxic substances. Certain types of household waste are also hazardous. Hazardous wastes could be highly toxic to humans, animals and plants; are corrosive, highly inflammable, or explosive; and react when exposed to certain things eg.,Gases. India generates around 7 million tonnes of hazardous wastes every year, most of which is concentrated in four states: Andhra Pradesh, Tamilnadu, Bihar and UttarPradesh. Household waste that can be categorized as hazardous waste include old batteries, shoe polish, paint tins, old medicines and medicine bottles.

Hospital waste contaminated by chemicals used in hospitals is considered hazardous. These chemicals include formaldehyde and phenols, which are used as disinfectants and mercury which is used in thermometers or equipment that measure blood pressure. Most of the hospitals in India do not have proper disposal facilities for these hazardous wastes. In the industrial sector the major generators of hazardous waste are the metal, chemical, paper, pesticide, dye, refining, and rubber goods industries. The wastes which comes from the industries when the byproduct or raw materials are converted into society acceptable materials. During this process the waste materials are released from the industries, they are of two types, namely, non-hazardous waste materials and hazardous waste materials.

14.2.3 Medical waste

Medical waste consisting of a needle, syringe with needle, surgical instrument or other article that is discarded in the course of medical, dental or veterinary practice or research and has a sharp edge or point capable of inflicting a penetrating injury on a person. Coming into contact with it; human tissue, bone, organ, body part or fetus; a vessel, bag or tube containing a liquid body substance; an animal carcass discarded in the course of veterinary research or medical practice or research; a specimen or culture discarded in the course of medical, dental or veterinary practice or research and any material that has come into contact with such a specimen or culture; or any other article or matter that is discarded in the course of medical, dental or veterinary practice or research and that poses a significant risk to the health of a person who comes into contact with it.

Hospital waste is generated during the diagnosis, treatment, or immunization of human beings or animals or in research activities in these fields or in the production or testing of biologicals. It may include waste like sharps, soil waste, disposables, anotomical wastes, cultures, discarded medicines, chemical wastes, etc. these are in the form of disposable syringes, swabs, bandages, body fluids, human excreta etc. This waste is highly infectious and can be a serious threat to human health if not managed in a scientific and discriminate manner. It has been roughly estimated that of the 4 kg of waste generated in a hospital at least 1 kg would be infected.

It includes the bio-medical waste which is generated during the diagnosis, treatment of immunization of human beings or animals or in research activities testing of biologicals during production.

14.3 Environmental Problems related with the Municipal Solid Waste

Solid wastes are all the wastes arising from human and animal activities that are normally solid and that are discarded as useless or unwanted. In early times, the disposal of human and the other wastes did not pose a significant problem. As the population was small and the amount of land available for the assimilation of waste was large. Now a days, we speak of recycling the energy and fertilizer value of solid waste, but the farmer in ancient times probably made a bolder attempt at this. Indication of recycling may still be seen in the premative at sensible, agricultural practices in many of the developing nations where farmers recycle solid waste for fuel or fertilizer value. The practice of throwing waste into the unpaved streets, roads and vacant lands led to the breeding of rats, with their attendant fleas carrying the germs of diseases and the out break of plague.

14.3.1 Vector of Disease

The lack of any plan for the management of sold waste led to the epidemic of plague, the black death. It was not until 19th Century that Public Health Control measures became a vital consideration and disposal of in a sanitary manner to control the vector of disease. The relation between public health and improper storage, collection and disposal of solid waste is quite clear as public health authorities have identified that rats, flies, and other disease vectors breed in open dumps as well as in poorly constructed or poorly maintained housing, in food storage facilities and in many other places. The improper disposal and management of solid municipal waste will cause 22 human diseases. It was published by the US Public Health Service study results. Data are also available to show that the illness, accident rate for sanitation workers engaged in the collection and disposal of solid waste is several times higher than for industrial employees.

14.3.2 Ecological Impacts

Ecological impacts such as water and air pollution have been attributed to proper management of solid waste, for instance, liquid from dumps and poorly engineered land fills has contaminated surface water and groundwaters in mining areas. The leached from waste dumps may content toxic elements such as copper, arsenic and uranium and may contaminate water supplies with unwanted salts of calcium and magnesium. Although residential solid waste accounts for a very small proportion of total waste generated, they are perhaps the most important because the generated in areas with limited storage space. As a result they can have significant public health.

Public Health concerns or related primarily to the infectious areas used for the storage of solid waste with vermin and insects. They often serves as potential reservoirs of disease. By far the most effective control measure for both rats and flies is proper sanitation. Typically, this involves the use of containers with tight lids, the period of washing of the containers as well as the storage areas, and the period removal of biodegradable materials (usually less than 8 days) which is especially important in areas with warm climates. Asthetic considerations are related to the production of odorous and the unslightly conditions that can develop when adequate attention is not given to the maintenance of sanitary conditions.

14.3.3 Brown agenda

Population increase and migration into the cities have created some of the most serious environmental problems associated with what is often characterized as the "brown agenda". This agenda lists the main shortcomings, such as lack of adequate waste management, lack of safety measures for mineral pollution control.

In more recent years there have been important changes in attitudes among stockholders such as local government NGOs/CBOs and citizens in the collection and processing of waste specifically in India.

14.4 Characteristics of Solid Wastes

Physico-chemical characteristics of municipal solid waste (MSW) depends upon various factors such as topography, season, social customs, economic status and food habits etc. The quality and quantity of municipal solid waste are the basic data for its proper management. This basic data will also helps in deciding the mode of processing and recovery and recycling.

The solid waste management in Indian megacities consists of biodegradable component, paper, plastics, metals and crockery, bio-resistant and inert materials. The biodegradable components mainly comes from households, markets, restaurants and hotels. The fraction of this components generated ranges between 30 - 45%, however 50% of solid waste generated in hotels is biodegradable, industrial areas will produce about 13-16% of biodegradable of the total waste generated, the paper and waste product varies between 5 -10% of the total municipal solid waste. (Table 14.1)

Table 14.1 Sources of solid waste and percent generation.

Source of solid waste	Percentage of their generation
Total MCH	30 – 45%
Hotels	50%
Industries	13 – 16%
Industries	8.5%
Commercial areas	7.4%
Hotels	12%
Hospitals	10%

The main produce of the paper waste is from industrial and commercial areas, hotels and hospitals the generate about 8.5, 7.4, 12 & 10% of their total solid waste generated. Metals and Plastics are mainly received from industrial and commercial areas. Bio-resistant materials includes leather, rubber and other synthetic materials and are of the order of 10-25% of the total waste generation. Industrial and commercial areas top in the list of largest produce of bio-resistant materials, above 35% of the solid waste received at the land filled site is inert. This consists of silt ash and constructional debris.

14.4.1 Physical characteristics of solid waste

(i) Size of components
(ii) Moisture content
(iii) Density

14.4.2 Chemical characteristics of solid waste

(i) Proximate analysis (Moisture, ash content, volatile matter, fixed carbon)
(ii) Fusing point of ash (Initial deformation temp., softening temp., Hemisperical temp., fluid temp.
(iii) Ultimate analysis (Carbon, Hydrogen, Nitrogen, Sulphur, Oxygen and Mineral matter)
(iv) Heating value

14.5 Solid Waste Collection System

Collection is the first fundamental function of solid waste management. Solid waste collection refers to the gathering of solid waste from place to place such as residential, commercial, industrial and institutional areas as well as public parks. There are generally two methods of collection namely Hauled container system and Stationery container system

(i) ***Hauled container system***

In this system the container is hauled from the collection point to the final point of disposal, processing facility or transfer station.

(ii) ***Stationary container system***

In this system the container is emptied into the collection vehicles at the point of collection.

14.5.1 Layout of collection Route

To manage a solid waste collection system effectively, the community, industrial park and the like should be divided into collection districts. In which a number of collection Routes must be laid out.

Laying out of Routes should follow a minimum of guidelines, among which could be the following.

(i) Route should not overlap and should be continuous.
(ii) On the job time of the collection Routes should be made equal as possible.
(iii) Congested areas should not be collected during rush hours.
(iv) To service a dead end street, collect it when it is on the right of the main street, to avoid making a left turn.
(v) As practicable as possible, collect a street on both sides.
(vi) As practicable as possible, select a consistent collection pattern.
(vii) For collection on steep streets, it is best for the vehicle to be collecting while it is moving down slope rather than up slope.

This is safer and will save energy and time.

14.5.2 Solid waste transfer stations and transportation

As the distance from the collection system to the processing facility or disposal site increases, the cost of hauling or transportation also increases. There will be eventually a certain transport distance, where management must decide whether or not a transfer station is to be built.

A transfer station is a facility where the wastes collected may be stored temporarily or transferred from the smaller collection vehicles to bigger transport vehicles for transportation to the destination point. There are two general types of transportations namely Direct discharge transfer and Storage transfer station.

(i) Direct discharge transfer : In this collection vehicles dump their loads directly into the larger transportation vehicles.

(ii) Storage transfer station : In this the solid wastes are emptied into storage pits or platforms. The wastes are then loaded into big transport vehicles for hauling to the destination point.

14.6 Factors Affecting the Solid Waste Generation Rates

The factors which effects the quality and quantity of solid waste generated includes.

(i) Geographic location
(ii) Seasonal factors
(iii) Collection frequency
(iv) Use of waste grinders
(v) Characteristics of population (standard of living their cultural habits)
(vi) Extent of salvaging and recycling
(vii) Public attitudes
(viii) Legislation

The industrial solid waste may cause acid mine drainage, ground water pollution due to leaching of toxic metals, loss of fertile top soil through erosion from OB, dumps, land damage through open cast operations and subsidence due to underground mining, consequently causing air, water and soil pollution.

14.7 Solid Waste Sampling Techniques

The technique is recommended to assess the individual components within a given domestic waste:

1. Unload a truckload of wastes in a controlled area away from other operations
2. Quarter the waste load.
3. Select one of the quarters and quarter that quarter.
4. Select one of the quartered quarters and separate all of the individual components of the waste into pre-selected components.
5. Place the separated components in a container of known volume and tare mass and measure the volume and mass of each component. The separated components should be compacted tightly to simulate the conditions in the storage containers from which they were collected.
6. Determine the percentage distribution of each component by mass and the discarded density.

Typically, from 100 to 200 kg (200 to 400 lb) of waste should be sorted to obtain a representative sample. To obtain a more representative distribution of components, samples should be collected during each season of the year. Clearly, no matter how many samples are analysed, common sense is needed in selecting the loads to be sorted, in analyzing the data, and in preparing projections.

14.7.1 Management Practices

The management procedure should consists of:

1. Reduction in MSW quantities
2. Recycling of materials
3. Energy recovery
4. Day-to-day solid waste management
5. Technological interventions
6. Efforts towards institutional and regulatory reforms

14.7.2 Targeting Waste Reduction at Source

Waste reduction can be accomplished in 3 ways:

1. Fees and tax incentives to promote market mechanisms to effect source reduction.
2. Mandatory standards and regulations: The seller of consumer goods must take back all the package waste that is produced. This is implemented in Germany. In India the regulatory agencies should take the lead in setting up rules prescribing targets for waste reduction in various manufacturing sectors.
3. Education and voluntary complains with policies by business and consumers.

Recycling of materials : About 45% of the component in MSW is organic and 55% inorganic. The materials, such as plastic, glass and metals can be recycled and re-used.

Energy recovery : Considering 45% (approx.) organic component present in any municipal solid waste, the possibility of energy recovery is very high. The energy recovery can be in the form of CH_4 (combustible gas) or can be converted into heat.

Day-to-day municipal SWM : Direct activities that must be considered and coordinated on a daily basis include waste-generation rates, on-site storage, collection, transfer and transport, processing and disposal. Indirect activities that are also an important part of a solid waste management programme include financing, operations, equipments, personnel, cost accounting and budgeting, contract administration, ordinances and guidelines and public communications.

Technological inventions:India has lagged behind in terms of adopting technologies for solid management. In particular, collection, treatment and disposal of waste require urgent consideration.

(i) Collection of waste: The preferred option would be to revamp the existing collection service structure to provide community with waste bins, conveniently placed for the people to deposit domestic waste, and door to door collection of waste. This along with separation of waste, at source, into biodegradable & non-biodegradable components would not only reduce the cost of transportation for final disposal but also provide segregated organic waste stock for waste to energy activities.

(ii) Treatment and disposal: Proper segregation of waste would lead to better options and opportunities for its scientific disposal. Recyclables for example, could be straightaway transported to recycling units, which, in turn, would pay the corporations for it, thereby increasing their income. Finally, the inert material that will be required to be sent to landfill would be of much lower quantity compared to unsegregated waste, consequently increasing the life of our existing disposal facilities.

14.8 Statistical Picture

Solid waste management has become a major environmental issue in India. The per capita of MSW generated daily, in media ranges from about 100 gms in small towns to 500 gms in large towns.

In India, around 3 lakhs tonnes of urban solid waste (USW) is generated every day from major urban centre. In Delhi alone, more than 5000 tones of Municipal solid waste is generated every day. It is expected to arise to 12,750 tones per day by 2015. Kolkata, is the world's second most thickly populated metropolis and India's biggest city. The Kolkota Municipal Corporation has an area of 104 Kms., population density over 31,615 Kms., and the total population of Kolkota Municipal Corporation is 43,88,000 according to 1991 census and produces 1800 to 2000 metric tones of solid waste per day (1991 data). Kolkota Metropolitan District cover an area of 1450 Kms., the population according to 1991 census 10.3 Million and produces 4000 metric tones of solid waste. Mumbai, the population of it grew from around 8.2 millions in 1981 to 12.3 million in 1991, registering a growth of around 49%. On the other hand, the MSW generated in the city increased from the 3200 tones per day to 5355 tones per day in that same period, registering a growth of around 65%. This clearly indicates that the growth in MSW in our urban centers has out placed the population growth in recent years. This trend can be ascribed to our changing life style, food habits, change in living standards.

Hyderabad is the 5th major Indian city with a total urban population of more than 5.0 million and produces 1500 to 2000 tonnes of urban solid waste daily. In India, the amount of waste generated per capita is estimated to increase at a rate of 1% to 1.33% annually. Fig.14.1 depicts the rising quantities of MSW from 1997 to 2047 assuming the daily per capita waste generation in 1995 as 0.456 kgs., and the per capita increase in waste generation has 1.33%. The calculated value of daily per capita waste generation in 1997 is 0.468 kgs. Fig. 14.1 is evident that the total waste quantity generated in 2047 would be approximately above 260 million tones. More than 5 times of present level. This enormous increase in solid waste generation will have significant impact in terms of the land required for disposing the waste as well as on methane emission. (Tables 14.2 to 14.6 show the broad spectrum of solid waste generation.

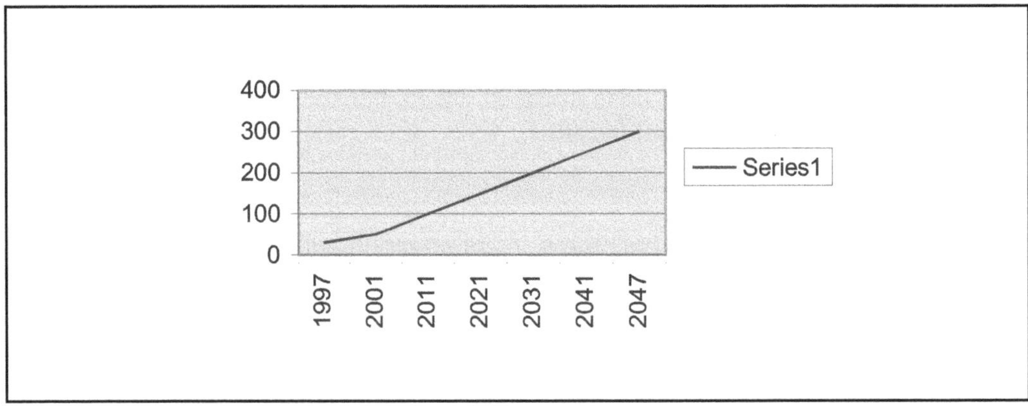

Fig 14.1 Projected trends in the generation of municipal solid waste in million tones/year. according to (BAU) business in the usual scenario.

Table: 14.2 Quantities of solid wastes generated (million tons/year)

Country	Agriculture	Mining	Construction & Demolition	Sewage Sludge	Energy	Industry Production	Municipal
U.K	260	240	3.5	27	13	62	110
U.S.A		1400	31.5	8.4	63	430	133
India		700 - 900	7.2		60		24

Table 14.3 Quantities of per capita solid wastes generated in different countries

Country	Kg/Person/day
India	0.25 to 0.33
Srilanka	0.40
Singapore	0.85
U.K	0.95 to 1.00
Japan	1.12
U.S.A	1.25 to 2.25

Table 14.4 Quantities of Municipal solid waste generated in some Indian cities

City	Tons/day
Ahmedabad	1600
Bangalore	2700
Kolkata	5000
Delhi	4600
Hyderabad	2800
Jaipur	1021
Kanpur	1314
Lucknow	1043
Chennai	3500
Nagpur	1100
Surat	1000

Table 14.5 Generalized composition of solid waste likely to end up at a disposal site

Type of waste	Average %
Paper products	43.8
Food wastes	18.2
Metals	9.1
Glass & ceramics	9.0
Garden wastes	7.9
Rock, dirt & ash	3.7
Plastics, rubber & leather	3.1
Textiles	2.7
Wood	2.5
Total	100.00

Table 14.6 The estimated total waste arising in Hyderabad

Composition	Weight (tons)	Per capita (gms)
Street and Household	2300	265.3
Glass and Iron pieces	40	4.4
Industrial Refuse	310	44.5
Hospitals	30	2.4
Others	120	1.78
Total	280	333.2

14.9 Case Study - Hyderabad

Accumulation of large quantities of solid waste, at urban centers, industrial sites and thermal power plants sites, is an adverse impact on the surrounding environment. The accumulation of such waste posses serious environmental problems. (Plate 14.1). Solid waste has a prevailing characteristic, which sets them apart from liquid and gaseous waste produced by the society. This characteristic is that the waste remains highly visible in the environment where they are disposed. Keeping the importance of proper and effective management of solid waste material, this project case study is designed. In this project, a detailed methodology is proposed to develop using very new and upcoming technology, namely, GIS and Remote Sensing and Analytical Hierarchical Process (AHP). The proposed study area is Hyderabad city and its surroundings. A Decision Support System is developed. In this project, it can be applied as in any Indian city to locate a favorable site for solid waste disposal. This is very much useful for decision makers who are involved in municipal corporation and in urban developmental activities.

The present study will provide a methodology for finding a suitable waste disposal site in any urban locality. It will give some information regarding the existing solid waste disposal site in and around Hyderabad and it will also provide probable sites suitable for waste disposal in Hyderabad.

The study aims to assess the capability and utility of GIS technology for Solid Waste disposal site selection. Satellite data has been used to identify various land use practices and wastelands. Other relevant data has been used in Arc Info GIS for the selection of suitable sites.

In this project, it is proposed to develop a generalized methodology and management technique for selection of suitable site for solid waste disposal. The computer based Decision Support System based on GIS and Remote Sensing techniques will be useful for the decision-makers who are heading the municipal corporation of Indian cities. The infrastructure developed in this program can be used to take up consultancy work for solid waste disposal siting, treatment of different types of industrial effluents, soil analysis for industrial estates etc.

14.9.1 Objectives of the Study

- To study the existing solid waste disposal sites in and around Hyderabad city.
- To find probable solid waste disposal sites in and around Hyderabad city keeping the growing demand in view.
- To create spatial digital database consisting of land use / land cover, drainage, road network maps with the help of IRS-IC PAN & LISS-III merged data & LANDSAT, IRS–1B Survey of India toposheets and ground data on Arc Info GIS platform.
- To evaluate Analytical Hierarchy Process (AHP) technique for its utility in selecting favourable site for disposal of solid waste.
- To generate and study the existing landuse and landcover map from the analysis of Remote Sensing data for identifying probable waste disposal sites with a special reference to the protection of urban environment by conducting site specific analysis.
- To integrate the data derived from different sources with GIS for site suitability and development of a Decision Support System (DSS)
- To provide a generalized methodology for finding a suitable waste disposal sites in any urban locality.

14.9.2 Description of Study Area

The local topography has completely guided the physical growth and the ecological structure of the city. Hyderabad city, which extended in 17^0 22' N latitude and 78^0 27'E and longitude situated on the banks of river Musi. The land surface in the Hyderabad is used for a variety of activities nearly 2/3 of Hyderabad is involved in residential and commercial use, while the rest area comprises of Industrial, semi-natural grass lands, vegetation types. The Hyderabad City according to Municipal Corporation of Hyderabad has been divided in eleven zones. The total extent of the area is 171 sq km.

In Hyderabad, solid waste is a heterogeneous liquid material. The weight/volume ration is 0.8g/cc at the dump yard, and density of fresh garbage is 0.43 tons/m3. The average moisture content of the garbage is 33% whereas the dump yard it is 11.2%. The net calorific value of Hyderabad house hold garbage is 1223 cal/gm. The municipal waste arising in Hyderabad consists combustible paper, plastic bags combustible vegetable/good leaves, non-combustible glass and metal, Incinerate bandages and toxic lead battery cells.

Climatic Conditions

The climate of the city is equitable. The earlier studies on climatological data for a period of 50 years shows that the daily mean maximum temperature varies from a minimum of 11.6^0 C during the month of December to 45^0 C in the month of April. Hyderabad gets the rainfall mainly from the southwest monsoon from the month of June to October, amounting to 125 mm on an average.

Population

According to 1991 census Hyderabad is one of the largest metropolis of India with a population of 4,334,437. The population of Hyderabad has increased from 0.448 million in 1901 to 1.429 million in 1961, between 1981 to 1991 the population went up to 4.34 million and the rate of the growth so far is 67.04%. The total population of the study area comprises, according to 1991 census, is 45 lakhs.

Geology

The city stands on gray and pink granites as foundation materials, which is suitable for building construction. The soils of Hyderabad are mainly brownish sandy and black soil, the northern most bit of the study area starts with a contour of 570 m above mean sea level gradually dipping down southwards up to the river Musi where the lowest contour reaches 450 m above mean sea level can also be said that the gradually falls from west to east almost creating a trough near the river bed on the eastern side of the city. The highest level is on the Jubilee Hills where the contour varies from 590 m to 607.5 m above mean sea levels. These changes in contour and distribution of rocks have their impact on the climate of the region.

Flora and Fauna

Flora : There is no forest area in Hyderabad district. However some scrub jungles and greenery was the home of vasugated flora. Consequent to the general hot and dry climate the flora of the district exhibits xerophytic adaptations. Since the district is urbanized there is no space for developing forests.

Fauna : It was recorded that during the beginning of 19th century, the district with its low scrubby jungles was home of leopards, bears, hyenas and occasionally tigers while in the more open plains the antelope was found in plenty. Hunting game was reserved for the ruling family of erstwhile Hyderabad State exclusively for enjoyment of the nobles.

Due to rapid urbanization, the forests have disappeared in the district. But there is a famous Nehru Zoological Park spread over 123 hectares. This is one of the biggest zoological gardens in the country. The main object of the park was to provide natural habitat to the various species of animals of wild and mild types, secured and reared by the zoo authorities. There are nearly 1100 animals, carnivorous as well as herbivorous and birds in large moated enclosures. The adjoining Mir Alam tank was convened into a charming drive lake for the visitors. Prehistoric animal park, ancient life museum, natural history museum, lion safari park are some of the attractions.

4.9.3 Significance of the study

In early times, the disposal of human and other wastes did not pose significant problem as the population was small and the amount of land available for the assimilation of wastes was large. Problems with the wastes started when humans first began to congregate in tribes, villages and communities and accumulation of wastes became a consequence of life. Improper disposal led to the breeding of rats, with their attendant fleas carrying the germs of disease and the outbreak of plague, the black death that killed half of the Europeans in the 14 century and caused many subsequent epidemics and high death tolls. The very recent incidents associated with the disposal of refuse are September 1994, bubonic plague : 50 persons died in Latur district, Maharastra and October 1994, bubonic plague : 75 persons died in Surat, Gujarat.

The reasons of taking up of this study in the context of existing conditions of solid waste management in Hyderabad City are:

(i) In the Hyderabad city, the existence of city controls on rubbish dumping and burning can be traced back to the 13th century, although it was industrial revolution and health problems. In view of growing population, rapid industrialization and increased urban land use, there is a need to select safe waste disposal sites.

(ii) The potential impacts of waste dumps on the environment are air pollution, surface & ground water pollution and soil pollution. Improper and unscientific disposal of urban solid wastes pollute air, water and soils in several ways.

(iii) One of the critical public health hazards from landfills, is due to the gases emanating from the decomposition of organic matter under partially anaerobic conditions of gases generated. CO can cause severe corrosively problems. Hydrogen sulfide is known as corrosive and when produced in high quantities can cause irritation to birds and animals.

(iv) The urban waste either fresh or composed is being used increasingly in areas adjoining the municipal or urban surroundings for agricultural purposes. Municipal solid waste offer many benefits for application to solids including soil fertility, increased water retention in soils, decreased soil erosion and decreased fertilizer requirements but soils retain the heavy metals in excessive concentrations decreasing crop yield.

14.10 Research Method

The following steps are adopted for selecting a disposal site using remote sensing and GIS.

Step 1 : Generating land use and land cover map from remote sensing data for identifying probable sites.

Step 2 : Studying the existing solid waste disposal sites in and around Hyderabad.

Step 3 : Carrying out site-specific analysis to verify the ground truth.

14.10.1 Data Sources and Products Used

The main source of spatial data to identify the solid waste disposal site is IRS IC – LISS III satellite digital data. The secondary data is drawn from survey of India toposheet and other thematic maps available in other organizations. This land use/ land cover data is then updated using the satellite data of IRS IC. (Plate 14.2). Table 14.7 shows the data products and corresponding organizations from which the required data have been generated.

Table 14.7: Data Products and Data Sources

SI.No	DATA PRODUCT	ORGANISATION FROM WHICH DATA COLLECTED/DERIVED
1.	Satellite data (LISS III of IRS IC)	NRSA, Hyderabad
2.	1:50,000 scale toposheet 56k7, 56k10, 56k11	Survey of India
3.	Other Thematic Maps & land Use/Land cover Map	Derived from satellite digital data

The present study is based on the Analytical Approach and ranking method for finding the best disposal site for dumping the solid waste. In the first part of the work the decision factors are divided into two categories namely exclusionary and non-exclusionary. The factors of exclusionary directly decides the suitability by YES or NO. The conditions of exclusionary category should directly satisfy the location otherwise that location will be discarded.

Non-exclusionary factors need ranking, and each individual factor will have the weightage of suitability. At the end of the assignment the sites are evaluated comparatively. The non-exclusionary weight age will be assigned only to the locations left after the exclusionary screening. Analytical Hierarchy Process (AHP) gives the facilitated way in decision making. The factors are sorted according to their relation and importance to the site of dumping which is called hierarchy of importance.

To derive the decision factors of exclusionary and non-exclusionary and to apply the AHP, the baseline digital database is necessary and this database is created on GIS platform.

14.10.2 Creation of Baseline Digital Database

Data types with their relevance in the digital database are classified, based on the source of acquisition and creation. For the preparation of the baseline digital data base as per objective, four different types of data are defined. They are as follows: topographical data, thematic data, collateral data and field data.

Topographical Data

The Topographical data comprise all the toposheets available from SOI on 1:25,000, 1:50,000 and 1:25,000 scales. The important topographical data layers are:

- Elevation points and contours.
- Road, Rail Network. (Transportation)
- Water Bodies
- Forest Boundaries.
- Administrative Boundaries.
- Drainage and Canal Network
- Settlements.
- Forest Areas.
- Scrubs/Waste lands.

Thematic Data

Thematic data as per requirements comprise the following themes, viz:

- Geographical map
- Land use/Land cover map (Kharif and Rabi) (Plate 14.1)
- Geomorphological map (Plate 14.2)
- Geotechnical map
- Slope map and soils map.

The Thematic data layers are prepared by digital cum visual image processing of satellite data.

Satellite data proposed to be used for this project: IRS ID LISS III

Collateral Data

Collateral data for the project consist of the following types.

- Depth to water level maps.
- Water quality maps and soil quality data
- Other miscellaneous data

Field Data

Data collection is one of the most important activity and such data required for planning and selection of probable solid waste disposal site. This is also required for verification and validation of thematic data layers.

14.11 Selection Criteria

The efficacy of any site for solid waste disposal depends on the selection of proper site and there are issues that have impact for site selection. Broadly they are divided into three categories i.e., economic, socio-economic and environmental. (Table 14.8)

Table 14.8 Factors of selection

ECONOMIC	SOCIO-ECONOMIC	ENVIRONMENTAL
Development Land		Wet Lands
Slope	Archeological & Historical Sites	Threatened or Endangered Species
Site Access	Land Use	Air Quality
Flexibility	Noise Impact	Odors
Capacity	Sensitive Reporters	Aesthetic Impact
Operation & Maintenance Cost	Social Impact	Flood Plains
		Soils
		Geology
		Surface & Ground Water Hydrology
		Topography

14.11.1 Excluding Criteria

The waste site-

- Must not be located on ground < 100m above sea level
- Must not be located within 160m of a wetland
- Must not be located within 160m of a sensitive area
- Must not be located within 200m of an existing waste site
- Must not be located within critical habitat area
- Must not be located within 3048m of any airport

14.11.2 Environmental Parameters Considered

The efficiency of any solid waste disposal site depends upon the selection of proper site and there are several issues that have impact for site selection. Broadly they can be divided into three categories i.e., economic, Socio-economic and environmental. The geological, geotechnical and hydro-geological parameters fall within the environmental category. The ultimate aim is to select a site where the greatest protection to the environment is provided. This process finally leads to one or more sites for detailed investigation and analysis.

14.11.3 Geological Parameters

Bed Rock Depth : It is the depth of the parent rock from the surface of the earth. Greater bedrock depths are preferable for siting. This prevents the contaminants from reaching the bedrock easily. Different types of bed rocks are susceptible to solution in varying degrees. Hyderabad has hard granite rocks ranging from 75-100 meters depth. (Table 14.9)

Structural features :

(i) **Fracture :** It's a small, weak plain along the layers of the bed.

(ii) **Joints :** These are the fractures of significant length.

(iii) **Faults :** Faults are the displacements of bed layers along the joints.

The presence of these structural features impacts the stability of the disposal facility and potential release of pollutants.

(iv) **Dyke :** It is an impervious barrier formed by the intrusion of lava into the fractures cutting across the sediment. All water bodies which encounter dykes in their flow path change their course. A special observation near the dykes is the level of groundwater on either side of the dyke. Areas with dykes represent more suitable locations, all else being equal.

14.11.4 Hydrological Parameters

(i) **Groundwater level:** It is the maximum level to which groundwater is expected to rise. The landfill may impact the groundwater to a great extent, the extent of this impact on ground water depends upon the level of water table. The higher the level of table more will be the impact.

(ii) **Surface water :** Disposal sites within the vicinity of surface water body pose a potential threat to the soil-water flora and fauna, also facilitates runoff of pollutants thus contaminating lakes/streams.

(iii) **Flood Plains :** All greater rivers annually flood portions of level land near their mouth and cover them with sedimentary deposits. The whole area thus flooded is called flood plain. Flood plains likely to be periodically inundated by surface water are not considered as acceptable sites for refuse disposal. Proximity of disposal to flood plains impacts transport of pollutants.

(iv) **Groundwater quality :** It is the natural quality of the groundwater as measured against drinking water standard. Areas with poor natural quality represent more suitable locations, all other conditions being satisfied. (Plate 14.3).

14.11.5 Soil Characteristics

(i) **pH :** pH characterizes the tendency of soil to absorb heavy metals, greater pH value is preferable for siting. pH value of soil indicates the logarithm of reciprocal of hydrogen ion concentration present in soil, it is thus an indicator of the acidity or alkalinity of soil. Acidic range: 0-7.0 and Alkaline range: 7.0-14.0

(ii) **Permeability :** It is the soil property that governs the rate at which water moves through it. Subsoil permeability impacts the release of pollutants to groundwater. Lower subsoil permeability is preferable for siting.

14.11.6 Critical Habitat Area and Airports

No landfill be constructed within critical habitat areas. A critical habitat area is defined as the area in which one or more endangered species live. It is sometimes difficult to define a critical habitat area. Siting a landfill within critical habitat areas is not suggested. No landfill should be constructed within 3048 m (10,000 ft) of any airport. This restriction is imposed to reduce bird hazard. Birds are attracted to landfills where food is available. Clearance from the proper authority should be sought if the proposed site is within the restricted area.

Table 14.9 The suitability of specific land forms at disposal sites

Land form	Comment	Suitability for land disposal site
Flood Plains	Frequent flooding	excluse
Drumlins	Limited land areas and steep slopes	excluse
Delta	Frequent flooding	excluse
Beach ridges	Limited area, high Ground water.	excluse
Coastal Plains	Highly heterogeneous Materials	detailed soil analysis required
Tidal flats	Frequent flooding	excluse
Sand dunes	Wind erosion, unstable	excluse

14.12 Unacceptable Characteristics

- **Geological**
  (a) Bed rock outcrops
  (b) Irregularities such as fissures or faults overlying groundwater.

- **Hydrological**
  - (a) Aquifer recharge zone
  - (b) Flood prone area
  - (c) Wetlands
  - (d) Seasonally high water tables
  - (e) Near private or community water supply wells or reservoirs
- **Climatic conditions**
  - (a) Extremely wet or cold conditions
- **Topographical characteristics**
  - (a) Overly steep slopes
  - (b) Broken terrain
- **Land use characteristics**
  - (a) Areas formerly used for landfills
  - (b) Areas contaminated with persistent residues from past chemical spills or waste treatment processing.

The Table 14.10 shows the list of differentiating parameters under exclusionary and non-exclusionary used in the execution of the project.

Table 14.10 Differentiation of factors

Exclusionary	Non-Exclusionary
Wetlands	Slope
Archeological and Historical Sites	Geomorphology/Geology
Settlements	Topography
Political Sensitive Areas	Landuse
Threatened and Endangered Species	Transportation Network
Flood Plains	

14.13 Converting Criteria into Data Layers

The data base created for solid waste disposal site selection consists of Landuse, water bodies, wet lands, major roads, data layers. The other data such as those on private water supply wells, areas of endangered and threatened species are not included, since the data do not exist for Hyderabad. To convert the criteria into a GIS layer data layers showing distances from water bodies were created then employing reclassification technique on the distance data layer areas unsuitable for siting landfill.

14.3.1 Creating Suitability Maps

Since all the criteria data layers are constraints, the process of combining the data layers is done with the use of the overlay technique in GIS. The multi criterion evaluation procedure will be adopted on a weighted linear combination of the factors in which each factor is multiplied by it's weight.

14.13.2 Satellite Data Processing

An area slightly bigger than Hyderabad and its environs will be splitted from LISS scene. First easily identifiable common points on image and map referred as Ground points (GCP'S) are collected so as to establish a geometric relationship between them. Then as affine transformation is applied to reallocate pixels in the uncertified image to corresponding locations in the rectified output image.

14.4 Analytic Hierarchy Process and Ranking

The Analytic Hierarchy Process (AHP) is a methodology to systematically evaluate, often conflicting, qualitative criteria. It is a relatively new addition to the family of multi-attribute decision models and was developed. Like other multi-attribute decision models, AHP also attempts to resolve conflicts and analyse judgments through a process of determining the relative importance of a set of activities or criteria. The methodology is based on the concept of trade-off and enables the decision-maker to develop the **trade-off** implicitly in the course of structuring and analyzing a series of reciprocal pair-wise comparison matrices. The AHP can be succinctly summarized in terms of its three principal components.

First, the principal problem is decomposed into a hierarchy. Each level of hierarchy consists of a set of elements and each element, in turn, is broken into sub-elements for the next level of the hierarchy. The final level consists of the specific courses of action that are being contemplated for adoption. Structuring any problem hierarchically is an efficient and intuitive way of dealing with the complexity and identifying the relevant components of the problem.

Second, within each hierarchical level, relative weights for the various elements are established using a measurement methodology. Use of the methodology requires the decision-maker to evaluate the elements in a particular level in a pair-wise fashion sing a 9-point scale. The pair-wise comparisons indicate the degree to which one element dominates another element of the same level, with respect to each element of the preceding level.

Third, the pair-wise comparison matrices are evaluated using a measurement theory. The basic assumption underlying the measurement theory in the AHP is that relative dominance can be measured by pair-wise comparisons. A pair-wise comparison of a set of 'n' attributes can be conducted and their relative importance can be established as follows;

14.14.1 Measurement Theory

If the attributes are denoted by $O_1, O_2, \ldots, O_n$ and their relative importance by $w_1, w_2, \ldots, w_n$, the pair-wise comparison matrix O may be expressed by the reciprocal matrix as shown below.

		O_1	O_2	....	O_n
O =	O_1	w_1/w_1	w_1/w_2	....	w_1/w_n
	O_2	w_2/w_1	w_2/w_2	....	w_2/w_n
	....	...	....	....	....
	O_n	w_n/w_1	w_n/w_2	....	w_n/w_n

Where, w_i/w_j reflects the relative importance of element i over element j. If O is multiplied by the transpose of the vector, $w^T = (w_1, w_2, \ldots, w_n)$, we obtain the vector nw and the problem takes the form

$$Ow = nw$$

It is implicitly assumed, in the case above, that the weights, w, are known. In case only O is known and the weights are to be recovered, the system (O-nI)w = 0, needs to be solved for the unknown w. This has a non-zero solution if and only if n is an eigen-value of O, i.e., it is a root of the characteristic equation of O. But O has unit rank since every row of O is a constant multiple of the first row. Thus, all the eigenvalues, $\lambda_{i\,,}$ i = 1, 2,.......,n, of O are 0 except one (Kreyzig, 1983). It is also known that

$$\sum_{i=1}^{n} \lambda_i = \text{trace (O)} \quad \text{sum of the diagonal elements} = n$$

Therefore, only one of the $\lambda_{i,}$ λ_{max} (the maximum eigenvalue, equals n and all the other $\lambda_i = 0$. The solution w of the above eigenvalue problem is any column of O. These solutions differ by a multiplicative constant. However, for operational reasons, it is desirable to normalize the solution so that the weights sum to unity. The result is a unique solution regardless of which column is used.

The matrix O satisfies the cardinal consistency property $O_{ij} \,.\, O_{jk} = O_{ik}$. Thus given any row of O, the rest of the entries can be determined from the relation. However, the scale w is unknown and thus, only the estimates of the ratios in the matrix can be made. In this case, the cardinal consistency relation need not hold, nor need an ordinal transitivity relation of the form:

$O_i > O_j \cap O_j > O_k \rightarrow O_j > O_k$ (where, O_i are the rows of O). since qualitative judgment remains inconsistent and intransitive, despite best efforts, the inconsistency in the comparison data needs to be considered. It can be the eigenvalues. It is also known, from the Perrorn-Frobenius therem, that a matrix of positives elements has a real positive eigenvalue whose modulus exceeds those of all other eigen values (Encyclopaedia of Mathematics). The problem Ow = nw, can thus be transformed to O'w' = λ_{max}w' , where, O' represents the matrix, which does not satisfy the cardinal consistency property; $\lambda_{max,}$ the maximum eigenvalue of matrix O', and w' the eigenvectors, corresponding to $ë_{max}$. The corresponding eigenvector solution is unique when normalized.

The question that comes up then is how close, in such cases, is λ_{max} to n and w' to w. has shown that for all possible states and that $(\lambda_{max} - n)/(n-1)$ serves as an index measure of consistency. The index indicated the departure from consistency of the comparison ratio, w_i/w_j, and the ratios are deemed consistent, if $\lambda_{max} = n$. has established, for different order random entry reciprocal matrices, an average consistency index which ranges from 0 for 1 to 2 element matrices, to 0.9 for 4 element matrices and 1.49 for 10 element matrices.

Thus, the pair-wise comparison matrix O is used to elicit the judgment of decision-makers or a group of experts with regards to the relative importance of a set of attributes for each element of the preceding level. The principle eigenvector (corresponding to the maximum eigenvalue is then extracted and normalized to yield local priorities for the elements of the matrix. Several methods can be used for deriving priority vectors from pair-wise reciprocal comparison matrices. Local priorities are then transformed to global priorities by weighting them with the global priorities of the elements of the preceding level. Continuing this process of eigenvector extraction and weighting through the levels of the hierarchy leads to a uni-demensional priority (weight) scale for the elements in the final level.

14.15 Decision Support System

For the present study, the spatial-AHP technique, was applied to identify and rank potential sites for solid waste disposal. The AHP decision making method, as used in spatial-AHP involves the following five steps

1. Identifying the decision factors associated with the problem.
2. Structuring them in a decision hierarchy.
3. Judging the relative importance of the decision-hierarchy elements.
4. Aggregating these measures in order to calculate a suitability index of the alternatives.
5. Ranking the elements according to the suitability indices.

In the present study, following step 4, the regions restricted by legal jurisdictions, physical obstructions, etc., were eliminated from the ranking operation. These steps are described subsequently.

Decision factors are used to relate attributes to suitability concerning a particular goal, as in this case, selecting potential sites for solid waste disposal. First, the major decision factors are identified. The primary factors which contribute significantly to the site selection criteria include the terrain information, information about the hydrogeological/geological parameters, land use/land cover information, the economic viability of the proposed site and factors based on legal, social, environmental or political restrictions and physical feasibility (presence of water bodies, proximity to roads, etc.).

The decision factors comprised of terrain information, hydrogeological/geological factors, factors ensuring environmental acceptability, physical feasibility and political restrictions. These factors were classified under two groups, namely, the exclusionary criteria, including the factors based on physical feasibility and political restrictions and the non-exclusionary criteria, consisting of the remaining decision factors (Table 14.11).

Table 14.11: Analytical Hierarchy Measurement Scale

Reciprocal Measure of Intensity of Importance	Definition	Explanation
1	Equal importance	Two activities contribute equally to the objective
3	Weak importance of one over another	Experience and judgment slightly favour one activity over another
5	Essential or strong importance	Experience and judgment strongly favour one activity over another
7	Demonstrated importance	An activity is strongly favoured and its dominance is demonstrated in practice
9	Absolute importance	The evidence favouring one activity over another is of the highest possible order o affirmation
2,4,6,8	Intermediate values between two adjacent judgments	When compromise is needed
Reciprocal of the above		If activity I has one of the above non-zero numbers assigned to it when compared with activity j, then j has the reciprocal value when compared with i.

Regions in the study area which are limited by physical restrictions, for instance ponds, streams, rivers and other water bodies with surface water, were also excluded.

Pixels representing road and rail networks were excluded for consideration, for obvious reasons. These three considerations essentially constitute the exclusionary criteria that where incorporated in the present study. The application of the exclusionary criteria involves a simple modeling operation in the GIS environment.

The non-exclusionary criteria are the ones for which the pair-wise comparison of AHP is applicable. Once the non-exclusionary decision criteria are selected, sub-factors and even sub-sub-factors are identified to better describe these criteria. For instance, terrain information can be categorized into information about the elevation above the mean sea-level and the slope. The information about elevation can be further grouped within selected class-intervals based on the elevation values.

For the present study, the decision hierarchy used, is given in the schematic diagram. The range of values, that were used for some of the Level IV parameters, are based on judgment. Any other justified range of values can be used. Other reasonable decision hierarchy can be incorporated within the same framework. For example, other decision factors can be used, depending on the decision-maker's perspective. Data availability, of course, is an important pre-requisite. Each factor used in the decision hierarchy must have appropriate data for the entire study area under investigation in a format that can be incorporated into the GIS. Decision hierarchies used in the present study uses data that were readily available.

14.15.1 Relative Importance Weights and Suitability Index

The Suitability Index for each call is determined by aggregating the Relative Importance Weights (RIWs) at each level of the hierarchy. The RIWs are the normalized eigenvectors corresponding to the maximum eigenvalues of the pair-wise comparison matrices constructed at each level of the decision hierarchy. The RIW assigned to each hierarchy element is determined by normalizing the eigenvector of the decision matrix. Eigenvector values are estimated by multiplying all the elements in a row and taking the nth root of the product, where n is the number of row elements. For example, for the first row in Table, where n is 3 the eigenvector value is given by normalization of the eigenvector is accomplished by dividing each eigenvector element by the sum of the eigenvector elements of the decision matrix.

The calculation of the RIWs are shown in Tables where the weights are calculated for the different levels (levels II, III and IV) of the decision hierarchy (Table. 14.12). The numbers used in the decision matrices (shown in the tables), signifying the relative importance of the competed elements, are also based on judgment, like the range of values used for some of the Level IV parameters shown in the tables. (Table 14.13)

Table 14.12 Schematic Diagram of Decision Hierarchy

Level I	Level II	Level III
Terrain Characteristics	Slope	Nearly level-Very gently Very gently-Moderate Moderate-steep
	Elevation above MSL(in meters)	110-115 115 – 125 >125
Hydrogeological Characteristics	GWI	Low Moderate High
	GWP	Poor-Nill Moderate-Good Good-Very good
	GWB	Very Low Low Moderate
Geological Characteristics	Cule	BSA KU Grass Land
	Geology	Basalt AAG/AMG/AT AAH
	HGN	Pedimat/RM PPS PPM

Table 14.13 Estimated Eigen Element and Relative Importance Weightage EEE & RIW :

Level I

	Terrain	Hydro	Geo	EEE	RIW
Terrain	1	1/3	1/5	0.405	0.105
Hydrological	3	1	1/3	1	0.258
Geological	5	3	1	2.466	0.637

Level II (a)

	Slope	Elevation	EEE	RIW
Slope	1	6	2.449	0.857
Elevation	1/6	1	0.408	0.142

Level II (a)

	GWP	GWI	GWS	EEE	RIW
GWP	1	1/3	1/5	0.405	0.105
GWI	3	1	5	2.466	0.637
GWD	5	1/5	1	1	0.258

Level II (a)

	Cule	Hgm	Geology	EEE	RIW
Cule	1	1/3	1/3	0.480	0.126
HGM	3	1	1/5	0.843	0.222
Geology	3	5	1	2.466	0.650

(a) Slope :

	a	b	c	EEE	RIW III	RIW II x (0.857)	RIW I x(0.105)
a	1	5	7	3.271	0.714	0.611	0.064
b	1/5	1	5	1	0.218	0.186	0.019
c	1/7	1/5 1		0.305	0.637	0.056	0.005

(a) = Nearly level-very Gently sloping

(b) = Gently-moderate sloping

(c) = Steep slope

(b) Elevation :

	A	B	C	EEE	RIW III	RIW II x(0.142)	RIW I x(0.105)
A(110-115)	1	5	7	3.271	0.714	0.101	0.010
B(115-125)	1/5	1	5	1	0.218	0.030	0.003
C(7125m)	1/7	5	1	0.305	0.066	0.009	0.0009

(c) Groundwater Potential :

P-N	M-G	G-V	EEE	RIW III	RIW II x(0.105)	RIV I x(0.258)
Poor-Nill	1	5	7	3.271	0.714	0.074
Moderate-Good	1/5 1	5	1	0.218	0.022	0.005
Good-Very Good	1/7	1/5	1	0.305	0.066	0.0069

(d) Groundwater Infiltration :

Low	Moderate	High	EEE	RIW III	RIV II	X(0.637)	RIV I x(0.258)
Low	1	5	9	3.556	0.735	0.468	0.121
Moderate	1/5 1	5	1	0.206	0.131	0.033	
High	1/9	1/5	1	0.281	0.058	0.036	0.009

(e) Groundwater Depth :

Very low	Low	Moderate	EEE	RIW III	RIW II(0.258)	RIW I(0.258)
Very Low	1	1/3	1/5	0.405	0.104	0.026
Low	3	1	1/3 1	0.258	0.066	0.017
Moderate	5	3	1	2.466	0.637	0.164

(f) Land Use/Land Cover :

BSA	KU	Gr.L	EEE	RIW III	RIW IIx(0.126)	RIW Ix(0.637)
BSA	1	5	7	3.271	0.714	0.089
KU	1/5	1	5	1	0.218	0.027
Gr.L	1/7	1/5	1	0.305	0.066	0.008

(g) Hydrogeomorphology :

PD	PPS	PPM	EEE	RIW III	RIW II (0.222)	RIW I (0.637)
PD 1	5	7	3.271	0.730	0.162	0.103
PPS	1/5	1	3	0.843	0.188	0.041
PPM	1/7	1/3	1	0.362	0.080	0.017

(h) Geology :

Basalt	AAG/AMG/AT	AAH	EEE	RIW III	RIW II(0.650)	RIW I(0.637)
Basalt	1	5	3	2.466	0.572	0.371
AAG/AMG/AT	1/5 1	3	0.843	0.195	0.126	0.080
AAH	1/3	1/3 1	1	0.232	0.150	0.096

14.16 Integration of AHP and GIS

The accuracy of the final modeling, integrating the GIS database with the AHP, depends on the quality of the database and on the fundamental assumptions of the AHP technique. The construction of the decision matrices (Tables 14.2 through 14.11), largely determines the final output. However, it must be realized that the construction of these matrices are essentially subjective, as in any other decision-making process, based on judgment.

The quality of data remains the limiting factor in any database-dependent modeling operation and is one of the possible sources of inherent errors. The quality of data, available for the scale, used for the present study, provide the information about the elevation above the mean sea-level at the smallest scale. Moreover, information from bench marks and triangulated heights were also incorporated in the elevation database. The rail and the road networks too, were digitized from the same maps. Hence, these informations are expected to be reasonably accurate as well.

By applying Analytic Hierarchy Process (AHP) and Geographical Information System (GIS) various layers are categorized according to Relative Importance Weightage (RIW). For final weight all individual weightages are summarized and evaluated. As each of the RIW includes its importance from the first level the final cumulative values will give more accurate results. According to the weightages allotted the entire area is categorized into five classes as excellent, good, moderate, poor and unsuitable.

14.16.1 Suitable Sites for Solid Waste Dumping

Excellent sites

1. The space between Konukunta, Ireddy Pally and Kotha Pally villages. The site is nearly three to four kilometers apart from each village. It is 18 km. From Dhulapalli forest department and 25 km. Far from Kukatpalli. This site is situated in the north-west direction of the city. It can serve the north and also west parts of the city. (Plate 14.4)
2. The space between Chevella and Malkapur villages, 3 km apart from each. It is 17 km from Rajendra Nagar and 20 km far from Nehru Zoological Park. It is situated in the South-West side of the city. It may be taken as the site to serve the southern part of the city.
3. The space between Chippalipalli and Phulmamdi villages and two and three kilometers far the villages respectively. It is at a distant of 7 km from Mysaram village, 20 km from Shamshabad and 33 km from Charminar.

Within 15 km from the Outskirts :

Northern side of the city

1. Between Kazipalli and Sultanapur villages
2. Between Borampet and Dhulapalli villages
3. North side 2 km from Bollaram village

West Direction

1. Between Narsingi and Gandipet villages

South-West Direction

1. Between Gaganpahad and Kothwalguda villages
2. East 2 km apart from Mamidipalli

South-East direction

1. Between Brahmanapalli and Koheda
2. Between Saroornagar and Kotlapuram
3. Between Abdulpur and Narayangiri

East direction :

1. East 2 km apart from Rampalli village. 7 km from Charlapalli.

Good sites above 15 km from the outskirts

East direction :

1. Near Rahimkhangudem (near Bibinagar)

South-East direction :

1. Between Nerapalli and Mylaram

South direction

1. Between Ghatpalli and Tulasinagar

South-West

1. Between Palligutta and Konkamamidi

West Direction :

1. Nearby Chillaguda and Kasaram (near Shankarpalli)

North West direction

1. Between Kodalanchi and Midheshwaram

North direction

1. Nearby Nawapalli and Kothapalli

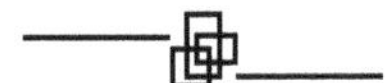

Plate 14.1 Disposal of waste in Open dumps

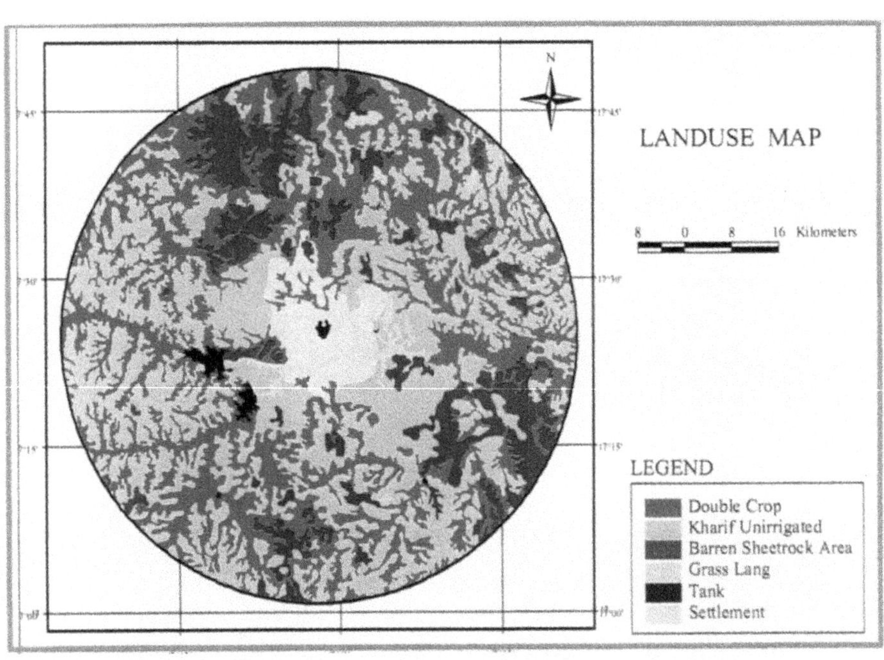

Plate 14.2 Land use Land cover of study area

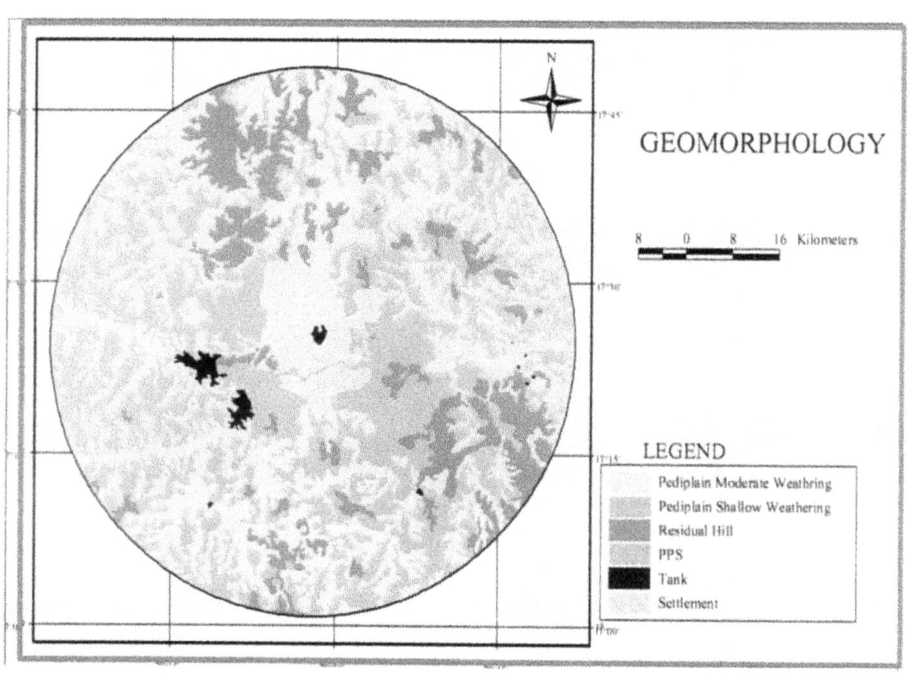

Plate 14.3 Geomorphology Map

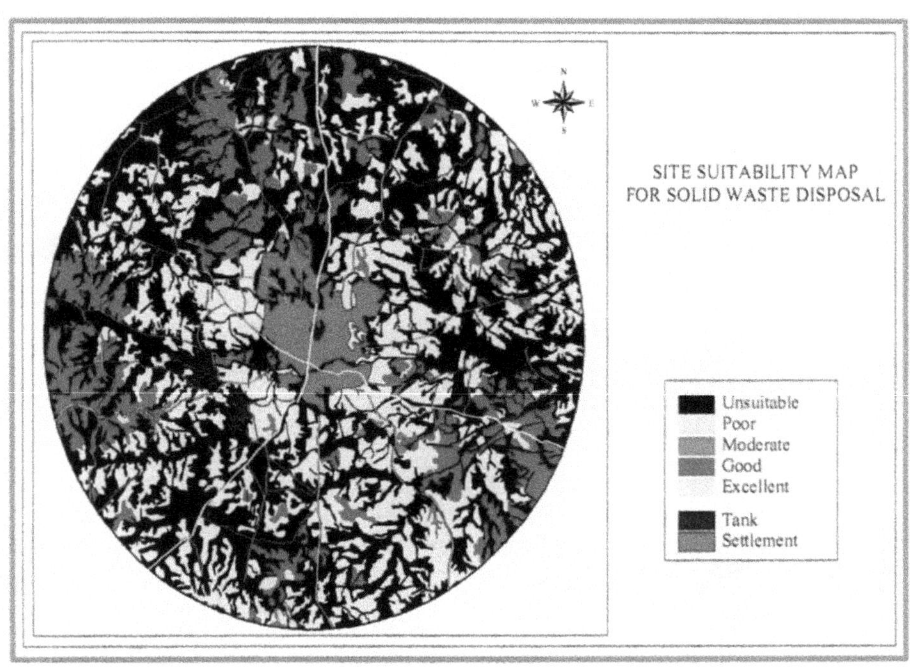

Plate 14.4 Site Suitability map for Solid Waste Disposal

15 Natural Disaster Management : Landslides

15.1 Introduction

The term landslides has been used to designate the movement of surface and near surface earth material down slopes and towards streams valleys and coastlines. Landslides are natural phenomena such as mudflows, earth flows, snow or debris avalanches and subsidence or natural events would occur with or without human activity. However, human activity has increased some of these events.

The same 'Landslides' is used both for the geomorphic process, which involves rapid gravity movements and for the resulting landform that is created by the displaced material. Furthermore, the term 'landslide' is also a misnomer because the process of landsliding has come to include a broad range of different types of motion whereby earth material dislodged by falling, sliding and flowing.

Natural disasters happen every year and their impact and frequency seem to have greatly increased in recent decades, mostly because of environmental degradation, such as deforestation, intensified land use, and the increasing population (Vincent, 1997). The urban growth is extending into hazard prone regions such as low elevation areas more prone to flooding. The result is that the magnitude of the impact of a disaster increases since it is dependent on the susceptibility of the land the vulnerability of the society (Verstappen, 1995). The vulnerability is also compounded with the increased complexity of interdependent systems (Comfort, 1999). Given that it is the transportation domain which interconnects those systems, interest has risen in determining methods which can be used to (1) measure transportation's vulnerability before such a disastrous event, (2) monitoring its availability during an event, and (3) assess the damages after an event. The occurrence of Natural Hazards is a serious constraint on economic development, particularly in developing countries, where the economic loss due to the impact of natural hazards often makes the difference between economic growth and stagnation. On the other hand practice has shown that adequate hazard mitigation is possible. Slope instability hazard zonation is defined as the mapping of areas with an equal probability of occurrence of landslides within a specified period of time. Slope instability hazard assessment is based on the analysis of the terrain conditions at sites where slope failures occurred in the past. Many landslides are natural phenomena that occur independently of human actions and many of these have enacted large human costs. There are also many landslides that have inflicted losses that have been used by the action taken to make the land suitable for some human purpose.

Population pressures and increasing urbanization along with local zoning laws have been forcing man flat lying areas and flood plains into the adjacent hill slopes. The hill slope as an environmental setting can help us to avoid the severe flooding population of the low lying areas, but can be equally hazardous because of potentially unstable earth materials. Therefore, a recognition of landslide - prone topography is becoming increasingly important in land use decisions.

Landslides can affect most hillslopes. The three principal causes for landsliding are excessive rainfall, human activity and earthquakes. Even nearly flat terrain, when undeclained by fine grained unconsolidated sediments is vulnerable. Even prolonged precipitation produce disaster on the leveled ground.

Thus, there are few areas that are completely safe from landslides although the probability for their occurrence is greatly enhanced by such factors as steep slopes, weak bedrock, unstable regolith, absence of vegetation and high moisture content and forces seeking to change hill side equilibrium.

15.2 Major Types of Landslides

The names of the major types of landslides are assigned on the basis of the type of displacement. Based on the mass movement, landslides are divided into four major groups.

(i) Slow flowage : Rock Creep, talus creep and soil creep
(ii) Rapid flowage : Earth flows/movements, mudflows, debris avalanche
(iii) Sliding : Slumps, rock slides, rock falls, and landslips
(iv) Subsidence : Sinking of mass

Fall : A fall may be defined as a free fall or rock fragments of various sizes. They take place very rapidly and consist of materials of which cliffs and bluffs are made.

Slide : Downward movements of intially intact masses of earth materials on well defined surfaces are called slide. There are two major types of slides such as rotational slides and transitional slides. Rotational slides are characterized by movement along curved surfaces that are concave upwards. The rupture surface may resemble the bowl of a spoon. The transitional slides entail movement on planer surfaces of repture that may be nearly horizontal or steeply inclined.

Lateral spreads represent a further development of nearly horizontal transitional slides, differing from them in the extent to which they consist of separate blocks that are broken and divided or of material that collapses and moves uniformly as it moves laterally.

Flows : Comprise a variety of slides that are distinguished by the chaotic movement of the material within the moving mass. This material especially sand, silt and clay, tends to appear during flow as if it were a viscous liquid. The surface of the flow has an irregular character, retaining no vestige of the original land surface.

Creep : It is a form of slope movement that takes place in a blanket of material that lies relatively close to the land surface. Creep is imperceptible except through observation of its long - term cumulative effects such as the formation of curved tree trunks. The average rate of downslope movement in creep is greatest near the surface and decreases with depth, so no sharp discontinuities separate the creep material from the material remaining in place.

Slump : A slump is a rotational slide in which the materials commonly retain coherence and move along a shear plane, which is concave upward.

Subsidence : Subsidence may be defined as the downward movement of the natural ground surface. Subsidence may be instantaneous and violent or gradual and may be caused by natural forces or by the activities of man.

15.3 Common Features of Landslides

The following features are to be considered in characterizing particular landslide

(i) The material of which the landslide consists

(ii) The shape and location of the rupture or slip surface

(iii) The displacement or distance travelled during the slide event.

In interpreting the processes of landslide movement, it is known the rate of movement, the behavior of the materials within the landslide during movement, and the record of any pre-landslide phenomena include earthquake precipitation and human modification of the terrain. Keeping in view of its importance, causes of hazards and controlling measures of landslides are discussed in the following.

15.4 Causes of Landslides and Related Phenomena

In many cases the real cause of landslides is the increase in driving force or decrease in resisting force are marked by immediate causes such as earthquake shock vibrations, or a sudden increase in the amount of moisture in slope material.

Landslide causes may be grouped into two, as external and internal. External causes produce an increase in the shear-stress (driving force per unit area) at relatively constant shear strength (resisting force per unit area). External causes include loading of a slope steepening of a slope by erosion or excavation, and earthquake shocks. A slight vibration or jerk to the mass would greatly add up against the frictional resistance and the mass would become unstable. The heavy traffic on hill roads is of great contributing factor towards causing the imbalance of the masses. Internal causes include changes due to increase in water pressure or decrease in cohesion of the slope materials. The steeping of the slope, water content of the stratum and mineralogical composition and structural features, which are tending to reduce the shearing strength of the rocks. (S.S.Ramakrishnan, et al.).

In addition to the above some other causes of landslides are intermediate having both external and internal causes. For example, rapid drawdown involves an increase in the shear stress and accompanied by decrease in shear strength other intermediate causes include spontaneous liquefaction and sub-surface weathering and erosion. Landslides occur when there is a loss of support under or infront of the shear surface, or when there is a change in the physiochemical constitution of the earth's materials. The three factors that influence stability of hill slopes are :

(i) Internal properties of the earth materials
(ii) The geomorphic setting and environment
(iii) Independant external factors.

(i) Internal properties of the earth materials : These include the type of material and its structural character such as degree of compaction and cementation thickness and arrangement of different rock types, strike and dip of zone of discontinuity, size and distribution of fault joints and other features and bedding surfaces.

(ii) The geomorphic setting and environment : Hill slope characteristics include the amount of relief, steepness of slope, shape of the land surface, slope orientation and aspect, nearness to rivers, vegetation, antecedent moisture conditions etc.

(iii) Independant external factors : These are termed the triggering machanisms because they provide the immediate stress that initiates the movement of mass. Prior to displacement, there is an accommodation equilibrium that exists the movement of mass. Prior to displacement, there is an accommodation equilibrium that exists between the earth materials and the geomorphic and environmental setting. The three most common disturbances that cause the critical threshold of cohesion to be exceeded are excessive precipitation, human activities and earthquakes.

15.5 Landslide Analysis

Analysis of a landslide is important to determine what factors have controlled the formation of the rupture surface and the movement on that surface. This requires an engineering analysis of the stability of the landslide mass and a geologic analysis of the changes in geologic and meterologic conditions that are correlated with landslide activity. Earth materials on slopes may fail and move or deform in several ways that includes flowage, sliding, falling and subsidence.

15.5.1 Slope Stability

The nature and extent of slope stability for the various types of downslope movements generally explained by examining the forces on slopes and their interrelated, variable types of earth materials, slope, topography, climate, vegetation, water and time.

15.5.2 Forces on Slopes

The causes of many landslides are examined by studying relations between driving forces, those which tend to move earth materials down a slope and resting forces those which tend to oppose such movement.

Slope stability is evaluated by computing a safety factor. Safety factor may be defined as the ratio of the resisting forces to the driving forces. If the safety factor is greater than one, the resisting forces must exceed the driving force and the slope is considered stable. If on the other hand the safety factor is less than one, then the driving forces exceed the resisting force and a slope failure is expected. Anything that reduces the resisting forces or increases the driving forces will lower the safety factor and thus increase the chances of a landslide.

15.5.3 Role of Earth Material

Depending upon the type of the earth material the frequency and magnitude of the landslides may be more or less for instance when a permeable rock such as sandstone overlies a weak rock such as shale rotational slides are most common. However, a resistant rock overlies a rock of very low resistance, than rapid undercutting of the resistant rock may cause a slab failure.

15.5.4 Role of Slope and Topography

Angle greatly influence the driving force on slopes as the angle of potential slip plane increases the driving force also increases. Thus everything else being equal landslide should be more frequent on steep slopes.

15.5.5 Role of Climate and Vegetation

Climate and vegetation may influence the type of landslide. Climate is significant in influencing the downslope movement because it controls the nature and extent of the precipitation and thus the moisture content of slope materials. For example, both earthflows and mudflows involve downslope movement of water saturated earth materials, but earthflows are common and mudflows relatively rare in humid regions. This results due to the infiltration of good amount of water into slopes in the humid areas facilitating earth flows.

The role of vegetation in landslides and related phenomena is complex because the vegetation in an area is a function of several factors including climate, soil type and topography. Vegetation is a significant factor in slope stability. Vegetation provides a cover that cushions the impact of rain falling on slopes thus facilitating infiltration of water into the slope while retarding grain by grain erosion on the surface. Vegetation has root systems that tend to provide an apparent cohesion to the slope materials. Lastly vegetation adds weight to the slope.

Problem concerned with slope stability and vegetation result from disturbance or removal of vegetation from slopes. In some cases however, vegetation increases the probability of landslides. Disturbance or removal of vegetation by logging has also been associated with an increase in landslides. Clear cutting or removal of all trees has caused several problems. First the rate of transpiration is greatly reduced thus soil moisture increases, sending to reduce the stability of the scopes leading to landslides.

15.5.6 Role of Water

Water is always directly or indirectly involved with landslides and its role is thus particularly important. Water is the universal solvent, and much weathering of rocks takes place. The weathering contributes to the instability of rocks which inturn leads to landslides.

The effects of water on slopes and landslides are quite variable, saturation of earth material causes a rise in more water pressure as the water pressure increases, the shear strength (resisting force) of the slope decreases, and the weight (driving force) increases, thus, the net effect is to lower the safety factor.

15.5.7 Role of Time

The forces on slopes often change with time. For example both driving and resisting forces may change seasonally as the moisture content or water table position alters.

Furthermore, these changes are greater in especially wet years, as reflected by the increased frequency fo landslides during wet years. A slope safety factor may also decrease with time because of progressive wetting which causes disarrangement of soil particles in the slope, lowering internal friction and thus the strength of the slope material.

15.6 Human Causes of Landslides

There has been a sharp increase in the number of man - induced landslides during the twentieth century. Because of expanding population, urbanization and construction, we have been rapidly modifying the landscape on a massive scale and have triggered through our own carelessness.

New engineering techniques, heavy land moving equipment have permitted the metamorphosis of slopes throughout the world on a scale of extraordinary dimensions. The nearly universal construction of dams and buildings of immense highway systems have added their toll to increase landslides. Most man made landslides are caused by changes in the slope configuration or by activities that permit additional moisture to enter the substrate material. Thus, any activity that increases hillside gradient, undercuts earth materials, adds weight to the slopes, or produces more water content can lead to instability and get the stage for landsliding.

15.6.1 Highway Construction

Road building is one of the most common ways that have directly caused landsliding. The majority of the landslides occur where a hill side cut is made and the design of slope stabilization structures is inadequate. When the underground water flow is change and the shear resistant of earth materials lessen, such artificial excavations become prime candidates for earth's movement. The construction of the roads in areas to be logged is an especially a serious problem because roads may interrupt surface drainage, alter sub surface movement of water, and adversely chage the distribution of mass on a slope by cut and fill operations.

15.6.2 Hillside Development

Building on hill side is a risk to structures, especially when there are improperly design or poorly located. Although single buildings rarely cause significant or large scale landsliding, house developments on this slopes lead to landslide. The most probable caused would be the increase in pore water pressure from lawn watering and septic tank effluent or cess pools.

15.6.3 Dams and Reservoirs

The construction of dams and development of artificial lakes and reservoirs create another type of conditions that can provide fertile settings for landslides. The primary cause of this new instabilities is the artificially induced changes in contiguous water tables.

15.6.4 Mining

Mining operation also provide triggering mechanism for landslide. In mining opertaions the over steepening of a cut into a terrace may produce failure in slope stability and leads to landslides. Other causes for landslides in mining area is due to the deforestation, transportation of ore materials and use of heavy machinery for mining operation in open cast and underground mines. Apart from this, landslides are also trigger by blasting involved in mining operations.

15.6.5 Lumbering

Timber operations especially clear cuttings may accelerate mass wasting process and produce a variety of different types of landslides. Landslides produce by multiple events such as removal of forest cover and excavation and dumping of the excavated materials in and around accompanied by rains produced landslides. Thus the shear charge of addition load coupled with diminishing of shear resistant of trees and the increased water content exceeds the strength of the earth materials, and failure is the result.

Earthquakes are the major immediate cause for some landsliding and produces disastrous effects.

15.7 Remote Sensing for Landslide Mapping

The purposes for using remore sensing include (Verstappen, 1995) : "to investigate the susceptibility of the land and the vulnerability of the society, to construct hazard zoning maps and potential damage maps, to monitor potential hazards, and to deal with emergency situations after a disaster". Before discussing remote sensing application in predicting monitoring, and assessing hazards and disasters, we should first define these terms. A disaster is "an event, concentrated in time and space, in which a society (or a community) undergoes severe danger and incurs such losses to its members and physical possessions (or appurtenances) that the social structure is disrupted and the fulfillment of all or some of the essential functions of the society is prevented" (Fritz 1961). A hazard is " the danger that a disaster event may develop or ensue" or "a threat to humans and what they value" (Kates, et al. 1985).

The use of remote sensing depends highly on the type of disaster or hazard of concern. Disasters or hazards can be divided into two groups, slow and rapid. In slow, "creeping" ones (e.g., desertification), remote sensing is genrally used for the purpose of monitoring and low temporal resolution images are often sufficient. In rapid, "instantaneous" disasters, the required capability of remote sensing varies largely. Due to the temporal requirements of the technology, it is usually not possible to monitor rapid disasters, e.g., landslide, earthquake, etc. Sometimes there is no clear distinction between slow and rapid when labeling a disaster. For example, a flood could either be classified as slow or rapid disaster depending on the conditions under which the event occurs. No matter what type of disaster occurs, rapid or slow, there are always some clues that remote sensing system can capture and use to predict a disaster (asses the hazard), monitor its progress, and assess the impact afterwards.

Satellite imagery has been used in the analysis of landslide occurrence primarily through the analysis of colour composites. Several studies have experimented with the use of a true colour composite (TCC) (Sauchyn and Trench 1978; Greenbaum et al. 1996). In most cases the primary restriction has proven to be spatial resolution, with only landslides of approx. 50 m × 50 m or larger being easily resolved. Rather better results were achieved using the 5.8 m spatial resolution of IRS-1 instrument however (Nagarajan et al. 1998).

False colour composite (FCC) images have proven useful in some cases, especially where a landslide scar provides a clear change in the surface properties, such as the removal of forest to expose bare soil. Greenbaum et al. (1995) successfully used this technique for the examination of landslides in Pupua New Guinea for example, similar results were also achieved by Rothery (1987) to identify rock avalanche deposits.

Mixed success has been achieved with the use of more complex analysis techniques. For example, Huang and Chen (1991) reported considerable success using a principal components analysis. Kusaka et al. (1996) on the other hand used the LANDSAT thermal bands to identify areas of perennially wet soil, which were linked with potential landslides. In all cases, spatial resolution has proven to be a key limitation, and shadows cause considerable problems.

Remote sensing has been used to study characteristic properties of ground surface due to the advantages of its broad area observation and periodicity. The integrataion of GIS with remote sensing data and thematic map data may facilitate greatly the assessment and estimation of regional landslide hazards.

15.8 Landslide Analysis in Geographic Information Systems

Geographic Information Systems (GIS) have become the promising tool for an effective analysis associated with the study of geologic hazards. The increasing computer - based tools are found to be useful in the hazard mapping of landslides. One of such significant tools for hazard mapping of landslides is Geographic Information Systems (GIS). GIS is an indeal tool for landslide modeling owing to its versatility in handling a large set of data, providing an efficient environment for analysis and display of results with its powerful set of tools for collecting, storing, retrieving, transforming and displaying spatial data from the real world. The key factor in landslide hazard mapping is the assessment of risk associated with the failures.

Landslides are one of the complex analyses, involving multitude of factors and need to be studied systematically in order to evaluate the hazard. One of the main advantages of the use of this technology is the possibility of improving hazard occurrence models, by evaluating their results and adjusting the input variables. An important aspect of landslide investigations is the possibilities to store, treat, and analyze spatiotemporal data that are available. Better interpretation and understanding of the failures could be derived from them in a GIS environment.

The unique capability of GIS to work with data referenced by vast spatial or geographic coordinates and its ability to incorporate appropriate engineeing models, have caused its proliferating application across the wide sections of engieering, especially in civil engineering where management of spatial-data is pivotal for the analysis. Any spatially-distributed data with a geo-reference to real world could be stored as points, lines and polygons (vector model) or as continuous fields (raster data model). Beyond GIS being used as a spatial database, it assists in modelling applications through handling a special form of data that would otherwise be compromised in conventional analysis (Miles et al. 1999). Also, GIS does not only serve as a database for parameter data, rather qualitative data can be integrated through spatial relationships rather than through relationships between attributes that may not exist (Frost et al. 1997). The other facilities such as : Query languages and user interfaces permitting rapid modification of parameter values : Convenient and quick updating of model parameters; Overlay function, where multiple maps are either visually or topologically combined and its potential in visualization of data using the graphic features, that assists the engineer in verifying data and information pertaining to the model and its application; and developing elevation maps and subsequent slope, aspect and hillshade themes (which are useful in the landslide analysis) are worth mentioning, as they are common to many applications of engineering models (Miles et al. 1999). In particular, the ability of GIS to present the data and analysis results in map forms play a key role in identifying the critical areas (where more rigorous analysis and improved solution is required) by its interactive visualization in a spatially optimized mode.

For the analysis of the causative factors the application of geographic information systems (GIS) is an essential tool in the data analysis and the subsequent hazard assessment. When making use of GIS techniques, the following methodological approaches can be differentiated.

- **Heuristic qualitative approach,** particularly suited for small - scale regional surveys. The scale of such surveys is in the order of 1:100.000 to 1.250.000. They are used mostly by regional planning agencies;

- **Statistical quantitative approaches** for medium scale surveys. These are in the range of 1:25.000 - 1:50.000, and are used by consulting firms or planning agencies for the preliminary planning of infrastructural works, such as the definition of road corridors. The methods at this scale can be sub-differentiated as follows :
- data driven multivariate statistical analysis;
- experience driven bivariate statistical analysis;
- predictive modelling through the application of probability and favourability functions.
- **deterministic approach** for detailed studies at large scale (1:2.000 - 1:10,000), without entering at the level of the engineering geological site investigation. Such small scale studies are used by consulting firms or local planning agencies for the detailed planning of infrastructural works.

All three methodologies require a good idea on the spatial distribution of landslides as the essential element for the analysis. The professional expertise, leading to a classification, is essential. Historic databases are most useful for this work. However, most of the landslide distribution mapping are to be obtained by aerial photo interpretation, and image processing.

16.9 Hazard mapping of landslides

Landslide hazard maps can be constructed by different methods, such as, from inventories; by consideration of site conditions including geology, hydrology, topography, and/or geomorphology; by statistical correlation of landslide frequency with geologic and geomorphic factors; using safety factors from stability analysis. Mechanics based models have also been used to estimate failure probability based on uncertainties about infiltration of rainfall, or pore pressure, and soil strength (Wu et al. 2000). A prior study of causal factors of landslides indicate, the ground conditions, geomorphological processes, physical processes and man made processes as the significant contributors. The pertaining data needed for landslide analysis for a particular slope are :

- Geologic and Geomorphologic features/setting
- Types and quantification of soil and/or rock properties
- Details of vegetation cover, folds, faults, etc.
- Past records of rainfall and earthquake incidences
- Appropriate hydrologic and stability models

Natural hazard is defined as the probability of occurrence of potentially damaging phenomena within a specified period of time and within a given area. Zonation also refers to the division of the land in homogenous areas or domains according to the degree of actual or potential hazard (Varnes, 1984). The analysis of hydrologic and stability models could be readily incorporated in the Arc View either using map calculator or avenue scripts. The results give the factors of safety in the form of maps or contours. The task of engineer becomes easier to identify, visually which areas are critical that deserve further attention. Hazard rating could be assigned based on suitable criterion. Table 15.1 gives such a rating criterion, based on a quantitative estimation of landslide occurrence over a given region, without mentioning the time period.

Table 15.1 Quantitative hazard rating of landslides (Miles et at., 1999)

Stability Criterion	Hazard rating
F > 1.5	Stable slope
1.25 > F < 1.5	Low Hazard
1.00 > F < 1.25	Medium Hazard
0 > F < 1.00	High Hazard

F is the factor of safety

15.9.1 Infiltration and Drainage Model :

The rainfall-induced failures are the common occurrences in soil slopes. Depending upon the intensity and duration of rainfall and permeability characteristics of the slope, infiltration causes development of pore pressures. Rise in piezometric level in hill side slopes, decreases the effective stress, eventually leading to failure of soil mass. Hence prediction of piezometric levels constitutes an important element in the evaluation of landslide hazards. Suitable hydrologic model predicting the level of pore pressures induced for a particular intensity of rainfall and for the given soil, hydraulic and slope conditions should be coupled with the stability model. The lumped parameter model given by Wu et al (1996), gives the recharge of the groundwater as a function of rainfall characteristics and soil properties. Then a finite-difference solution be used to calculate the pressure head at different points in a slope.

15.9.2 Stability Model

Slope failures triggered by any causal factors, involves movement of mass or slide. The exact configuration of the movement and its mass is difficult to predict. Though, depending upon the material characteristics, slope conditions, suitable theoretical models could be used for analysis. Limit equilibrium solutions are available for stability analysis, depending on the assumption of the failure plane, whether it is plane failure of finite/infinite slope or circular failures. However, in actual slope failures, the assumption of single slope failure is not realistic. It may include a slide of mass with circular failure surface and also with plane surface failure, as multiple slope failures are possible. Hence, the stability model should capture both the types of failures according to the conditions of its occurrences. Appropriate criterion could be established for identifying which type of failure would occur for the given conditions.

15.10 Case Study : Kohima Area

Natural disasters are brought about by forces of nature that cause a great damage to human civilisation. The Himalayas are one of the world's most tectonically active areas. Landslides occur frequently and represent a very real hazard to life and property. Since study of landuse/landcover is a part of management of natural hazards, study of landuse/landcover has been carried out in Kohima area, Nagaland, using multidated remote sensing data of 1989 and 1999. The changes in landuse/landcover classes are classified using maximum likelihood classifier applied on IRS IC LISS III digital data through image processing software, namely, EASI/PACE. In the study area, it is mainly found below the settlement areas, rolling Jhumland, slopes and road construction sites. Lack of proper drainage accounts for forty percent. Jhumland areas and deforestation are also responsible for around twenty percent of the landsliding while road construction site is responsible for another 40% occurrence. Kohima, the state capital of Nagaland, is located in the North-Eastern region of India between the latitudes of 25^0 30' N to 25^0 45' N and longitude 94^0 0' E to 94^0 15' E. It is bounded in the north by Wokha district, in the south by Manipur, in the east by Zunheboto and Phek district, in the west and north-west by Assam. The area shows a rugged topography of high hills, narrow valleys, steep cliffs, deep gorges are common features. The maximum and minimum heights encountered are over 3700m and below 360m respectively above mean sea level. The slope in the study area ranges between 15^0 and 36^0. Areas close to drainage and streams are comparatively steep. It has a variety of climatic conditions ranging from warm sub-tropical over the foothills and the plains to cool and temperate in the high hills. The average annual rainfall is in between 200 and 250 cm. Two groups of rocks are found, namely, the disang and barails. The causes observed are increase in population, illegal felling of trees, and wasteland management. Based on the analysis of the result at various stages of work, it can be said that remote sensing method has been able to detect the changes in landuse/landcover classes during the selected time period. Using Remote Sensing and Geographical Information System (GIS), natural disasters can be identified and preventive measures undertaken to minimise the losses. Certain types of disasters cannot be predicted directly through Remote Sensing but potential zones can be identified fairly accurately. Landslides are an inherent part of the environment that required control and management strategies. Hence, GIS and Remote Sensing tools can be administered for disaster management and preventive measures for the study area.

The human transition to urban settlement is a continuing phenomenon creating a complex environmental problem on global climate, the intricate web of forest, ecology, and diversity of living beings. It is also manifested in the form of soil erosion, climatic changes, landslide and so on which are a common phenomenon, possessing a great menace to the economic development of the study area. The study has been carried out to identify various changes in landuse/landcover pattern over a selected period of time using IRS IC LISS III digital data. The spatial information on landuse/landcover and their pattern of change are essential for planning, utilisation and management of the area because they permit better understanding of land utilisation aspects on cropping pattern, fallow land, forest, grazing land, wasteland, and surface water bodies which are very vital for developmental planning.

Landslide which is a recurrent phenomenon, is mainly found below settlement areas, terrace fields, rolling Jhumland slopes and road construction sites. Landslide occurrence around settlement areas is mainly due to lack of proper drainage to convey high runoff water during the rainy days and this accounts for forty percent of landslides. Road construction is also quite significant and it accounts for forty percent. Jhumland areas and deforestation are also responsible for about twenty percent of the landslide occurrence. Slope stability (Plate 15.1) for the various types of down slope movement is generally explained by examining the forces on slopes and their interrelated, variable types of earth materials, slope, topography, climate, vegetation, water and time. Climatic condition is aslo significant in influencing the down slope movement because it controls the nature and extent of the precipitation and thus the moisture content of slope materials. The role of vegetation in landslides and related phenomena is complex because the vegetation in an area is a function of several factors including climate, soil type and topography.

Because of expanding population, urbanisation, and construction, we have been rapidly modifying the landscape on a massive scale and have triggered through our own carelessness. The nearly universal constructions of buildings, immense highway constructions and hillside development systems have added toll to increase landslides. Most man made landslides are caused by changes in the slope configuration or by activities that permit additional moisture to enter the substratic material.

15.10.1 *Objective of the study*

(i) To study the existing land use pattern of Kohima area, Nagaland, with special reference to the evaluation of landslides.

(ii) To find the changes in the land use pattern since 1989 with the help of remote sensing data.

(iii) To map the zone of landslides using GIS and to suggest remedial measures for prevention of landslides.

15.10.2 Slope Analysis

Slope plays a major role in destabilizing the slope besides other factors, so slope analysis is one of the factor used in hazard analysis. Four categories of slope representing Escarpment/Cliff, very step slope, moderately step slope, and gentle slope have been used.

Slope in the study area ranges in between 15^0 to 45^0. There are few fairly flat areas on the top of the knolls and ridges.

Slope Classes	Slope in Degrees
Gently slope	$15\text{-}25^0$
Moderately steep sloping	$25\text{-}35^0$
Very Steep Sloping	$35\text{-}45^0$
Escarpment/Cliff	$>45^0$

15.10.3 Drainage and its Density

Drainage is the life line of any given area which determines the productive capability of the region in terms of its water resources. The main drainage systems present in the study area are Dzuna, Sidzu, Dzurii, Dzulakie and Kehorii. The drainage pattern observed in the area is dendritic type with varying density. A number of Perennial and Ephemeral streams with 'V' shaped valleys and gorges are common.

Drainage density is an indicator of linear scale landform elements that reflect permeability, rock types and local relief. The drainage density in the study area ranges between 0.5 to 3.2 Km/Sq Km.

15.10.4 Land use/Land cover Analysis

Digital image processing involves the manipulation and interpretation of images obtained in the digital form using various modules like GCP'S, geo-referencing and registration with the aid of computers. One of the most often used methods of information extraction is multispectral classification. Most of the information extraction techniques rely on analysis of the spectral reflectance properties of imagery and employ special algorithms designed to perform various types of "spectral analysis". The process of multispectral digital classification is performed using supervised maximum likelihood Classifier.

The total master plan area of Kohima is around 3855.62 hectares out of which 1080.52 Ha is vacant land. The total developed area is 2436.47 Ha. Table 15.2 shows the existing landuse of Kohima. Plate 15.2 (a) shows the enhanced satellite digital image and Plate 15.2 (b) depicts the corresponding colour coded output of classified image for landuse/landcover of a part of Kohima.

Table 15.2 Existing land use of Kohima

Land use	Area in hectare	% of the total land
Residential		
Government	496.25	12.87
Private	746.93	19.37
Commercial	232.83	6.03
Administration	454.42	11.78
Industrial	13.56	0.35
Recreational	8.70	0.22
Transportation & Communication	93.24	2.41
Security uses	56.65	1.46
Agricultural uses	338.63	8.78
Other uses	11.56	0.29
Vacant land	1080.52	28.02
Total	**3855.62**	**100.00**

Source : Land use survey, town-planning organisation. Kohima, 1998

Based on the land use/land cover classified system, the different landuse/landcover patterns are classified for the years 1989 and 1999 as shown below in the Table 16.3 and pie chart. (Fig. 15.1)

Table 15.3 Land use/ land cover of Kohima area for 1989 and 1999

Land use	Areal extent (Ha)				Change
Classes	1989	Percent (total area)	1999	Percent (total area)	Percent from '89 to '99
Built up	3242.72	5.93	3933.05	7.19	21.28
Dense Forest	14020.08	25.65	11424.25	20.90	18.5-
Open Forest	5925.08	10.84	5299.37	9.69	10.57-
Land with scrub	8020.30	14.67	7298.48	13.35	8.99-
Land without scrub	10115.90	18.51	9399.93	17.20	7.07-
Grassland	7561.57	13.83	6898.48	12.62	8.76-
Crop land	1194.03	2.18	1328.84	2.43	11.29
Shifting Cultivation	3716.19	6.80	4993.89	9.13	34.38
Seasonal follow	1820.11	3.33	1783.11	3.26	2.02-
River	N.A.	498.91	0.91	0.00	
Landslide	N.A.	574.07	1.05	0.00	
Total	**54645.36**		**54645.36**		

* -ve: indicate decrease in landuse classes.
Source : NRSA, Hyderabad.

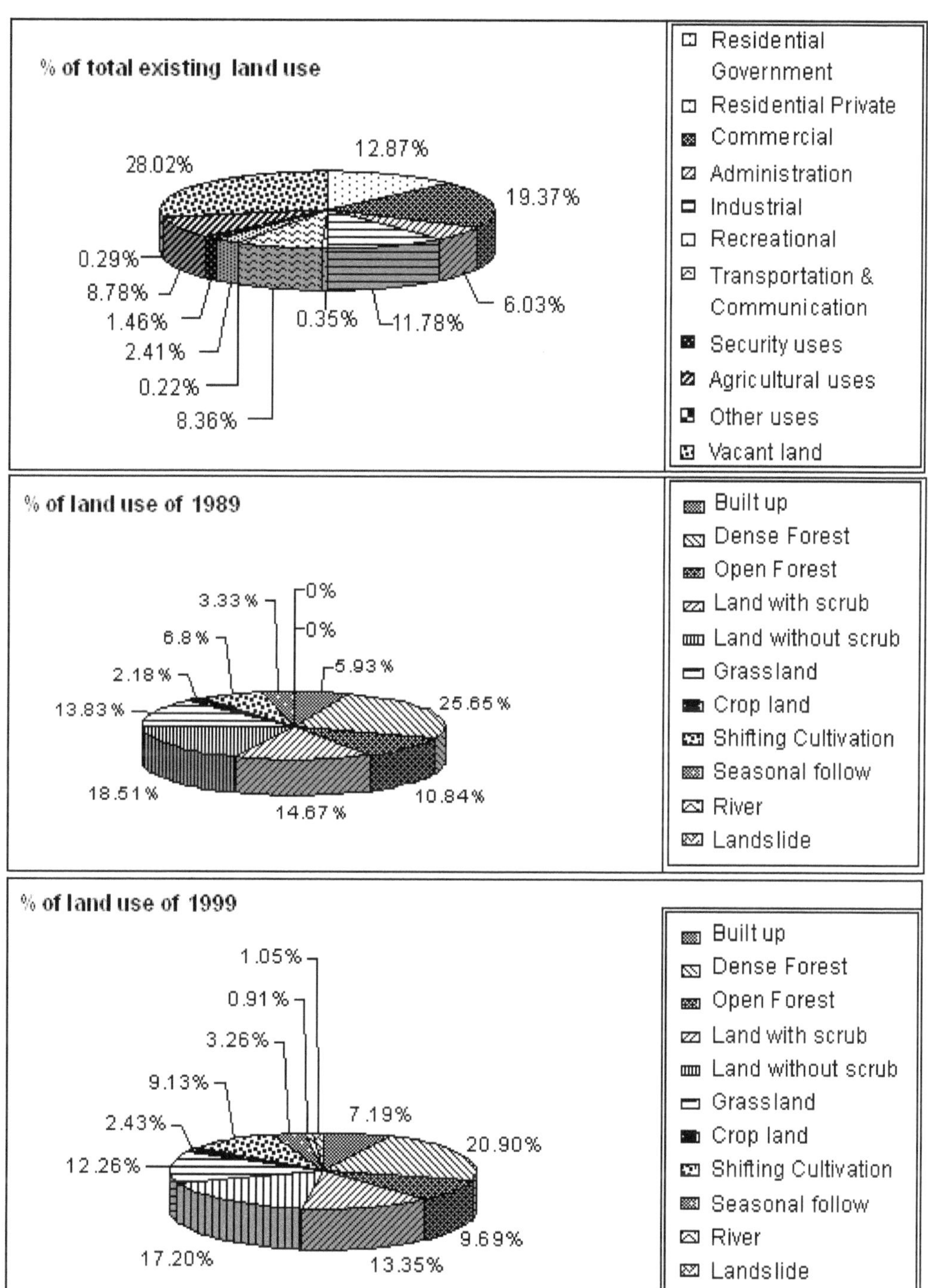

Fig. 15.1 Pie-Chart Showing Existing land use of Kohima, Nagaland.

15.10.5 Land Use Change Analysis

From the above analysis it has been shown that the land use pattern has been changing. It is observed that the dominant change which occur in the study area, is the decrease of forest area by 3944.02 hectares. This change may be due to tremendous increase in shifting cultivation and felling of trees. Increase in built-up land, which increases to 690.33 ha, is due to increase in population and economic activities. Since the people depend mostly upon agriculture, and the land not being very suitable to terrace cultivation, shifting cultivation is adopted as a major practice which increases to 1277.7 ha. The decrease in the area under wasteland, which is about 715.77 ha, may be due to transformation of land to built-up and waste management. Major landslide activities (Plate 15.3) seen along the road cutting and settlement areas covering an area of 574.07 ha are due to man's activity, lack of proper drainage and geological instability of land.

15.10.6 Results of the GIS Analysis System

The best viable solution is complete avoidance of the area and this strategy has the advantage of being the safest alternative. However, as urbanisation is in its development, prevention and management of the area have been a key factor. So some of the alternative prevention practices are slope reduction, retention structures like ground covers or other vegetation development, retaining walls or other stabilised structures and hillside benching. Surface water may be diverted around the slope by a series of gutters, roadcuts, and surface ditches, and collector systems may also be used to drain ponds and remove standing water. Groundwater may be inhibited from entering a slope by excavating a cut of trench, galleries, and tunnels. Control of toe and surface erosion are caused by using concrete or crib walls, or spur dykes. Control surface erosion due to runoff by use of surface drains, vegetation cover and gabion. People should be made aware of consequences of shifting cultivation.

As observed from the study undertaken, there have been rapid changes in landuse pattern like decrease in forest area, increasing shifting cultivation and so on which have serious consequences on the environment as well as on climatic condition. Hence awareness should be infused into the people as to the sustainable management of the land. Prevention of large natural landslides is nearly impossible but good geological engineering design could be identified and practised for preventive measures and control exercises to minimise the losses. Furthermore GIS software can be used to creat operational database for taking remedial measures to protect the environment in which the buffer zone can also be created around sensitive areas for future planing.

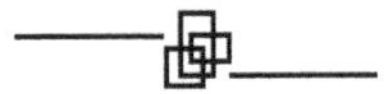

16
Pipeline Alignment

16.1 Introduction

The siting of pipeline and highway routes are similar problems. A recent example of pipeline siting that employed both remote sensing and GIS methods offers insight into siting procedures that could be applied to either of them. A reported case study of the routing of the Caspian pipeline is, as follows :

The Caspian Pipeline Consortium (Sultanate of Oman, Republic of Kazakhstan, and the Russian Federation) sponsored a pilot project that employs remotely sensed data and GIS to assist the routing process for a pipeline that will carry oil from the Tengiz oil field in Kazakhstan, on the Caspian Sea, to Novorossiysk, Russia, on the Black Sea. Though the new pipeline will be 500 miles long and linked with 500 miles of existing pipeline, the pilot study area for the route siting involved only the last few per cent of the total length of the new construction. The consortium has already employed general, political and economic guidelines to constrain the area of study to an inverted L-shaped corridor, and identified four points (A, B, C, and D) through which the pipeline was constrained to pass. The object of the study was to find a least-cost pathway and to compare its cost with the likely cost of three straight-line segments through these four points.

The idea employed was to extend the algorithms traditionally used in GIS for defining the paths of stream channels in a drainage basin, where a DEM is the input and water is always assumed to flow downhill. Feldman et al. (1985) extended the algorithms to define the least-cost pathway of a pipeline, with the path constrained to pass through specific nodes to specified end points. They generated a single weighted surface layer, based on input layers representing factors that contribute to the cost of traversing between two points with a pipeline. This single layer, called a combined weighted surface layer, is analogous to a topographic surface, in that it has peaks (areas of relatively high cost) and valleys (areas of relatively low cost). Therefore, a two-dimensional grid of this combined weighted surface layer replaced the DEM as input to the drainage algorithm, and the route of the pipeline has a specified beginning point and a specified endpoint. A cumulative cost surface layer was generated from the combined weighted surface layer, consisting of the sum of all cells that had been as "channel" grid cells between the points of origin and ending.

In order to select a probable corridor for gas or petroleum or waste material transportation (liquid) pipelines, extensive data is needed. Such data are used to assist in locating a cost effective pipeline route based on land cover type, wet land distribution, the proximity to other environmentally sensitive resources. The international scenario on such studies says that the organisations like Reliance Petrochemical Ltd., Caspian pipeline consortium, Destin pipeline, and the Gulf of Mexico, have faced similar problems in searching for a cost effective methodology for Environmental Impact Assessment (EIA) for siting pipeline routes. The tools that are available for such studies are not cost effective and time effective. Therefore all the companies which are involved in developing environmentally friendly pipeline alignments for any given area are searching for new methods. These new methods, using remote sensing data and other collateral data products with the integration of data through GIS, need a research methodology, that should focus attention on the creation of a database by which siting of pipeline routes should be performed. Such a database can also be used for the study of EIA for siting pipeline alignments.

A query based decision support system is developed to answer a number of queries within the GIS database. This requires the creation of a positional accuracy of the database. The database creation is based on image processing of satellite digital data, existing maps, data collected from organisations and GPS. This database is utilised to devise an automated method to provide data more accurate for EIA. All these studies lead to establish a detailed instrumental procedure to carry out the environmental impact assessment studies for siting pipeline alignments.

16.2 Conceptual Design

The Remote Sensing and GIS method has obvious advantages for the siting of pipeline alignment and the monitoring of natural petroleum and gas leaks from pipelines, both onshore and offshore. This method can also be used for environmental monitoring of pipeline corridors with the help of satellite data and integration of such data with conventional data through GIS technique. From theoretical considerations, the plumes could be detected from space borne hyperspectral infrared sensors. Earlier investigations on related studies show this method is more potential as a safety tool for environmental monitoring of pipeline alignments and leaks.

A recent experimental study of pipeline alignment, which has employed both Remote Sensing and GIS methods offer insight into the siting procedures. This study aims at central India pipeline alignment survey using Remote Sensing data, and is funded by Reliance Petroleum Ltd. and executed by private consultants. During the execution of this project, the project personnel have come across so many problems relating to the methodology, type of data products, analysis methods, and more specifically, environment related problems. The weighting factors that made up each of the individual component layers are actually cost factors associated with the pipeline crossing of slopes, streams, wetlands, roads, rail, rocky land, agricultural land, and urban/ industrial areas. The methodology that is being used should consider all these individual layers weighted and added to obtain a combined weighted surface layer on which the pipeline alignment is to be delineated.

16.2.1 Methodology

The following steps are to be adopted to develop a methodology for siting pipeline alignment:

(i) ***Satellite Data collection***

In developing an RS & GIS based methodology for siting pipeline alignment, the satellite digital data are necessary for obtaining information about land use/land cover, roads, rail roads, drainage crossings, stream network, wetland distribution, and marshy land, and proximity of environmentally sensitive resources. Therefore, the satellite data obtained by IRS-1D (Pan + LISS III) sensor in digital format are collected from National Remote Sensing Agency (NRSA), Balanagar, Hyderabad. Collection of maps namely, SOI toposheets, slope maps, and geology maps are obtained from National Atlas and Thematic Mapping Organization (NATMO) and Geological Survey of India. (GIS)

(ii) Scanning of all maps for GIS database creation, geocoding, and geo-referencing.

(iii) Identification of sufficient number of well distributed Ground Control Points (GCPs) on base map

(iv) Digital analysis of remotely sensed data and transforming of GCPs on satellite digital image for geo-referencing

(v) Classification of digital data for the generation of land use/land cover layer and contour map for delineation of pipeline alignment.

(vi) Delineation of pipeline routes based on classified Imagery, contour information, land use\land cover, demographic, and socio-economic data.

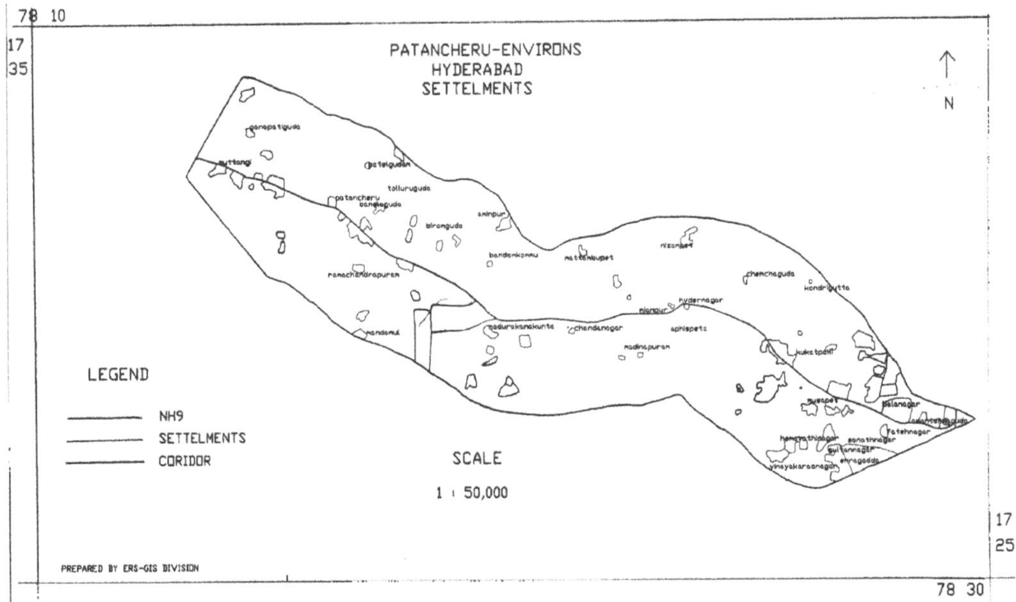

Fig. 16.1 Line buffer used in pipeline alignment study of patancheru industrial area.

(vii) GIS database generation and its analysis for the selection of the best pipeline alignment along with the generation of DEM for hydraulic gradient of proposed pipeline alignment.

More specifically, all the above operations are grouped in as follows :

* Acquisition and Processing of Satellite Data
* GIS Database Creation
* GIS Data Analysis
* Map and Table output production
* Integration for the development of generalised Decison Support System (DSS).

Processing of Satellite Data :

* Image restoration and registration
* Geocoding/Geo-referencing
* Data merging (PAN + LISS)
* Image classification/analysis
* Generation Ground Control Points (GCP)
* Training sample database creation

* Supervised classification for land use/land cover map generation
* Visual analysis of classified output
* Modifications for final map preparation

GIS Database Creation (Spatial and Attribute Database) :

A large number of data layers are to be created to support environmental analysis. Standard Data conversion methods to create a single consistent ARC/INFO GIS database (ARC/INFO is database software produced by ESRI) are used through scanning of all maps and digitisation/conversion methods to create the digital database.

Deliniation of pipeline routes is based on classified imagery, contour information, land use/land cover, demographic, and socioeconomic data. A consistent land-use map for the entire area under study is also necessary to evaluate pipeline routing alternatives.

GIS Data Analysis System (4 Analysis Methods) Consists of

* Buffer Analysis
* Overlay analysis
* DTM generation and analysis and
* Route Analysis

16.2.2 Buffer Analysis

Whenever information regarding proximity to the pipeline is needed, buffer zones around the pipeline data layer will be created at a user-specified distance. For overlaying the buffer with the data layer in question (wells, agricultural land uses, sensitive species locations and so on), it is possible to determine the number and identity of these features or areas within the buffer distance with the help of land use/land cover maps. Fig. 16.1 shows an example of line buffer used in the study of siting pipeline alignment for transportation of wastewater.

16.2.3 Overlay Analysis and DTM

This type of analysis will be employed whenever questions arise regarding the location of two or more types of data. An overlay analysis within a GIS creates a new layer of data which are the combination of pipeline arcs and polygon arcs and polygons. A quick summary analysis on this layer then yields the length of pipeline within each polygon.

DTM can be generated based on elevation data from topographic maps and spot heights. Data is prepared using 'Fly Model' of EASI/PACE. Image Processing Software. DTM with Remote sensing data can be draped to get a virtual view of the terrain. Plate 16.1 is an example of DTM.

16.2.4 Route Analysis System

Most of the questions posed related to the length that each pipeline route traversed various environmentally sensitive resources, such as, wetlands and stream crossing (land use/land cover map). Some of the questions that have to be answered within the GIS database include:

* For each pipeline alignment, how many miles of land use are traversed ?
* Are there any drinking water wells within 150ft of each alignment (centerline)? if so, at what milepost?
* Does the pipeline cross any wetlands? If so, what is the beginning and ending milepost for each wetland?
* Are any streams or rivers crossed? If so, what is the milepost of the crossing ?
* Are any highways, roads, or utilities crossed? If so, what is the milepost of each crossing?
* What is the total length of each alignment?
* What is total length within each revenue division? At what milepost does the pipeline cross the boundary?
* At what milepost does the pipeline cross into or out of public lands?
* At what milepost does the pipeline cross within a mile of a listed special status species? How many times has this occured along each alignment?
* At what milepost does the pipeline cross within a mile or hazardous waste sites and underground storage tanks? How many times has this occured along each alignment?

It is possible to work with attributes of the pipeline in terms of mileposts rather than x, y coordinates. Furthermore, attributes may be modified without underlying pipeline coordinates.

16.2.5 Map and output Production

A series of maps showing the study area, pipeline route, and environmental features will be produced to the calculated lengths of traversed features, and milepost information will be generated. The route is then segmented on a milepost basis according to land use/land cover type. Finally, the mileposts/ landuse tables will be summerised and printed, allowing comparison between the pipeline route alternatives.

16.2.6 Data Integration

More accurate information will be generated regarding the location of the pipeline and the environmentally sensitive features potentially affected. GPS will be used to obtain precise coordinates of the pipeline centerline wetlands crossed by the pipeline. Route analysis allowed the production of milepost information to indicate where the pipeline crossed wetlands, streams, or other environmentally sensitive features. Buffer analysis located environmentally sensitive features, such as threatened and endangered species, wells, and other environmental sensitive sites within a particular query distance.

16.2.7 Tools to be Used

Hardware

* A Pentium III Computer System (Processing of Satellite Data and GIS Data Analysis)
* A0 Colour Scanner (Scanning of Maps)
* GPS (Field Work-Locations)

Data Products and Sources used for the pipeline study with special reference to Indian conditions are provided in Table 16.1.

Table 16.1 Data Products and Sources

S.No.	Data Product	Source
(i)	Satellite Data (IRS-PAN, LISS)	National Remote Sensing Agenacy
(ii)	Toposheets	Survey of India
(iii)	Other Thematic Maps Land use/Land cover Geology Forest Cover Soil type Demographic information Rain fall Natural Hazards	National Atlas and Thematic Mapping Organisation (NATMO) Satellite Data Field Work India Meteorological Department (IMD)
(iv)	Geological Maps	Geology Survey of India, Satellite Data
(v)	Socioeconomic Data	Bureau of Statistics and Economics, India

The application of remote sensing and GIS has demonstrated the capability and potential of high resolution satellite data in the inventory and assessment of natural resources for environmental impact assessment studies. Similar studies can be carried out using IRS-IC/1D high resolution satellite data of the latest time frames for urban studies in the form of sprawl analysis, encroachment studies, urban infrastructure studies and so on. Increase or decrease of saline soils, land degradation, degradation in forest cover can also be monitored using remote sensing satellite data.

16.3 Case Study - Patancheruvu Industrial Area

The case study deals with the application of IRS - LISS data and GIS techniques for the siting pipeline alignment from Patancheru to Balnagar of Hyderabad district, Andhra Pradesh using the methodology.

The study area covers 18km length alignment passing through different areas. Satellite data of IRS - LISS III on 1 : 50,000 scales have been interpreted for this study. The thematic information required for suggesting pipeline alignment are the occurrence and distribution of various landforms, mining/industrial belts, densely/ sparsely habituated areas (villages, towns etc.,) forest cover, changing river courses, flood prone areas, erosion prone areas, marshy/water logged areas, stream/canal/ railway line/road network etc.,.

Satellite data with multi spectral capability and repetitive coverage have provided such integrated information required for pipeline alignment. For easy approachability and maintenance, infra structural facilities like road/railway network, important towns/cities etc., have also been considered in deciding alignment. After conducting field traverses along in the pipeline corridor maps have been prepared finally on 1 : 50,000 scale for regional appreciation showing that tentative alignment by shortest root with detailed information and alignment with actual number of turning point, road/railway line/stream/canal etc., In addition pipeline alignment is also proposed nearer to the national highways and important cities to facilitate easy maintenance and monitoring.

16.3.1 Land use/ Land cover Assessment

The total geographic area of the present study is distributed in Nine scenes of IRS LISS III data and these data sets have been acquired form NRSA (National Remote Sensing Agency) Hyderabad.

Image processing software namely ENVI 3.5 is used for all preprocessing steps to get a composite image, showing the total geographical area which is used land use/ land cover assessment. Using Envi 3.5 classification routines, land use/ land cover maps are prepared.

Preprocessing steps :

(i) Image Interpretation

(ii) Image Enhancement

(iii) Image Rectification

Comparison Matrix

	Barren	Water	Vegetation	Agriculture	Built - up
Barren	1	9	7	7	9
Water	0.111	1	0.5	1	0.5
Vegetation	0.143	2	1	5	
Agriculture	0.143	1	0.20	1	2
Built - Up	0.111	1	0.167	0.5	1

Normalized Matrix

	Barren	Water	Vegetation	Agriculture	Built-up	Total	Weight (Tot/5)
Barren	0.663	0.643	0.799	0.483	0.487	3.066	0.6132
Water	0.074	0.0714	0.057	0.069	0.028	0.2994	0.05988
Vegetation	0.095	0.143	0.113	0.345	0.324	1.02	0.024
Agriculture	0.095	0.0714	0.023	0.069	0.19	0.4484	0.08968
Built - up	0.074	0.0714	0.019	0.0345	0.054	0.2529	0.05058

	Barren	Water	Vegetation	Agriculture	Built-up	Total	Weight (Tot/5)
Barren	0.6132	0.53892	1.428	0.62776	0.45522	3.6631	5.974
Water	0.06806	0.05988	0.102	0.0896	0.02529	0.38215	6.39
Vegetation	0.0876	0.1197	0.204	0.448	0.3034	0.9901	4.853
Agriculture	0.0876	0.0598	0.408	0.0896	0.1011	0.3881	4.328
Built - up	0.6806	0.0598	0.03406	0.448	0.0505	0.2574	5.09

$$\lambda = \frac{5.94 + 6.39 + 4.85 + 4.33 + 5.09}{5}$$

= 5.327

CI = I – n / n – 1

= 5.327 – 5 / 5 – 1

= 0.0819

Finally consistency ratio = CI / 1.24

= 0.0819 / 1.24

= 0.06604 which is < 0.1

Hence, it is good consistency in pair wise comparison

	Rank 1 (meters)	Rank 2 (meters)
Least	74 meters	105 meters
Moderate	182	177
Good	1545	409
Very Good	18212	17574
Excellent	7654	10284

Classification

In the present study both supervised and unsupervised classification routines are used for land use / land cover assessment. The classification proceeds through selection of training areas, calculation of statistics for the identified training areas and decision boundary of maximum probability based on mean vectors, variance, covariance and correlation matrix of the pixels.

16.3.2 Unsupervised Classification

In ENVI 3.5 unsupervised classification there are two algorithms namely

(i) Iso Data Classification

(ii) K - Means Classification

For the present study K - Means unsupervised algorithm is used.

K-Means unsupervised classification calculates initial class means evenly distributed in the data space and then iteratively clusters the pixels into the nearest class using a minimum distance technique. Each iteration recalculates class means and reclassifies pixels with respect to the new means. All pixels are classified to the nearest class unless a standard deviation or distance threshold is specified, in which case some pixels may be unclassified if they do not meet the selected criteria. This process continues until the number of pixels in each class changes by less than the selected pixel change threshold or the maximum number of iterations is reached.

16.3.3 Supervised classification

In ENVI 3.5 supervised classification there are seven algorithms namely ;

(i) Parallelepiped Classification

(ii) Minimum Distance Classification

(iii) Mahalanobis Distance Classification

(iv) Maximum Likelihood Classfication

(v) Spectral Angle Mapper Classification

(vi) Binary Encoding Classification

(vii) Neural Net Classification

For the present study Maximum Likelihood Classification supervised algorithm is used.

Maximum likelihood classification assumes that the statistics for each class in each band are normally distributed and calculates the probability that a given pixel belongs to a specific class. Unless you select a probability threshold, all pixels are classified. Each pixel is assigned to the class that has the highest probability (i.e., the "maximum likelihood").

The pipeline passing through the various landforms and have been categorized accordingly as shown in the above chart.

The pipeline passing through the rank 1 is 27, 667 meters wheras Rank 2 is 28, 550 meters and hence concluded that Rank 1 is the suitable in all aspects for the alignment of the pipeline.

- This works the utility of remotely sensed data for oil pipeline alignment study. The survey of India toposheets prepared two decades ago have helped us to take the base information whereas the recent topographical and terrain information which are the most important for this study such as various landforms, slope, geological formations, streams, canals, forest cover, soil type etc., have been collected from the satellite imagery as it has the temporal capability.
- The pipeline corridors have been selected in such a way to minimize the sensitive areas along the route like reef, vegetated areas etc.,
- It is seen from the imagery for rank 1 the pipeline is passing through the sand, mudflat with sparse vegetation and barren mud flats up to the landfall point.

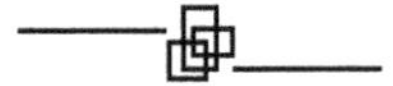

17
Potential Fishing Zone Mapping

17.1 Introduction

India, with a long coastline, is one of the major marine fish producing countries and occupies seventh position in the world in regard to the quality of fish harvest. The declaration of Exclusive Economic Zone (EEZ) in 1977 for exploration, exploination and utilisation of the living resources in the sea up to 200 nautical miles from the shore has given exclusive rights to the country to formulate its policies for the development and use of the fisheries resources in an area of 2017900 sp.km (NRSA, 1993). This extended jurisdiction, provides a challenge for a new and improved basis for rational exploination and utilisation of the resources and to enhance opportunities for fisheries to play a greater role in increasing food production so as to help minimise protein malnutrition and contribute to the betterment of the socioeconomic conditions of the poorest sections of the population.

Though it is nearly one and half decade since this extended jurisdiction has become available to the country, for various reasons the progress in the exploitation of the resources in the entire EEZ region has not been satisfactory. The fishery has only been taking the resource available in the inner continental shelf up to about 50 m depth. By and large the resources available in the areas beyond 50 m depth still remain unexploited. The situation, hence, calls for a reassessment of the strategies and policies for marine fisheries development and management in India.

Commercial fishermen are always been concerned about safety at sea and in making the best catch for the amount of time spent in search of productive fishing areas. In the last decade, commercial fishing on the high areas become an increasingly competitive and economically risky business. To ensure a profit, it has become necessary for the fishermen to utilise available technological and scientific knowledge to improve their catches.

Accordingly, modern commercial fishermen require timely, reliable and accurate Meteorological and Oceanographic information. Although the use of satellite imagery in the preparation of fisheries-aids products has been limited, it has the potential to contribute significantly in meeting the increased needs of fishermen for environmental information. Satellite Oceanic remote sensing is to play an important role in fishery research and fishery management by providing synoptic and Oceanic measurements for use in evaluating environmental effects on the abundance and availability of fish populations.

17.2 Factors Affecting the Availability of Fish

The factors affecting the availability of fish are Sea Surface Temperature (SST), temperature, currents, salinity, ph, upwelling and abundance of food. These factors are described in the following paragraphs.

17.2.1 Sea Surface Temperature

The Sea Surface Temperature (SST) can be monitored by remote sensing technology. Radiometer sensitive to the thermal infrared range of the spectrum can, under cloudy condition, be utilised to map the SST. Further relations between the SST and the potential fishery resources have been subject to investigations before the existence of remote sensing technology. Ocean frontal zones, as resolved by strong gradients in the surface temperature distribution, have historically been important as indicators for fishing localities over all the oceans.

17.2.2 Temperature

Sea water temperature may influence the activity, movement and metabolic process of the fish. Every stock of fish has a normal temperature range for maturation and spewing activities. Water temperature may delay or hasten the onset of spawing and the fishes migrate to favourable places for ensuring the required temperature range for the above purpose. This may also coincide with a particular season such as spring, summer, monsoon, post monsoon etc. incubation of eggs, development and growth of larvae and young fish are also influenced by favourable temperature ranges.

The temperature has also its effect on the food organisms of the fish and this in turn will trigger seasonal migration of fish in search of food another thermal feature which affect the vertical migration of fish is the thermocline which is defined as a layer within the sea where rapid decrease in temperature occur in the vertical downwards. Above this thin layer, water are usually warm and rich in dissolved oxygen content, which conditions are favourable for the pelagic fishes. Temperature in the sea at different depths is governed by several factors such as solar radiation, summer heating, winter cooling, tides, currents, upwelling and divergence.

17.2.3 Currents

Fishes generally respond directly to the environmental changes to the current. Determination of temperature, oxygen content, salinity and plankton composition are used as tools for identifying water masses and their movement. For instance, along the West - coast of India, it has been observed that the current are mainly southwards during the monsoon and post-monsoon and mainly Northwards from November to April. Along the east coast of India, a north-easterly current flows during January to July and south-westerly currents during August to December.

17.2.4 Salinity

Salinity variations at surface levels mainly depend upon rainfall, run off from the rivers and evaporation at the sea surface. Salinity variations in offshore waters are relatively less compared to estuarine and near shore coastal waters. The salinity maximum found along the west coast during winter and summer seasons may act as a barrier for the vertical migration of eggs, juveniles and fish.

17.2.5 The pH

The pH of seawater generally observed in the sea range from 7.5 to 8.4. This means that the sea water is generally alkaline. Fishes in general are able to detect slight changes in pH values and they migrate accordingly.

17.2.6 Upwelling

Upwelling is a vertical current system from deeper depths (from about 400-500 m depth) towards surface. In coastal waters, upwelling is caused by of shore winds (winds from land to sea) or coastal (horizontal) current with the coastline to the left of current and/or by a combination of wind and surface currents under favourable circumstances. It is established that, most of the very good fishing grounds of the world oceans are located in the areas where upwelling is found to occur. Since subsurface water rises to the surface during upwelling a marked decrease in sea surface temperature is observed. The surface temperature in these regions will be less than that at the non-upwelling regions of the same latitude.

In the normal case, sea water temperature decreases with depth. During the process of upwelling the cooler waters from the depths are transported to subsurface and surface level. Upwelling increases the salinity of top layers and due to the combined effect of low temperature and greater salinity the density of subsurface waters increases in the areas of upwelling. As regards chemical conditions during upwelling the oxygen content shows marked decrease whereas the nutrient contents increases.

17.2.7 Abundance of Food

The food organisms consist of phytoplankton, zooplankton, nekton and benthic organisms. Their production in turn is determined by solar energy received, availability of nutrients, currents and other conditions such as weather, rainfall etc.

The availability of food suitable of larvae at the appropriate time is related to the phytoplankton production, which, again is closely related to the seasonal changes of temperature and the amount of light (Reddy M P M 1993). Also the abundance of zooplankton, important food for the larvae, is related to the abundance of phytoplankton. In these above sentences, the important point is that the food for larvae is plankton. Even it is linked to zooplankton which is also a food which is directly related to the abundance of phytoplankton. So we can say that where there is lot of phytoplankton we can also find lot of zooplankton and where we find lot of zooplankton we can find fish at that place (Reddy M P M, 1993). So in the present chapter more importance is given to the link between phytoplankton and fish. The essence of the work is to say that chlorophyll can be used as an indication of fish areas with the help of remote sensing technology.

17.3 Potential Fishing Zone (PFZ)

The area in the water body where the availability of finding fish is more is referred to as Potential Fishing Zone. These zones can be located using conventional methods and also by using Remote Sensing technology. Some of the important points of this PFZs are given below.

17.3.1 Potential Fishing Zone Map

The Indian subcontinent is divided into nine sectors with the help of Naval Hydrographic Office (NHO) maps. This base map will have coastal reference points, fishing villages and important fishing harbours and land marks and bathymetry. The SST is analysed to delineate the thermal gradients and the thermal features are optically transferred on to the corresponding sectors.

Finally the distance of Potential Fishing Zone (PFZ)s from the coastal reference point, depth direction at which PFZ is available (for location) and the direction at which the PFZ is present with reference to the coast is read from the PFZ map.

17.3.2 Conventional Methods of Identifying Fishing Zones

From the decks of their boats, commercial fishers observe target species, habitat, prey and changes through time to the environment. Their observations, perceptions and impressions are coloured by their values, culture and interactions with other fishers. Fishers who are at sea every day hold a vast amount of knowledge, however, their information is considered anecdotal.

As targeted fish species have increasingly become over or fully exploited, awareness of the need to use all available information as the basis for management decision making has grown. Thus, local knowledge, or what has come to be known as traditional ecological knowledge that could be valuable to scientists. Local knowledge in fishing communities has been described in various places in the Pacific, South America, the Caribean and India, but rarely has this knowledge been assessed with regards to management and sustainability (Ruddle, K 1994).

17.4 The Need of Remote Sensing Technology

At present the fishing activities are concentrated to a narrow belt of inshore waters up to a depth of about 50 m. This shows that even the productive fishing areas along the continental shelf beyond 50 m is under exploited. The deep sea fishing venture is in the infant stage and information on fish stocks beyond 50 m depth zone is still elementary. Inspite of revolutionary change due to the impact of mechanisation of the fishing crafts and by use of synthetic fibers for gear, the marine fish production is far short of the estimated potential of 3.92 million tons from the Exclusive Economic Zone (EEZ).

17.4.1 Remote Sensing and Fishery

Satellite remote sensing is becoming routinely used in marine fisheries applications, including fisheries research, resource management, and marine fish harvesting. Fish schools generally are not directly detectable from space.

The rationale of applying satellite remote sensing to marine fisheries questions is based on the principles that :

1. Variations in ocean conditions cause natural fluctuations in the distribution, abundance and availability of marine fish stocks,
2. Variations in ocean conditions can influence the vulnerability of fish to harvesting.

17.4.2 Remote Sensing and Oceanography

The increasing population and far dwindling land resources all over the world are causing concern and a need is felt to exploit marine resources. Most of the Island and coastal nations have large dependency on Ocean resources primarily on sea food. Though India is predominantly an agrarian nation, off late realised its vast potential with respect to Oceanic resources. However, the vagaries of monsoons largely influence agricultural/marine fish production in India. More than 90 % of the food consumed by the human race is produced by the farmers, while the fisherman provide about 10 % of it. Our endeavor is to take the Remote Sensing Technology directly to fishermen for their betterment. The use of satellite remote sensing in oceanography expanded considerably in the 1970's and satellite with dedicated oceanographic sensors were first launched in 1978.

17.5 Coastal Sensing Systems

17.5.1 IRS P4 - Oceansat

Most of the existing missions cater to land applications. IRS P4 satellite, which is also called as Oceansat, will primarily cater to oceanographic applications. This second generation satellite was launched by the indigenous Polar Satellite Launch Vehicle (PSLV) on 26th May, 1999 from Sriharikota, India, into a polar sun - synchronous orbit at an altitude of 720 Km. The satellite has a high repetivity of two days. The payload includes a Ocean Colour Monitor (OCM), a Multifrequency Scanning Microwave Radiometer (MSMR) and solid state memory for recording data outside the visibility of a ground station. The OCM has six spectral bands in the visible and two in the near infrared region of the electromagnetic spectrum. The specifications of OCM are presented in Table 17.1 and Table 17.2 shows the applications of OCM.

Table 17.1 Specifications of IRS - P4 OCM

S. No.	Specifications			
1.	Ground Resolution (m)			
	Real-Time Mode (Full Resolution Mode)		: 360 (across - track)	
			: 236 (along - track)	
2.	Swath (km)		: 1420	
3.	Instrument Spectral Bands	Wavelength Range (nm)	SNR at Ocean Radiance*	Radiance's Expected SNR
	C - 1	402-422	9.1	307.6
	C - 2	433-453	8.4	440.7
	C - 3	480-500	6.6	415.0
	C - 4	500-520	5.6	408.8
	C - 5	545-565	4.6	412.2
	C - 6	660-680	2.5	345.6
	C - 7	745-785	1.6	393.7
	C - 8	845-885	1.1	253.6
	*In mw/(cm^2-Sr-mm)			
4.	MTF (at Nyquist)	> 0.2		
5.	Quantisation	12 bits'		
6.	Data Rate (Mbps)	20.8 Mbps		
7.	Transmission Frequency	X - band		
8.	Along Track Streeing (To avoid sunlight)	+ 20^0		

Table 17.2 Applications of OCM bands

Band No.	Band Centre (nm)	Band width (nm)	Applications
1.	412	20	Yellow substance and turbidity
2.	443	20	Chlorophyll absorption maximum
3.	490	20	Chlorophyll and other pigments
4.	510	20	Turbidity, suspended sediment
5.	555	20	Chlorophyll, suspended sediment
6.	670	20	Chlorophyll absorption
7.	765	40	Oxygen absorption R - branch
8.	865	40	Aerosol optical thickness, vegetation, water vapour Reference over the ocean

17.5.2 CZCS - NIMBUS Satellite

When the Nimbus - 7 satellite with CZCS was launched in 1978 there was increased enthusiasm and interest by fishery scientists to utilise satellite ocean colour measurements in fisheries research especially the pelagic fishery. Ocean colour measurements from the CZCS are being used in fishery resource applications for the location of water fronts, circulation features, quantitative determination of ocean colour that are directly related to chlorophyll concentrations (Laurs et al. 1984). Laurs et al (1984) has discussed the use of SST for identifying Potential Fishing Zone (PFZ). They even emphasised on ocean colour and wing. Ocean colour, which is mainly due to the presence of chlorophyll which in turn, indicates the presence of phytoplankton. This chapter mainly provides insight on identification of chlorophyll and linking this to the availability of fish. As it is a fundamental knowledge that chlorophyll which is present in phytoplankton is a primary link to zooplankton and fish in the food chain.

Satellite-derived ocean colour data are probably more useful in fisheries applications than any other oceanic measurement made from space. This was shown during the operation of the Coastal Zone Colour Scanner (CZCS) on board the NASA research satellite NIMBUS 7. This satellite provided useful and excellent information on global ocean colour for approximately 7 1/2 years during 1978-1986 (Michael Laurs R 1993). In this study importance on satellite derived ocean colour data are given which is probably very useful in fisheries application. Colour to oceans is basically impounded by phytoplankton due to the chlorophyll in the plankton using remote sensing technology.

These principles are established on the theory that information is necessary on the "changing ocean" instead of "average" ocean conditions, to understand, model, and predict the effects of the ocean environment on marine fish stocks. Monitoring of the marine environment is done for studies on the different aspects of the sea including biological, physical, chemical, geological and meteorological parameters in order to estimate the productivity of the sea and its relationship directly and indirectly to the fish potential and other environmental factors. These studies which are carried out giving thrust to different aspects by different agencies, obtained through satellite sensing and photography covering wider area of the ocean help fishery forecasting much more effective and meaningful.

17.6 Indicators of Fish Potential

17.6.1 Ocean Colour

In recent years, study of ocean from satellite in terms of temperature, sea level and ocean colour has given as insight into the changing climate of the world. An understanding of ocean water properties and processes can be obtained more effectively through orbital Remote Sensing as compared to conventional ship based observations. The conventional methods lead to point measurement whereas satellite sensor with a wide area synoptic coverage aids in collecting spatial as well as temporal data. Use of appropriate algorithms along with relevant sea truth data from ship based sensors can be used to complement the data from satellite sensor.

Ocean colour, in particular, phytoplankton pigment play a vital role in the oceanic system. Almost it works like "Biological Pump" to maintain the ratio of CO_2 and O_2 in the atmosphere. Remote sensing in optical region is found useful in understanding the spatial distribution of ocean water constituents both particulates as well as dissolved materials. Amongst these the one which has definite response in the visible region are the phytoplankton pigments (Chlorophyll - a and pheaopigment-a) which impart in general a green colour to the sea water, chlorophyll - a is the pigment present in all living plants which is responsible for photosynthesis. In the ocean this pigment is contained in microscope free floating organisms (expect Sargassam Sp.) called phytoplankton which is the first link in the ocean food chain (Aishwarya Narain, 1993). The above study has given in detail how chlorophyll is identified using a different technique through Remote Sensing. An estimate of ocean colour provides an useful input about the primary productivity due to phytoplankton. Primary productivity estimation leads to the third level productivity i.e. fish in the ocean food chain. The requirements of a satellite sensor for oceanic regime is quite different from these designed for meteorological or land applications.

The first and foremost is the strength of the apparent signal received at the satellite sensor that has information about the ocean water which constitutes only 20 per cent of total signal and the remaining 80 per cent is due to the intervening atmosphere resulting from molecular scattering, namely Rayleigh scattering and scattering due to suspended dust particles like aerosols caused by Mie scattering.

Atmospheric correction schemes have been suggested to account for the removal of these atmospheric effects (Gordon 1978, 1980, 1981; Gordon and Clark, 1981) so as to arrive at the upwelling subsurface radiance which contains information about the ocean water constituents, particularly, phytoplankton pigment.

The change in the ocean colour from blue to green may also serve as an indicator of increasing plankton abundance. The green colour is associated with the presence of chlorophyll. Area with high biological productivity is indicated by high chlorophyll biomass. This area can be said to attract fish and where fish school known to congregrate (Abdul Hamid Yasin and Mansor Mat Isa, 1993). As the paper discuss about phytoplankton and fish, where fish can be located. OCM data can be used where its bands have special application for identification of chlorophyll, which in turn indicates the presence of phytoplankton.

17.6.2 Phytoplankton and Fish

A distinction among different substances has to rely on characteristic spectral signatures within the visible spectrum. Phytoplankton is the only substance for which such a signature exists, the flurorescence of chlorophyll a, l = 685 nm. Phytoplankton respond to variations in light intensity as they more mixed through the water column.

Remotely sensing plankton's optical properties, using the signature of chlorophyll as an index of abundance. The emphasis on chlorophyll is inevitable, given chlorophyll's distinctive optical characteristics, which enables plant material to be distinguished from the other suspended matter. The relationship between chlorophyll and organic carbon is highly plastic, varying with growth, irradiance, nutrient availability and temperature. Because of water's capacity for high levels of latent heat, temperature is more likely to vary on daily to seasonal scales, and in a manner predictable form the balance of radiative transfer and evaporative cooling. By contrast, phytoplankton are subject to relatively rapid changes in both the intensity and the spectral quality of light as they move vertically in a water column.

The light attenuation within anybody of water follows (approximately) the Lambert - Beer law, decreasing exponentially with depth. Phytoplankton is low and the dominant attenuation is water itself, which is highly absorbent in red light. The dominant penetrating wavelengths are therefore blue light. Coastal waters are intermediate with respect to the magnitude and spectral dependence of attenuation. The light regime experienced by phytoplankton also depends on the rate of mixing, which is much more rapid in estuarine waters than coastal or open ocean waters.

The spatial and temporal variability in phytoplankton abundance is a major factor determining the productivity of mussels. During periods of low phytoplankton abundance, food depletion may be a significant factor in the productivity of mussel farms (Shaun C Ogiluie et al, 2000). This of course, forms a strong base, that phytoplankton can be used to identify fish area, and Remote Sensing technology is applied in identifying phytoplankton.

17.6.3 Sea Surface Temperature and Fish

Among the various oceanographic parameters the two viz,. ocean colour and temperature have been widely used in locating marine fishery resources. Fish like other animals survive best in particular ocean environmental conditions, which are characteristic for each species. They react to change within the environment and undertake migrations to those areas where the conditions, such as proper temperature, salinity, nutrients, food and other factors are favorable to them.

Fish and sel - fish are cold blooded animals and, unlike mammals and birds, their body temperature is not internally regulated but approximate to that of their surrounds. Fish have the ability to perceive and select a limited thermal range in which its offers the opportunity for maximum expression of activity is ultimately revealed in their abundance and distribution.

The sense of temperature in fish seems to be well developed. Temperature of sea water bears a marked effect on fish by affecting their metabolic rate which increase with rising temperatures, so that in warmer waters it is higher, while in cooler waters it is lower. Thus, temperature affects fish growth, multiplication and development. It also influences the migration of fishes and their geographic and ecological distribution.

The sense of temperature in fish seems to be well developed. Temperature of sea water bears a marked effect on fish by affecting their metabolic rate which increase with rising temperatures, so that in warmer waters it is higher, while in cooler water it is lower. Thus, temperature affects fish growth, multiplication and development. It also influences the migration of fishes and their geographic and ecological distribution. Fish respond to minimal changes in temperature, unnoticeable to man. Fish responses to a 0.03^{O} C temperature change are noticed. In addition to direct influence, changes in seawater temperature may indicate changes in other environmental factors. The idea of remote measurement of the surface temperature of an object stems from Plank's radiation Law.

Fisheries forecasting using satellite derived SST has been implemented by the NRSA, Hyderabad since 1990, for the first time in India. Under the national project on "Ocean Related Remote Sensing Applications", the SST over the Arabian Sea and the Bay of Bengal on the thermal gradients, the eddies and thermal fronts which are indicative of fish school aggregation are detected. These are called the "Potential Fishing Zone" (PFZ) which are intimated to the fishermen and Governmental and Non-Governmental agencies in near real time as Fisheries Forecast, 3 to 5 days in advance.

17.6.4 Phytoplankton and Sea Surface Temperature

Rhe availability of food suitable for larvae at the appropriate time is related to the phytoplankton production, which again is closely related to the seasonal changes of temperature and the amount of light. Also the abundance of zooplankton, important as food for the larvae, is related to the abundance of phytoplankton, and to the spawing period of adult plankton animals, which again is controlled by sea water temperature. The warm surface layer of sea water is separated from the colder deep water by a thin layer of rapid temperature change called "thermocline" or "temperature discontinuity later". The temperature in the surface layer is uniform due to mixed layer. The mixed layer supports pelagic fisheries, passively floating plankton, fish eggs, and larvae, whereas the cold waters below the thermocline support the benthic and deep water fauna.

From the above, it is clear that sea water temperature plays a very important role in fish abundance. It is an environmental factor of outstanding significance. Knowledge of the surface and subsurface temperature of water and their relation to local distribution patterns of fishes is important from the commercial fishing point of view.

Among the various oceanographic parameters the two viz., ocean colour and temperature have been widely used in locating marine fishery resources. Ocean colour in general and chlorophyll - a present in free floating plants (phytoplankton) in particular gives an indication about the favourable environmental conditions. In recent past, experiences were drawn by various investigators towards mapping the phytoplankton pigments using CZCS on board NIMBUS 7 satellite. However, in absence of ocean colour scanner, largely the experiences have been towards using the SST derived form NOAA AVHRR for different target species all over the world (Aishwarya Narain, 1993). This author emphasised that, mostly given on SST, by using different thermal bands and how to correlate SST for fishery exploration. It has been clearly mentioned that due to the absence of ocean colour scanner most of the research work has shifted to SST. OCM sensor has eight bands each of it having different applications as the study also works on chlorophyll identification using OCM 2, 3, 5 and 6 bands which are specially applicable for identification of chlorophyll concentration.

Shipboard collection of water and plankton specimens is both expensive and time consuming, and interpolation to regional scales is also difficult. Because oceanic phytoplankton provide food source for zooplankton, both phytoplankton and zooplankton, in general, are associated spatially and temporally. Given that relationship, both phytoplankton and zooplankton can be monitored by using remote sensing data because oceanic phytoplankton concentrations can be estimated by using Remote Sensing data. SST was used because it has been related to phytoplankton concentration. The relational between SST and cholera out breaks was expected because plankton blooms depend on warm ocean temperatures (Bard Lobitz et al, 2000). Although other variables, such as nutrients load associated with run-off, mixing, upwelling, light availability etc., may be related to plankton blooms, the relationship between SST and *Vibro cholerae* is more direct. Furthermore, it is now known that plankton serve as a reservoir for Vibro cholerae. It is established that through SST and plankton, the presence of Vibro cholerae which caused cholera out break may be detected. The identification of plankton using chlorophyll is been studied, using Remote Sensing technology by establishing the relationship between SST and plankton.

17.7 PFZ Method and Process

It is well known that various environmental factors greatly affected the behavior of fish. Fisheries forecasting can be taken into practice. The relations between the oceanographic factors and the behaviors of adult fish would give out sound bases for short term forecasting of fishing zone. Though the whole life circle of fish would be studied for full knowledge of the fish stock, the adults are the target for fishing and subsequently, the forecasting of short term fishing zone (Han Shixin, 1993). For each species to be forecasted the following biological features and their relation with the oceanographic conditions ought to be studied :

1. Distribution
2. Population Structure
3. Spawning
4. The target ages
5. the migration
6. Feeding habit
7. Fisheries
8. Abundance of stock and exploitation level
9. The social - economic importance

Three factors would be selected before any forecasting is really taken into action, namely, target species, Oceanographic factor and the schedule and area for forecasting.

17.7.1 The Method of Forecasting

Using basic knowledge about the target fish, the potential fishing zone could be drawn into the map through the steps listed below :

1. Identify the critical oceanographic condition in the map;
2. Roughly delineate the area where the target fish is expected;
3. Determine the zones in expecting area where critical condition is favorable to target fish;
4. Compare with practical fishing operation and last issued chart;

5. Discuss the difference come from comparison to see whether the change is regional, if not find out the reason then go back to the first step and if the answer is positive then;
6. Draw the potential fishing zone into the map.

After charting, some analysis and comments on the oceanographic features and stock state as well as prediction of expected change in near future could be attached.

17.7.2 Mode of Dissemination

PFZ information is sent to the grass root fishermen, Governmental and Non - Governmental agencies both in the form of text by telegram/telex as well as by fax depending on the facilities at the receiving end. The facilities of central telegraphic office can be utilised by the users. Information is also passed on to those who call over telephone. The PFZ information is translated into local language and broadcast over the all India radio and television in the form of post system, broadcast system, telephone system and radio facsimile system.

17.7.3 Forecasting and Management

Fish being a dynamic aquatic resource, the ability to forecast the catches or variations in the seasonal abundance of the stocks in different areas assumes great importance in the development and management of fisheries of exploited stocks principally two types viz., 1. Short term and 2. Long term forecasts are recognised.

Short term forecasts are made using different approaches :

(i) Analytical approach (Virtual Population Analysis - VPA)

(ii) Analysis of time series data on catch

(iii) Influence of environment on the availability or abundance

(iv) Changes in fishing pattern and

(v) Remote sensing and PFZs.

Among the short term forecasts, two approaches need serious consideration in India :

(i) to improve the methods of understanding the influence of environmental characteristics on the abundance or availability of fish in different areas in different periods and to make the forecasts of the same,

(ii) to make analysis of time series catch data to make forecasts of catch in the next year or in a particular period during next year.

Certain environmental parameters as limiting factors on the availability of stocks for exploitation is well-known and the study of environmental factors for the prediction of availability or abundance of fish stocks in the fishing grounds is gaining momentum.

There is some evidence of suitability of these approaches to Indian marine fisheries but attempts aiming at comprehensive studies should be made. In the area of long term forecasts, considerable work si done in India on single species assessments but in context of multi - species, multi - gear nature of Indian marine fisheries, assessments of all species together in a mixed fishery are urgently required for effective management of fisheries.

A long term forecast such as estimating maximum sustainable yield (MSY) depends on the factors such as recruitment, growth, mortality etc. which maintain the stock in the long run. Such an assessment is relevant to the strategy of management and is generally a part of an attempt to determine some long term goals or targets for management.

17.7.4 Need for Mass Awareness Among Fishermen Communities

Traditionally fishermen have been venturing into the sea for fishing to areas that are known to be good fishing grounds based on their past experience. They also make use of their knowledge on ocean colour, wind direction, surface current direction for paying their nets. However the catches cannot always be assured as the fish stock availability is dependent on several physical, chemical and biological factors.

In the case of PFZ fisheries forecasting fish stock occurrence is predicted with reasonably good success ever since experimental trails were conducted in the year 1990 and 1991 along the Karnataka coast with the participation of fishermen. Fishermen were convinced that this model is quite helpful as it showed increase in fish catches in the areas mentioned in the forecast. Even though this is a proven technology, it takes an exclusive and dedicated effort to take this technology to the grass root level (fishermen) for application in the field. As the fishermen communities take a very practical assessment before accepting any new technology, a greater responsibility lies in convincing them with practical demonstration.

17.8 Potential Fishing Zone Mapping Using NDVI

Vegetation indices are built as linear combination of albedos or bi - directional reflectance which depend on radiation properties of the surface and can be effectively utilized for discriminating different vegetation types. Normalised Difference Vegetation Index (NDVI) is sensitive to the presence of green vegetation and has been successfully used in numerous regional and global applications for studying the state of vegetation canopy reflectance, photosynthesis and transpiration (Seller P. J. 1985).

Inspite of variations in atmospheric effects (Kaufman 1984, Fraser and Kaufman 1985) and radiometric degradation in the red and near infrared bands (Holben et al 1990) the success of NDVI as a descriptor of vegetation lies in its normalization (Fraser RS and Kaufman Y. J. 1985).

17.8.1 Generation of NDVI

For the land (agriculture) applications, the so-called NDVI is commonly used. The NDVI is sensitive to the presence of green vegetation on land, based on the spectral characteristics of chlorophyll in the vegetation. Normally, since water is purely a black body for near infrared, the NDVI is negative for water surfaces with blue - green algae floating at the very surface, the reflectance in channels will be increased dramatically, due to the abundant presence of chlorophyll giving the NDVI a positive value. (G. J. Prangsma and J. N. Roozekrans, 1989). Here the use of near infrared bands have been used. So based on this scientific study, where it says that, NDVI can be used in identifying the presence if phytoplankton in a water body the present study has been carried out. Earlier studies conducted by various investigators. AVHRR-NOAA satellite data, using NDVI to identify plankton on water body. This case study uses OCM-IRS-P4 data to identify plankton, using NDVI. Photosynthetically active vegetation typically has a reflectance < 20% in the narrow band visible (0.5 – 0.7 mm) but a much higher reflectance up to 60% in the near infrared (0.7 – 1.3 mm).

17.8.2 Ranges of NDVI

The NDVI is a bounded ratio that ranges between-1 to +1. Clouds, water and snow have negative NDVI since they are more reflective in visible than near infrared wave lengths. Soils and rock have broadly similar reflectance giving NDVI close to '0'. Only active vegetation has positive NDVI being typically between about 0.1 and 0.2 values at the higher end of the range indicates increased photosynthetic activity and a greater density of the phytoplankton biomass. To the raw image (Plate 17.1) of the study area NDVI is applied. Through various computational procedures a grey coloured image of the study area is prepared. After applying the computational procedure, which is discussed in the next paragraph, the NDVI ranges have been taken and classified (Table 17.3). This classified ranges are then added to the image by applying various pseudocolours for each range of values of NDVI. According to this colouring the abundance of chlorophyll has been indicated (Plate 17.2).

The ranges between – 0.2 to 0.025 is classified as nil because the negative value indicates that chlorophyll is completely absent in that region. As only water body is reflected the NDVI value is negative for that area. The range form 0.025 to 0.1 is classed as Low. Which indicates that chlorophyll is very less or it has been scattered, or the growth is very low. The range from 0.1 to 0.175 is classified as medium class. Here the growth of phytoplankton may be is the early stages. The NDVI values between 0.175 to 0.4 is classified as High class. Mostly this refers tot he land area where vegetation is more when compared to the water body of the phytoplankton abundance. From the NDVI classification, it is said that the medium class range is the one which is showing the highest value of chlorophyll in the water body. So the presence of phytoplankton is more in this region.

17.8.3 Computational Procedure for NDVI

The remote sensing data used extensively for large area vegetation monitoring. Typically the spectral bands used for this purpose are visible and near infrared bands. Various mathematical combinations of these bands have been used for the computation of NDVI, which is an indicator of the presence and condition of green vegetation. These mathematical quantities are referred as vegetation indices. There are three such indices : Simple Vegetative Index (SVI), Rational Vegetative Index (RVI) and Normalised Differential Vegetative Index (NDVI).

These indices are computed from the equations :

$$VI = \text{Near IR - Red}$$

$$RVI = \text{Red/Near IR}$$

$$NDVI = (\text{Near IR - Red})/(\text{Near IR + Red})$$

Based on the formulate the ranges of the NDVI values are computed for all pixels of image. These values represent different grey levels in the image. Then the image is classified using the grey level values, and pseudo colouring is given to the grey levels. Basing on this pseudo coloured image different polygons are created on a tracing sheet. This map is then scanned, digitised in Auto CAD R 14 and edited in ARC/INFO software. This edited map is then exported into Arcview and the polygons are represented in the way we needed. And the final out put map showing different layers of the NDVI values is presented (Plate 17.2).

Table 17.3 NDVI ranges of the area

S. No	NDVI Range	Remarks
1	– 0.2 to – 0.1	NIL
2	– 0.1 to 0	NIL
3	0 to 0.025	NIL
4	0.025 to 0.05	LOW
5	0.05 to 0.075	LOW
6	0.075 to 0.1	LOW
7	0.1 to 0.125	MEDIUM
8	0.125 to 0.15	MEDIUM
9	0.15 to 0.175	MEDIUM
10	0.175 to 0.2	HIGH
11	0.2 to 0.3	HIGH
12	0.3 to 0.4	HIGH

17.8.4 Application of NDVI

NDVI is applied to the raw image using EASI/PACE image processing software. To the output image, classification is done and a pseudocolour image showing NDVI values has been developed (Plate 17.2). This image shows the different NDVI values where the photosynthetic activity is high or low or nil. Taking this image as a base, different polygons have been delineated. These polygons are delineated according to the pseudocolour given to each polygon. A base map of this image showing NDVI values, as polygons, is prepared. Each polygon or each pseudocolour has been given a number for easy identification.

This base map showing different polygons is then scanned. This scanned image is attached to the Auto Cad for digitisation. To work in a GIS software the image should be in the vector form. The image what we have scanned is in the raster form. To convert this raster image to the vector form digisting is done. In autocad different layers are created with different colours, for each polygon having different identification numbers. After digitisation of all the polygons, the image is then exported into Arc/ Infor GIS software.

17.8.5 GIS Data Preparation

GIS, which is a computerised database system, is easy for storage and analysis of the spatial data. GIS is a general purpose technology for handling geographical data in digital form and satisfying operations such as reformatting, change of projection resampling, generalisation, calibration of models, forecasting, prediction, tabulation, report generation and mapping.

GIS offers a multi dimensional, that is, attribute dimension, spatial dimension, and temporal dimension as they integrate multi-sector, multi-level and multi-period database. For better understanding of the image, GIS helps a lot as, we can in this context present well all the different layers or zones making use of the NDVI values. Care is taken for the data quality. Here each pseudocolour, which is represented as a polygon, is given an individual label with identification number.

17.8.6 PFZ Map Preparation

Editing is done to the image in ARC/INFO software. In editing at first the whole image is verified for dangles. These dangles are the major problem creators in the image. After processing of these dangles, labels are given to each polygon. Then the map is exported to Arcview software. Here in the image, arcs, labels, and polygons, which are imported, are created as themes. Each theme consists of different layers such as arcs, polygons, labels and annotations. There is an option of loading individual layers or any number of layers at a time. In this work only polygons have been loaded. At first whole image is displayed with a single colour. According to the preference, individual colours are given to each individual number of the polygons. This process is called classification. Classification can be done based on many parameters such as on system identification numbers, user identification (id) area, perimeter of the polygon. For the identification of different zones, classification is done by using Id's. As said before each individual polygon has been given a particular number, according to these numbers the classification is done. Each number which represent the level of chlorophyll content and which inturn represents the level of presence of fish in each polygon.

All these processes are done Arcview software. For the purpose of displaying, they are arranged in the layouts. In the layout the image, its title, north indication, legend and scale are attached and displayed. There are many models for each of these options and can be selected any of them. Then the final output of the map showing the potential fishing zones is displayed. (Plate 17.3).

In the final output map all the land features have been given a particular number and colour, all the areas having no phytoplankton biomass, all the zones having low phytoplankton biomass, and all the zones having high growth of phytoplankton biomass are given a different number and colour, for easy identification of the zones of fish potential.

Chlorophyll identification can be related to phytoplankton abundance and phytoplankton abundance can be inturn related to locating of fish species at that region in the map. It shows the zones of chlorophyll presence and its levels. Based on these results the map showing PFZ is prepared. It is interpreted that where the presence of phytoplankton is more the availability of finding fish is more in that region. Where the phytoplankton is less, the availability of finding fish will be less in that region. This whole picture is based on the fact that chlorophyll can also be used as a base for the presence of fish.

Based on the foregoing study, it is concluded that NDVI, which is mainly used on land applications for the identification of the vegetation index can be used on water surface. The main concept of NDVI is based on the chlorophyll absorption. Phytoplankton which is also having chlorophyll, had given positive NDVI values. In a food chain the primary link is of phytoplankton to zooplankton or to fish. Hence it is concluded that presence of phytoplankton can be used as a base for the presence of fish, for which NDVI can be used in identifying PFZs as an indicator.

17.9 Coastal Change Detection

Coastal environment, which is very dynamic in nature, is the place where several processes like fluvial, marine and natural processes occur and which are the main reasons for formation of present day landforms. In recent years aquaculture has grown in many folds in coastal Andhra Pradesh resulting in conversion of agricultural fields into aqua farms which in turn have adverse effect on soil fertility due to increase of saline compounds. In India total aquaculture shrimp production increased from 30,000 tonnes in 1990 to 1, 02, 000 tonnes in 1999. Keeping in view the environmental degradation, salinization of drinking water and reduction of regular crop yield, an attempt is made by the author of this book to study the temporal changes on land use/land of Kakinada Bay and its environs due to aquaculture.

Shrimp aquaculture is responsible for about 80% of the conversion of mangrove land. The decrease in the area of sparse mangrove cover is partly reversed by the conversion of dense into sparse mangroves probably through fuel wood collection and grazing. The rate of conversion of mangroves into shrimp ponds increased in the period 1997 to 1999, suggesting that shrimp pond construction started in fallow land and crop lands but then encroached on mangroves in the absence of suitable fallow land. Policy regulations banning the conversion of mangroves to shrimp ponds and the protected status of the Godavari forest have not been able to prevent the conversion of mangroves into shrimp ponds. In addition to the conversion of mangroves, shrimp aquaculture contributed to a number of other environmental and social conflicts. Protests from land villagers, supported by various social and environmental NGOs culminated in 1996 supreme court judgement and the 1997 aquaculture Bill which established a regulatory framework for shrimp aquaculture in India.

A comprehensive change detection analysis of a part of east coast of India i.e. Kakinada Bay is carried out based on temporal satellite data. Temporal changes in the land use/land cover for a period of 10 years was determined using IRS IB LISS II and IRS ID LISS III data and a change detection map was prepared. From the remote sensing images it was apparent that in the Godavari delta area, about 14% of the aquaculture farms have been constructed on mangrove lands. The mangrove coverage in the Kakinada Bay was reduced by 0.12 times with respect to 1989-90 to 1999-2000, which is due to the conversion of mangrove forests to aquaculture. The aquaculture coverage increased by 8.4 times, vegetative coverage by 2.3 times, fallow land by 1.7 times, dry land by 2.1 times, wetland by 0.13 times and water body 0.8 times in the decade.

A number of policy measures have to be recommended to increase the enforcement and to decrease the environmental and social costs of shrimp aquaculture in India. All farms including the existing farms need to be obliged to obtain a license to continue operations and the state level committees should be given sufficient means to carry out site investigations of all new applications. The state level committees should also give more consideration to the coastal zone management to prevent any further damage to the coastal environment.

Table 17.4 Mangrove areas by state (Sq. km)

State	Sidhu (1963)	Blasco (1977)	India Forest Resource Assessment (1987-89)	India Forest Assessment (1993-94)
Andhra Pradesh	184	100	400	380
Orissa	120	50	200	210
Tamil Nadu	26	15	50	20
West Bengal	4189	2 000	2120	2120
Total East India	**4 519**	**2 165**	**2770**	**2730**

Sources : Sighu, 1963; Blasco, 1977; Government of India, 1991a, 1997.

Table 17.5 Mangrove areas by state (Sq. km)

State	1990	1994	1999
Andhra Pradesh	6000	34 500	84 269
Goa	525	600	650
Gujarat	125	700	997
Karnataka	2 500	3 500	3 540
Kerala	13 000	14 100	14 595
Maharashtra	1 800	2 400	970
Orissa	7 075	8 500	11 332
Tamil Nadu	250	2 000	2 670
West Bengal	33 815	34 400	42 525
Total	65 090	100 700	161 570

Sources : 1990, 1994 : MPEDA in James, 1999; 1999: India Hatcheries Organization, unpublished data, 1999.

Table 17.6 The impact of shrimp aquaculture in the Godavari delta (ha)

Land use Patterns	Land use area			Converted to shrimp		
	1989	1997	1999	1987 - 1997	1997 - 1999	1989 - 1999
Crop land				4543	2324	6903
Fallow land				3149	1327	4497
Dense mangrove	16586	15987	15318	433	471	1137
Sparse mangrove	4530	3786	3199	604	666	1030
Total mangroves	21116	19773	18 517	1307	1137	2167
Other						
Aqua farms	2006	13 032	19239			

17.9.1 Coastal Zone Management Policies

For regulation of development activities, the coastal states were requested to prepare coastal zone management plans through the 1991 CRZ Notification. The Coastal Zone Management Plans require the classification of all coastal land within 500 m land-wards of the high tide line into four categories.

CRZ-I includes ecologically sensitive areas, including parks and mangroves and areas between the high tide line and the low tide line. In this zone no new construction is permitted landwards of the high tide line. Between the low tide line and high tide line the only construction activities permitted are those per-taining to facilities for carrying treated water discharge into the sea, facilities for carrying seawater for cooling purposes, oil, gas, etc.

17.9.2 Statistical Details

The area in hectares of each category from classified data sets are calculated and shown in Table 17.7 and Fig. 17.1. The graph showing the comparison of areas is shown in Fig. 17.2. The final change detection map is shown in plate 17.4.

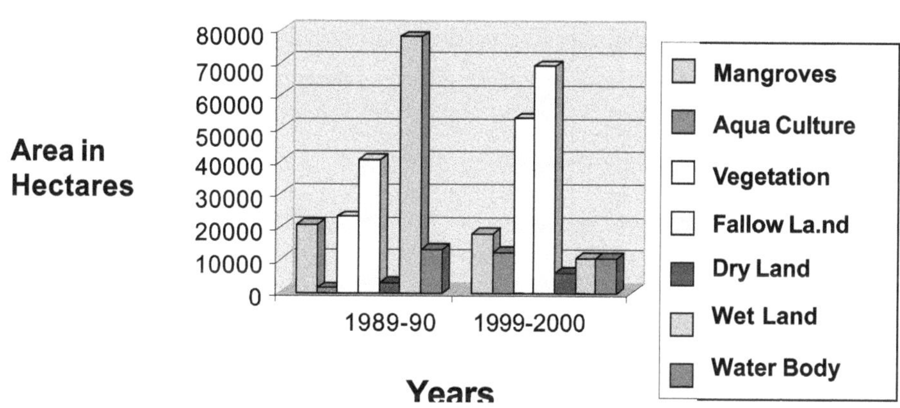

Fig. 17.1 Graph showing the area in hectares

Table 17.7 Areas in hectares

Areas in hectares of each category from classified data sets :

Year	Mangroves	Aquaculture	Vegetation	Follow land	Dry land	Wet land	Water body	Ocean
1989-90	20810.749	71445.8750	23054.8249	405.152507	2894.1250	7550.438	213017.499	49043.8241
1999-2000	18040.8700	12174.8800	52958.7510	68791.6262	6060.0625	10448.8751	10488.3749	49667.0616

*1989-90 there is clouds cover of 193.3300 Hectares

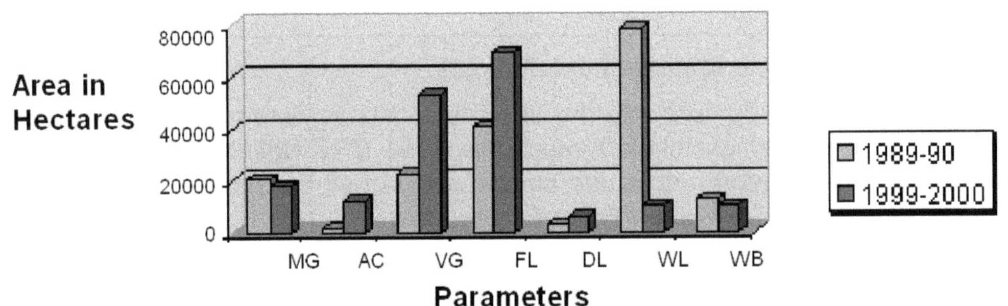

Fig. 17.2 Graph showing the comparison of areas

Table 17.8 Area land use patters changes (hectares)

1999-2k / 1989-99	Ocear	Vegetation	Fallow land	Dryland	Wet land	Mangroves	Aqua culture	Water Body	Total areas in 1989-90
Ocean	4788.5561	1.0625	1.5825	208.3760	184.6875	68.4375	0	895.9375	49043.6241
Vegetation	20.4375	12873.6248	8310.9378	506.6875	546.6875	37.8750	472.0000	286.8215	2305.46249
Fallow land	173.3750	10080.00	25480.5832	1802.6250	1241.1875	78.1250	912.5826	678.8125	40515.2507
Dry land	193.3125	159.5825	299.6875	1449.8125	250.5625	149375	110.5000	609.0625	2894.1250
Wet land	273.3750	28803.2512	33971.7504	1068.1875	4472.6876	661.2500	7656.5000	848.7500	77754.4392
Mangrove	288.3750	478.4375	139.1250	145.8125	1333.8125	16622.1872	949.5625	653.4375	206107.497
Aquaculture	0	121.3125	232.3750	0	79.7500	0	1012.4375	0	1445.8750
Water body	1030.9375	433.5000	355.6250	830.0000	2337.5000	550.3125	986.3125	6483.5624	13017.7499
Total areas in 1999-2k	49667.0616	52956.7510	68791.6262	6069.0625	10446.8751	18033.1247	12098.8750	10466.3749	+193.3300 clouds

Grand Total = 228529.72 Hec

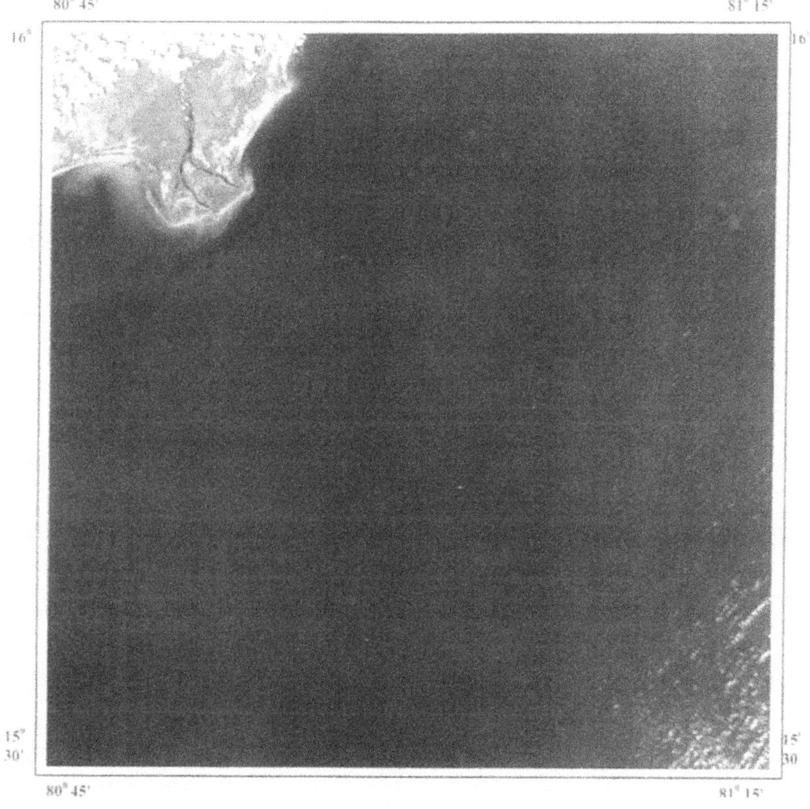

Plate 17.1 IRS-P4 (Oceansat-1) satellite image showing the study area (a part of Bay of Bengal)

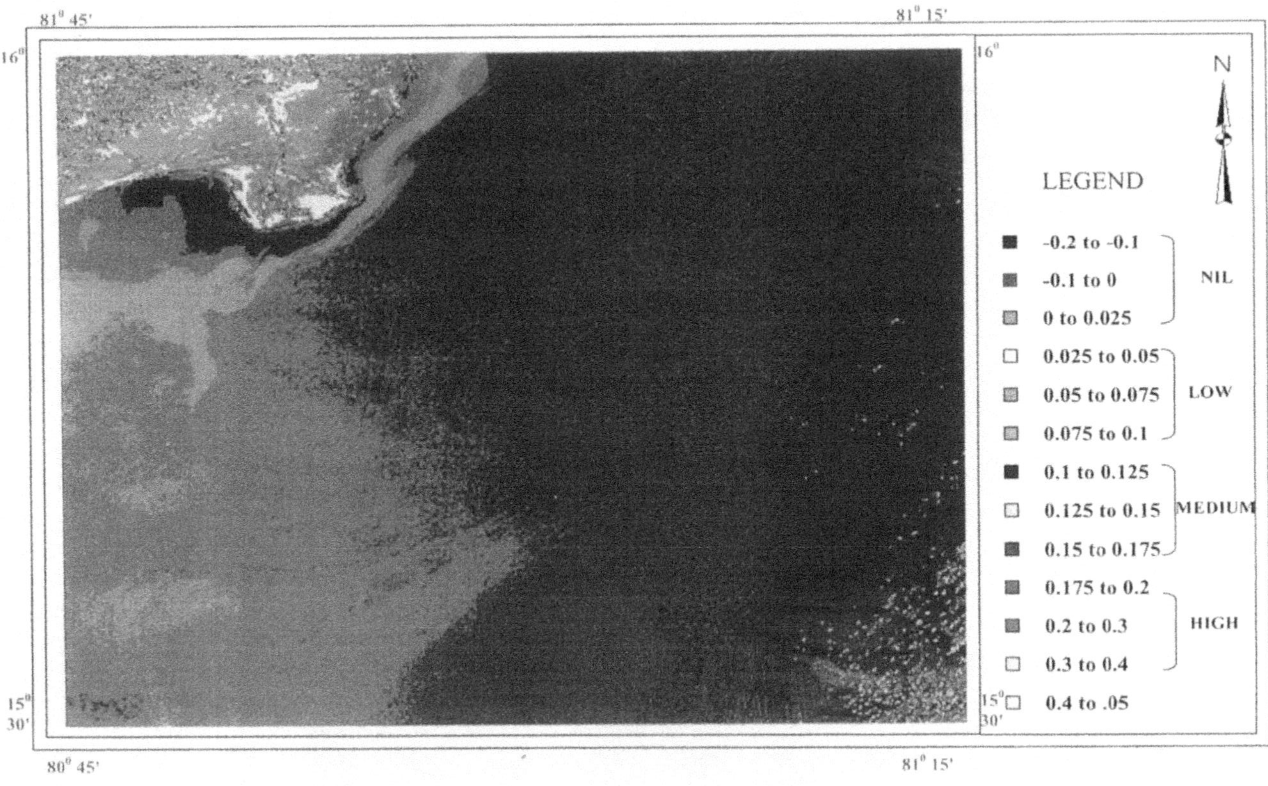

Plate 17.2 Pseduo colour image showing NDVI values as a base for indentification of fishing zones

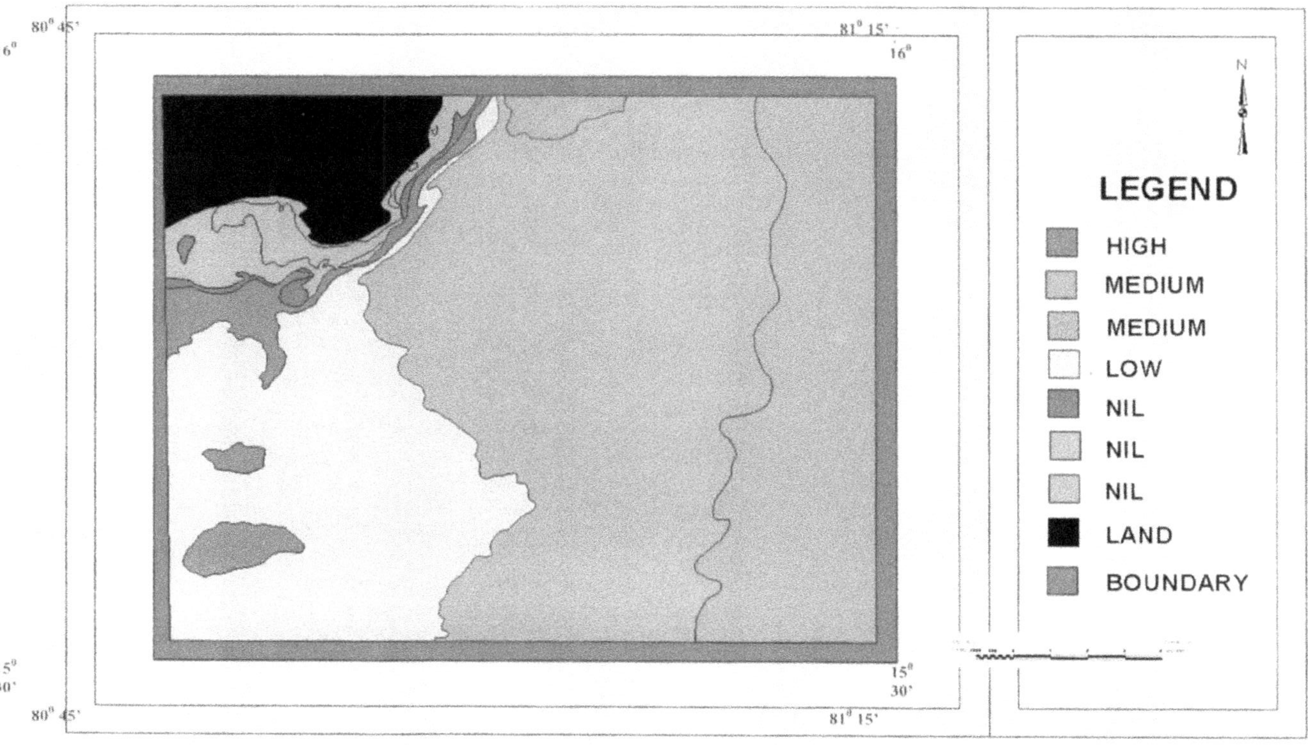

Plate 17.3 Map showing potential fishing zones

Plate 17.4 Coastal change detection map of Kakinada Bay

18

Urban Planning and Management

18.1 Introduction

Optimum utilization of land by human beings leads to his survival, development and progress of the region. Rapid industrialization causes haphazard and unplanned growth of urban centres, which becomes more complicated by the fact that it must take place within the build up area. These pressures of continually growing population result in over crowding and become burdens to limited civics. Amenities which forces the middle class and the builders to outlying suburbs, a phenomenon called as urban sprawl or growth.

This irregular growth of urban area consumes agricultural and flat upland on their periphery. Therefore, proper planning has to be done for controlled and regulatory growth of urban centres. Urban development policy requires an approach that aims at optimizing the productive advantages of cities and towns, while at the same time minimizing or mitigating the negative impacts of urbanization. In addition, new urban areas will have to be planned in adequate numbers to accommodate such a growth of urban population and provide them basic infrastructure services.

Urban projects range from urban transport planning to urban housing development, new-town planning and building design to the design and implementation of water supply and disposal systems. Innovative methodologies and technologies for solution finding are increasingly being adopted by institutions as well as companies to facilitate the planning, management and development process. Sophisticated computer-based data handling technologies make simple databases and drawing facilities integral to the capture, collation, manipulation and presentation of diverse data sets and allow the decision maker to combine conventional map-based information with diverse textual or attribute data sets.

To establish the base data set for any major development planning project requires considerable investment of time and resources in data capture: topographic data, drainage, soils, geomorphology, land use/land cover, water supplies, existing, planned and projected infrastructure, communications as well as up-to-date census data, development projections and so on are all vital to the establishment of coherent development priorities. Existing sources of data need to be identified, relevant data extracted and pertinent data gaps addressed through systematic data collection strategies that reflect the objectives and real needs of the client.

A large rural land use planning programme will frequently have a significant focus on the causes and effects of rural-urban migration, in particular the pressures on resources that result from urban growth. Data on health, education, water supplies, sanitation and population growth and movement are central to this and to the identification of options for change in urban areas and in preparing land use plans for rural areas.

GIS can be used to ensure that infrastructure development, land tenure and the supply of adequate services could be effectively planned and monitored. Particularly pertinent in developing countries are the adoption of good data collection and entry practice including not only GIS functionality but GPS, sampling, cartography, source selection and how these relate to the relevant urban planning issues.

GIS is a decision support tool that allows the planner to visualise real-world situations and to analyse and identify options to achieve stated objectives. In urban planning these may include infrastructure and utilities development, habitat protection, services location and investment needs. For utilities themselves this entails asset and facilities management through the integration of historical data with ever-changing data from systems monitoring both the demand and the supply of the relevant resource. Thus the seasonal and diurnal demands of the population must be understood as much as those of physical and environmental factors to contribute to a holistic management information system.

18.2 Role of Remote sensing and GIS

Remote sensing techniques which gather data about earth's surface features in raster form on a periodic basis and GIS which helps in analysis and integration of relation database containing both spatial and aspatial details through automated machine processing system provides an alternative means of mapping of dynamic features at regular intervals. By such integrated study it is possible to take relevant decisions at the appropriate time by the urban planners.

The significant benefit of GIS technology is its ability to visualize spatial data and to interpret information visually. A computer aided database management system (DBMS) in GIS environment is a very efficient tool for quick management decision and can also be used in urban infrastructure management. There is a high potential of integrating GIS within the application area planning and management layout of pipelines for water supply, sewage, gas supply, drainage, cable network for electricity and telecommunications, transport network, locations of facilities and other distributions.

GIS application in urban infrastructure management allows systematic collection, updating, processing and distribution of related information. Thematic overlays contained within GIS could also be used as reference source for defining the spatial location of phenomenon (urban sprawl) and also comparing such phenomenon against identified features employing digital processing of RS images and selected thematic overlays integrated within GIS, where digital enhancements are apprised with greater capability than by use of RS or GIS alone.

18.2.1 Satellite Remote Sensing Data for Urban Planning and Development

Over the years, satellite-based remote sensing data have been successfully utilized for mapping, monitoring, planning and development of urban sprawl, urban land use and urban environment. Satellite data complements and supplements data available from both aerial photographs and conventional ground sources for urban planning. Though data from large-scale aerial photography are desirable for urban surveys, it is slow, becomes outdated fast, is costly and difficult to obtain at regular intervals of time. Although small format aerial photography (SFAP) provides details of urban areas and is 5 to 10 times cheaper than normal photography, it is precluded due to operational and other data limitations. The biggest strength of satellite-born-sensor data is its periodicity, fastness and economic efficiency. Also, the data being available in analogue and digital form, it is amenable to both visual and digital analysis for extraction of information. The current satellites allow data to be generated up to 1 : 50,000 scale using IRS (LISS-11) and up to 1: 25,000 scale using SPOT data. Today, with the availability of IRS-IC data, it is not only possible to generate data on 1 : 50,000 / 1 : 25,000 scales using LISS-111, but it has been seen with the PAN data that it is possible to generate up to the desired scales of

1 : 20,000, 1 : 15,000 and 1 :12,500, useful for urban planning.

Using satellite data, different urban features can be identified and mapped using a false color composite (FCC) of three bands or single band data wherever applicable. Also, the digital data can be enhanced or merged with high resolution data for better feature identification and mapping. Earlier studies have proven the usefulness of visible and near infrared multispectral bands for urban feature identification and mapping. Also, the use of thermal band data available from Landsat (TM) data has also been demonstrated in mapping 'urban heat islands'.

18.2.2 Applications Scenario

With the successful launch of IRS-IC satellite and the availability of high spatial resolution data in 5.8 m in single panchromatic band and 23m (LISS-III) resolution data in four multispectral bands along with the stereo and revisit capability of 5 days, it would now be possible to explore its potential either singularly or in combination in the different areas of urban land use survey and planning. There are a variety of potential areas of urban applications which can be possible using IRS-IC data. The few important such applications are:

(i) Urban housing
(ii) Urban transportation and traffic planning
(iii) Urban cadastral and real estates
(iv) Urban ecology and hazards
(v) Urban fringe area land use dynamics
(vi) Urban land use change detection and updating
(vii) Urban base map preparation
(viii) Urban green belts or open space mapping
(ix) Urban encroachment of slums onto vacant lands and urban land use zoning
(x) Urban utilities and infrastructure
(xi) Urban water supply and sanitation
(xii) Urban tourism and recreation
(xiii) Urban census
(xiv) Urban landscape design

18.3 GIS in Urban Planning

The significant benefit of GIS technology is its ability to visualize spatial data and to interpret information visually. A computer aided database management system (DBMS) in GIS environment is a very efficient tool for quick management decision and can also be used in urban infrastructure management. There is a high potential of integrating GIS within the application area planning and management layout of pipelines for water supply, sewage, gas supply, drainage, cable network for electricity and telecommunications, transport network, locations of facilities and other distributions. GIS application in urban infrastructure management allows systematic collection, updating, processing and distribution of related information. Thematic overlays contained within GIS could also be used as reference source for defining the spatial location of phenomenon (urban sparwl) and also comparing such phenomenon against identified features employing digital processing of RS images and selected thematic overlays integrated within GIS, where digital enhancements are apprised with greater capability than by use of RS or GIS alone.

Geographical Information Systems are computer aided decision support and planning tools which integrate data from maps (spatial data) and other auxiliary data (attribute data) for a geographical area of interest. They can be used to create and maintain geographic databases and are eminently suited for what-if-analysis in any planning related activity.

Once the GIS package integrates the spatial and attributed data, the planner has a powerful tool, which can be used for information dissemination or analysis. By querying the geographic database in several ways the planner can present the available information in a variety of formats: as printed tabular reports, graphically as map display and map outputs in paper.

The analysis capabilities of a GIS package allow the planner to address what-if questions and work out a variety of action plans in a scientific manner. A number of problems can be solved by geographic analysis. Typical examples are : Town ship development, relationships between agricultural parameters such as yield and salinity, Land capability analysis, site locations for facilities, environmental problems such as animal migration.

18.3.1 Urban Planning, Implementation and Monitoring

At the threshold of 21st century it is an interesting question to ask ourselves, if 20th century belonged to technological advancement, then what will the major event of 21st century be ? Last century witnessed a trickle of urbanization and emergence of metropolises. Never in the history of cities, have stakes related to controlling the world's urban development, been as crucial for the future of humanity. This is true not only in terms of quantity, as city dwellers are expected to account for roughly half the worlds population by the year 2000, but also in terms of quality, since the bulk of economic, social, scientific, technological and cultural development will take place in cities and in particular in largest conurbation known as Metropolises. The increasing urbanization of the world's population is inescapable and irreversible. This goes to show that it is necessary and fundamental for policy makers to make technologies like GIS and Remote Sensing imperative for the urban planning.

Biggest challenge of 21st century will be to understand phenomenon of urban agglomeration. Most of the 20th centuries concept and visions about cities are being questioned. Traditional approaches and techniques designed for town and cities may turn out to be inadequate tools to deal with metropolises. New approaches, need to be invented, other actions/methods need to be incorporated in the existing work methods. From the vast experience of the past, a solution for the future planning methods has to be derived. The objective is to confine the search for new techniques to deal with large urban agglomeration and application of GIS and Remote sensing technique at the various stages of planning, implementation and monitoring of the urban projects. One of the established practice to search for new techniques is to identify the problems faced by the planners in the past.

From the experience of last two decades in planning and development of large urban projects, it is learnt by the planners that transformation of paper plan on the ground takes much more than simply preparation of the plan. Identification of land for land acquisition, actual physical possession of land and development of roads etc., takes considerable time. During this incubation time ground realities have changed and planning proposals do suffer. Therefore, rate of implementation of the plans is below the desired level. One of the major lacuna identified by the planners is non-availability of base-maps, absence of any method for updating the base maps and the absence of any monitoring mechanism.

18.3.2 Nature of Problems in Using Geoinformatics

Broadly the problems can be described in the following manner :-

(a) In most of the metropolitan areas Survey of India base maps toposheets are available. However , these maps are 5 to 10 years old which create problem for making accurate planning proposals and predictions.

(b) Even after the base maps are prepared and proposals are placed on it , the difficulty arises due to absence of any reliable method for updating available base maps .

(c) The fringe area where most of urban development is taking place i.e. Urban Fringe area is very dynamic is terms of land development, so a device / method is required to detect changes on continuous basis, so that the planning proposals can be re-adjusted and re-shaped as per the ground realities .

(d) Physical planning largely depends upon inputs from the physical survey, various social and physical requirements and site conditions. The planners based on his training and intuition, works out the planning proposals. However it has been observed that GIS and Remote Sensing can provide intelligence for formulation of the proposals.

(e) The most important aspect which is generally overlooked is to co-relate or transfer planning proposals on revenue maps. The planners face difficulty in understanding the plans prepared, by revenue staff as these plans are on different scale and projection. Similarly revenue staff is unable to superimpose planning proposals on revenue maps manually. However it has been observed that superimposition of any two 'digital maps' is easy as digital maps are scale free. Thus an interface between planning basemap and revenue maps can be established.

18.3.3 Search for Solutions

Once problems faced by the planners while preparing basemaps etc. are identified, then search for solutions leads to mapping related technologies which are at our disposal. GIS and Remote Sensing are mapping related technologies and it is capable to overcome problems. Application of these technologies has given way to innovative methods in planning process -

(a) Digitisation of planning base maps and various layout plan has facilitated updating of basemaps wherever changes have taken place in terms of land development etc. Digital maps provide flexibility as digital maps are scale free. Superimposition of any two digital maps which are on two different scales is feasible. This capability of digital maps facilitate insertion of fresh survey or modified maps into existing basemaps. Similarly superimposition of revenue maps on basemaps with reasonable accuracy is great advantage compared to manually done jobs.

(b) Computerization within the planning department helps to disseminate information from one section to another section also . This may result in better horizontal co-ordination between various sections and also remove redundant data from the database .

(c) Experience shows that Remote Sensing techniques are extremely useful for change detection analysis and selection of sites/areas for specific purposes like land acquisition etc. The use of satellite imageries in Rohini Project has demonstrated systematic analysis of land available for acquisition and monitoring of areas which are prone to encroachment etc.

(d) Since all this information/maps is available in digital format, co-relating various layers of information i.e. satellite imagery, planning maps and revenue maps on common feature is feasible, with the help of GIS & Image processing software available in the market. Such superimposed maps provide valuable information / inputs for planning, implementation and land management

GIS and Remote Sensing technology are useful to detect -

(a) Large scale encroachments.

(b) Large scale conversion of agriculture land into non-agriculture land.

(c) Vacant pockets of agriculture land for acquisition.

(d) Identification of revenue boundaries of villages on satellite imageries.

(e) Large scale change detection in status of terrain is possible with help of satellite imageries.

GIS and Remote Sensing technology have capability to provide necessary physical input and intelligence for preparation of basemaps , formulation of planning proposals and act as monitoring tool during implementation phase . Satellite imageries helps to maintain truthful record of terrain during that period . Thus GIS and Remote Sensing are emerging as a powerful land related technologies for monitoring and management of land . Thus GIS and Remote Sensing has capability to provide FOURTH dimension to the city i.e., TIME.

18.4 Issues in Urban Planning

It is now being recognized that cities are the engines of growth at both regional and national level. To facilitate and sustain this growth, cities have to provide both a high quality of life and an efficient infrastructure for economic activities. Environment management and protection strategy addresses the critical environmental problems, which mainly concerns preservation of lakes and water bodies, its catchment area and its water quality and land use management in catchment areas. The other environmental issues relate to the disposal and treatment of urban waste and its recycling and the socio-economic problems caused by the displacement of population, in context to Future City spread. Environment management of lakes and water bodies are vulnerable to urban pressure in its close vicinity. It is essential to enforce land use control measures in the catchment areas to prevent further environmental degradation and thereby achieve desired level of sustainability. The sustained efforts are needed for plan implementation to improve the quality of city life. Hence an effective plan implementation strategy needs to be evolved to achieve the following objectives

(i) Protect natural environment.
(ii) Conservation of areas of cultural heritage.
(iii) Optimize land use and land utilization
(iv) Provide services and infrastructure
(v) Participatory approach for supply of land and infrastructure development.

Urban planning is basically resource generation, resource development and resource management exercise. The efficiency of urban settlements largely depend upon how well they are planned, how economically they are developed and how efficiently they are managed. Planning inputs largely govern the efficiency level of human settlements. There is a widely held view that the master planning methods adopted over the last few decades have not produced a satisfactory physical environment. The urban development planning process in the past has been unduly long and has been largely confined to the detailing of land use aspects. The plans have paid inadequate attention to the provision of trunk infrastructure, environmental conservation and financing issues. They have been unrealistic and have not been accompanied by investment programmes and capital budgets. Integrated urban development planning approach, taking into account regional, state and national strategies, and spatial, functional and other linkages between human settlements, has not been given much recognition. Also the planning and plan implementation processes have not paid adequate attention to the integration of land use and transport planning. The fact that transport is a key determinant of land use and "leads" development is sometimes ignored.

The Five Year plans laid stress on the need to undertaken town planning and evolve a National Town Planning Act so as to provide for zoning and land use, control of ribbon development, location of industries, clearance of slums, civic and diagnostic surveys and preparation of Master Plans. Although a significant step in urban development was undertaken in the Plan in the form of Central assistance to the states for the preparation of master plans for selected areas, comprehensive action was not taken by the states for the adoption and implementation of the plans. The urban development planning should, essentially, be supportive of the economic development in the country. At present, hardly 20 per cent of the urban centers have some sort of a Master Plan, which is many cases is just a policy document. It is estimated that there are about 1200 master plans prepared by various Agencies responsible for plan preparation but their implementation is not encouraging. The implementation of master plan facilitates the orderly and planned development of cities in a sustainable manner, which would ultimately help in good governance.

18.5 Urban Growth Management : A Case Study

Urban growth is taking new forms characterized by heavy dependence on transportation, communication and multiple activity centre. It becomes necessary to understand the mechanisms at play to capture aspects associated with this complex process to develop frame works in the context of increasing importance of a National Urban System The type of urban growth proliferation that witnessed today calls for policy orientations that will not only contain the parasitical tendencies of metropolitan growth but also integrate growth of lower order settlements with development of metropolitan areas. In this study an attempt is made to employ recent tools and techniques associated with mathematical models, remote sensing and GIS to verify a growth strategy that is formulated to contain metropolitan dominance and the problems associated with such a phenomenon. Land use-transportation planning models that establish the interaction between the land development and transport requirement are employed to verify the growth strategy formulated for the Hyderabad city's influence region. Appropriate land use structures and compatible transport networks are generated using remotely sensed data in a GIS environment.

The basic objective of this research effort was not only to formulate growth strategies that are innovative and responsive to the current needs as part of the urban growth management but also demonstrate the usage of GIS which is becoming a powerful application tool that can be employed in diverse areas ranging from planning cities to evolving business strategies.

18.5.1 The Role of Technology Total Stations, GPS and GIS

India having 26 States and 4 Union Territories and each state has been divided into Districts and then to Villages and Thesiles and so on. There is Land, which belongs to Government, land which is Private Property, land which is totally unused. As population grows, infrastructure tends to grow and property value starts increasing. India is also known for family disputes where close knit family with members as big as 100 number has become past. The family disputes in villages make more and more divisions on the land. So keeping Land Recor ds has become a necessity.

The function of Land Records can be described as follows :

To make Property Maps and to do settlement.

(i) To do Topographic survey for Infrastructure development.

(ii) To keep database records related to property.

(iii) To evaluate the property value.

(iv) To divide the property in case of disputes.

(v) To work in tandem with Board of Revenue, Public works departments, Utility companies.

18.6 Disadvantages of using Conventional Method

Land Records have realized the following disadvantages.

1. Chains data can be taken only in perpendicular direction, while actually land/ fields is of Zig Zag shape.
2. Chain data is always inaccurate. The land cost of today demands a more accurate method.
3. The maps generated by manual methods by draftsman cannot be reproduced at user required scale. If scale has to be changed then whole map has to be redrawn.
4. Attribute data for a field cannot be attached on a paper map. For example The name of the owner, the area, the cost etc. has no relation with the map.
5. Decision making process using old paper maps becomes very cumbersome and some times it leads to wrong decision.

18.6.1 How technology can help Department of Land Records

Today's technology like Computers, Total Stations, GPS, Mapping Software, GIS Software etc., can do the following for Land records.

1. Making a map using Coordinates rather than by angles and distances. The advantage is just by press of a key you have the position on the ground.
2. Area computation can be done using Total Stations as well as GPS then and there itself. This feature can be used for solving property disputes.
3. Using Total Stations and Mapping Software the maps can be created automatically. No need of manual data entry, manual line joining and manual creating a lot of problems.
4. Map at any scale can be created because of availability of data in electronic format. In a very short time user or specific maps can be generated.
5. Existing old data (Ladder Diagram) can be inputted into the same mapping software to bring years old data to the same level of today.
6. Using Total Stations, Mapping Software and Geographical Information Software a full-fledged GIS can be generated so that just by click of a key, all the attributes corresponding to a land piece can be obtained. The decision making process becomes very simple and cost effective.

18.6.2 Total Solutions for Land Records

Total Stations are surveying instruments used for land surveying and Mapping. Since Total Stations are capable of measuring distances upto 2.5 Km to 3 Km, these are, most of the time, used for Plot/Lot survey, Detail Mapping instruments and establishing control points or temporary points over a small area. Global Positioning System (GPS) as the name stands can give position of a point anywhere on the globe with high accuracy. GPS can measure base lines for High accuracy control points without any line of sight requirement. Since Total Stations work on the principle of signal reflection, line of sight must be there between Total Station and Prism reflector. This makes GPS more effective tool for control points establishment. However GPS cannot be used in a area with lot of trees, High rise buildings because of satellite signal interference. The BEST way is to mix Total Stations and GPS together for effective survey of Land. Following solutions are offered by Leica Geosystems, for Land Records:

(i) To do Topographic survey for Infrastructure development.

(i) Cadstral/land records mapping using Total Stations: (TPS300 series, TPS700 Series, TPS1100 Series)

(ii) Differential GPS for control point establishment.

(iii) Real Time GPS for Cadstral mapping wherever obstructions are minimum and in open area.

(iv) Mapping Software for Land Records " LISCAD"

(v) Local Language software for customization of maps and database. (For eg. Tulika Software for Malayalam Language)

18.6.3 Solution based on real time GPS

GPS is recommended to be used for establishing Control points in the villages. Those control points should be used in Total Station Survey. So combination of Total Station and GPS will be the correct combination.

18.6.4 Use of GPS for Land Records

The following are the major usage of GPS for Land Records :

- To establish Control points to be used for Total Stations.
- To do detail Mapping.
- To do Land/Plot survey where there are no serious obstructions like Trees, Buildings etc.
- To connect one Village with other using long Base lines.

18.7 Urban land use and infrastructure identification and delineation

Urban land use reflects the morphology, economic and socio-cultural framework of a city or a town. Here, IRS-IC LISS-III and PAN data can be digitally merged and the enhanced standard FCC can be generated through Laplacian filtering technique and edge enhancement operations using the EASI-PACE image processing software. Further, the merged (LISS-III and PAN) data sets can be generated by taking intensity from PAN data and hue and saturation from LISS-III data, after converting blue, green and red channels of LISS-III into IHS channels. Subsequently, using these data sets various details of urban land use classes of level-HI can be identified such as low/ medium and high rise structures associated in the residential use, slum areas, industrial complexes, major and minor roads and bridges, education institutions, open spaces such as parks, garden, playground, stadium and race course, utilities such as sewage-treatment plant, agricultural research farm, bus depot, vacant lands and newly developing layouts, etc. as seen in Hyderabad.

Further, plot-wise urban development was attempted by registering (using GCP from SOI map sheet) the cadastral parcel boundary map of Jodhpur gram panchayat onto the IRS-IC (PAN) data. It has been observed that both voids (filled with structures) and non-voids (vacant lands) can be delineated on a plot by plot basis.

18.8 Urban transport network identification and mapping

Transport network forms an important infrastructural element of the whole urban area. It allows connectivity and movement of people, traffic and goods from both intra (within) city and inter (outside) city hinterlands. Here, the IRS-1C (LISS-III and PAN) merged image data set is converted to TIFF and GRID format and made compatible to ARC-INFO GIS environment. This GRID is geometrically corrected by selecting GCP on the image and SOI map sheets. The geometrically corrected grid data can then be imported in ARC/INFO GIS and on screen digitization attempted for drawing the existing and newly identified major and minor roads and intersections, crossings and bridges and railway lines.

18.9 Urban city guide map change detection and updation

SOI city guide maps form the largest land use and topographic maps available on 1 : 20,000 scales for a few selected cities in the country today. Although these maps provide the detailed possible information on urban land use, majority of these maps are outdated (surveyed up to 1970s) and are restricted in areal coverage to a limited city area and often do not show the extended coverage and the recent developments within and near fringe areas of the cities. City guide map can be digitally merged with the IRS-1C (PAN) data, which clearly brings the different land use information such as roads, bridges, residential colonies, industries, institutions, new layouts, development onto the vacant lands/tanks and historical monument sites, etc., as surveyed in the city guide map of 1973-74 period and recent developments noticed in the IRS-IC (PAN) data of 1996, over a period of more than 20 years.

References

A.K.Sanyal, Geometric Geodesy, (1991), Survey of India Publication.

Adler., H.H. 1968, Infrared Spectra of Phosphate Minerals: Splitting and Frequency Shifts Associated with Substitution of PO^3_4 for AS O_4^{-3} in mimetite American Mineralogist, vol.53,pp. 1740-1744.

Agarwal, N.K. and Sharma, M.K.(1983) Application of Remote Sensing Techniques of Forest Resource Assessment with particular reference to NOAA-7 satellite of USA, Forest Survey of India, Dehradun.

Akea.,A.(1971) Identification of land use classes and forest types by means of microdensitomenter and discriminant analysis, Application of Remote Sensing in Forestry, Report, University of Freiburg, FRG, 147-64. American Society of Photogrammetry : Manual of Photogrammetry (1980) George Bank Co.

Anbalagan R., " Landslide hazard evaluation and zonation mapping in mountainous terrain", Engineering Geology, Elsevier Science publishers B.V., Amsterdam, 32(1992) 269-277.

Anji Reddy., M. 1997, A detailed Statistical Study on Selection of Optimum IRS LISS Pixel Configuration for Development of Water Quality Models, International Journal of Remote Sensing, Vol.18, No.12, pp. 2559-2570.

Anji Reddy., M. and K.M.Reddy, 1996, Performance Analysis of IRS Bands for Land Use/ Land Cover Classification System using Maximum likelihood Classifier, International Journal of Remote Sensing, Vol. 17, No.13, pp.2505-2515

Anji Reddy., M. and S.Srinivasulu, 1994, Comparison of IRS-1B LISS-11A Pixel Array Sizes for Estimating Suspended Solids Concentration in Hussain Sager Lake, Hyderabad, India – A Statistical Approach, International Journal of Remote Sensing, Vol. 15, No.18, pp.3693-3706.

Anji Reddy., M. 1995, Ph.D Thesis, Jawaharlal Nehru Technological University, Hyderabad, India

Aronoff., S.1989, Geographic Information Systems : A Management perspective, WDL Publication, Ottawa, Canada.

Barrow.,G.M.,1962, Introduction to Molecular Spectroscopy, New York, Mc Graw-Hill, 318p.

Bernhardsen., T., 1992, Geographic Information Systems, VIAK IT, Norway.

Berry., J.K., 1982, Cartographic Modelling Procedures for Extending the utility of Remotely Sensed data: In B.F. Richason (ed.) Remote Sensing : An input to Geographic Information Systems in the 1980's, Proceedings of the Pecora 7 Symposium.

Berry., J.K., 1987, Fundamental Operations in Computer-Assisted Map Analysis, International Journal of GIS, 2, pp. 119-136.

Berry., J.K., 1993, Beyong Mapping : Concepts, Algorithms and Issues in GIS. GIS World Inc. Colorado.

Bracken., I., and Webster, C., 1989, Information Technology for Geography and Planning, London, Routeledge.

Burrough, P.A., 1986, Principles of Geographical Information Systems for Land Resources Assessment, Clarendon Press, Oxford.

Burrough, P.A., and Frank., A.V., 1996, Geographical Objects and Indeterminate Boundaries, Taylor and Francis, London, 345p.

Burrough., P.A., 1983, Geographical Information Systems for Natural Resources Assessment, New York, Oxford University Press.

Burrough., P.A., 1983, Multiscale Sources of Spatial Variation in Soil 1. The Application of Fractal Concepts to Nested Levels of Soil Nariation, Journal of Soil Science, 34, 577-97.

Burrough., P.A., and Masser., F.I., 1988, European Geographical Infrastructures, Opportunities and pit falls, Taylor and Francis, London.

Carrara Alberto, " Gis-Based Techniques For Mapping Landslide Hazard", Cnr-Cioc, Bologna

Chrisman., N.R., 1982, A Theory of Cartographic Error and its Measurement in Digital Databases, Proceedings Auto Carto, 5, pp. 159-68.

Chrisman., N.R., 1994, Beyond the Snapshot, Changing the Approach to Change, Error and Process, Unpublished Manuscript.

Chrisman., N.R., 1997, Exploring Geographic Information Systems, Wiley, New York.

Clair E. Ewing and Michael M.Mitchell, Introduction to Geodesy, (1970), American Elsevier Publishing Company, Inc. New York.

Comfort, L. K., "The impact of Information Technology upon Disaster Mitigation and Management," Proceedings of the Second Conference on the Applications of Remote Sensing and GIS for Disaster Management, Washington D. C., Jan. 19-21, 1999.

Curran., P.J., 1988, Principles of Remote Sensing, ELBS., pp 282.

David Clark, (1983) Plane and Geodetic Surveying for Engineers Vol I CBS Publishers & Distributors, New Delhi.

E.Faisz General Cartography

Fisher., J., 1989, The pixel, a snare and a delusion, International Journal of Remote Sensing, 18, pp. 679-685

Frost, J.D., Caroll, D.P., and Rochaway, T.D. 1997. Spatial liquefaction analysis, Spatial analysis in soil dynamics and earthquake engineering, Geotech. Spec. Publ. No. 67, ASCE, Reston, Va., 70-86.

George P.Kellaway : (1949) Map Projections, E.P.Dutton & Co. Inc.

Goodchild, M.F., 1991 a Issues of Quality and Uncertainty: In.Miller, J.C., (ed), Advances in Cartography, Oxford, Elsevier Science, pp. 111-139.

Goodchild., M.F Parks., B.O. and Stayaert., L.T., (ed) 1993, Environmental Modelling with GIS., Oxford University Press, Oxford and New York.

Goodchild., M.F., 1980, Fractals and the Accuracy of Geographical Measures, Mathematical Geology, 12, pp, 85-98.

Goodchild., M.F., 1992, Geographical Information Science, International Journal of Geographical Information Systems 6(1), pp 31-45.

Greenbaum, D., McDonald, A. J. W. and Marsh, S. H. 1996. Rapid methods of landslide hazard mapping. Proceedings of the Thematic Conference on Geologic Remote Sensing, 11, I.287-I.296.

Hand book of Topography – Plane Tabling (1967), Survey of India Publication.

Hand book of Topography – Theodolite Traversing (1967), Survey of India Publication

Hand Book of Topography – Triangulation and its Computation (1967), Survey of India Publication

Heller,R.C and Ulliman, J.J.(1983) Forest Resources Assessments, Manual of Remote Sensing, American Society of Photogrammetry, Washington DC, chapter 34.

Henderson, S.T., 1970 Daylight and its Spectrum, Adam Hilger Ltd., London, pp 363.

Huang, S.L. and Chen, B.K. 1991 Integration of Landsat and terrain information for landslide study. Proceedings of the Thematic Conference on Geologic Remote Sensing, 8, 743-754

Hunt., G.R., Salisbury, J.W., and Lenyoff, C.J., 1973, Visible and Near Infrared Spectra of Minerals and Rocks. V11. Acidic Igneous Rocks, Modern Geology, Vol. 4, pp 217-224

Kusaka, T., Ootsuka, M., Shikada, M. and Kawata, Y. 1996 Estimation of landslide areas using satellite imagery and spatial features of watersheds. Proceedings of the Thematic Conference on Geologic Remote Sensing, 11 II.415-II.420.

Laurini., R.Thompson., D 1992; Fundamentals of Spatial Information Systems, Academic Press, London.

Lawrini., R. and Thompson. D., 1992, Fundamentals of Spatial Information Systems, Academic Press, London.

Lillesand., T.M. and R.W. Keifer1979, 1994 and 2000, Remote Sensing and Image Interpretation, John Wiley and Sons, New York.

MacEahren., A. M., 1994, Some Truth with Maps, Association of American Geographers, Washington D.C.

Mandelfort., B.B., 1982, The Fractal Geometry of Nature, Freeman., New York.

Marble., D.F. and D.J. Peuquet, 1983, Computer Software for Spatial Data Handling, Current Status and Future Development Needs, Buffalo, New York, Geographic Information System laboratory, State University of New York.

Mark., D.M., 1986, The use of Quadtrees in Geographic Information Systems and Spatial Data Handling : In M.Blankemore(ed.) Proceedings, Auto Carto London, vol. 1, pp 517-526.

Mather, P. M., 1987, Computer Processing of Remotely sensed Images : An Introduction, John Wiley & Sons, 352 p.

Miles, S.B. and Ho. C.L.1999. Applications and issues of GIS as tool for civil engineering modeling, Journal of computing in civil engineering, ASCE, Vol.13, No.3, 144-161.

Nagarajan R., Mukherjee A., Roy A. and Khire, M.V. 1998. Temporal remote sensing data and GIS application in landslide hazard zonation of part of Western Ghat, India. International Journal of Remote Sensing 19 (4), 573-585.

Peter Richardus / Ronk.Adler (1974) Map Projections, North-Holland Publishing Company – New York.

Petley David N. *, Crick William D.O.**, Hart Andrew B. ** "the use of satellite imagery in landslide studies in high mountain areas". *Lecturer, Department of Geography, University of Durham, United Kingdom,**Engineering Geologists, United Kingdom.

Porer W. McDonnell, (1991) Land Mark Enterprises, Cordova.

Prof.R.P.Misra, Fundamentals of Cartography

Ramakrishnan S.S, Sanjeevi Kumar V, Zaffar Sadiq M.G.S.M, Venugopal K. "landslide hazard zonation map of Marappalam using Remote sensing technique and GIS",Anna University, Chennai-25.

Rhind., D.W., 1988, Personality as a Factor in the Development of a Discipline, American Cartographer, 15, pp 3277-89.

Robinson & Sale : Elements of Cartography

Sabins, F.F.,1987, Remote Sensing-Principle and Interpretation, Freeman and Co., New York, Chapter 7, pp 246-60.

Sivakumar Babu G.L., Mukesh M. D."Landslide analysis in Geographic Information Systems", Department of Civil Engineering, Indian Institute of Science, Bangalore.

Star, J.L., Estes., J.E., McGore, K.C., 1997, Integration of Geographic Information Systems and Remote Sensing, Cambridge University Press, New York.

Susan Cole, Early Maps of India

Text Book:

* Environmental Geology by Strahler & Strahler.

* Landslide and related phenomena by Sharp, C.F.S.

* Mass Movement in Man and Physical Environment by Morton, D.M & Streitz.

Tina Cary, Digital Photogrammetry (1996) American Society for Photogrammetry & Remote Sensing

Tomlin., C.D., 1990, Geographic Information Systems and Cartographic Modelling, Englewood Cliffs, N.J, Prentice Hall.

Tomlin., C.D., 1991, Cartographic Modelling : D.J. Maguire, M.F. Good Child and D.W. Rhind (eds) Geographical Information Systems : Principles and Applications, Harlow, Longman.

Tomlin., C.D., and Berry., J.K., 1979, A mathematical structure for Cartographic Modelling in Environmental Analysis., Proceedings of the American Congress on Surveying and Mapping, Falls Church VA. American Congress on Surveying and Mapping.

Townshend, J.R. and Justice, C.(1981) Information Extraction from remotely sensed data, Introduction of Remote Sensing, 2, 313-29.

Trench, N. R. and Sauchyn, D. J. 1978. Application of Landsat data to the identification and delimitation of landslides in Colorado. Abstracts with Programs - Geological Society of America, 10 (7) 506.

Varnes, D.J. 1984. Landslides hazard zonation: A review of principles and practice, Natural Hazards No.3, Commission on Landslides of the Int. Assoc. of Engrg. Geology, UNESCO, Paris, 63.

Vincent, R. K. Fundamentals of Geological and Environmental Remote Sensing. Upper Saddle River, NJ: Prentice-Hall, Inc., 1997.

Vincent., R.K., Rowan, L.C.Gillespie.R.E., and Knapp, C., 1975, Thermal Infrared Spectra and Chemical Analysis of Twenty-Six Igneous Rock Samples. Remote Sensing of Environment, 4.4, pp 199-210.

Wolf, Paul R, Elements of Photogrammetry (1974) Mc Graw Hill Inc.

Wolfgang Torge, Geodesy (1991), Walter de Gruyter & Co., Berlin.

Wu Tai-Chi, Usher John M., "Application Of Remote Sensing for the Prediction, Monitoring, and Assessment of Hazards and Disasters that Impact Transportation ", Department of Industrial Engineering, Mississippi State University.

Wu, W., Abdel-Latiff,M.A., and Wieczorek, G.F, 1996. Landslide hazard prediction. In Proc. Of the 7th Int. Symp. On landslides, Trondheim, Vol 1, pp. 423-428.

Index

C

D

H

I

J

K

L

M

N

O

P

Q

R

S

T

U

V

W

Z

www.ingramcontent.com/pod-product-compliance
Ingram Content Group UK Ltd.
Pitfield, Milton Keynes, MK11 3LW, UK
UKHW061534190726
13845UKWH00005B/100